THE CHANGING GLOBAL ENVIRONMENT

South and Central America, the Atlantic Ocean, and portions of Africa and North America are seen on this photograph from NASA's Applications Technology Satellite III. Satellites have the capability of monitoring many kinds of pollution on a world-wide basis and can detect incipient changes in world climate.

THE CHANGING GLOBAL ENVIRONMENT

Edited by

S. FRED SINGER

D. REIDEL PUBLISHING COMPANY

DORDRECHT-HOLLAND / BOSTON-U.S.A.

Library of Congress Catalog Card Number 73–86096

ISBN-13: 978-90-277-0402-3   e-ISBN-13: 978-94-010-1729-9
DOI: 10.1007/978-94-010-1729-9

Published by D. Reidel Publishing Company,
P.O. Box 17, Dordrecht, Holland

Sold and distributed in the U.S.A., Canada, and Mexico
by D. Reidel Publishing Company, Inc.
306 Dartmouth Street, Boston,
Mass. 02116, U.S.A.

TABLE OF CONTENTS

EDITOR'S PROLOGUE

An increased scale of human activity has brought with it pollution, defined as "an undesirable change in the physical, chemical, or biological characteristics of our air, land, and water that may or will harmfully affect human life or that of any other desirable species, or industrial processes, living conditions, or cultural assets; or that may or will waste or deteriorate our raw material resources".* Under certain circumstances, the natural processes are unable to keep pace with the increase of pollutants, and then serious problems arise, which are usually on a local scale. On occasion, however, pollution effects may persist long enough so that the atmosphere or the ocean circulation may spread them over the whole earth.

One classic example, well-documented, is the rise in the concentration of atmospheric carbon dioxide produced by the intensive burning of fossil fuels during the past few decades. The buffering action of the ocean has not been able to keep pace with the increased rate of production, and we now find the CO_2 content increased by about 12% and still rising. But while there is little argument concerning the existence of such an increase, there is no agreement as to the consequences of an increase on the radiation balance of the Earth and, therefore, on world climate. There are now many other examples of worldwide effects: pesticides in the world's oceans, for example. Only within the past year has it been recognized that chlorofluoromethanes used as inert propellants in aerosol spray cans can have far-reaching effects on stratospheric ozone which in turn would increase levels of biologically harmful ultraviolet radiation at the Earth's surface. And there may be pollution effects extant which we have not yet recognized or whose consequences we cannot yet ascertain.

It is important therefore to examine the situation at frequent intervals, to determine whether a pollutant released in our environment could have far-ranging effects in the biosphere, and to probe particularly all interconnections in order to expose any weak link in the ecological chain. Most worrisome are situations in which a triggering action sets off a feedback mechanism that preexists in nature. Such feedback mechanisms are believed to be responsible for major changes in the climate and for the production of the Ice Ages. It behooves us therefore to examine very carefully, and even conservatively, all pollution effects arising from human activities. It is important also that this examination involve scientists from different specialties but with broad interests. Many disciplines must be represented, including geophysics, geology, biochemistry, biology, medicine, and ecology. The subject matter has ob-

* 'Waste Management and Control', Committee on Pollution, National Academy of Sciences/ National Research Council, Washington, D.C., 1966.

vious interest to the general public, to policymakers in the government, and to all who are concerned about the effects of man's activities on the environment.

This is a most appropriate time for reviewing our changing global environment, and especially the effects caused by human activities which accelerate geochemical natural cycles or introduce new chemicals or new effects, or just large quantities of heat. The last few years have seen an increasing awareness and sensitivity about global pollution effects, even those potential effects of supersonic transports on the stratosphere. The latter problem has triggered active experimental and theoretical work, and there have been new scientific insights and results on photochemistry, on the natural origin of carbon monoxide, and on atmospheric particulates.

In addition to keeping all topics updated, a special effort has been made to keep the volume integrated and coherent, to aid the student who wants to gain a comprehensive view of global pollution effects.

I am indebted to the authors and many colleagues for critical and useful comments. I owe particular thanks to Mr John Stanley for his scientific and editorial assistance throughout this volume.

University of Virginia
Charlottesville, Va.
December, 1974

S. FRED SINGER
Professor of Environmental
Sciences and Member of the
Center for Advanced Studies

PART I

OVERVIEWS ON GLOBAL EFFECTS OF
POLLUTION

SPECIFIC AND GENERAL EFFECTS OF
POLLUTION

INTRODUCTION

We know a great deal about historical climate and its variations from various geological studies. There are two points worth remarking on. One is that the climate changes frequently and radically, but that the degree of variation and even sense of variation depends on the time scale which we are considering. Secondly, that this is a most unusual geological period for the Planet Earth; we are living in a period of mountain building and glaciations, whereas during most of the last 250 million years (m.y.) there was little ice and little topography.

A good view of climate change of the last hundred m.y. can be gained by looking at the paper of Kellogg. We are now in a period of extensive glaciations. The previous interval occurred 300 to 250 m.y. ago, when even the Sahara was glaciated. (Of course, it was at that time near the position of the South Pole; we know that 300 m.y. ago the continents had not broken apart and formed one land mass.) Apparently between 250 and 20 m.y. ago there was little ice on the Earth, even at Antarctica. Continental basins were flooded by shallow seas. This was the period when plant life and marine life proliferated and when most of our fossil fuels were laid down.

We suspect that the onset of glaciations has something to do with the presence of continental masses at high latitudes. In any case, once glaciation is started, then there exists a positive feedback: the high albedo of the ice rejects solar energy and therefore keeps cooling the earth. Eventually, however, glaciation is limited, we believe, by the drop in ocean temperatures which slows down evaporation and therefore stops feeding the ice.

Deep sea cores taken by the Glomar Challenger expedition found that Antarctic glaciations started about 20 m.y. ago. It is perhaps significant that 4 m.y. ago the Central-American isthmus formed, separating the Atlantic and Pacific. At about that time glaciation began in the Northern hemisphere. Since then, there have been rapid cycles of glaciation with the last one occurring only 20000 yr ago.

On the time scale of millenia, therefore, we are today in an unusually warm period, while on the time scale of millions of years this is a cool period. On shorter time scales our present period can be considered either warm or cold depending on the length of the time scale considered. An idea may be gained by looking at the paper of Mitchell (see esp. p. 152).

Having briefly considered historical climate, we are not yet able to identify the reasons for climatic change with certainty. Most likely, there are a number of causative factors, depending on the time scale of change considered.

There are three questions we can ask:

(1) Have we considered all of the effects and factors that can produce a change in climate?

(2) Is any change in climate necessarily bad?

(3) Can we compensate for climate changes?

1.   It is clear that we do not fully understand all of the human actions that can influence climate. The four most important factors are: chemical changes in the atmosphere, particularly changes in CO_2 concentration; the presence of dust and aerosols; changes in surface albedo, including ice and snow, clearing of land, inundation, building of cities, etc.; and generation of heat. Their effects are indicated briefly in the survey papers of Part I and discussed in more detail in Parts II and III of this volume.

But what about the endogenous climate changes produced by volcanic eruptions, by mountain building, by continental drift, etc.? What about the fluctuations in the output of solar ultraviolet which produces changes in the ozone concentration of the stratosphere; solar particulate radiation; variations and indeed reversals of the earth's magnetic field; and astronomical events such as the impact of meteorites or the passage of comets? We simply do not have the knowledge to evaluate these factors fully, nor have we necessarily exhausted all of the possibilities.

We do believe, however, that climate changes are generally *triggered*. By this we mean that a small change of a crucial parameter, at the proper time and at the proper place, can cause a large climate change by initiating or triggering an inherent feedback mechanism which exists in our natural environment. This possibility of course must be a source of constant concern.

Since we do not understand all of the pathways of climate change, nor even the inherent feedback mechanisms, we can never be quite sure about the consequence of any particular human activity. Some comfort can be gained from the fact that human interventions up until now have not been large compared to natural disturbances. But this situation may change in the future. Hence it is incumbent upon us to be ever watchful and to maintain a global monitoring network.

It is also incumbent upon us to understand better the causes and possibilities of climatic change. One approach is through a general circulation model using the dynamic equations of meteorology such as the one described by Manabe in this volume.

But there is also the possibility of climatic models, whose development has been carried out by W. Sellers, B. Saltzman, M. Budyko, and others.* Unlike the highly developed general circulation models, the climatic models are usually only one or two-dimensional and use extensive parameterization. This of course is artificial but makes them manageable. With more time and more research these parameterizations can be made more certain, particularly with the use of weather satellites whose observations relate more directly to climatic models.

* See, e.g., Sellers, W.D.: 'A New Global Climatic Model', *J. Appl. Met.* **12**, 241–254 (1973).

One interesting result of climatic modeling has been to show the existence of several different states of climate corresponding to the same parameters. There exists a kind of 'hysteresis effect', indicating that climate state depends on the path taken in its development.

2. We can now turn to the second question and ask whether changes in climate are necessarily bad. Since throughout history, climate changes have been the rule rather than the exception, and since the biosphere has survived and evolved, one might be tempted to make light of those who decry a warming of the climate, while others worry about bringing back the ice ages. I am persuaded to think that any climate change is bad because of the investments and adaptations that have been made by human beings and all of the things that support human existence on this globe.* Even minor fluctuations of climate could change the distribution of fish (the cod has moved further north in the last few decades from Iceland to Central Greenland), upset agriculture (by forcing it onto soils that are not suitable), and inundate coastal cities (by raising sea levels). Such changes could occur at a faster rate perhaps than human society can evolve.

3. But the possibility of adjusting climate artificially remains a fascinating one. Of the various methods available, it seems most likely that changes in surface albedo or the introduction of particulates into the stratosphere might be most cost-effective. One can even visualize a space-borne 'sun shade' which intercepts solar radiation in a variable but controllable manner. These are fascinating possibilities for the future; the basic principles are discussed in this volume.

---

* See, e.g., Newman, J. E. and Pickett, R. C.: 'World Climates and Food Supply Variations', *Science* **186**, 877–881 (1974).

POLLUTION EFFECTS ON GLOBAL
CLIMATE – AN INTRODUCTION

S. FRED SINGER

University of Virginia, Charlottesville, Va., U.S.A.

Abstract. Neglecting human heat production, the Earth's climate can be modified by three different methods:
(i) a change in the albedo of the Earth's surface;
(ii) a change in the albedo of the atmosphere;
(iii) a change in the atmospheric content of gases having strong radiative properties.
The most important mechanisms are briefly described here. Other articles in this volume give a more detailed treatment.

1. Introduction

Before discussing how the Earth's climate can be modified by pollution, it is first necessary to summarize our understanding of the present surface and atmospheric temperatures.

The only heat source of any consequence is the Sun's radiation* which corresponds to that of a black body of about 6000 K emitting its maximum energy in the visible region of the spectrum near 0.47 μ. At the distance of the Earth, one astronomical unit from the sun or 1.5×10^{13} cm, the energy flux S is 1400 W m^{-2}. The Earth reflects a portion of this energy back into space corresponding to its optical albedo. The rest is absorbed and turned into heat and re-radiated as infrared or thermal radiation with a maximum wavelength in the vicinity of 10 μ. If the Earth had no atmosphere but a good surface conductivity of heat, then its surface temperature would be

$$T = [(1-A)S/4\sigma]^{1/4},$$

where T is the planetary temperature, A the albedo of the Earth and σ is the Stefan-Boltzmann constant. (The factor 4 appears because the Earth absorbs according to its cross-section which is a circle, but radiates according to the surface of the sphere). For an average planetary albedo of approx. 33 %, the temperature would be 253 K ($-20\,°C$).

The Earth's atmosphere changes this simple picture considerably. A clear atmosphere is reasonably transparent to solar visible radiation; hence does not absorb much of the incident solar radiation. But even a clear atmosphere contains water vapor (H_2O), carbon dioxide (CO_2), and ozone (O_3) which gases are strong absorbers and radiators in the infrared region of the spectrum. This makes the atmosphere essentially opaque for the infrared radiation that is re-emitted from the Earth's surface, except for a 'window' which extends approximately between 8 and 12 μ.

For a clear atmosphere containing an average amount of dust about 65% of the

---

* In the future, heat released by human activities may become of some importance: specifically, energy production by nuclear and geothermal methods, and by the increasing use of fossil fuels. See the discussions by Singer (pp. 25–44) and by Kellogg (pp. 13–23).

S. Fred Singer (ed.), The Changing Global Environmental, 7–12. All rights reserved.

incoming solar radiation reaches and is absorbed in the Earth's surface. However, only about 10% of the total (infrared) radiation leaving the surface is directly transmitted to space.* The remainder is absorbed by the radiative gases in the atmosphere and re-emitted successively at the (lower) temperature of the gas layer. Part of it is re-emitted downward to the ground and part of it upward to the next layer, and finally to space. The downward radiation helps to keep the surface temperature relatively high by what is termed the 'greenhouse' effect. The radiation emitted back into space can be thought of as consisting of two components (See Figure 1): (i) that emitted from the very cold outermost layer of the atmosphere; and (ii) the component directly emitted from the Earth's surface and reaching space through the 'window'.

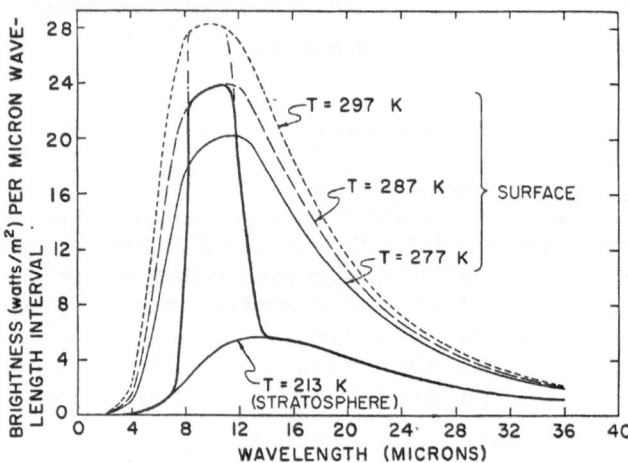

Fig. 1. Schematic diagram showing the two components of infrared energy flux leaving the Earth's atmosphere. The flux in the 'window' region corresponds to an average surface temperature. The second component can be thought to originate from an upper and colder atmospheric layer, essentially from the highest 0.3 mm of precipitable water vapor.

The atmosphere also has the important function of distributing the energy from the Sun, cooling the equatorial region and heating the polar regions, in a complicated process of energy transfer which is made more difficult by the rotation of the Earth.

The distribution of the radiating gases is largely responsible for the layering of the Earth's atmosphere. The troposphere extends from sea level up to approximately 12 km at which point the temperature has dropped to approx. −60°C. Water vapor is about 10 times more important than carbon dioxide, both for radiative heating by absorbing solar radiation and for radiative cooling by emitting more infrared than is absorbed. In the stratosphere however, radiative heating by the absorption of solar radiation by ozone is dominant. The radiative cooling that does occur is carried

---

* For an average cloudy atmosphere the values are 45% and 5%, resp.

on primarily by CO_2, with water vapor playing a minor role above about 20 km.*

This brief discussion sets the stage for understanding changes in global climate by changes of the Earth's environment. There are four general methods:

1. A change of the Earth's surface albedo.

2. A change in the albedo of the atmosphere.

3. A change in the chemical composition of the atmosphere.

4. Releases of heat which are large in relation to solar heating. (This mechanism is discussed in pp. 25–44.)

2. Changes in Surface Albedo

The natural albedo of the Earth's surface varies from extremely low values (almost black or totally absorbent), like the ocean surface, to extremely high values (very white and therefore nearly completely reflecting), such as freshly fallen snow. In between, we have low-albedo irrigated crop land and high-albedo deserts. From this preliminary discussion it can be seen that human activities, principally urbanization and agricultural development, will have little effect on the overall surface albedo which is largely given by the oceans, and the polar snow and ice covers.

Natural variations occur throughout the year which would dwarf any human effects, especially the natural growth and retreat of the snow and ice cover. Herein, however, lies a point of possible instability whereby human activity, if properly or improperly applied, can influence climate as a whole. If, for example, the temperature in the arctic regions is lowered, then ice sheets will spread. Since snow and ice have a higher albedo than any other ground cover, the result would be a further reduction of temperature, resulting in a further increase of the ice sheet. This suggests an instability which, unless checked by other factors, could lead to an equator-to-pole ice cover. We know however that during the ice ages the ice sheet did not penetrate beyond middle latitudes. Hence there must be factors in the atmosphere which counteract this 'albedo instability'.

In this connection, ambitious operations have been proposed from time to time. For example, if the Bering Straits were dammed and the arctic ocean were to become ice-free, then it might become possible to start another ice age (if as Ewing and Donn [3, 4] suggested in 1956 the principal requirement for another ice age is the supply of water made available by an ice-free arctic ocean). It is believed that once the arctic ice sheet is removed the arctic ocean would remain ice-free. Some intriguing suggestions have been made by J. O. Fletcher [5] for removing the ice sheet. One of them requires simply an albedo change of the ice, spraying it with soot or another dark material which could absorb more solar radiation.**

* The reasons are complex; ozone is most plentiful in the stratosphere (where its production takes place); also, it absorbs strongly in the ultraviolet region of the spectrum where much of the solar energy is concentrated. On the other hand, water vapor is unimportant because its concentration is so low; unlike CO_2 which is uniformly mixed with the major gases N_2 and O_2, the water vapor mixing ratio drops to the low value of 2–3 ppm at the tropopause. As the atmospheric temperature drops with altitude, the H_2O 'freezes out'. See the article on SST effects (pp. 125 to 134) for further discussion.
** For further elaboration, see the papers of Kellogg, Bryson and Wendland, and Landsberg. The possibility of climate effects from oil spills in the arctic ocean has been discussed by several authors: see *Science* **186**, 843–846 (1974).

3. Changes in Atmospheric Albedo

The reflecting properties of the Earth's atmosphere with respect to incoming solar radiation are determined mainly by the amount and distribution of water and ice clouds, and less so by particulate material. Water clouds form at lower altitudes, ice clouds (cirrus) in the upper troposphere. Water droplets and ice crystals mainly scatter the visible radiation rather than absorb it. On the other hand, dust and aerosols partially reflect and partially absorb incident solar radiation, depending on their exact composition. It is possible that the small absorptivity of clouds is produced by 'dirtiness' introduced by a dust nucleus.

The average cloud albedo varies from 10 to 25% for cirrus to about 70 to 80% for thick cumulus clouds. It should be pointed out that 'white' clouds, i.e. with a high albedo in the visible, can be 'black', i.e. efficient absorbers and therefore radiators, in the infrared region of the spectrum.

The effect of atmospheric particles in the size range of 0.1 to 2.5 μ is complicated by their optical properties and their exact location. Particles which absorb in the infrared part of the spectrum (but not in the visible) may cool the upper atmosphere, while particles containing absorbing material in the visible may heat it.* For example, it is believed that high-altitude cirrus will cool the upper troposphere, while the eruption of the volcano Agung which put a great deal of dust into the stratosphere produced worldwide heating of the layer.

Conceivably human influences can be important here. High altitude aircraft laying down condensation trails produce artificial cirrus clouds, while at the same time ejecting smoke particles which simulate volcanic dust. At lower altitudes, much particulate material is produced in industrialized areas by stationary power plants, metal smelters, etc., and in the tropical agricultural areas by slash-and-burn agricultural procedures.

At present, the injection of particulate material (dust, smoke and aerosol particles) by human activity is small compared to the natural injection of particles which come on a steady basis from forests in the form of terpenes, from oceans in the form of salt particles, from deserts and other wind-blown dust, and from occasional eruptions of volcanoes. The latter can and do put large quantities of particulates into the stratosphere where they remain for several years and add to the already present stratospheric aerosol layer.**

4. Changes in Atmospheric Composition

The major concern is the steady increase in the carbon dioxide content of the atmosphere produced by the rapid burning of fossil fuels. Also of concern are changes in the water vapor concentration and ozone concentration in the stratosphere. Of lesser

* For a more detailed discussion see Ensor, D. S. et al., J. Appl. Met. 10, 1303–1306 (1971); and Weare, B. C. et al., Science 186, 827–828 (1974).
** Many papers in this volume elaborate on this theme; see especially the papers of Mitchell, Bryson and Wendland, Manabe, and Ellsaesser.

concern for global climate change are sulfur dioxide, nitrogen oxides, carbon monoxide, methane and other minor gases, although they may produce intense local problems.

Since late in the 19th century, man has released carbon dioxide into the atmosphere at an ever accelerating rate through the combustion of fossil fuels.* About 50% of the CO_2 released has remained in the atmosphere, the rest being absorbed into the ocean. The present rate of CO_2 production by man is estimated at 16×10^9 metric tons per year. The average planetary atmospheric CO_2 concentration in 1972 was very close to 324 ppm, or 12% above the generally accepted 19th century base level of 290 ppm. From a variety of observations one may conclude that the secular rate of increase of CO_2 is now about 0.3% per year. By 2000 A.D. the increase will be roughly 30% over base level.

According to the numerical investigations of Manabe and collaborators (see pp. 73–77) increases in CO_2 result in a warming of the entire lower atmosphere, with the amount of warming dependent on the humidity, but with cloud distribution, surface albedo, etc. remaining constant. Under those conditions, a 10% CO_2 increase can lead to a warming of 0.2 to 0.3 °C.

Mitchell (pp. 149–173) concludes that probably not more than one-third of the planetary temperature disturbance of the past century is attributable to variations of atmospheric CO_2. Other mechanisms are evidently required to account for part of the warming observed between 1880 and 1940, as well as for the cooling observed since 1940. It is generally believed that atmospheric turbidity (due to particulates, mostly natural but partly manmade) could be responsible for the difference. On this basis, Mitchell has projected temperature trends for the next 100 yr, attributable to anticipated future increases of CO_2 and particulates, both from human activities (see figure on p. 170). The different curves correspond to different estimates of the future rate of increase of particulates and their effect on temperature.

It should be emphasized that such extrapolations cannot be considered realistic unless the complete atmosphere is modeled at the same time. Increased temperature will lead to increased evaporation from the ocean and therefore an increased water content in the atmosphere. If, therefore, cloudiness increases, then this will offset temperature increases. For the time being we must rely on careful monitoring over the complete Earth's surface by means of weather satellites, and on improvements in the numerical modeling of the atmosphere by using high-speed computers.

Another area of concern is a possible general change in the ozone content of the atmosphere. Since ozone has important radiative properties and controls the heating in the stratosphere, a general concentration change could have further consequences on a global scale. Interest in this problem has been focused recently through discussions on the environmental effects of the supersonic transport. Water vapor ejected by the burning of aircraft fuels, and especially nitrogen oxides produced in high temperature combustion, can remove some of the stratospheric ozone, or possibly

* A number of papers in this volume discuss this matter in detail, as well as some of the consequences. See especially the papers by Singer (pp. 25–44), Mitchell (pp. 149–173) and Manabe (pp. 73–77).

even a large fraction. Since the water vapor content in the stratosphere is low, only two parts per million, injection from SST's or from other human activities could lead to important consequences.*

5. Other Chemical Emissions

The emission of other gases may have important local effects and lead to intense air pollution problems, but on a global scale they are not considered to be serious. The breakdown shown in the Table (p. 33) gives emissions for the United States for various air pollutants. The major aspects of the circulation through the atmosphere of SO_2, NO_x, and CO have been discussed by Robinson and Robbins (pp. 111–123). They estimate the source magnitude, the residual atmospheric concentrations, and discuss the scavenging processes.

One third of the sulfur reaching the atmosphere comes from pollutant sources, namely as SO_2 (see Table of p. 112). Within the atmosphere there is a net transfer of sulfur from land to ocean areas. The average tropospheric concentrations of sulfur compounds are shown in the Table (p. 114). Most of the scavenging takes place through precipitation and dry deposition as shown in the flow diagram of Figure 1 on p. 114.

The circulation of nitrogen compounds through the Earth's environment is quite complex. Of the seven significant atmospheric forms of nitrogen the only important pollutants emitted by human activities are NO and NO_2. (See Table on p. 117). The atmospheric concentrations are shown in the Table (p. 118) and a simplified flow diagram is given in Figure 2 on p. 119.

The major significant source of human-produced carbon monoxide is the gasoline engine (See the Table on p. 85). However, the average CO concentration in the atmosphere is only on the order of 0.1 ppm. The major scavenging process may be biological, probably through soil bacteria, although some of the carbon monoxide that is transported into the stratosphere may be oxidized there. In any case, the residence time in the atmosphere is only between 0.1 and 0.3 yr.

References

1.   *Inadvertent Climate Modifications*, Report of the Study of Man's Impact on Climate (SMIC), MIT Press, Cambridge, Mass. (1971).
2.   *Man's Impact on the Global Environment*, Report of the Study of Critical Environmental Problems (SCEP), MIT Press, Cambridge, Mass. (1970).
3.** Ewing, M. and Donn, W. L.: 'A Theory of Ice Ages, I', *Science* 123, 1061–1066 (1956).
4.** Ewing, M. and Donn, W. L.: 'A Theory of Ice Ages, II', *Science* 127, 1159–1162 (1958).
5.** Fletcher, J. O.: 'Polar Ice and the Global Climate Machine', *Bull. Atomic Sci.* 26, 40–47 (1970).

* It has only recently been recognized that oxides of chlorine can destroy stratospheric ozone catalytically like NO_2 (see p. 130); see Stolarski, R. S. and Cicerone, R. J., *Can. J. Chem.* 52, 1610–15 (1973). Freons used in aerosol spray cans seem to provide an important global source of chlorofluoromethanes; nearly 10^6 ton will be produced in 1974, and 5 m.t. has already been released into the atmosphere. Being inert they accumulate in the troposphere with a concentration of $\sim 10^{-4}$ ppm, but are photodissociated into free chlorine in the stratosphere; see Molina, M. and Rowland, F. S., *Nature*, 240, 810–912 (1974).
** Indicates references actually cited in the text.

CLIMATE CHANGE AND THE INFLUENCE OF MAN'S ACTIVITIES ON THE GLOBAL ENVIRONMENT

WILLIAM W. KELLOGG

National Center for Atmospheric Research*, Boulder, Colo., U.S.A.

Abstract. The history of changes in the Earth's climate are traced from the earliest times to the present as recorded in the rocks and ice caps. Man's potential impact on the climate is discussed including the impact of carbon dioxide, particulate matter, 'albedo' changes, irrigation, and the direct release of heat. The results of man's activities are uncertain at this moment, but some preliminary computer calculations indicate that man can influence the climate of the Earth and that the direction of this influence in the decades to come must be that of a warming essentially in the Northern Hemisphere. The Arctic Ocean pack ice represents an unstable part of the ocean-atmosphere system in that a warming that would remove the pack ice would produce a major one-way transition. The impacts of such a transition would be very grave for some regions of the Earth, but cannot be spelled out to become a disaster for mankind.

1. Introduction

The climate has changed many times in the past. The evidence for this is all around us, written in the shapes of the land, the composition of the ocean sediments, and the structure of the great ice caps that cover Greenland and the Antarctic.

We are now coming to the realization that man can influence the planet's environment on a grand scale. Though he may have influenced the climate already, it has so far probably been in a *small* way. Our concern is that he may be able to do it in a *larger* way, and could begin to match Nature's forces with his own in the future.

What can we say about man's influences relative to Nature's? This is the central question that we face when we consider the future of mankind. I will try to explain some of the things that we think we know about it at the end of this paper.

To a certain extent the conclusions that I will present are those of an international Study of Man's Impact on the Climate (SMIC) [1] that met near Stockholm in the summer of 1971, a study in which I participated, augmented by some additional ideas that have been introduced since then. One overall conclusion that emerged from the SMIC was that we have a long way to go before we can give very many definite answers, but the path to a better understanding of the causes of climate change is now more clearly marked than it was a few years ago (See also references [2] and [3]).

2. A Brief Look at the History of Climate

Looking at the record of past climatic conditions, it becomes so fuzzy as we turn back the clock beyond 500 m.y. that very little can be said about the conditions on Earth

---

* The National Center for Atmospheric Research (NCAR) is sponsored by the National Science Foundation.

before then. We believe that life came on to the dry land about a billion years ago, and we know that there was a kind of ice age at about 500 m.y. ago, and then another at about 250 m.y. ago, and that between these relatively short periods of glaciation of the poles the Earth was warmer than it is now. In fact, only about 10% of the time during this 500 m.y. period, during which mammals evolved on the Earth, were the poles glaciated as they are now.

The ice age in which we now live probably started about 20 m.y. ago in the Antarctic, and then spread to the Northern Hemisphere two or three million years ago. Figure 1, by Flohn [4] gives a rough idea of how the temperature at mid-latitudes may have changed during the past 60 m.y. More recently, as shown in this figure, we have had a succession of warming and cooling trends lasting for a hundred to two hundred thousand years each. We have just emerged from the last cold spell, the Pleistocene Ice Age, when a great ice sheet sat over the North American Continent, and another smaller one sat over Scandinavia and Asia. This last period of glaciation only retreated 10000 yr ago, which is but a fraction of a second measured on the scale of time of the Earth's history.

Imagine the changes that must have been forced on primitive man by these glaciations of large parts of his continents, and all the changes that must have occurred in the rainfall and temperature patterns of the rest of the land! It is almost certain that many of the deserts that exist now were much smaller or were not even deserts during the last ice age, and the patterns of forests, lakes, and rivers must have been very different from those at present, even though the continents had essentially their present shape.

More recently, only 1000 yr ago, recorded history tells us that the Vikings were able to sail in their small longboats as far as the North American continent, and then a

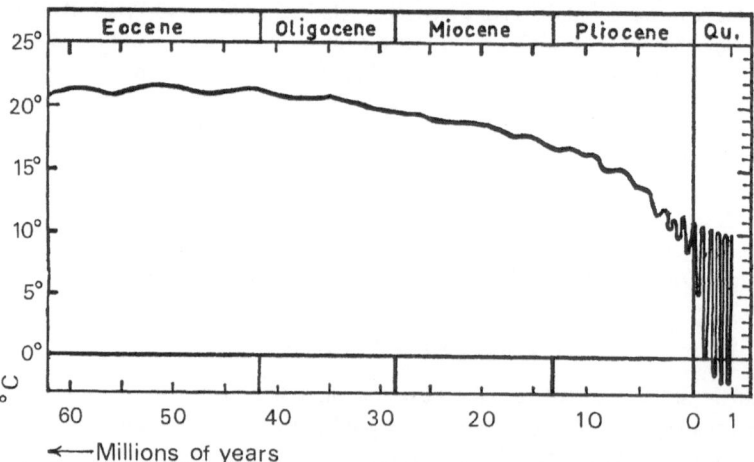

Fig. 1. An estimate of mid-latitude temperatures during the past 60 m.y. relative to the present mean temperature. Source: H. Flohn, Ein geophysikalisches Eiszeit-Modell, *Eiszeitalter Gegenwart* **20**, 204–231 (1969); also reproduced in SMIC, p. 31.

around 1100 A.D. the cooling trend that took place brought the polar ice further south over the Atlantic, and they were no longer able to reach their colonies. Eskimos from Greenland, hunting along the southern edge of the ice pack were reported in the Faroe Islands during this period of cold.

Even more recently, the cold of the 17th Century caused great hardship in Europe,

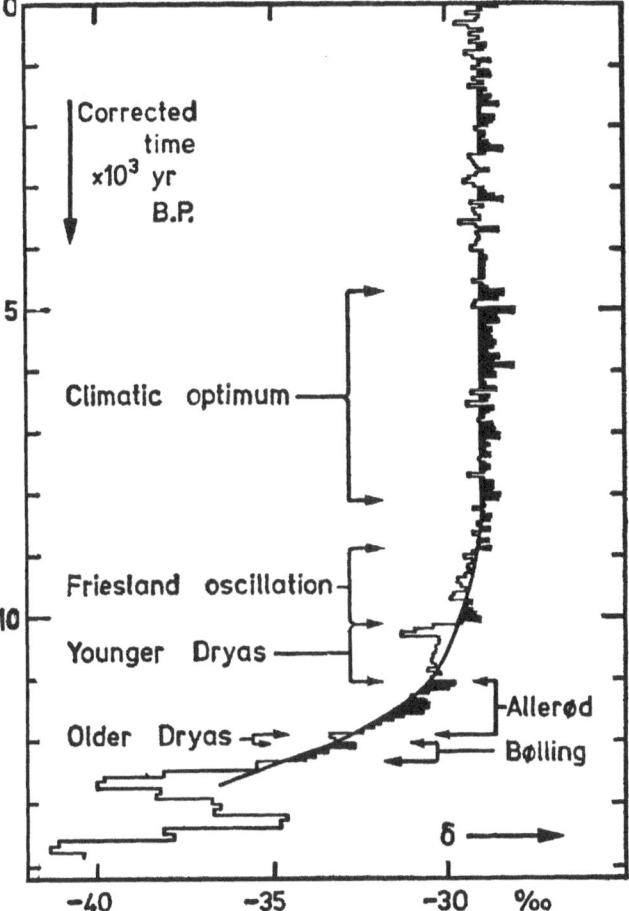

Fig. 2.   Variations of the O_{18}/O_{16} oxygen isotope ratio with depth below the surface (converted here to age of sample) in an ice core taken from the Greenland ice sheet at Camp Century. An increase in this ratio corresponds to an increase in temperature at middle and high latitudes in the Northern Hemisphere, and the values shaded in black that exceed the smoothed curve would therefore correspond to periods when the temperature was warmer than the long term trend. The names of the climatic periods shown refer to those assigned in Europe. Source: W. Dansgaard, S. J. Johnsen. H. B. Clausen and C. C. Langway, *The Late Cenozoic Glacial Ages*, symposium edited by K. K. Turekian, Yale University Press, New Haven, Conn. (1971); also reproduced in SMIC, p. 38.

and has come to be known as 'the Little Ice Age'. But these climate changes probably corresponded to mean temperature fluctuations of only one or two degrees in northern Europe, whereas the changes that accompanied the major glaciations of the more distant past must have been five or ten degrees at those same latitudes, perhaps more.

With this in mind it is interesting to study the Greenland ice core record shown in Figure 2, prepared by Dansgaard *et al.* [5]. Although the quantity plotted is oxygen isotope ratio, it can be considered as roughly proportional to mean temperature at the latitude of Greenland, and the various periods of relative warm and cold that I have just referred to (and others) can be discerned in this 14000 yr record. The total record from this core (not shown here) goes back about 150000 yr.

3. Theories of Climate Change

Why does the climate of the Earth change from time to time? This is of course the question that we would dearly like to answer, and many attempts have been made to do so. The theories that have attempted to account for climate change range from those which invoke a completely external influence, such as a change in the output of the Sun, to an almost toss-of-the-coin theory which tries to describe the ocean-atmosphere system as one which has several stable regimes so that it can, as it were, flip from one regime to another. This is the theory of 'almost intransitivity' best enunciated by Lorenz [10] of MIT. In between are other theories which are undoubtedly all parts of the picture, but we do not yet know the relative importance of such things as volcanic activity (which could darken the sky for decades or centuries at a time), the drift of continents (which on a very long time scale can certainly alter the relative behavior of the oceans and the atmosphere), the fluctuations in the ocean circulations themselves, etc. A complete theory of the climate will have to take all of these various influences into account in order to reconstruct what has happened in the past. The ocean-atmosphere-land system is a complicated one indeed, and no theoretical model has yet been constructed which pretends to include all the important factors. In fact, even the modeling techniques which have been used successfully to simulate the day to day behavior of the general circulation of the atmosphere will not work for climate study models, a point that is discussed at some length in the SMIC report.

4. The Concept of Feedback Mechanisms

Even though we really do not have a satisfactory theory or model for the climate yet, we do see several very important sets of relationships that can govern this quasi-equilibrium that I have referred to. There are several examples of feedback mechanisms that are probably operating in our atmosphere right now.

The first of these is the carbon dioxide that man has been putting into the atmosphere.* The carbon dioxide content of air has indeed increased in the past few decades,

* See p. 36.

and we are reasonably certain that this is due to the large amount of fossil fuels that have been burned during this period, thus releasing the carbon locked beneath the surface of the Earth and putting it into the atmosphere and the oceans. Machta [9], shows the past increase and an estimate of the future trend up to 2000 A.D. One of the effects of this carbon dioxide in the atmosphere is to trap some of the heat which would otherwise escape to space in the form of infrared radiation – it serves as a kind of blanket. The sunlight can still come through and heat the surface, but the infrared radiation from the surface cannot escape as easily when there is more carbon dioxide in the atmosphere. This is known as 'the greenhouse effect', and adding more carbon dioxide simply increases the temperature at the surface of the Earth.

Now, granting this, what other things will happen when we increase the surface temperature? One thing is that more water vapor will be evaporated from the surface of the oceans. Water vapor acts very much as carbon dioxide does with regard to the greenhouse effect, so the extra water vapor added to the atmosphere further enhances the greenhouse effect, and further raises the surface temperature. It seems as though this complementary effect could go on indefinitely, with a small increase in carbon dioxide causing a small increase in water vapor, etc. This *positive* feedback mechanism, or amplification, has sometimes been referred to as 'the runaway greenhouse', and our sister planet Venus is cited as a sad example of the complete end of such a runaway greenhouse effect – a miserable planet where the temperature at the surface is hot enough to melt lead, and where all the water that might have been in the oceans and all the carbon dioxide that might have been in the rocks has remained in the atmosphere, adding to a super-greenhouse effect.

Another example of a feedback mechanism, one that also happens to be a positive one, is the polar ice. Consider the fact that the polar ice reflects much more sunlight than the land or the open seas, and therefore, if we increase the area of the polar ice, more sunlight will be reflected. A small decrease in temperature of a degree or two (particularly in summer) can cause the polar ice cap to spread towards the equator, this reflects more sunlight, this causes less heat to be available to warm the planet, the temperature falls still further, the polar ice creeps still further towards the equator and so on until (one might expect) the polar ice would reach the equator and the planet would become a solid cake of ice over its entire surface. This mechanism, which has intrigued meteorologists in both the U.S. and the Soviet Union, can be called the 'runaway ice cap'.

Obviously, the Earth has not become frozen, but we can learn from this example how sensitive the polar regions are to a small change in the heat balance of the planet as a whole. We can see that this is so by the fact that, while the equatorial regions seem to have had very little change of temperature during the ice ages, the polar regions undergo very large changes in temperature. During the past century, when the mean temperature of the Earth has been changing by something under 1 °C, as shown in the figure by Mitchell (p. 152), the polar regions have changed by over one degree Centigrade in just the last decade, as shown in Fig. 3 by Flohn [1].

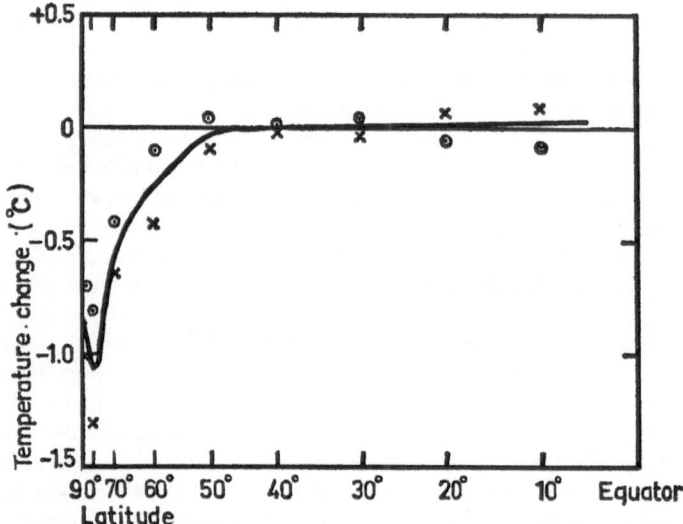

Fig. 3.   Mean temperature in the Northern Hemisphere for 10° latitude belts, expressed as deviations from the average for the period 1931 to 1960. The solid line gives the change in the decade 1961 to 1970, the circles the changes 1961 to 1965, and the crosses 1966 to 1970. Source: Prepared for SMIC by H. Flohn using data published by F. R. Pütz, Klimatologische witterkarten der nordhemisphere für den temperature im dezenius 1961/70, *Beilage zur Berliner Witterkarte* **106**, 23 (1971).

The conclusion that one draws from these rather simple considerations is that the ocean-atmosphere system can certainly change its balance, and sometimes quite suddenly. The remarkable thing is that there must be a kind of shakey equilibrium, and that the Earth has remained habitable by life forms more or less as we know them for at least a billion years. Somehow, these runaway mechanisms have not been allowed to spoil things irrevocably for the planet Earth, as they must have for the planet Venus. There are obviously stabilizing factors that are strong enough to keep our global climate within reasonably narrow bounds, permitting ice ages to come and go but damping out any large fluctuations.

5. Civilized Mankind as a Factor in Climate Change

Now man enters the scene, and we must ask whether he can reach any of the lever points on this gigantic mechanism and influence it. If there are lever points that he can reach, history has shown that he will probably be tempted to tamper with them, whether for better or for worse.

All the lever points that we have thought of that are accessible to man (except one important one) operate through the *heat balance* of the atmosphere-ocean system. The main human influences, more or less in order of their current importance as we see them now, are:

(1)   Carbon dioxide – We have already mentioned this. The effect of increasing carbon dioxide in the atmosphere is to cause a *heating* at the surface.

(2)   Particles added to the atmosphere – These particles both scatter and absorb radiation, and the net effect of an increase in particulates is *probably a cooling* influence, though we are not entirely sure. Particles only remain in the lower atmosphere for a week or less, and tend to remain in the latitude belt in which they are created.*

(3)   Changing the reflectivity or absorptivity of the Earth – its 'albedo'. Most of the things that we can think of, such as plowing grassland, cutting down forests, building cities, etc., cause more heat to be retained by the surface and therefore cause a net *heating*.

(4)   Irrigation of farmland and evaporation of water – Although the immediate and local effect of evaporating water in an irrigated region may be a cooling, the ultimate effect is a *warming*, since more sunlight is absorbed and the heat is released back into the atmosphere when the water condenses as rain or snow at some other place.

(5)   Direct release of heat from the generation of energy – This obviously results in a *heating*. In the decades to come, this will undoubtedly move to the top of the list if we continue to generate energy at an ever increasing rate.

Looking back at this list, it seems clear that as we continue to burn carbon fuels, the carbon dioxide that we add to the air will continue to increase. So will the other factors, and most of them are in a *direction* of increasing the surface temperature – all but one, the dust.

The last one on the list, the direct release of heat to the atmosphere from man's insatiable need for energy, is also continuing to go up, and will do so after he runs out of fossil fuel and resorts to a nuclear fuel. Estimates of the increase in the amount of energy that man will require vary greatly when we go beyond about the year 2000 A.D., but here are some interesting statistics about the release of heat now and estimates up to the year 2000 A.D., taken from the SMIC Report [1].**

The flux of solar radiation at the earth is 1400 W m^{-2}, and the average net radiation in the course of a day and night at the top of the atmosphere is about 350 W m^{-2}, which obviously depends on latitude. Of this, about 30% is reflected, and less than half of the remainder is absorbed by the surface, so the solar radiation that can be used to heat the surface and the lower atmosphere is, on the average, about 100 W m^{-2}.† Now, if we consider that the industrialized area of the world (about 0.5×10^6 km^2) accounts for 75% of the world's energy production and release in the form of heat, the average flux of artificially generated heat is about 12 W m^{-2}. The energy production divided by urban area for some cities is considerably greater than this average, being about 630 W m^{-2} for central New York City, 127 for Moscow, 21 W m^{-2} for West Berlin and Los Angeles, and 7.5 W m^{-2} for all of Los Angeles

---

* See also others papers in this volume, esp. Mitchell, Bryson and Wendland, and Landsberg.

** See also discussion on pp. 40 to 43.

† Since the continental land masses are mostly at mid-latitudes, the average for the continents is 67 W m^{-2}.

County, an area of 10^4 km^2. It is therefore evident that in several areas of 10^4 to 10^5 km^2 the manmade energy production is 5 to 10% of the net solar radiation absorbed.

Since a reasonable estimate for the increase of energy production gives about a factor of five increase by the year 2000 A.D. (5.5% yr^{-1} for 30 yr), it is likely that over such areas in the developed countries energy production will be 25 to 50% of the net solar radiation absorbed, and the average for all the continents if it were to be spread evenly would be nearly 1%.

A widely quoted vision of the future by Herman Kahn foresees a world in 2100 A.D. with 20 billion people, each family using an amount of energy equivalent of a U.S. family with a $20000 yr^{-1} income (Kahn's '20–20 vision'). One can make an estimate of the total energy production by such a society and compare it with the energy received from the Sun, and the human energy flux comes to about 2% of the net absorbed solar energy for the planet as a whole – and 4 to 5 times this for the continents alone. Such an artificial source of heating would cause several degree rises in the mean temperature of the continents, probably more at the poles – our knowledge of how the non-linear atmosphere-ocean system would respond to this input suggests that it would be highly non-uniform and complex, and still largely unpredictable with our present models. (See below.)

There is one part of the system that might be particularly sensitive to such a mean warming of the Northern Hemisphere: the Arctic Ocean. This point is discussed in the SMIC Report [1] and the essentials are as follows: The Arctic Ocean ice pack is about 3 m thick on the average in winter, and less than 2 m thick by the end of the summer melting season. Although there are leads with open water, especially in summer, they freeze over again quite readily in winter, the freezing encouraged by the low salinity of the top 10 to 30 m of the Arctic Ocean. There is, in other words, a relatively fresh water layer that floats at the surface under the ice cover, not mixed by the wave action that would be present in an open ocean.

We do not know how much mean temperature rise would be required to melt the ice pack to the point that it would not be able to reform the following winter, but it must be just a few degrees. Summer temperatures are more influential here than those in winter. After some appreciable opening of the Arctic Ocean to wave action the fresh water layer would be mixed with the salt water below, and the same wave action would prevent ice from reforming except in bays and estuaries. The lower albedo of the open water would cause more solar radiation to be absorbed, thereby further warming the Arctic Basin. The added evaporation would cause more rainfall in summer and more snowfall in winter.

One study of the regime that would exist with the Arctic Ocean ice removed was made by the RAND Corp., using the Mintz-Arakawa 2-level general circulation model. Figure 4 from the report by Warshaw and Rapp [7] shows the temperature *difference* and the zonal wind *difference* between the condition with the 'ice out' and the 'ice in'. The circulation patterns were considerably changed by the addition of the oceanic heat source that became available when the sea ice was removed, in particular

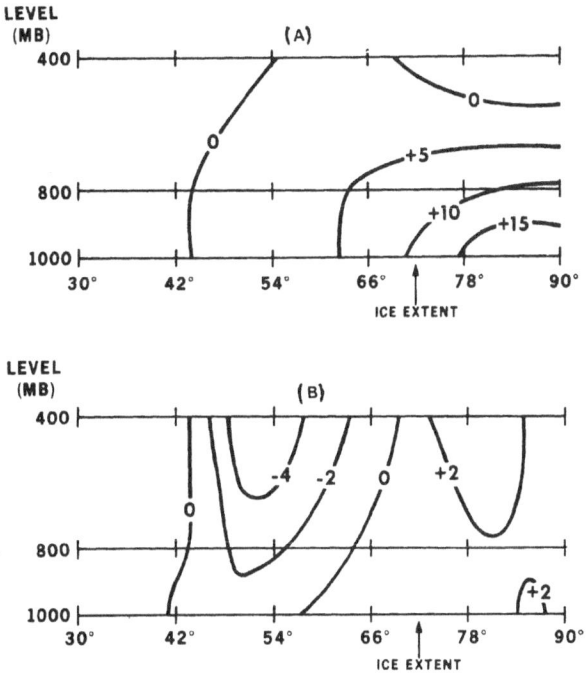

FIG. 4. At top (a) are shown the temperature differences between runs of the Mintz-Arakawa general circulation model with the Arctic Ocean ice out and the ice in, as a function of latitude and height averaged for a latitude zone; and the lower diagram (b) shows the corresponding differences in the west-east (zonal) wind in m/s^{-1}. Removing the arctic ice caused a weakening of the mid-latitude westerlies and a slight increase in the higher latitude westerlies. Source: M. Warshaw and R. R. Rapp, *An experiment on the sensitivity of a global circulation model.* Studies in climate dynamics for environmental security, RAND Report No. R–908–ARPA, Santa Monica, Calif. (1972).

a weakening of the cold-core low vortex over the Arctic Basin. Unfortunately, the model did not, apparently, reveal the altered precipitation patterns.

Thus, the fact is that man by sheer brute force may be able to begin to add enough heat to the atmosphere to be a definite factor in the climate; and in the Northern Hemisphere the Arctic Ocean might boost the effect after it reached a certain limit.

The next thing that should be considered is what would happen then to the ice caps of Greenland and the Antarctic Continent. It is possible that a warming of the polar regions would result in a gradual melting of these ice caps with a corresponding rise in the ocean level. (Melting the Greenland ice cap alone would cause a rise of about 7 m). However, this outcome is certainly not obvious, since there would probably also be an increase in snowfall over Greenland due to water vapor added by the open Arctic Ocean. In any case, the rate of change of these vast ice caps would be very slow indeed, and the shrinking or growth would be measured in centuries. This is clearly one of the problems that requires more work.

Some experiments have been done at the National Center for Atmospheric Research by Dr Warren Washington [8] to see how much the climate does change when additional heat is added to the atmosphere in the populated regions of the Earth. As one approaches a point where something like 50% of the solar energy is matched by man-made energy over large areas of the Earth (probably an unrealistically large amount, chosen to demonstrate the qualitative changes), one can see definite changes in the circulation pattern of the atmosphere, changes that can be interpreted as changes in the climate. The changes are remarkably nonuniform, but our models are not entirely suitable for this kind of experiment, and we must continue to study the matter further.

The one human environmental factor that I mentioned that was *not* directly related to the radiation balance is the possible influence that large-scale operations of jet aircraft in the stratosphere might have on the ozone layer. This was quite an important issue in the U.S. during the debate on the SST prototype development appropriation about 1970-71. The point was first raised by the Study of Critical Environmental Problems (SCEP) [2], a summer study organized by MIT in 1970, and was further looked at by the Study of Man's Impact on the Climate (SMIC) [1] in 1971, and by a number of other individuals. The idea is that the water vapor and the oxides of nitrogen released by these high-flying jets could react with the ozone in the stratosphere, and in such a way as to reduce the amount of ozone. A reduction in the amount of ozone would cause somewhat more ultraviolet sunlight to reach the surface of the Earth.

While this is probably an effect which can eventually be predicted, some of the factors that we need to know in order to make a definite estimate of the effect of SST's are still being sought, and the main hope will be in developing models of the atmosphere and specifically of the ozone region that are adequate to do experiments with. By this I mean that we will use a numerical model, run on a large computer, that will simulate the atmosphere adequately to demonstrate the probable effects of large numbers of SST's. At the moment, we do not know what the answer will be though some early estimates have shown that very large numbers of SST's could indeed cause an appreciable change in the ultraviolet radiation reaching the ground. This would, in turn, be an environmental change on a global scale that would have quite an influence on us. But I must emphasize that we have not finished our research on this yet, and the influence of even very large numbers of SST's may turn out to be so small that we should not be concerned about them.*

6. Conclusion

From the above one can, and probably *should*, conclude that man can influence the climate of his planet Earth. The direction that this influence will take in the decades to come, if man continues to demand more energy, coupled with his increasing

* See also discussion on pp. 125–134.

population, must be that of a warming, especially in the Northern Hemisphere. The Arctic Ocean pack ice represents an unstable part of the ocean-atmosphere system, in that a warming that would remove the pack ice would produce a major one-way transition, and the northern polar regions would then include a large open body of water the year around. The implications of such a transition are very grave for some regions of the Earth, but cannot be said (as some have) to spell disaster for mankind.

While we do not know how to predict such a transition, nor indeed whether it will take place at all in the foreseeable future, the distinct possibility that it could occur as a result of human influences warrants every effort to understand the mechanisms that govern climate and climate change. In fact, this is likely to be one of the major future thrusts of atmospheric research, one that has already begun.

References

1. *Inadvertent Climate Modification*, Report of the Study of Man's Impact on Climate (SMIC), MIT Press, Cambridge, Mass. (1971).
2. *Man's Impact on the Global Environment*, Report of the Study of Critical Environmental Problems (SCEP), MIT Press, Cambridge, Mass. (1970).
3. *Man's Impact on the Climate* (ed. by W. H. Matthews, W. W. Kellogg and G. D. Robinson), MIT Press, Cambridge, Mass. (1971).
4. Flohn, H.: 'Ein Geophysikalisches Eiszeit – Modell', *Eiszeitalter Gegenwart* **20**, 204–231 (1969).
5. Dansgaard, W., Johnsen, S. J., Clausen, H. B., and Langway, C. C.: in K. K. Turekian (ed.), *The Late Cenozoic Glacial Ages*, Yale University Press, New Haven, Conn. (1971).
6. Putz, F. R.: 'Klimatologische Witterkarten der Nordhemisphere für Temperature in Dezenniuse 1961/70', *Beilage zur Berliner Witterkarte* **106**, 23 (1971).
7. Warshaw, M. and Rapp, R. R.: 'An Experiment on the Sensitivity of a Global Circulation Model', *Studies in Climate Dynamics for Environmental Security*, Rand Report No. R-908-ARPA Santa Monica, Calif. (1972).
8. Washington, W. M.: 'On The Possible Uses of Global Atmosphere Models for the Study of Air and Thermal Pollution', in W. H. Matthews, W. W. Kellogg and G. D. Robinson (eds.), *Man's Impact on the Climate*, MIT Press, Cambridge, Mass. (1971).
9. Machta, L.: in D. Dyrssen (ed.), 'The Changing Chemistry of the Ocean', *Nobel Symp.* **20** (1972).
10. Lorenz, E.: 'Climatic Change as a Mathematical Problem', in W. H. Matthews, W. W. Kellogg and G. D. Robinson (eds.), *Man's Impact on the Climate*, MIT Press, Cambridge, Mass. (1971).

ENVIRONMENTAL EFFECTS OF ENERGY PRODUCTION

S. FRED SINGER

University of Virginia, Charlottesville, Va., U.S.A.

Abstract. The production and transmission of concentrated amounts of energy are the keystones for our rapid rise in standard of living. The major source of energy comes from fossil fuels, a resource which was accumulated over hundreds of millions of years but which is rapidly being used up: oil and gas in a matter of decades, coal in centuries: a brief episode in human history, but with a profound impact. Hopefully, nuclear breeders and fusion reactors will be developed in good time to supplant fossil fuels.

We describe here the uses of various forms of energy, historical and future trends, and especially the various environmental effects. Chief among these are the increase in global carbon dioxide and the generation of waste heat. Their effects are judged to be noticeable but not serious at this time.

1. Introduction

The most characteristic aspect of the industrial revolution certainly is the large production of energy. And, in the process of increasing the amount of energy produced and consumed, the character of the energy sources has changed and so has the possible impact of this energy release on the biosphere.

Before 1800, human beings throughout the world used mainly solar energy in various forms. In some cases it was the kinetic energy of the air or water, through wind motion or water falls, which drove mills or pumps. Or it was the burning of wood or various oils of animal or vegetable origin which provided heat and light. This involved the oxidation of organic materials which were created by the photosynthetic actions of plants, and thus constituted a direct conversion of *contemporary* solar energy. Or human or animal power was used to do work, and here the energy came from the metabolism of food – again an oxidation of organic material or the conversion of solar energy. This is not an all-inclusive list, for there were minor releases of energy from such sources as gunpowder, and there are historical records of the use of heat energy from the Earth's interior – geothermal energy. Another characteristic was that natural sources of energy were exploited where they were. Little attempt was made to expand them artificially. And, of course, the transmission of energy was strictly limited.

The situation became quite different after the year 1900, and has assumed completely new dimensions in the last two decades with the advent of nuclear energy. Its most striking aspect is the increased per-capita use of energy in the developed countries; in fact, the correlation between energy use and degree of development and standard of living is not accidental (Figure 1). A human being expends about 2000 kcal per day or about 100 W (thermal);* the per-capita use of energy in the United States is

* See Table I for energy conversion factors, and other useful energy data.

S. Fred Singer (ed.), The Changing Global Environment, 25–44. All rights reserved.
Copyright © 1975 by D. Reidel Publishing Company, Dordrecht–Holland.

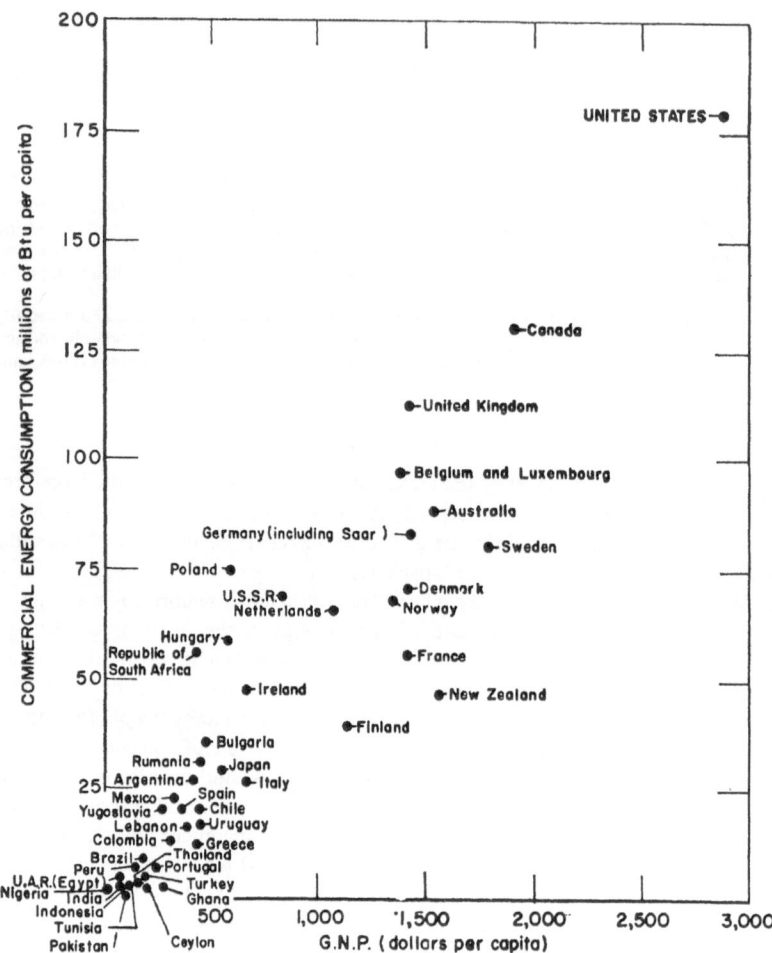

Fig. 1.   Per capita energy use and per capita gross national product selected for countries, 1961.
Source: *Energy R & D and National Progress*, Interdepartmental Energy Study, Office of Science and
Technology, Washington, D.C. (1964).

10000 W and growing at the rate of 2% to 3% yr^{-1} [21]. This increased use of energy
has had an impact on all areas, but consider for a moment the ancient occupation of
agriculture. Until recently, with very few exceptions, most of humanity spent most of
its energy and labor in providing food to sustain itself. A considerable amount of
effort and energy went into providing food for draft animals – a most uneconomical
process. In striking contrast, the farming population in the United States has dropped

TABLE I

Energy data

1 nt m (newton-meter)	$= 1$ J (joule) $= 1$ W s (watt-sec) $= 10^7$ erg	
1 gm cal	$= 4.184$ J	
1 BTU	$= 1.055 \times 10^3$ J $= 252$ gm cal $= 778$ ft-lb	
1 kWh (kilowatt hour)	$= 3.6 \times 10^6$ J $= 3.412 \times 10^3$ BTU	
1 W yr (watt-year)	$= 3.155 \times 10^7$ J $= 3 \times 10^4$ BTU	

2×10^3	kg cal	$=$ energy/day needed to sustain 1 man
50×10^3	BTU	$=$ daily per capita consumption by primitive agricultural man
800×10^3	BTU	$=$ daily per capita consumption by 1970 American
60×10^6	BTU	$=$ annual consumption by American family car
70×10^6	BTU	$=$ annual consumption for heating typical American house
64×10^{15}	BTU	$=$ U.S. gross consumption in 1970 $= 2.2 \times 10^{12}$ W yr
200×10^{15}	BTU	$=$ world annual gross consumption (1970)
1.03×10^3	BTU;	in one cu ft natural gas
9.55×10^4	BTU;	in one gal LP gas or ethane
1.24×10^5	BTU;	in one gal gasoline
4.1×10^6	BTU;	in one ton of TNT
5.8×10^6	BTU;	in 1 barrel (bbl) of petroleum (1 bbl $= 42$ gal $= 310$ lb)
25.8×10^6	BTU;	in one short ton of coal (1 short ton $= 907.2$ kg)
77.8×10^6	BTU;	in one gm U235
13.5×10^{15}	BTU;	annual domestic coal production (1970)
17.1×10^{15}	BTU;	annual domestic petroleum production (1970)
23.4×10^{15}	BTU;	annual domestic natural gas production (1970)
63.8×10^{15}	BTU;	annual world oil production (1970)

1Q $= 10^{18}$	BTU	and is approximately:
33×10^{12}	W yr	
1.0×10^{21}	J	
250×10^{15}	kcal	
3.0×10^{15}	cu ft of hydrogen	
1.0×10^{15}	cu ft of natural gas	
300×10^{12}	kWh – (thermal)	
8.0×10^{12}	gal of gasoline	
7.0×10^{12}	gal of crude oil	
170×10^9	barrels of crude oil	
50×10^9	short tons sub-bituminous coal	
40×10^9	short tons bituminous coal	
4.0×10^9	lb of uranium (burnt in light-water reactor)	
50×10^6	lb of uranium (breeders – 60% eff.)	
24×10^4	million tons of TNT	

4 Q $=$	cumulative US demand (1970–2000)
6 Q $=$	economically recoverable world petroleum resource[a]
30 Q $=$	economically recoverable world coal reserves
41 Q $=$	economically recoverable world fossil fuel reserves
9 Q $=$	economically recoverable US fossil fuel reserves
90 Q $=$	world nuclear LWR (U235)
9 Q $=$	US nuclear LWR (U235)
9000 Q $=$	world nuclear breeders
5400 Q $=$	yearly amount of energy from the Sun intercepted by the Earth

[a] The term 'economically recoverable' leads to considerable confusion if not qualified, and care should be exercised in its interpretation and use. The values used here are those of Starr [35], and are defined to be those resources available at no more than twice the current cost. A value commonly seen, 3×10^{18} gm carbon, is the N.A.S. estimate of recoverable fossil fuels. See also Table III.

from 70% (in 1820) to its present 5%, and we are better fed on the average then we ever have been.

Hand-in-hand with the introduction of energy in greater amounts has been the introduction of new sources of energy; fossil fuels, particularly oil and gas – and, of course, nuclear energy.

At first glance, the fossil fuels which represent after all solar energy which has been stored up over 600 m.y., appear to be quite different from nuclear power derived either from the fission or the fusion of atomic nuclei. Yet in a deeper sense they are related. The solar radiation which we are receiving today and have received for billions of years in the past is derived ultimately from nuclear energy sources in the Sun's interior. But for the purposes of our essay, it will be convenient to treat contemporary solar energy, fossil fuels, and nuclear energy separately. Although every energy production and consumption process releases waste heat into the environment, the different energy sources each have a different and peculiar pollution effect.

2. Fossil Fuels

The production of fossil fuels is based on the well-known carbon cycle. Through photosynthesis, plants use solar radiant energy to convert atmospheric carbon dioxide and water into carbohydrates and release oxygen into the atmosphere. Then, when the plant materials decompose or burn or are consumed by animals, the process is reversed. Oxygen is used to convert carbohydrates into energy plus carbon dioxide and water. These are the basic steps in the carbon cycle. Under normal conditions the amounts of CO_2 and O_2 in the atmosphere remain approximately in equilibrium on a year-to-year basis.*

But there are small long-term imbalances in the carbon cycle; and thus the fossil fuels, which include coal and lignite, oil shales, and tar and asphalt, as well as petroleum and natural gas, have all had their origin from plants and animals which have existed on the Earth during the last 600 m.y. Over a period of geological history extending back to the Cambrian, a small fraction of these organisms has become buried in sediments or muds under conditions which have prevented complete deterioration and complete oxidation. And so, after various chemical transformations, these organisms have been preserved as our present supply of fossil fuels. Although the same geological processes are still operative, the amount of new fossil fuel that is likely to be produced during the next few thousand years is inconsequential. Therefore, one can assume that the existing fossil fuels will be progressively exhausted and constitute a nonrenewable resource.

The present quantity of CO_2 in the atmosphere is 2400 billion tons, corresponding to a concentration of about 320 parts per million (450 ppm by weight) [21]. We will

---

* The annual amount of carbon dioxide involved in photosynthesis is about 110 billion tons per year [13], roughly 70% by land plants and 30% by marine plants [29]. (This is about 1/20 of the amount of atmospheric CO_2.) It is matched by the annual release from organic matter through oxidation to within 1 part in 10^4 (see Table II).

TABLE II

Distribution of planetary carbon

	Per unit area	Over the Earth's surface $(5.11 \times 10^{18} \text{ cm}^2)$
Mass of atmosphere	1030 g cm^{-2}	5.3 $\times 10^{21}$ gm
Mass of atmospheric CO_2 (present)	0.45 g cm^{-2}	2.38 $\times 10^{18}$ gm
CO_2 in ocean water	27 g cm^{-2}	1.4 $\times 10^{19}$ gm
CO_2 produced geologically over 4.5 b.y.	5×10^4 g cm^{-2}	2.5 $\times 10^{23}$ gm
Fossil fuel reserves (economically exploitable)[a]		3 $\times 10^{18}$ gm
Total coal[b]		15 $\times 10^{18}$ gm
Carbon in biosphere	0.29 g cm^{-2}	1.5 $\times 10^{18}$ gm
Organic C in sedimentary rocks	10^3 g cm^{-2}	5 $\times 10^{21}$ gm
CO_2 consumed in photosynthesis		1.1 $\times 10^{17}$ g yr^{-1}
CO_2 released in burning of fossil fuels		1.5 $\times 10^{16}$ g yr^{-1}
CO_2 increase in atmosphere = 1.0 ppm yr^{-1} or 0.3% yr^{-1}		5 $\times 10^{15}$ g yr^{-1}

[a] See footnote to Table I.
[b] A fraction of which might be recovered.

describe later how this concentration has been increasing as a result of the burning of fossil fuels. But it is important to realize at this stage that the origin of the CO_2 in the atmosphere is from the Earth's interior and that it has been released, like other atmospheric gases and water vapor, from volcanoes. As a matter of fact, more than 100000 times the present quantity of CO_2 has been released [13]. Nearly all of it has combined with elements like calcium and magnesium to form carbonates which have been precipitated as limestones or as dolomites (calcium-magnesium-carbonate). Sometimes this precipitation has been accelerated by marine organisms, such as those that form coral reefs or by foraminifera. It is interesting to note that, when limestone is converted into lime in the process of making cement, then CO_2 will be released back into the atmosphere. With a current production of around 500 million tons of cement, about 5×10^{14} gm of CO_2 is released back into the atmosphere per year. This figure is about 3% of the amount released through the burning of fossil fuels, namely about 1.5×10^{16} gm yr^{-1} [21] (See also Table II).

Until about the 13th century, the human race was solely dependent on solar energy which was essentially contemporary, stored at the most for 100 yr or so in the form of wood. Then there was discovered in northeast England that certain black rocks found along the seashore and thereafter known as 'sea coles' would burn. From this discovery dates the mining of coal and the systematic exploitation of the Earth's supply of fossil fuels. For the first time in human history man had found a huge supply of concentrated energy which could be commanded. The exploitation of petroleum began only about 100 yr ago; but from small beginnings there has grown a tremendous energy industry.

The relative ease with which this concentrated energy could be harnessed removed an important constraint to population growth. It has allowed the human population to increase and achieve a new (and precarious) ecological balance. Over most of the last

TABLE III

Energy contents of the world's initial supply of recoverable fossil fuels.[a]
From: M. K. Hubbert, *Energy Sources for Power Production*, International Atomic Energy Symposium, 1970

Fuel	Quantity	Energy Content (in various units)				Percent
		$(10^{21}$ J)	10^{15} kWh	Q Units 10^{18} BTU	Trillion W y	
Coal and lignite	7.6×10^{12} metric tons	201	55.9	191	6378	88.8
Petroleum liquids	2000×10^9 bbls $(272 \times 10^9$ metric tons)	11.7	3.25	11.1	371	5.2
Natural gas	10000×10^{12} ft^3 $(283 \times 10^{12}$ m^3)	10.6	2.94	10.0	335	4.7
Tar–Sand oil	300×10^9 bbls $(41 \times 10^9$ metric tons)	1.8	0.51	1.7	58	0.8
Shale oil	190×10^9 bbls $(26 \times 10^9$ metric tons)	1.2	0.32	1.1	36	0.5
Totals		226.3	62.9	214.9	7178	100.0

[a] See also footnote to Table I. Hubbert estimates recoveries of 50% of the total coal resource, 6% of the tar sands, and only 0.01% of the shale oil. Note that tar sand and shale may well produce larger amounts if oil prices rise. The heat values used here are averages for naturally occurring coals, petroleum and gas.

century, world consumption of energy from fossil fuels has increased at about 4% yr^{-1} [12]. The human population is now growing at a rate of about 2% yr^{-1}. Therefore, at present, the world's average non-nutrient energy consumption per capita is increasing at about 2% yr^{-1}.

But since the Earth's fossil fuels are strictly limited and non-renewable, it follows that they will eventually be depleted. M. King Hubbert [11] has estimated that the Earth's coal supply can serve as a major source of industrial energy for another 3 or 4 centuries, but petroleum, both because of its small initial supply and because of its more rapid rate of consumption, will last only about 70 yr (see Table III). Nearly 2% of the coal supply, and nearly 14% of the available crude oil have already been consumed.

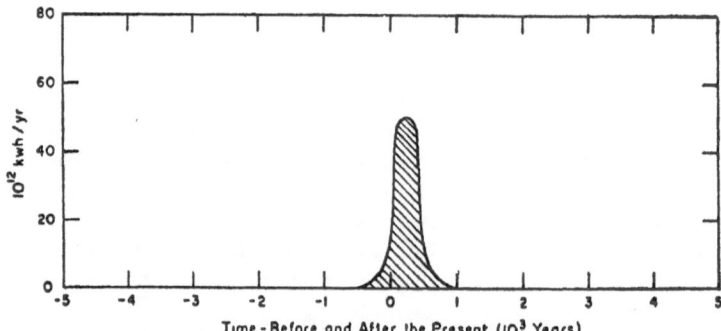

Fig. 2.   Epoch of exploitation of fossil fuels in historical perspective from minus to plus 5000 yr to present. Source: *Resources and Man*, Committee on Resources and Man, (Preston Cloud, Jr., Chmn.), National Academy of Sciences, Washington, D.C. (1969) [11].

As seen in historical perspective, the exploitation of fossil fuels is indeed only a short pulse in human history (see Figure 2).

Coal has been losing ground since the turn of the century to oil and natural gas. A more detailed breakdown is shown in Figure 3 which gives amounts used for various energy sources. As can be seen, by the year 1960, fuelwood, which was the most important energy source a hundred years ago, has disappeared and nuclear energy is beginning to take its place. Hydro-energy which has always been a small fraction of the total is becoming less and less important as its share of the total diminishes.

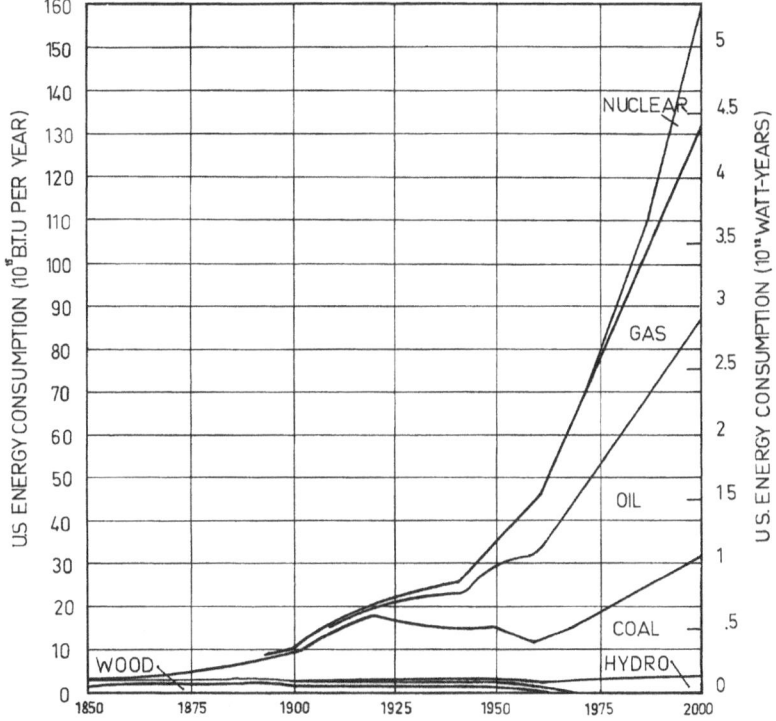

Fig. 3.   Past and projected energy use in the United States. Since 1850 consumption has increased by a factor of 30 and is projected to more than double over the present values by the year 2000. The sources of this energy has and will continue to change, with coal, hydro and wood yielding to nuclear, gas and oil. Source: Chauncey Starr 'Energy and Power' *Sci. Am.* **225**, 37 (1971) [35].

Some attention, of course, has to be paid to the actual way in which the energy is used. Nuclear, hydro, and most of the coal is first put into the form of electrical energy. Oil and natural gas are also used directly, particularly for space heating in homes, for industrial heat and, of course, for transportation.

In 1970, the U.S. economy consumed some 63×10^{15} BTU [7]. This energy cor-

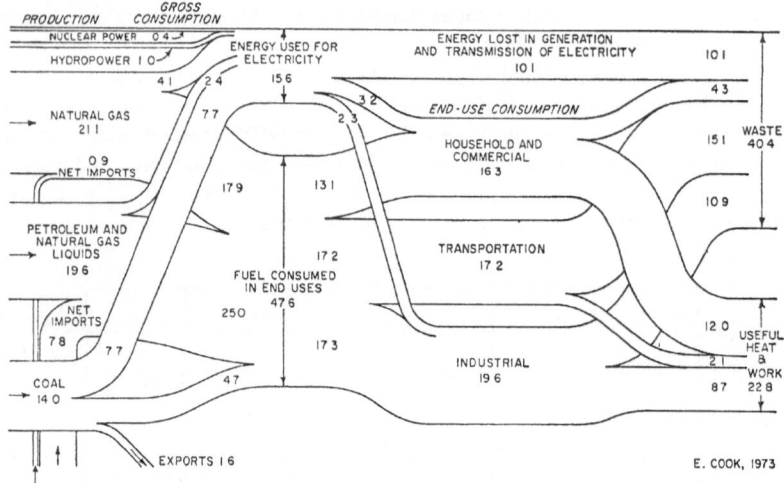

Fig. 4.   Use of energy in the United States for 1971 expressed in terms of 10^{15} BTU. The total consumption was 63.2×10^{15} BTU or an average per capita usage of 10 kW which is 100 times the food intake level of 100 W. The overall thermal efficiency of the whole system was about 40%, with transportation being least efficient. Source: Earl Cook (Texas A & M).

responds to that produced by 2.5 billion tons of coal – equal to about 5 yr U.S. coal production. About 1/4 of the energy is for the generation of electricity; most of the rest is used directly in the form of natural gas and petroleum.

This energy is put to a variety of uses (See Figure 4). Space heating is one of the largest. The average home requires as much energy as the average family car, about 70 million BTU or 900 gall of oil. Actually gas and electricity are becoming the preferred methods for the heating of homes.

The other domestic uses are for cooking and water heating, air conditioning, and lighting. Here, electricity supplies now more than half of the energy and is increasing its share gradually.

Around 10% of the national energy consumption goes to commercial uses, stores, offices, hotels, apartment houses, etc. Here coal and oil still play a more important role than in the individual household, but electricity is gradually gaining.

Industry takes more than 35% of the total energy used in the U.S. Agriculture probably consumes no more than 1% of all of the energy, mainly for the operation of tractors and for running of irrigation and drainage equipment.

As of 1970 at which time nuclear power was still quite unimportant, electricity was generated roughly 46% from coal, 16% from hydro, 24% from natural gas, and less than 13% from oil [7]. As is discussed below, in future years this ratio may be affected by the air pollution which the combustion of various fossil fuels produces. Fossil fuels are also used as the raw materials for the petrochemical industry. In spite of the extremely rapid growth, however, the total is less than 2% of the annual consumption.

The remainder of energy consumption is in the field of transportation and here it is mainly in the form of gasoline for cars. But taking all forms of transportation, including trucks, buses, rail and airplanes, the equivalent consumption in 1970 was 2.75 billion barrels or roughly 16×10^{15} BTU, about 25% of the total consumption.

3. Environmental Effects

We can now turn to a discussion of the environmental effects of the combustion of fossil fuels so that we can better judge its influence on the biosphere. As can be seen from Table IV, the five most common air pollutants, expressed in tons per year in the United States, are carbon monoxide, sulfur oxides, hydrocarbons, nitrogen oxides, and particles, in that order. The major sources are automobiles, industry, electric power plants, space heating, and refuse disposal. By far the greatest emission is carbon dioxide, but this will be discussed separately. The burning of fossil fuels also produces effects on water: chemical effects when the air pollutants are washed down by rainfall, and 'thermal' pollution, local heating effects which occur mainly when waste heat from thermal power plants is disposed. This will be discussed later also.

The simple listing given in Table IV unfortunately does not convey the importance of various air pollutants. Nor does it indicate their interactions or synergistic effects on the biosphere. The fossil fuels contain natural pollutants – radioactivity and possibly mercury in coals, vanadium in certain crude oils, whose effects are not always known. Furthermore, there are artificial additives, such as lead in gasoline, which may have pronounced effects.

The most important impurity in fossil fuel is *sulfur* and it causes some of the most troublesome effects. The only major source of sulfur dioxide appears to be pollution – 85% from fossil fuel combustion, the remainder from smelting and petroleum refining [28]. Natural sources, such as volcanoes, are responsible for but a very small portion of the atmosphere SO_2.* The sulfur dioxide may form sulfuric acid which

TABLE IV

National air pollutant emissions (millions of tons per year, 1965)[a]

	Totals	% of Totals	Carbon monoxide	Sulfur oxides	Hydro-carbons	Nitrogen oxides	Particles
Automobiles	86	60%	66	1	12	6	1
Industry	23	17%	2	9	4	2	6
Electric power plants	20	14%	1	12	1	3	3
Space heating	8	6%	2	3	1	1	1
Refuse disposal	5	3%	1	1	1	1	1
Totals	142 million		72 million	26 million	19 million	13 million	12 million

[a] Source: 'The Sources of Air Pollution and Their Control', Public Health Service Publication No. 1548, Government Printing Office, Washington, D.C., 1966.

* H_2S and SO_4 are however mostly from natural sources and account for approximately 65% of the total sulfur injected into the atmosphere.

often becomes associated with atmospheric aerosols, or it may react further to form ammonium sulphate. A typical lifetime in the atmosphere is about a week; but when the substances are removed by precipitation, they increase the acidity of the rainfall. Values of pH of about 4 have been found in the Netherlands and in Sweden, probably because of the high level of industrial activity in western Europe [15]. As a result, small lakes and rivers were beginning to show lowered pH values and endangering the stability of some ecosystems. A pH below 5.0 is considered inimical to all but the most tolerant freshwater fish [8]; the report of the committee on Water Quality Criteria recommends that no materials be added to water to give a pH below 6.0 [26].

Direct sampling of atmospheric particulates has revealed a stratospheric dust layer at 18 to 20 km with the majority of particles being sulfates [47]. These sulfates result from the oxidation and condensation in the stratosphere of H_2S and SO_2 which has been transported up through the tropical tropopause [12]. Such fine particles could have an effect on the radiation from the upper atmosphere and thereby affect mean global temperatures [12, 21, 22].

Larger *particles* are produced in the lower atmosphere by all kinds of sources, with the combustion of fossil fuels making an important contribution [5, 22]. Smoke from the combustion of petroleum products and coal, and fly ash are the main contributors. But pollution control technology is adequate to limit such emissions. Hopefully, therefore, they will diminish in importance in the future.

The major source of atmospheric *carbon monoxide* has long been considered to be almost exclusively pollution by man – especially from the incomplete combustion of fossil fuels. Certainly man is responsible for the very high concentrations occasionally exceeding 100 ppm, encountered around many population centers. Another source of carbon monoxide is the oxidation of methane; it seems to be the major source of the global carbon monoxide concentration of about 0.2 ppm [36]. Other natural sources exist [17]. Thus energy production seems to impose only a local perturbation on the global CO balance [12a].*

Nitrogen oxides occur naturally in the atmosphere in the form of nitrous oxide (N_2O), nitric oxide (NO), and nitrogen dioxide (NO_2). N_2O is most plentiful, relatively inert, and has a global concentration of about 0.25 ppm [28]. Nitrogen dioxide is a strong absorber of UV light and triggers photochemical reactions that produce smog [12]. In combination with water vapor it can form nitric acid.

The production of nitrogen oxides in combustion is very sensitive to temperature and is particularly important in explosive combustion which occurs in the internal combustion engine. If this engine is ever replaced by an external combustion engine which operates at a steady lower temperature, rather than at high peaks, then the emission of nitrogen oxides can be greatly reduced.

Hydrocarbons are emitted naturally into the atmosphere from forests and vegetation, and in the form of methane from bacterial decomposition of organic matter. Man produces only about 15% [12], but these contributions are concentrated in urban

* Man's activities are however responsible for an appreciable fraction of the atmospheric methane [31] and may therefore contribute additional atmospheric CO.

areas. The main contributor is the processing and use of petroleum, especially gasoline from the internal combustion engine. The reactions of hydrocarbons with nitrogen oxides and ultraviolet light produce the famous photochemical smog which was first encountered in Los Angeles. The biological effects of some of the products, such as ozone and complicated organic molecules, may be quite severe. Some of them are thought to be carcinogenic. Ozone has very detrimental effects on vegetation but, fortunately, these effects are localized. So far, at least, no regional or worldwide effects have been discovered.

There are ecological effects, however, that can be produced by *oil spills*. To put this matter in perspective, the Santa Barbara spill of 1969 involved some 10^{10} gm and the Torrey Canyon spill of 1967 about 10^{11} gm. These produced an intense local concentration of oil which is toxic to various marine organisms, particularly in its aromatic fractions. In addition, there is a yearly worldwide spillage which has been estimated as around 10^{12} gm which occurs in small amounts in various oil operations [21]; also there are natural oil seeps whose magnitude has not been estimated. About 10^{12} gm of waste motor oil is believed to be dumped every year in the United States alone and adds to the total load. So far at least there have been no detectable worldwide effects. It can therefore be assumed that bacteria degrade the oil fairly rapidly.

Carbon dioxide is the only combustion product whose increase on a worldwide basis has been documented. The injection of large quantities of CO_2 in the last few decades has been very sudden in relation to time scales in nature that are important. For example, while the top layer of sea water may take perhaps five years to adjust itself to the atmospheric carbon dioxide level [13, 33], the upper layers and especially the lower layers require some hundreds or thousands of years to adjust to a changed concentration. If the oceans were perfectly mixed at all times, then the carbon dioxide added to the atmosphere would distribute itself about 5/6 in the water and about 1/6 in the air. In fact, the distribution is about half and half.

The base CO_2 concentration of the late 19th century atmosphere is generally accepted to have been 290 ppm. Precise observations by C. D. Keeling *et al.* [24] of the Scripps Institution of Oceanography conducted at Mauna Loa Observatory placed the 1960 concentration at about 315 ppm, with an average annual increase of about 0.2% or 0.6 ppm (Figure 5)*. Thus in 1960 the CO_2 concentration was 8% above the base 1890 levels. Extrapolations to the year 2000 put the concentration at about 380 ppm or 32% above the base level [22].

The most widely discussed effect of increased CO_2 is the possibility of a general temperature rise. Carbon dioxide molecules have strong absorption bands, particularly in the infrared wavelength regions between 12 and 18 μ. It is in this region where most of the thermal energy radiating from the Earth into space is concentrated. By increasing absorption of this terrestrial radiation and by re-radiating it at a lower temperature, corresponding to that of the upper atmosphere, the carbon dioxide reduces the amount of heat energy lost by the Earth's surface.** This has been

* Recent data indicate an increase of 1 ppm yr^{-1} since 1968. See Mitchell pp. 149–173.
** See also Figure 1 on p. 8.

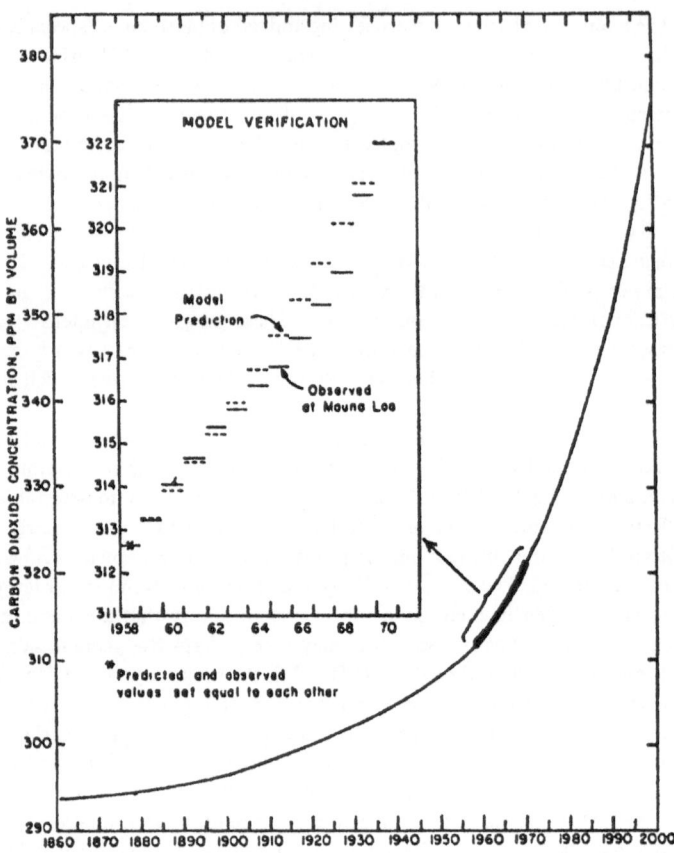

Fig. 5.   Model calculations and observed values for past and future carbon dioxide concentrations in the 'unpolluted' atmosphere. The increase is due to the worldwide combustion of fossil fuels [17].

termed the 'greenhouse' effect, although the operation of a real greenhouse depends only partly on the radiation property of the glass and more particularly on cutting down convective heat transfer.

The possibility that additional CO_2 from the burning of fossil fuels could produce a worldwide increase in temperature seems to have been posed first by the American geologist P. C. Chamberlain in 1899. In 1956, Gilbert Plass [25], a physicist, calculated that, if the CO_2 content of the atmosphere were to double, the worldwide mean surface temperature would rise 6.5°F. In 1963, Fritz Möller [23], a German meteorologist, calculated that a 25% increase in CO_2 would increase the average temperature by 1° to 7°F depending on the effects of water vapor in the atmosphere. The most extensive calculations have been performed by S. Manabe and R. T.

Wetherald [20, 18], who calculated that doubling atmosphere CO_2 from 300 ppm to 600 ppm would increase surface temperatures by 4.25 °F assuming average cloudiness, and by 5.25 °F assuming no clouds, all other factors being equal; i.e., constant relative humidity. Unfortunately, the problem is more complicated. Increasing surface temperature and lower level atmospheric temperature not only increases evaporation of water but changes cloudiness and, therefore, the albedo (average reflecting power) of the Earth. The normal average albedo is of the order of 35 % which means that 35 % of the incident sunlight is immediately reflected back out into space [19]. Changes in cloudiness, therefore, can have a very pronounced effect on the atmospheric temperature and on climate.

The situation is further complicated by atmospheric turbidity. Mitchell [22] has determined that atmospheric temperatures have risen 0.6 °C since 1860 until about 1940 (See Fig. 1 in Mitchell). Between 1940 and 1967, while warming occurred in Northern Europe and North America, there was a global temperature decline of 0.3 °C. Mitchell finds the cooling trend which has set in to be due to dust or smoke from volcanoes and partly from human activities; e.g., from contrails of jet planes.

To summarize, the increase in CO_2 in the atmosphere is quite certain and supported by reliable measurements. Its effect on climate is uncertain,* partly because no good worldwide radiation measurements are available and partly because of the interfering effects of changes in cloudiness and changes in the turbidity of the atmosphere. One of the exciting possibilities before us is the use of a weather satellite as a means for keeping track of the actual energy radiated back into space by the Earth, and therefore the possibility of performing for the first time reliable and standardized measurements of the 'global radiation climate'.

The higher levels of CO_2 may not maintain themselves for very long. For one thing, the oceans which contain 60 times the amount of CO_2 as compared with the atmosphere will begin to absorb the excess as mixing of the intermediate and deeper layers proceeds, and precipitate out a certain amount. For another, the increased CO_2 content of the atmosphere will stimulate a more rapid plant growth, a feature that has been experienced and utilized in experimental greenhouses. Of course, the CO_2 is returned to the atmosphere when the plant decays. But since forests constitute about $\frac{2}{3}$ of the world's photosynthesis on land and, therefore, roughly $\frac{1}{2}$ of the world total, their absorption of the excess CO_2 will spread out the peak over a longer time interval.

Above all, supplies of fossil fuels are limited (Table III); even if all known economically exploitable reserves $(3 \times 10^{18}$ gm C) [13] are burned suddenly, the CO_2 content would increase to a 3000 ppm peak but then immediately diminish.

The oxygen consumed in the combustion of the fossil fuels would, of course, come from the atmosphere and result in a worldwide decrease in oxygen. If all fossil fuel reserves were to be burned, the oxygen decrease would be substantially less than 3 % of the present value [4], or an absolute decrease of about 1 %. This is considered to be an inconsequential change.

* It is perhaps premature to add here to the many existing accounts, some of them quite lurid, of melting ice caps and rising sea levels, as well as of glaciation. A sober account is given by Kellogg [14a].

4. Nuclear Energy

The discovery of atomic energy processes and their introduction into the power generation picture is changing the dimension of man's energy generation and also the effects on the biosphere. Two processes are of concern to us: (1) the fission of heavy nuclei such as uranium; and (2) the fusion of light nuclei such as deuterium and tritium.

Nuclear energy can be considered as a heat source different from coal or oil; but once the energy has been released in the form of heat, it can be used in a very similar way to generate electricity. Therefore, the waste heat problem is very similar. The pollution characteristics, however, are different – radioactive rather than chemical.

The fission reaction has to start with uranium-235 since it is the only naturally-occurring isotope which is fissioned by the capture of slow neutrons. It therefore supplies the neutrons needed to carry out other reactions.

Each fission event of uranium-235 releases approximately 200 MeV or 3.2×10^{-11} J. One gram of uranium-235 therefore corresponds to 8.19×10^{10} J, an energy equivalent of 2.7 metric tons of coal or 13.7 barrels of crude oil [11]. It is also approximately equivalent to 1 thermal MW-day. A nuclear power plant producing 1000 electrical megawatts with a thermal efficiency of 33 % would consume about 3 kg of uranium-235 per day.

A nuclear 'burner' uses up large amounts of uranium-235; it is in short supply since it has an abundance of only 0.7 % in natural uranium ore. If reactor development proceeds as the AEC expects, inexpensive reserves of uranium (at less than $10 per pound of U_3O_8) would be used up in the mid '80's, and medium-priced fuel (up to $30 per pound) would be used up by 2000. Hence there has been a great fear that the present reactors would deplete our supplies of inexpensive and medium-priced uranium before converter and breeder reactors are developed to make fissionable plutonium 239 and uranium 233. Either one of these can be used as a catalyst to burn the highly-abundant uranium 238 or thorium 232 (Figure 6). Since thorium plus uranium have an abundance in the Earth's crust of about 15 ppm, we have here an immense energy source, millions of times greater than all the fossil fuels.

Further off is the possibility of generating energy by nuclear fusion. Of the two processes considered – the deuterium-deuterium reaction or the deuterium-tritium reaction – the latter is somewhat easier since it requires a lower reaction temperature. Essentially, lithium 6 is burnt, using tritium as a catalyst. The reactions proceed as shown in Figure 7. The amount of energy is limited by the abundance of lithium 6 in the Earth's crust, about 2 ppm.

As for the deuterium-deuterium reaction, it is a practically inexhaustible energy source, since 1 part in 6700 of the hydrogen in the sea is deuterium [11].

If we are lucky, then breeder reactors, and perhaps even fusion reactors, will be developed to a commercial stage before we run out of fossil fuels, particularly oil, and before we run out of uranium 235. Looking back at the situation thirty to fifty years from now, the epoch of fossil fuels and nuclear burners will seem a transitory event, but one which has profoundly influenced the course of human events. With inex-

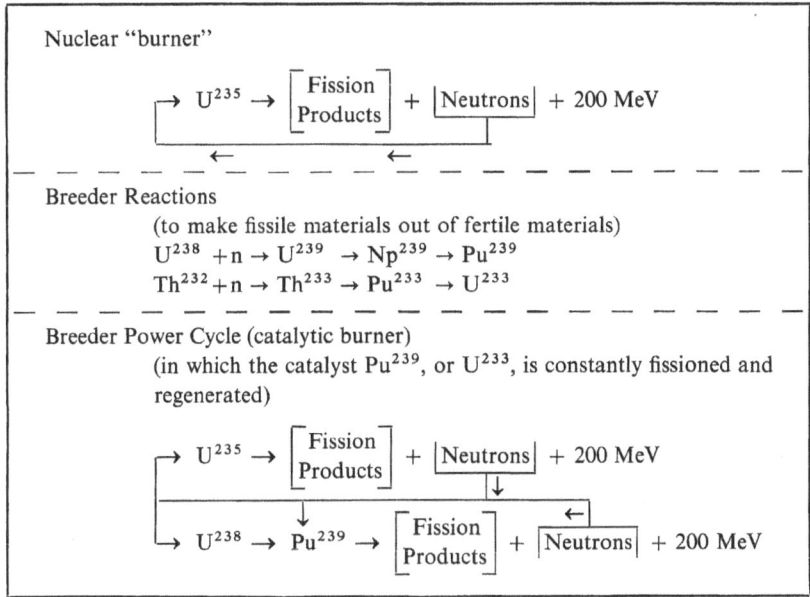

Fig. 6.   Power production by nuclear fission.

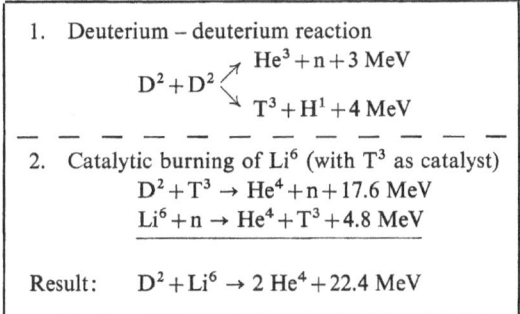

Fig. 7.   Power production by nuclear fusion.

haustible supplies of nuclear energy (inexhaustible, but not cheap!), cars may be running on ammonia or methane, artificially produced; coal and oil shale will be used as the basis for chemicals; and electric power, of course, will be generated directly in large nuclear catalytic (breeder or fusion) reactors. Electricity will be used to produce electrolytic hydrogen which will then be used for the manufacture of ammonia, methane, the reduction of ores, the production of fertilizers, and directly as a fuel.

It is difficult at this stage to predict the effects of large-scale uses of nuclear energy on the biosphere. We must make certain assumptions concerning the disposal of radioactive wastes. We must assume that the presence and storage of radioactive wastes from reactors as liquids in underground tanks will be replaced by the technique in which the non-volatile, long-lived activities are solidified and stored permanently in underground salt mines, and where the noble gas krypton 85, and tritium as tritiated water are held up until they decay.

5. Thermal Environmental Effects

This leaves dissipation of waste heat as the major problem as well as the major impact of human energy production. It has both direct effects on the biosphere, and it could have indirect effects by way of modification of the climate. It is useful to make this distinction here between local problems of thermal pollution, i.e., in the immediate vicinity of the power plant, and the global problems of the thermal balance created by the use and release of ever-growing amounts of energy.

The efficiency of power plants is determined by thermodynamic considerations; no matter whether they are fossil-fueled or nuclear-fueled, one tries to create very high temperature steam to drive the turbines and then condense this steam at the lowest practicable temperature. Water is the only practical medium for carrying the heat away and, as a consequence, more than 80% of the cooling water used by U.S. industry is used by electric power plants [32]. For every kWh of energy produced about 6000 BTU (about $\frac{2}{3}$ of the heat energy) is rejected in fossil-fuel plants and about 8500 BTU in present-day nuclear plants. This then is the energy that has to be dissipated in the immediate vicinity of the power plant. The remainder, of course, is carried in the form of electrical energy and then consumed and dissipated over a much wider geographic area – usually directly to the atmosphere without the intermediary of cooling water.

6. Local Effects

The local problem of thermal pollution can be viewed from the U.S. perspective where power consumption has been doubling every eight to ten years. The rapid increase in electric power plants as well as their increase in size is putting a real strain on the available amounts of cooling water. For example, by 1980 the electrical needs of the U.S. will require the use of $\frac{1}{6}$ of the total available fresh water run-off in the entire nation for cooling purposes. Now if we discount flood flows which usually occur about $\frac{1}{3}$ of the year and account for $\frac{2}{3}$ of the total run-off, it becomes apparent that the power industry will require about half the total run-off of the remaining $\frac{2}{3}$ of the year [32]. Even though some 95% of the water used in once-through cooling is returned to the stream, the increased temperature of the water has a potentially dele-terious effect on other water uses. The higher temperatures decrease the amount of dissolved oxygen and therefore the capacity of the stream to assimilate organic wastes. Bacterial decomposition is accelerated, which again depresses the oxygen levels. This oxygen reduction in turn decreases the viability for aquatic organisms while at the same time the increased temperature raises their metabolic rate and need for oxygen. Temperature changes, either increases or decreases, have pronounced effects on the life cycle of aquatic organisms and high temperatures specifically are lethal to fish [6].

In the face of stringent water quality requirements by the States and by the U.S. Department of the Interior, power companies are installing cooling devices before returning the water to the stream. These devices may be cooling ponds or spray ponds or cooling towers [1]. Usually some of the cooling water is evaporated, and the excess heat is therefore dissipated into the atmosphere rather than returned to the stream. There is no problem here of technical feasibility, but it does raise the cost of producing electricity which should result in an increase to the domestic consumer of 1 % or 2 %.

7. Regional Effects

This strategy of 'dilution' or spreading of the waste heat has to be re-examined as the scale of the problem increases. We are already aware of the 'heat islands' created in metropolitan areas and are beginning to examine their meteorological effects which may not be altogether bad, but they certainly will change the pre-existing climatology. It is well known, of course, that cities are warmer than the surrounding countryside, and this has a definite effect on the ecology and biospheric activity in metropolitan areas. The release of heat in a relatively small local area also causes a change in the convective pattern of the atmosphere. The addition of large amounts of particulate material from industry, from space heating, and from refuse disposal provides cloud condensation nuclei [30]. One documented study in the State of Washington shows an increase of approximately 30 % in average precipitation over long periods of time as a result of the air pollution produced from pulp and paper mills [10].

8. Global Effects

The worldwide effects of energy consumption can be estimated as follows: At present (1970), the U.S. population of 204×10^6 is estimated to consume 63.2×10^{15} BTU yr^{-1} or 2.2×10^{12} W (installed electrical capacity is approximately 3×10^{11} W in the U.S.). Thus, U.S. energy consumption per capita is a little over 10 kW while for most of the world it is still just barely above the food intake level of 100 W. At the present time, the U.S. consumes roughly $\frac{1}{3}$ of all the energy [7], so that total world consumption is of the order of 6.6×10^{12} W (see Table V).

Future projections depend, of course, very much upon assumptions. If we assume that in fifty years the rest of the world will have reached an energy consumption level equivalent to present U. S. consumption, and if we assume a population of 10^{10}, then the total man-made energy would be 10^{14} W, distributed however in some patchy manner which depends on the actual distribution of population centers as well as on the distributing effects of the atmosphere and oceans.

These numbers must be compared with the energies which the Earth receives from

TABLE V

Global energy balances

U.S. energy consumption rate (1970)	2.2×10^{12} W
World consumption (1970)	6.6×10^{12} W
World consumption assuming population of 10 billion and 10 kW (thermal) per capita	1×10^{14} W
Local energy dissipation rate in a city (assuming 20000 people per sq mile and 10kWt per capita)	78 W m^{-2}
Solar constant	1400 W m^{-2}
S: Solar energy absorbed (averaged over entire Earth)	220 W m^{-2}
E: Infrared energy emitted (averaged)	220 W m^{-2}
Energy storage (typical value, either positive or negative) computed as $\frac{1}{2}$ (F-Fy), where F is radiation balance (half-yearly average), and Fy is yearly average; radiation balance is (S-E) computed over latitude zones of the Earth. (Ref. J. C. Johnson, *Phys. Meteorology*, p. 171)	30 W m^{-2}
Earth's surface area	5.11×10^{14} m^2
Land area	1.48×10^{14} m^2

the Sun. The solar constant of 1400 W m^{-2} applies to solar radiation outside of the atmosphere. Going from a cross-section to a spherical surface introduces a factor of $\frac{1}{4}$. Roughly $\frac{1}{3}$ of this energy is reflected back into space and an additional small amount is abstracted in the upper atmosphere. Therefore, over the total Earth's surface, about 220 W m^{-2} is absorbed and converted into infrared energy which is radiated back

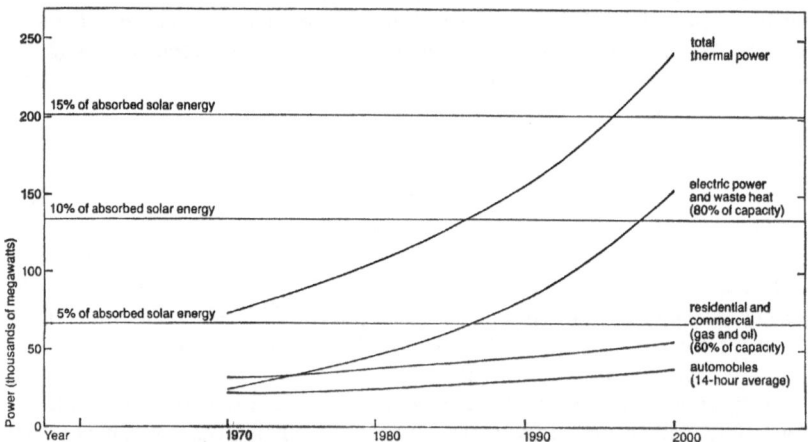

Fig. 8. Thermal power generation in the Los Angeles Basin. (Calculated by L. Lees [21]). This area of 4000 sq miles generates thermal power equivalent to more than 5% of the solar energy absorbed at the ground. The year 2000 value is expected to be 18%, based on extrapolating electrical power consumption with a doubling time of 10 yr, and extrapolating other consumption of energy with a lower rate. Even if extrapolation is not quite correct, it can be seen that the waste heat loads are very large and can certainly lead to changes in local climate.

into space, thus preserving the Earth's radiation balance.* Over the whole Earth, the heat loss is about 1.2×10^{17} W or about 1000 times the energy which would be dissipated by human activity.

It would be premature, however, to assume that the energy input from human activities can be neglected – just because the average heat input is so much less than the solar heat input. The atmospheric engine is very subtle, and energy inputs in particular places can change weather patterns. (see, e.g. Figure 8)

References

1. Belter, Walter G.: 'Thermal Effects – A Potential Problem in Perspective', in D. A. Berkowitz and A. M. Squires (eds.), *Power Generation and Environmental Change*, MIT Press, Cambridge, Mass. (1971).
2. Berkowitz, D. A.: 'Global Energy Balance', *Science* **169**, 426 (1970).
3. Bolin, B. and Bischof, W.: 'Variations of the Carbon Dioxide Content of the Atmosphere in the Northern Hemisphere', *Tellus* **22** (1970).
4. Broecker, Wallace S.: 'Man's Oxygen Reserves', *Science* **168**, 1537 (1970).
5. Bryson, R. A. and Wendland, W. M.: 'Climatic Effects of Atmospheric Pollution', this volume pp. 139–147.
6. Carlson, Clarence A.: 'Impact of Waste Heat on Aquatic Ecology', in D. A. Berkowitz and A. M. Squires (eds.), *Power Generation and Environmental Change*, MIT Press, Cambridge, Mass. (1971).
7. Cook, E.: 'The Flow of Energy in An Industrial Society', *Sci. Am.* **224**, 134, 1537 (1970).
8. Doudoroff, P. and Katz, M.: 'A Critical Review of Literature on the Toxicity of Industrial Wastes and Their Components to Fish', *Sewage Ind. Wastes* **22**, 1432 (1950).
9. *Energy R & D and National Progress*, Interdepartmental Energy Study, Office of Science and Technology, Washington, D. C. (1964).
10. Hobbs, P. V., Radke, L. F., and Shunway, S. E.: 'Cloud Condensation Nuclei from Industrial Sources and Their Apparent Influence on Precipitation in Washington State', *J. Atmospheric Sci.* **27**, 81 (1969).
11. Hubbert, M. K.: 'Energy Resources of the Earth', *Sci. Am.* **224**, 3 (1971).
12. *Inadvertent Climate Modification*, Report of the Study of Man's Impact on Climate (SMIC), MIT Press, Cambridge Mass. (1971).
12a. Jaffe, Louis: 'The Global Balance of Carbon Monoxide', this volume, pp. 83–110.
13. Johnson, Francis S.: 'The Oxygen and Carbon Dioxide Balance in the Earth's Atmosphere', this volume, pp. 49–56.
14. Junge, C. E. and Manson, J. E.: 'Stratospheric Aerosol Studies', *J. Geophys. Res.* **66**, 2163 (1961).
14a. Kellogg, William: 'Climate Change and the Influence of Man's Activities on the Global Environment', this volume, pp. 13–23.
15. Lundholm, Bengt: 'Interactions Between Oceans and Terrestrial Ecosystems', this volume, pp. 329–336.
16. MacDonald, Gordon, J. F.: 'Climatic Consequences of Increased Carbon Dioxide in the Atmosphere', in D. A. Berkowitz and A. M. Squires (eds.), *Power Generation and Environmental Change*, MIT Press, Cambridge Mass. (1971).

* Another estimate for power flux is 68 W m^{-2} which refers to the "mean annual difference between absorbed solar radiation and long wave radiation into space" (MacDonald) [16]. or to the "difference between the absorption of solar radiation at the surface and the net radiative emission from the surface" (Berkowitz) [2]. A preferred estimate, and at the same time the lowest flux figure, would be 30 W m^{-2} which is the energy storage, computed as one-half the difference between the half-yearly and the yearly average of the radiation balance computed over latitude zones of the Earth. Use of this value would lead to an extreme projection for 2020, whereby the 'solar effect' would be 140 to as little as 40 times human energy dissipation, depending on whether we take total earth surface, or just land area.

17. Machta, L.: in D. Dyrssen (ed.), 'The Changing Chemistry of the Oceans', *Nobel Symp.* **20**, Almqvist and Wiksell, Upsala. (1972)
18. Manabe, Syukuro: 'The Dependence of Atmospheric Temperature on the Concentration of Carbon Dioxide', this volume, pp. 73–77.
19. Manabe, Syukuro: 'Cloudiness and the Radiative, Convective Equilibrium', this volume, pp. 175–176.
20. Manabe, S. and Wetherland, R. T.: 'Thermal Equilibrium of the Atmosphere with a Given Distribution of Relative Humidity', *J. Atmospheric Sci.* **24**, 241 (1967).
21. *Man's Impact on the Global Environment*, Report of the Study of Critical Environmental Problems (SCEP), MIT Press, Cambridge, Mass. (1970).
22. Mitchell, J. Murray, Jr.: 'A Reassessment of Atmospheric Pollution as a Cause of Long-Term Changes of Global Temperature', this volume, pp. 149–173.
23. Moller, F.: 'On the Influence of Changes in the CO_2 Concentration in Air on the Radiation Balance of the Earth's Surface and on the Climate', *J. Geophys. Res.* **68**, 3877 (1963).
24. Pales, J. C. and Keeling, C. D.: 'The Concentration of Atmospheric Carbon Dioxide in Hawaii', *J. Geophys. Res.* **70**, 6053 (1965).
25. Plass, G. N.: 'The Carbon Dioxide Theory of Climatic Change', *Tellus* **8**, 140 (1956).
26. *Report of the Committee on Water Quality Criteria*, Federal Water Pollution Control Administration, U.S. Government Printing Office (1968).
27. *Resources and Man, Committee on Resources and Man*, (Preston Cloud, Jr. Chairman), National Academy of Sciences, Washington, D. C. (1969).
28. Robinson, Elmer and Robbins, Robert C.: 'Gaseous Atmospheric Pollutants from Urban and Natural Sources', this volume, pp. 111–123.
29. Ryther, John H.: 'Is the World's Oxygen Supply Threatened?', *Nature* **227**, 374 (1970).
30. Schaefer, Vincent J.: 'The Inadvertent Modification of the Atmosphere by Air Pollution', this volume, pp. 177–196.
31. Singer, S. F.: 'Stratospheric Water Vapour Increase Due to Human Activities', *Nature* **233**, 543 (1971).
32. Singer, S. F.: 'Environmental Quality and the Economics of Cooling', in D. A. Berkowitz and A. M. Squires (eds.), *Power Generation and Environmental Change*, MIT Press, Cambridge, Mass. (1971).
33. Sisler, Frederick D.: 'Impact of Land and Sea Pollution on the Chemical Stability of the Atmosphere', this volume, pp. 57–71.
34. *The Sources of Air Pollution and Their Control*, Public Health Service, Publication No. 1548, Government Printing Office, Washington, D. C. (1966).
35. Starr, Chauncey: 'Energy and Power', *Sci. Am.* **225**, 37 (1971).
36. Weinstock, B. and Niki, H.: 'Carbon Monoxide Balance in Nature', *Science* **176**, 290 (1972).

PART II

CHEMICAL BALANCE OF GASES IN THE EARTH'S ATMOSPHERE

INTRODUCTION

Johnson reviews our current thinking on the origin and evolution of the Earth's atmosphere. It is now widely accepted that the primeval atmosphere contained no free oxygen, and that appreciable oxygen concentrations developed only because of the evolution of living organisms which, in turn, had to adapt themselves to the environmental changes caused by the oxygen.

Today, as was first pointed out by L. V. Berkner and L. Marshall, the balance between gain and loss of oxygen is extremely close and very fragile. Ecological effects of pollution could well lead to a decrease of oxygen on a global scale. This problem is discussed critically by Sisler who considers also the fate of the other major gases in the atmosphere.

Carbon dioxide plays an important role: it furnishes the carbon for plant growth, thereby supporting all forms of life. It also exercises a moderating effect on the world's climate by retaining some of the heat radiation which would otherwise escape into space. The concentration of CO_2 has increased suddenly and markedly since the beginning of the industrial revolution – because of the rapid burning of coal, gas, and oil. The concentration will continue to increase, in spite of the introduction of nuclear energy. As a result, atmospheric temperatures may increase all over the world.

The calculations by Manabe predict an increase of 0.8 °C, which could have appreciable environmental consequences and change the world's climate. But CO_2 is not the only factor which affects the long-term state of climate; other types of atmospheric pollution play an important role (see Part I and Part III).

In his final paragraph, Manabe points to the necessity of considering the exchange of CO_2 between atmosphere and ocean. As Johnson has pointed out, the ocean contains 60 times the amount of CO_2 in various dissolved forms and acts as a buffer for changes in atmospheric CO_2 concentration. But, as develops from the report of Berger and Libby, the problem may be quite complex, with the exchange rate possibly controlled by an enzyme. This, of course, raises the possibility again that worldwide marine pollution may control the concentrations of atmospheric CO_2 by affecting the organisms which produce the enzyme.

One of the major pollution inputs to the atmosphere from human activities is in the form of carbon monoxide, almost all of it produced by motor cars. We have only recently become aware of the fact, however, that there are several natural sources, which jointly are about an order of magnitude stronger than the anthropogenic sources. The characteristics of the various sources and sinks of atmospheric carbon monoxide are described by Jaffe. Robinson and Robbins consider the sources and fate of natural and man-made pollutants of sulfur and nitrogen.

The final paper by Singer reviews the various effects of SST's on the stratosphere. At present, the major uncertainty relates to how much ozone would be destroyed by the nitric oxides released in the combustion process.

THE OXYGEN AND CARBON DIOXIDE BALANCE
IN THE EARTH'S ATMOSPHERE*

FRANCIS S. JOHNSON

University of Texas at Dallas, Dallas, Tex., U.S.A.

Abstract. The scarcity of noble gases on Earth, compared to cosmic abundance, constitutes powerful evidence that the Earth was formed without an atmosphere and that the atmosphere has evolved by release of gases from the Earth's interior. These gases consisted mainly of carbon dioxide and water vapor, and they contained no free oxygen. The water vapor has gone mainly to form the oceans, and the carbon dioxide has gone mainly into carbonate rocks, and a minor constituent, nitrogen, remains the principal atmospheric constituent. Oxygen has been released mainly by photosynthesis, which involves the consumption of carbon dioxide. Most photosynthesis does not make a lasting contribution to atmospheric oxygen, because the products of photosynthesis undergo decay and oxidation, consuming as much oxygen as was produced in their production. A small fraction (about 10^{-3}) of the products of photosynthesis escapes decay and provides a lasting contribution to atmospheric oxygen. This natural conversion of carbon dioxide from the Earth's interior to oxygen is completely overwhelmed today, by a factor near 10^3, by the burning of fossil fuels, and the carbon dioxide content of the atmosphere is therefore increasing at a relatively rapid rate. Though detailed studies are lacking, the possibility exists that world climate may be affected. The risk of a serious perturbation appears small, but the problem is only poorly understood and the confidence level in such a prediction is low.

1. Introduction

The Earth's atmosphere is not a residual of an early atmosphere that was formed along with the Earth. Instead, it has largely arisen by continuous, though not necessarily uniform, degassing of the Earth's interior throughout geological time [12]. The gases released from the Earth's interior contained no free oxygen; its development in the atmosphere is a notable occurrence on Earth, and it results from the presence of life. The release of gas from the Earth's interior has probably resulted from local heating associated with tectonic activity. It appears most likely that the release commenced about 4.5×10^9 yr ago and that it was initiated by the capture of the moon by the Earth [14].

The release of gas from the Earth's interior has been much more massive than the present atmosphere would suggest. First of all, water should be considered as an essentially atmospheric gas. The total quantity of water released from the Earth's interior approximates 3×10^5 g cm^{-2} averaged over the Earth's surface, and most of this is now found in the oceans, simply because the Earth is cool enough so that it has condensed. The next most important constituent has been carbon dioxide, for which the release has been about 4×10^4 g cm^{-2} [7]. However, most of this has been removed from the atmosphere by dissolving in water followed by its precipitation from the ocean in the form of calcium carbonate. The quantity remaining in the

* This research was supported by the National Aeronautics and Space Administration under grant NGL 44-004-001.

atmosphere today is 0.45 g cm^{-2}, while the amount in the ocean amounts to 27 g cm^{-2} averaged over the Earth's surface, or 60 times as much. Thus, all but a few thousandths of the carbon dioxide released from the Earth's interior has been extracted from the atmosphere-ocean system and locked up in geologic deposits. Nitrogen has been a rather minor constituent among the gases released, but it is chemically inert and, unlike the other gases, it has accumulated in the atmosphere. The total release has been close to 10^3 g cm^{-2}, and about 10% of this has been removed and locked up in geologic deposits [5]. A fraction 10^{-8} of the atmospheric nitrogen is fixed each year [4] but most of this is returned to the atmosphere within a few years through decay of plant materials.

2. The Oxygen Balance

There are two significant sources of atmospheric oxygen: the photodissociation of water vapor, followed by the escape to space of the hydrogen that is released; and photosynthesis, followed by accumulation of unoxidized deposits of organic materials, mainly carbon.

In the case of the photodissociation of water vapor as a source of oxygen, the hydrogen that is produced must escape to space in order to leave the oxygen as a persistent constituent of the atmosphere. If the hydrogen did not escape from the upper atmosphere, it would eventually become oxidized, using as much oxygen as was released during its formation. Since hydrogen is a light gas, it can escape from the upper atmosphere by virtue of its thermal energy. However, under present conditions at least, the rate of escape is rather slow, not over 10^8 atoms cm^{-2} s^{-1} [3]. Over geologic time, approximately 10^{17} s, this amounts to 10^{25} atoms cm^{-2}, or 17 g cm^{-2}. The amount of water which must be dissociated to produce this amount of hydrogen is 150 g cm^{-2}. Unless conditions at some earlier time were very different from those prevailing now, the rate was probably no larger in the past. The limitations on the rate of escape are provided mainly by the rate at which atomic hydrogen diffuses out of the atmosphere from its region of formation, and the rate at which the atoms enter into chemical combination with other constituents of the atmosphere. To increase the escape rate, the atomic hydrogen concentration in the photodissociation region would have to build up above the present values – something that the chemistry might not permit. The low temperature of the upper atmosphere is a secondary factor; it makes the upper atmosphere very dry and reduces the number of photodissociation events undergone by water molecules, but this is not of much significance since the hydrogen atoms mainly recombine into water anyhow. Another secondary factor limiting hydrogen escape is the protection from photodissociation provided by oxygen molecules, which absorb the dissociating radiation. In the absence or near absence of oxygen, carbon dioxide would assume this role. However, if neither gas were present and the rate of photodissociation of water vapor were to increase, hydrogen recombination would also increase and the escape rate would not necessarily be affected very much unless the recombination processes involving hydrogen were in some way inhibited at the same time.

The second source of oxygen is photosynthesis, which of course was not active at the beginning of geologic time. The first life apparently originated over 3×10^9 yr ago. Cloud [2] associates this earliest life with the banded iron formations, assuming that an oxygen receptor was needed to avoid oxygen poisoning of the organism and that it was provided by ferrous ions. The last of these banded iron formations is about 2×10^9 yr old, and this probably indicates the time when oxygen and peroxide mediating enzymes arose and permitted green plant photosynthesizers to release free oxygen. Thereafter there was probably a generally increasing rate of oxygen production, and finally some significant accumulation in the atmosphere. Berkner and Marshall [1] identify the oxygen concentration at the beginning of the Cambrian era 6×10^8 yr ago as 1% of the present atmospheric value. At this level of oxygen concentration, organisms change their mechanism of energy release from fermentation to respiration, which is far more efficient. Further, the oxygen concentration in the atmosphere at this point is adequate to form a low level ozone layer. This ozone layer would not be sufficient to completely screen out biocidal radiation in sunlight ($\lambda \approx 2500$ Å), but when augmented by a thin layer of water, it would attenuate the radiation to the point where life could exist. Consequently, the oceans became capable of supporting life almost everywhere at this point in geologic time.

Berkner and Marshall further identify the end of the Silurian era 4.2×10^8 yr ago as the point in time when the atmospheric oxygen reached 10% of its present concentration. Ozone then provided protection from lethal radiation even over land areas, and life emerged ashore.

Most of the oxygen that has been introduced into the atmosphere by photosynthesis in excess of decay of organic materials has been consumed by oxidation of surface geologic materials and volcanic gases. There are two ways to estimate the total quantity of oxygen involved. One is to examine sedimentary rocks and compare their degree of oxidation to that of the igneous rocks that were weathered to form them. The other is to determine the quantity of unoxidized organic remains; these include coal and oil deposits, of course, plus a much greater amount of highly dispersed organic carbon in sedimentary rocks. Neither method is very exact.

The total mass of sedimentary rocks is estimated to be about 2.4×10^{24} g [7] which amounts to an average of about 5×10^5 g cm^{-2} over the Earth's surface. The igneous rocks that weather to produce sedimentary rocks contain about 3.80% ferrous oxide. The sedimentary rocks contain 0.9% ferrous oxide. To produce the increased oxidation, assuming that all of the missing ferrous oxide has been oxidized to ferric, 80×10^{20} g oxygen is required. In addition to this, some other oxidation has occurred. The volatiles that have escaped from the Earth's interior include 130×10^{20} g sulfur, 420×10^{20} g chlorine, and 40×10^{20} g nitrogen [7]. The sulfur was probably emitted as H_2S or SO_2, and it has been oxidized to SO_3, utilizing from 280×10^{20} to 70×10^{20} g oxygen, depending upon whether it was emitted as H_2S or SO_2. The nitrogen was probably released either as N_2 or NH_3; in the latter case 84×10^{20} g oxygen would have been required to free the nitrogen. The chlorine was probably emitted as HCl, which would react with metal oxides or hydroxides without consumption of oxygen.

Finally, the 1900×10^{20} g carbon dioxide that has been introduced into the atmosphere may have been emitted as carbon monoxide, in which case the amount of oxygen consumed in oxidizing it would have been 700×10^{20} g. Thus, the oxygen required for oxidation of crustal materials and volatiles emitted from the Earth's interior must be at least 150×10^{20} g; if the gases released from the Earth's interior were strongly reducing, the quantity could be much greater.

The coal, oil and other hydrocarbon reserves are products of photosynthesis that have not become oxidized. They are estimated to amount to 3×10^{18} g carbon (National Academy of Sciences, 1962), or 0.6 g cm^{-2} averaged over the Earth. In addition, the carbon in living matter and undecayed organic material amounts to about 1.5×10^{18} g [12]. However, these represent only a small fraction of the surviving unoxidized products of phtosynthesis. Of all carbon that has gone into sedimentary deposits, about 20% is of organic origin [16]. The total carbon associated with 1900×10^{20} g CO_2 is 500×10^{20} g, so 100×10^{20} g is in organic fossil form. This amount of stored carbon implies that 270×10^{20} g oxygen has been made available to the atmosphere. This suggests that the total oxygen production beyond the amount consumed in decay of organic materials has been almost thirty times the amount now in the atmosphere. It also indicates that the gases released from the earth's interior have consisted predominately of carbon dioxide rather than monoxide; otherwise the oxygen source would have had to have been about a factor of three larger than can be justified in terms of fossil organic carbon. However, such an amount is not outside the range of speculation [12].

Table I summarizes the amounts of the four principal gases introduced into the atmosphere, and the amounts remaining in the atmosphere today.

During the processes of geologic evolution sedimentary rocks become exposed and weathered. This recycling of sedimentary rocks is important in the chemical balance of the geologic system; for example, it provides an important portion of the minerals constantly carried to the sea by river water. The cycle time involved has been estimated by Li [7] to be 2×10^8 yr based on the age distribution of exposed sedimentary rocks; a similar estimate can be made on the basis of calcium in river water and the total content of calcium in sedimentary deposits. This weathering of sedimentary rocks exposes some of the organic fossil carbon that was associated with the release of

TABLE I

Atmospheric constituents

	H_2O	CO_2	N_2	O_2
Total supply	1.5×10^{24} g	1.9×10^{23}	5×10^{21}	2.7×10^{22}
Total now in atmosphere	10^{19}	2.25×10^{18}	4×10^{21}	10^{21}
Total supply per unit area	3×10^5 g cm^{-2}	4×10^4	10^3	5×10^3
Total in atmosphere per unit area	2	0.45	8×10^2	2×10^2

Note: There is sixty times as much CO_2 in the ocean than in the atmosphere but only about 1% as much oxygen and nitrogen in the ocean as in the atmosphere.

oxygen to the atmospheric reservoir. This constitutes the most important drain on the oxygen reservoir today. The release rate, corresponding to the exposure by weathering of the entire reservoir of 100×10^{20} g fossil carbon in 2×10^8 yr, is 5×10^{13} g yr^{-1}, thus consuming 1.3×10^{14} g yr^{-1} oxygen. This is over an order of magnitude greater than the average rate over geologic time of oxidation of geologic materials and volcanic gases, which is 9×10^{12} g yr^{-1} oxygen. Thus, the net rate of oxygen production by photosynthesis and deposition in sedimentary deposits of unoxidized organic material goes 90 % to the oxidation of old organic carbon, leaving 10 % for the oxidation of newly weathered igneous rocks and volcanic gases.

It is speculative as to how uniform the rate of oxygen production has been over geologic time. We will assume that it has been uniform over a period of 3×10^9 years, in which case the average rate of release has been 9×10^{12} g oxygen yr^{-1} or 2×10^{-6} g oxygen cm^{-2} yr^{-1}. It is worthwhile comparing this with the rate of photosynthesis today, which is estimated to produce 3×10^{16} g C yr^{-1}, about half on land and half in the oceans [6], [13], and [15]. The associated rate of oxygen release is 8×10^{16} g yr^{-1}, about 4 orders of magnitude above the average net rate of oxygen production over geologic time and almost 3 orders of magnitude greater than the rate of oxygen consumption by weathering of sedimentary rocks today and the associated oxidation of fossil carbon. This implies that only about one part in 10^3 of the products of photosynthesis escape oxidation. However, it is that small part of photosynthesis that provides the real contribution to atmospheric oxygen and allows oxygen to persist in the atmosphere even in the face of losses to the earth's crust by oxidation of geologic materials.

3. Carbon Dioxide

The carbon dioxide annually consumed in photosynthesis is 11×10^{16} g. However, the rate of release of carbon dioxide into the atmosphere by oxidation of recently grown organic materials matches this within about one part in 10^3. Therefore, the removal of carbon dioxide each year to deposit new fossil carbon amounts to about 10^{14} g. The burning of fossil fuel now releases about 1.5×10^{16} g carbon dioxide yr^{-1}, two orders of magnitude greater than the rate of return to the fossil reservoir. In fact, it is an appreciable fraction ($\sim \frac{1}{7}$) of the carbon dioxide entering the photosynthesis cycle each year, although this is of little significance owing to the relatively short cycle time in living material. It is more important to compare the carbon dioxide released from fossil fuels with the carbon dioxide consumed to produce new fossil carbon each year; these figures are 1.5×10^{16} and 10^{14}, showing that we are depleting this resource over a hundred times faster than it is being renewed. The total reservoir of fossil carbon in forms suitable for exploitation has been estimated to be near 3×10^{18} g. When this resource has been fully expended, the oxygen consumption associated with this will amount to about 6×10^{18} g, or 1.2 g cm^{-2} averaged over the earth, not enough to seriously deplete the atmospheric oxygen reservoir.

Table II summarizes the time constants for atmospheric gases for some of the source

TABLE II

Time constants

	Rate	(Atmospheric content)/rate
CO_2		
Photosynthesis	1.1×10^{17} g yr^{-1}	20 yr
Combustion	1.5×10^{16}	150[a]
Fossilization	10^{14}	2×10^4
O_2		
Photosynthesis with no decay	8×10^{16} g yr^{-1}	10^4 yr
Net organic production	8×10^{13}	10^7
Photodissociation and hydrogen escape	2×10^{11}	5×10^9
N_2		
Bacterial action	4×10^{13} g yr^{-1}	10^8 yr

[a] Present observed rate of increase of atmospheric CO_2 is 0.7 ppm yr^{-1}, for which time constant is 450 yr. The difference is explained by a portion of the carbon dioxide becoming dissolved in the ocean.

and loss mechanisms. These figures should be used with caution for carbon dioxide because of exchange that can take place between the atmosphere and the ocean, which contains sixty times more carbon dioxide than does the atmosphere.

There has been much speculation on the effect of the massive release of carbon dioxide that is taking place as a result of burning fossil fuel. There has been a clear trend toward higher carbon dioxide content of the atmosphere, with an increase of about 10 % since the beginning of this century; the exact increase is uncertain because of inaccuracies in the early measurements. However, a clear upward trend is discernible also in recent years [10]. An important aspect of the carbon dioxide increase in the atmosphere from fossil fuel combustion is the degree to which it is shared with the oceans. Early estimates suggested that most of the carbon dioxide released by burning fossil fuels had remained in the atmosphere and that no significant portion of it had been shared with the oceans. However, the data of Pales and Keeling [10] are much more reliable than the early data and they indicate an exchange time from the atmosphere to the ocean of about 5 yr. Their indicated rate of increase of atmospheric carbon dioxide from 1958 to 1963 is 0.7 ppm yr^{-1}, or 0.5×10^{16} g yr^{-1}, which is just about $\frac{1}{3}$ the rate of release of carbon dioxide from fossil fuel burning during that period. However, the rate of accumulation of carbon dioxide in the atmosphere can also be influenced by other factors, such as long-term temperature trends for the Earth, changes in patterns of cultivation, etc.

4. Global Effects

We shall inquire as to the steps in the oxygen-carbon-dioxide cycle that are sensitive to perturbation or instability. For example, about half the carbon dioxide consumption each year in photosynthesis occurs in the ocean, where phytoplankton are the primary producers. The growing spread of pollution has already shown up indirectly in midocean area. DDT dust has been observed far out at sea [11]. Also, sea birds, the Bermuda petrel, have been found to have substantial concentration of chlorinated hydrocarbons in their eggs [17], and this has apparently deleteriously affected their reproductivity. Owing to their food habits – they are at the peak of the oceanic food chain pyramid – the chlorinated hydrocarbons have apparently been obtained from fish, whose food source in turn goes back ultimately to the phytoplankton, which presumably gathered the insecticide from dust settling into the ocean. Similar evidence comes from Clear Lake, California, where water containing less than 0.02 ppm DDT produced plankton containing 5 ppm, and fish near the top of the food chain contained enough insecticide to kill birds that ate them. The sensitivity of phytoplankton to such materials is probably variable, and one cannot immediately discount the possibility that insecticide and herbicide pollution may have widespread and unforeseen effects that could affect the plankton that converts a large proportion of our carbon dioxide to oxygen.

This problem is not alleviated by the fact that the net oxygen production – photosynthesis exceeding respiration and decay – is very small compared to the total production. A few ocean areas with anoxic bottom conditions and a few marshy areas in which peat is forming are presumably the key areas for maintaining our oxygen replenishment on a long term basis, because they are reservoirs in which unoxidized organic materials are accumulating. These limited areas are at least as susceptible to poisoning as are the open oceans. It is a matter of importance to man's future to recognize and preserve these areas.

Another alarming possibility is that of instability in the relationship between atmospheric carbon-dioxide concentration and average world temperature. An increase in carbon-dioxide, say from burning of fossil fuels, may increase the hothouse effect on Earth and increase the average temperature of the Earth a small amount; calculations indicate a factor-of-2 change in atmospheric carbon dioxide while maintaining constant relative humidity would produce a temperature change of 2.4 K [8]. Heating the ocean would tend to drive carbon dioxide out of solution into the atmosphere, further increasing the hothouse effect. Or one might imagine the effect going the other way as a result of increasing the flow of nutrients to the oceans. By stimulating the growth of phytoplankton, the rate of removal of carbon dioxide from the atmosphere might be increased, reducing the carbon dioxide content of the atmosphere and, with this, the hothouse effect. This could lead to world-wide cooling and an increase in the proportion of the carbon dioxide stored in the ocean relative to that in the atmosphere, further enhancing the effect. Conceivably, the ice ages could have resulted from such oscillations. Alternatively, they may have been produced by a variation in the rate of emission of carbon dioxide from the Earth's interior.

Because of the importance of these problems to man's future we should be very confident of our full understanding of them. At present, the ideas and even the numbers must be regarded as highly conjectural. Catastrophic problems appear to be in prospect for mankind because of the population explosion and its associated pollution explosion. One of the many possible ways in which pollution can precipitate a catastrophe is by upsetting the oxygen balance in our atmosphere. However, the time constant for this is sufficiently long that a destructive course may well have been followed beyond the point of no return before it is recognized.

References

1. Berkner, L. V. and Marshall, L. C.: 'The History of Oxygenic Concentration in the Earth's Atmosphere', Discussions of the Faraday Society, No. 37 (1964).
2. Cloud, P. E.: 'Atmospheric and Hydrospheric Evolution on the Primitive Earth', *Science* 160, 729–736 (1968).
3. Donahue, T. M.: 'The Problem of Atomic Hydrogen', *Ann. Geophys.* 22, 175–188 (1966).
4. Donald, C. M.: 'The Impact of Cheap Nitrogen', *J. Aust. Inst. Agric. Sci.* 26, 319–338 (1960).
5. Hutchinson, G. E.: 'The Biogeochemistry of the Terrestrial Atmosphere', in G. P. Kuiper (ed.), *The Earth as a Planet*, University of Chicago Press, Chicago, pp. 371–433 (1954).
6. Leith, H.: 'The Role of Vegetation in the Carbon Dioxide Content of the Atmosphere', *J. Geophys. Res.* 68, 3887–3898 (1963).
7. Li, Yuan-Hui: 'Geochemical Mass Balance among Lithosphere, Hydrosphere and Atmosphere', *Am. J. Sci.* 272, 119–137 (1971).
8. Manabe, S. and Strickler, R. F.: 'Thermal Equilibrium of the Atmosphere with a Convective Adjustment', *J. Atmos. Sci.* 21, 361–385 (1964).
9. National Academy of Sciences, Energy Resources Report 1000-D (1962).
10. Pales, J. C. and Keeling, C. D.: 'The Concentration of Atmospheric Carbon Dioxide in Hawaii', *J. Geophys. Res.* 70, 6053–6076 (1965).
11. Risebrough, R. W., Huggett, R. J., Griffin, J. J., and Goldberg, E. D.: 'Pesticides: Transatlantic Movements in the Northeast Trades', *Science* 159 (3820), 1233–1235 (1968).
12. Rubey, W. W.: 'Geologic History of Sea Water', *Bull. Geol. Soc. Am.* 62, 1111–1148 (1951).
13. Ryther, J. H.: 'Biological Oceanography, Geographic Variations in Productivity', in M. N. Hill (ed.), *The Sea*, Vol. 2, Interscience, New York, pp. 347–380 (1963).
14. Singer, S. F.: 'The Origin of the Moon and Geophysical Consequences', *Geophys. J. Roy. Astron. Soc.* 15, 205–226 (1968).
15. Strickland, J. D. H.: 'Production of Organic Matter in the Primary Stages of the Marine Food Chain', in J. P. Riley and G. Skirrow (eds.), *Chemical Oceanography*, Academic Press, pp. 477–712 (1965).
16. Wickman, F. E.: 'The Cycle of Carbon and the Stable Carbon Isotopes', *Geochim. Cosmochim. Acta* 9, 136–153 (1955).
17. Wurster, C. F. and Wingate, D. B.: 'DDT Residues and Declining Reproduction in the Bermuda Petrel', *Science* 159 (3818), 979–981 (1968).

For Further Reading

1. On geochemical processes, including quantitative estimates of the quantity of elements in various geological regimes:
   Brian Mason, *Principles of Geochemistry*, 2nd ed., John Wiley and Sons, New York, 1964.
2. On many assorted problems in the area of atmospheric evolution:
   P. J. Brancazio and A. G. W. Cameron (eds.), *The Origin and Evolution of Atmospheres and Oceans*, John Wiley and Sons, New York, 1964.
3. On chemical and life processes in the oceans and particularly the articles by E. Steeman Nielson and J. H. Ryther for world-wide rates of photosynthesis:
   M. N. Hill (ed.), *The Sea*, Vol. II, Interscience, New York, 1963.
4. A review article on evolution of planetary atmospheres:
   Francis S. Johnson, 'Origin of Planetary Atmospheres', *Space Sci. Rev.* 9 (1969), 303–324.

IMPACT OF LAND AND SEA POLLUTION
ON THE CHEMICAL STABILITY OF THE ATMOSPHERE

F. D. SISLER

Environmental Protection Agency, Washington, D.C., U.S.A.

Abstract. Gross pollution produced by modern technology and world population growth may precipitate changes in concentration of the major gases in the atmosphere which are essential for life
and human welfare. The hydrosphere, in particular the ocean, is in dynamic equilibrium with the
atmosphere and is largely responsible for controlling the chemical composition of the latter. The
equilibrium processes are biogeochemical as well as physical in nature. An upset in the ecological
balance, as typified by photosynthesis and respiration processes of marine plankton, is considered.
The possible consequences to atmospheric quality reflected in change of concentration of oxygen,
nitrogen, and carbon dioxide are discussed, but a major disruption on a global scale is not apparent
now. However, subtle changes may be taking place which may have far-reaching consequences in
coming years.

1. Introduction

The chemical composition of the Earth's atmosphere is controlled to a large extent by
biogeochemical cycles existing on land and in the seas. It is estimated that well over
one-half of the natural constituents of the atmosphere originate from, and are in
equilibrium with, the hydrosphere, which includes the oceans, estuaries, inland seas,
lakes and rivers. These water bodies are now faced with gross pollution from a vast
assortment of chemicals. Since some of these chemicals are toxic and cumulative, and
since most persistent pollutants discharged on land or to the atmosphere ultimately
accumulate in the hydrosphere, there is a distinct possibility that the ecological
balance in this region could be upset with serious consequences to the quality of the
atmosphere. The oceans, the major portion of the hydrosphere, now can no longer
be considered the ultimate sink for all types of modern pollutants, even though they
have served as a most efficient septic tank since Cambrian times.

In recent years, several provocative papers, notably those of Berkner and Marshall
[1, 2], Commoner [6, 7] and Cole [5], point to dangerous trends resulting from modern
technological practices affecting natural chemical balances of and between land, sea,
and air, as well as the climate in general. Effects from burning fossil fuels, widespread
use of inorganic fertilizers, depletion of forested areas with reduction of photosynthetic potential on land, release of toxic or harmful chemicals from industrial
factories and agricultural practices, and increasing population have been stressed.
Planned use of nuclear materials for power and for Earth-moving projects are also
considered. With respect to the atmosphere, the life-supporting gases of oxygen,
nitrogen, and carbon dioxide appear to be in danger of change in relative composition.

Oxygen and carbon dioxide are the principal gases involved in photosynthesis [35].
Keeling and associates [22, 23, 24, 25] have made extensive studies of the distribution

S. Fred Singer (ed.), The Changing Global Environment, 57–71. All rights reserved.

of carbon dioxide in the atmosphere and surface waters. Seasonal variations in CO_2 often relate to photosynthesis activity in the oceans. Since the carbon dioxide concentration in sea water is strongly influenced by pH, this physico-chemical parameter is important in oxygen equilibrium [8, 11, 20, 40].

Gambell and Fisher [14, 15] report acid rainfall along the United States east coast in the vicinity of Chesapeake Bay as low as pH 4.0. Rainfall samples obtained synoptically at a site on St. Thomas Island, representative of Atlantic coastal conditions reasonably free from pollution, usually gave pH values in excess of 8.0 and never less than 7.6. Apparently, this acid rain has been falling, at least sporadically, since the middle 1950's.

The Chesapeake Bay Institute of Johns Hopkins has been measuring the vertical and horizontal pH distribution throughout Chesapeake Bay since 1949 [19]. Beginning the summer of 1959, their published reports showed distinctly acid conditions in various sections throughout this large estuary. The reports covered the time period 1949–61 and low pH readings, below pH 7 (and reportedly below pH 4 at one station in the lower Bay in 40 and 60 feet of water), were observed to be widely distributed, both horizontally and vertically during the remaining years.

Gambell and Fisher postulated that the acid rain resulted from the distillation of acid surface waters in the vicinity. Sulfur acids were prominent in the collected samples of rain, but other acids may also have contributed to the low pH readings. That the Chesapeake Bay may have been a source of the acid rain has not been established, but the fact that highly buffered estuarine waters, as compared to fresh water, could become acidic seems quite unusual.

Besides acid, modern rain may contain a variety of chemicals which are potentially harmful to living organisms. The heavy metal lead is a case in point. In the past 30 + years, the exhaustion of lead from automobiles to the atmosphere as burned lead tetraethyl has increased linearly from zero to greater than 1.6×10^{11} grams per year [41]. Most of this ends up in the oceans. Lead is toxic to many organisms including man. It is now well known that living organisms in the seas have an amazing ability to concentrate trace elements and chemicals [17]. The continued pollution of the seas with toxic substances such as lead, mercury, refractile chemical pesticides and polychlorinated biphenyls may well pose a threat to human welfare. Goldberg [17] discusses this point in detail in this volume.

Harris and co-workers [43, 44, 46] have pointed out that bioassay techniques for evaluating the chronic effects of sublethal concentrations of heavy metals and organochlorine compounds are lacking. Environmental quality standards set on an antropomorphic basis may not provide adequate protection for lower trophic level organisms.

2. Oxygen

The oxygen balance of the atmosphere is a major concern. The importance of ocean photosynthesis to oxygen concentration in the atmosphere seems well established. Berkner and Marshall, who have made extensive studies of the genesis of oxygen in the Earth's atmosphere, have considered the potential degradation of this life-supporting gas as a consequence of man-made pollution [1, 2]. They consider the present O_2 equilibrium to be unstable. This assumption is based on a rather involved theoretical model of the genesis of the present atmosphere since pre-Cambrian time of -27×10^8 yr. Prior to this period, the authors reconstructed the secondary, abiogenic atmosphere from Urey's basic premise of the absence of a primordial atmosphere. The secondary atmosphere evolved from leaching of reducing gases and water vapor from volcanic processes. The rarified secondary atmosphere (0.001 % present atmosphere) permitted the penetration of short wavelength ultraviolet light (0.1–02 μ) causing photodissociation of water vapor to form H, O_1, O_2, and O_3.

Eventually the concentration of O_2 built up to a point where further dissociation of water was restricted because of the Urey 'shadowing' effect. The overlying O_2 filtered out sufficient short UV such that a self-limiting, steady state process prevented further buildup of oxygen.

With the appearance of photosynthetic life, the oxygen level increased sharply in a relatively short time since photosynthesis uses visible light (0.4–08 μ) which is not filtered out by increasing oxygen concentration.

Berkner and Marshall reason that the present O_2 equilibrium is unstable because a relatively small drop in production (as might be caused by biocides) will result in a sharp drop in concentration. They assume that the oxygen demand by reduced chemicals and respiration is precisely balanced by present Earth's oxygen production. Once the scale is tipped by lowered production, the consequent lowered O_2 concentration will permit more lethal UV to reach the Earth's surface, further stifling photosynthesis.

The strength of the oxygen model presented by the authors rests heavily on their evaluation of the limiting effects of ultraviolet radiation on photosynthetic plants. An important question concerns the matter of shielding. Ultraviolet at any wavelength is reflected, scattered or absorbed by thin sections of most substances more dense than gases. Many plants as well as animals are protected by an integument which is impervious to UV. The sunbather is protected from mutagenic UV by a thin coating of oil. Atmospheric dust should screen out much of the UV. Pure water, although more transparent to short UV radiation in a rarified atmosphere, rarely occurs in nature. Most ocean and lake waters contain sufficient impurities in the form of suspended and colloidal materials, such that absorption coefficients based on pure laboratory solutions do not apply.

Even assuming pure ocean water where most photosynthetic oxygen is produced, the effect of more intense short UV would be to depress the photosynthetic layer by some centimeters, possibly meters, but not to a considerable depth compared to the average depth of the oceans.

On land, a thin layer of dust will protect most biota which do not have a protective integument.

It is the writer's present opinion, therefore, that the instability of the present atmosphere oxygen concentration based primarily on the UV effect is open to some question.

3. Carbon Dioxide

Evidence that carbon dioxide is increasing in the atmosphere has been presented by various investigators, particularly Plass [31, 32, 33], Keeling [22], and Pales and Keeling [29]. According to Pales and Keeling, the concentration of this gas has been increasing steadily since 1958. However, the rate has been decreasing, even though the rate of fossil fuel utilization has been increasing [22, 29].

In 1965, 1.7×10^{10} tons of CO_2 were injected into the atmosphere. From C^{14} data, the atmospheric residence of CO_2 was estimated at five years [3, 22]. A substantial part of atmospheric CO_2 becomes dissolved in ocean waters in the five-year period.

According to Kuentzel [26] and Kerr [45], the rate of photosynthesis by algae increases in the presence of additional carbon dioxide. At least with fresh water algae, it appears that the concentration of CO_2 is the rate-limiting nutrient rather than phosphorus or nitrogen. Kuentzel points out that algae require very little phosphorus in their aqueous medium, of the order of 0.01 mg/L (ppm). Kuentzel attributes the presence of massive algae blooms in eutrophic lakes to additional CO_2 produced by bacteria which ferment the organic matter introduced by pollution sources such as sewage. Although nitrogen and phosphorus are important nutrients to the photosynthetic algae, they are usually present in excess, whereas carbon dioxide becomes rapidly depleted in the presence of active photosynthesis and algae growth becomes self-limiting if the main source of CO_2 is from the atmosphere.

It is interesting to speculate whether photosynthesis in the oceans by photoplankton might be increased as the result of human activity by introduction of additional carbon dioxide, either directly from the atmosphere, or indirectly from organic matter and sewage from land runoff, outfalls and barged wastes. The continental coasts of the United States, excluding Alaska, receive 10^{12} gall of water per day from land runoff. Substantial organic matter is introduced to coastal waters by this route. The major sources of the organic matter are soil humus, farm wastes, and human sewage. The latter has been increasing with the population which now exceeds 200 million. Since livestock is raised in proportion to the population, farm wastes are also increasing. No figures are available on the average organic content of waters reaching the coasts. Raw sewage has an oxygen demand of 0.0011 lb O_2 gall^{-1}. Farm wastes should have a similar oxygen demand and, volume-wise, should exceed human wastes by several times. Moreover, farm wastes are not usually treated in sewage disposal plants. It can be safely assumed, therefore, that surface water draining to the continental coasts contains a substantial amount of organic matter contributed directly or indirectly by human activity.

It seems quite evident that the ocean waters surrounding the continental United States are receiving more nutrients in the form of organic matter, factory-made chemical fertilizers, and carbon dioxide from the atmosphere at a rate exceeding that of a millenium ago. If a similar process is occurring on a global basis, a reasonable assumption, considering world population growth, is that photosynthesis may be increasing. If this actually is the case, then one might expect an increase in the rate of oxygen replenishment to the atmosphere from the ocean surfaces. Unless excess oxygen production from ocean waters is exactly compensated by combustion of fossil fuels and oxidation of recent organic matter on the continents, there should be an increase in the relative concentration of oxygen in the atmosphere. The residence time of oxygen in today's atmosphere should be less than that of a thosand years ago, in any case.

Domestic and industrial effluents also contain toxic substances which can inhibit photosynthesis [44, 46]. Further research will be required to establish the impact of human effluents on the metabolism of marine plankton communities.

We are now witnessing man-made eutrophication of large lakes; e.g., Lake Erie. Can Kuentzel's interesting observation on stimulation of photosynthesis from fresh water algae by additional CO_2 be extended to ocean photosynthesis? Compared with relatively unpolluted fresh water in equilibrium with the atmosphere, sea water contains dissolved CO_2 (as CO_2, H_2CO_3, HCO_3^-, and $CO_3^=$) in much higher concentrations (see Table I). The introduction of additional CO_2 to sea water might not have the same stimulating effect as described by Kuentzel's fresh water system. On the other hand, the introduction of any agent to sea water that might cause a shift in pH to the acid side might cause a dramatic increase in available CO_2 with consequent increase in the photosynthesis rate. The reserve of CO_2 in sea water as carbonate is not directly available to photoplankton.

TABLE I

Relative concentration of gases in the atmosphere and in sea water

Gas	A Atmosphere[a]	B Sat. sea water[b] ($S = 34.3^0/_{00}$) ($T = 10\,^{\circ}C$	$\dfrac{B \times 10^{-1}}{A}$
	Mol. fraction %	ml/L $^0/_{00}$	
Nitrogen (N_2	78.09	11.56	0.015
Oxygen (O_2)	20.95	6.44	0.031
Argon (A)	0.93	0.3	0.032
Carbon dioxide (CO_2)	0.03	34–56	113–188
Neon (Ne)	2×10^{-3}		
Helium (He)	5×10^{-4}	1.7×10^{-4}	
Krypton (Kr)	1×10^{-4}		
Hydrogen (H_2)	5×10^{-5}	3×10^{-3}	0.015
Xenon (Xe)	8×10^{-6}		
Ozone (O_3)	1×10^{-6}		
Radon (Rn)	6×10^{-8}		
H_2S	?	0–22	
CH_4	?	0–30	

[a] Values obtained from *Handbook of Chemistry and Physics*, 47th Edition, represent sea-level atmospheric composition for a dry atmosphere and do not necessarily indicate exact condition of atmosphere, e.g. air pollutants.
[b] Values of major constituents obtained from *The Oceans*. High values for H_2S and CH_4 found only in waters over anoxic basins.

Moreover, the total carbon dioxide concentration in all molecular, ionic, and mineral forms is decreased in sea water at higher pH values. There is usually an alkaline shift associated with active photosynthesis resulting in a tendency for carbonate to precipitate out of solution. Because of the unique equilibrium of the carbon dioxide system in sea water, the oceans' major buffering mechanism, the availability of this essential nutrient becomes self-limiting to photosynthesis. Keeling [22] assumes the rate of CO_2 plant uptake on land to be proportional to the CO_2 partial pressure of the atmosphere. This follows Kuentzel's observations of fresh water algae. Whether additional CO_2 reaching the oceans will increase the rate of photosynthesis remains to be verified.

As yet there appears no firm evidence of change in relative concentrations of oxygen or nitrogen in the lower atmosphere, although local transient and diurnal changes for oxygen have been noted by some observers [10, 20]. Up until the present time, the relative concentration of oxygen in the atmosphere of approximately 21 % appears to be regulated largely by the incident annual solar light reaching the Earth, rather than other factors such as local concentration of nutrients. Variations in primary productivity of the oceans show a wide range in regional and seasonal differences (see Table II). Broadly speaking, the photosynthetic activity, with consequent oxygen production, ranges from 5 to 5000 mg carbon per m^2 per day. Thus, oxygen production in the oceans is unevenly distributed and varies as much as a thousand times. Since this pattern is constantly changing, it is remarkable that the concentration of oxygen in the atmosphere remains constant. Johnson [20] in this volume estimates the time constant for O_2 in the atmosphere as produced by photosynthesis to be 10^4 yr, which

TABLE II

Oceanic productivity measurements, based on C^{14} fixation (Original sources compiled in Strickland (1960), Ryther (1963), except Equatorial Atlantic from IGY cruise data)[a]

Region	Time of Year	Primary productivity (mg C/m^2/day)
Atlantic Ocean areas		
Arctic		
Ice Island T3	Midsummer	0–24
Station Alpha	Late summer	0–6
Western Barents Sea, near Bear Island		
Arctic water	May	1300
Atlantic water	May	275
Norwegian Sea, near Spitzbergen		
North Atlantic water	June	2400
Arctic water	June	400–600
Northern North Atlantic		
Faroe-Iceland Ridge	Summer	650–2700
Near Iceland	Summer	530–1300
South of Greenland	Summer	550
Irminger Sea	Summer	150–250
North Sea		
Annual range	Yearly	100–1500

Table II (continued)

Region	Time of Year	Primary productivity (mg C/m²/day)
Northeast Coast of England	May	220
	October	110
	February	5
Danish Coastal waters	March bloom	300
	August (max)	700
	December (min)	10
Eastern North Atlantic		
15 miles off Oporto	September	100
200 miles off Oporto	September	150
Mediterranean Sea		
2 miles off French Coast	Midsummer	30–40
Western North Atlantic		
Continental shelf and inshore		
Spring flowering	Spring	1930
Mean rate		560
Weighted annual mean		330
Sargasso Sea	April bloom	890
	Summer	100–200
Caribbean Sea (10–20°N)		100–200
Equatorial Atlantic		
West Basin		
Near 20°N	February	60–160
At 8°15′N	April–May	230–300
At 8°15′S	March	70–280
At 15°45′S	April	20–130
East Basin		
Near 20°N	February	190–780
At 8°15′N	April–May	180–1480
At 8°15′S	March	100–370
At 15°45′S	April	90–420
Southwest Atlantic		
Walvis Bay and Benguela Current	December	500–4000
Pacific Ocean areas		
Equatorial Pacific		
Near coast of Ecuador	Autumn	500–1000
9°N, 90°W (Costa Rica dome)	November	410–800
11°N, 115°W	Autumn	10
Range, 30°N–30°S	March	100–250
Northwest Pacific		
Sea of Japan		
Kuroshio Current	Summer	50–100
Oyashio System	Summer	250–500
Sea of Okhotsk, North of Japan		
Range of values	May	6–5100
Mean value	May	2000
Indian Ocean areas		
Equatorial Indian Ocean		200–250

[a] From *Encyclopedia of Oceanography* (ed. by R. W. Fairbridge), 1966 (10).

may account for its present uniform concentration.* Whether this stability will be affected by man-made activity is still an open question.

Recent Weather Bureau reports (see also Bryson and Wendland in this volume), state that there is a worldwide increase in particulate matter in the atmosphere which filters out some solar radiation. This suspended material is believed to be largely smoke and dust particles from burning grass lands and denuded soil from man-made activities. Evidence for the rapid buildup of atmospheric particulate matter is found in recent deposits in glaciers and high altitude snow deposits. More recently, exhausts from jet engines, factories, and motor vehicles have added to the atmosphere's particulate burden. The impact of this man-made solar filter may cause a reduction in the Earth's temperature – a reverse of the 'greenhouse effect' [3]. Whether the Earth is cooling off or warming up is another matter open to question. Mitchell [27] discusses this point in this volume. Whether this increase is a temporary perturbation or a more prolonged effect resulting from man-made activity is not yet established. Solar flares and storms, for example, would be an obvious cause of a temperature perturbation which is not man-made.

4. Nitrogen

The situation with respect to nitrogen is even less clear than that of oxygen. Although the writer does not feel that an upset in the nitrogen cycle is as imminently serious as with oxygen, the two are interrelated along with, also, carbon and hydrogen. Nitrogen differs from the other elements mentioned, notably in its distribution. Most all of the Earth's nitrogen is in the atmosphere, with a smaller percentage tied up in the biosphere (including the hydrosphere) and some small deposits of biologic origin, such as saltpeter in Chile and oxidized organic minerals from guano deposits in a few isolated areas. Although volcanic gases show N_2, O_2, H_2S, SO_2, CO, CO_2, CH_4, HF, HCl, etc., the N_2 and O_2 are believed to be from either atmospheric contamination or from interaction with sea water. The analyses of unweathered sedimentary rock by Friedman, show very little nitrogen present [13]. The concentration of nitrogen in igneous rock, moreover, is 10000 times less than that of oxygen [36]**. In brief, the Earth's crust cannot provide a substantial reservoir of nitrogen as is the case with oxygen and carbon dioxide. This also applies to the hydrosphere where most of the dissolved nitrogen present originates directly from atmospheric equilibria, with a small fraction

---

* Uniform concentration refers to the lower atmosphere. The situation is apparently different at higher altitudes. Norton and Warnock [28] report a seasonal variation of atmospheric molecular oxygen in the 100–200 km region as $50\pm20\%$ lower in winter than in summer between latitudes 45 and 65°.

** To date, no one has obtained a positively identified uncontaminated sample of the mantle, which constitutes 67.2% of the Earth's volume [21]. Igneous rock, representative of the Earth's crust, accounts for less than 1% of the Earth's volume. Some traces of nitrogen have been found in diamonds which are believed to have originated in the mantle where required high temperatures and pressures prevail for their formation [9, 21]. The mantle would seem a logical reservoir or source of nitrogen if this element is of terrestrial origin. More than 90% of the mantle is believed to consist of compounds of four elements – magnesium, iron, silicon and oxygen [21].

TABLE III

Percent concentration
of oxygen and nitrogen in
igneous rock

Oxygen	46.42%
Nitrogen	0.00463%

(as gas, inorganic, and organic compounds) indirectly via the biosphere. In contrast, the oceans provide a vast reservoir of oxygen from water itself since O^{18} studies show that O_2 liberated in photosynthesis originates almost entirely from H_2O and not CO_2 [16].

Although nitrogen in its various molecular and ionic forms has a potential of eight electron steps, the most common forms found in nature are nitrogen gas, oxides of nitrogen gas, organic nitrogen, ammonia, nitrite and nitrate. Microorganisms play an important role in the nitrogen cycle, affecting reversibly all natural states in redox reactions.

A simplified version of the nitrogen cycle in the biosphere as presented in many texts (see, e.g., p, 571 of [30]) is as follows:

$$\rightarrow N_2 \rightarrow RNH_2 \rightleftharpoons NH_3 \rightleftharpoons NO_2 \rightleftharpoons NO_3^- \rightharpoondown$$

where R represents an organic complex such as in amino acids and protein. Those microorganisms that liberate free nitrogen from nitrite and nitrate are called denitrifiers. The literature is not too clear as to whether denitrification is the principal process for liberation of free nitrogen from the biosphere. This gas is observed to emanate from organically-rich, anaerobic, water-saturated soils and sediments where little or no nitrite or nitrate should exist [12, 34, 37, 39]. This point is mentioned because pollution of estuaries with organic-rich wastes covered by silt could create conditions speeding up release of nitrogen gas to the atmosphere with a subsequent reduction in the retention time of nutrient nitrogen available for desirable biological activity.

A few species of microorganisms fix* nitrogen gas while many species are involved in liberating free nitrogen from organic and inorganic materials. Since nitrogen fixation requires high free energy (endergonic), whereas nitrogen liberation yields energy (exergonic), a considerable conservation of energy is realized in the biosphere if the nitrogen liberation step can be slowed down. To accomplish this, it is necessary to understand more thoroughly the precise environmental conditions which tend to keep soil or sediment nitrogen in a fixed form. Commoner [6, 7] was perhaps alluding to this when he deplored the widespread practice of using inorganic nitrogen fertilizers in place of organic forms.

Unless plants can quickly absorb the inorganic nitrogen, the excess in the soil is

---

* Nitrogen fixation is defined as the process whereby free nitrogen combines chemically with other elements. This process occurs on land and is encouraged in organic farming. Although nitrogen fixation by microorganisms and phytoplankton of the oceans is known to exist, the total contribution to the ocean's nitrogen budget is not clearly understood [21, 38].

quickly washed into the nearest drainage system where it is not only irretrievaby lost
to the farmer, but where it creates local problems of eutrophication. Where organic,
urea, or ammonia nitrogen is used as fertilizer, the soil tends to retain these compounds
[39]. Excessive use of urea, since it is water soluble, may also create eutrophication
after land drainage. The use of manure and the practice of plowing under nitrogen-
fixing leguminous plants provides not only a desirable source of nitrogen, but also a
humus-type soil which acts as a blotter to retain nitrogen compounds, desirable
moisture content, and other essential nutrients. Commoner [6, 7] stresses the importance of humus soils in the N cycle. Also, such a composted soil is better aerated,
which should retard excessive nitrogen liberation as free gas.

In contrast, the incentives for quick cash crops have encouraged the use of inorganic
fertilizers with little or no composting. In fact, some farm practices during harvest
time remove the entire cash crop, roots, stems, and leaves from the ground leaving the
soil bare, subject to erosion by water and wind. As to the latter, the dust bowls of the
Earth apparently are on the increase, as mentioned above [3].

The nitrogen content (hence the protein content) of plants varies considerably
according to the species. In the corn plant, protein approximates 10% of the entire
plant [42]. Under controlled laboratory conditions, the protein content of algae
(chlorella) can be made to vary from 7% to as high as 88% depending upon the
nitrogen concentration used as an essential nutrient [4].

These laboratory findings suggest that natural environmental conditions affecting
the nitrogen balance may have an important role in determining the quality and
quantity of protein in other plants, on land and in the sea. Therefore, in considering
the importance of the nitrogen cycle in food production, proper control of the environment seems highly desirable. This can readily be accomplished on land. Control
is still possible in the estuaries. In the oceans, however, at the present time, about all
that can be done is to observe changes and continue research of the nitrogen cycle
in this environment. As with oxygen and carbon dioxide, studies of equilibria trends
and exchange processes at the sea surface should be stressed to predict possible
atmospheric changes.

The origin and fate of nitrogen on Earth is interesting and puzzling. As to its origin,
aside from what has been discussed above, it is assumed by Rubey, Urey, and others
that the present nitrogen is of terrestrial origin.

The fate of nitrogen, of course, is a question of some importance as discussed by
Cole [5]. Although Cole does not elaborate on how nitrogen can be lost to Earth, this
deserves consideration. There seem only two possibilities; loss to outer space and
nuclear transformation [16]. Both appear negligible, hence the persistence of nitrogen
on Earth seems assured in the foreseeable future.

5. General Considerations

The complexity of the bio-geochemical cycles in the oceans and their dynamic and
fluid nature makes long range forecasting extremely difficult. In any attempt to get a

TABLE IV

Environmental extremes for microbiological processes [47]

Temperature:

    -269 to $> +80\,°C$ survival range

    5 to $\sim +70\,°C$ metabolically active

    20 to $40\,°C$ optimum; most species

Hydrogen-Ion concentration:

    < 0.5 to ~ 10.5 pH survival range

    6 to 8 pH optimum range for most species; acid and alkali producing species are notable exceptions

Water:

    Liquid water essential for active metabolism. Many species will survive for various periods of time in media containing less than 1 % free or unbound water

Pressure, hydrostatic:

    0 psi to $> 10^5$ psi survival range

    $\sim$ 15 psi optimum; most species

Surface tension:

    $\sim$ 25 to 100 dyn cm^{-1} survival range

    $\sim$ 75 dyn cm^{-1} optimum; most species

    $<$ 50 dyn cm^{-1}, lethal for many species

Osmotic pressure:

    $\sim$ 0 to 20 000 psi survival range

    15 to 200 psi optimum; most species

Cations (other than protium):

    Hg highly toxic for most species. Ag, Cu, Cd and Ce toxic for vegetative cells. Mechanism of toxic action of most metal cations not established. Notable exceptions are specific enzyme inhibitors

Anions (other then hydroxyl):

    Halogens highly lethal for most species

    Toxic activity of most other anions not established

Organic toxins:

    Formaldehyde lethal for most species

    Phenolic compounds and detergents toxic under various conditions. Most organic toxic compounds, e.g. antibiotics are highly specific for individual species

Electricity:

    $>$ 10 000 V and 1000 A survival range

    Electrolyte products of gaseous halogens existing as impurities in water could be toxic if occurring in concentrations 71 %. This effect would result only with high sustained D.C. current

Sound energy:

    10 000 cps will disintegrate many species. Sound energy below and above this frequency not critical for survival. Low frequency sound energy stimulates metabolic activity

Magnetism:

    $0 > 10 000$ Gauss survival range. Some species exhibit paramagnetic inhibition and/or morphologic and physiologic change in sustained homogenous and nonhomogenous fields of $>$ 100 Gauss

Electromagnetic radiation:

    UV light critical between 2000 and 3000 Å for most species

    Highly bactericidal between 2500 and 2800 Å

    Visible light, i.e. 4000 to 8000 Å essential for photosynthetic species

    Infra-red energy, i.e. $>$ 8000 Å utilized by some species, i.e. purple sulfur bacteria

Radioactivity:

    Alpha radiation not critical under most circumstances, i.e. with light shielding provided by average environment. Lethal ionization results from exposure

    Soft beta radiation not critical

    Hard beta radiation may cause lethal ionization effects. Resistance to hard beta radiation varies greatly among species

    Gamma and X-ray radiation, true electromagnetic radiation are highly lethal to virtually all species, but lethal exposure time varies considerably among species. Gamma radiation is highly mutagenic

Gravity:

    $0 > 1000$ G survival range except at upper extremes. Some species inactivated by high artificial gravity; i.e. ultra centrifuge handle on these, it is imperative that one appreciates the importance of the role of plankton of microbial dimensions as geochemical agents: such would include bacteria, yeasts, unicellular algae and all other microscopic organisms. These self-propagating biological catalysts function in strict conformity to the environmental parameters of their particular ecological niches. Each particular species may be considered unique in its optimum requirements for growth and functions. All are subject to the antibiotic influences of neighboring organisms. All are subject to physical and physico-chemical influences and stresses which govern function. Generalizations can be misleading in predicting under what set of circumstances a given flora will flourish and affect its own environment. It is useful, however, to appreciate extreme boundary conditions. Table IV presents a partial list of environmental extremes which limit the activity of the microscopic plankton. The question of whether the world's oceans can undergo accelerated eutrophication as the result of human activity has been the subject of much recent speculation. Certainly some type of eutrophication is taking place at the margins, such as large estuaries and certain coastal regions. These regions would include ocean pockets adjacent to densely populated areas, e.g., the Baltic Sea, the Mediterranean Sea*, the Gulf of Mexico, etc. Transient eutrophication effects may also be manifest in the open ocean basins. These would be expected where ocean currents carrying continental pollutants would surface and produce sluggish eddy currents not vigorously aerated by natural mixing processes. Such a situation may have been witnessed by Thor Hjerdahl and crew during his cross Atlantic voyage of RAII.

Considering the vastness of the worlds oceans, there is a need to focus attention on those places where troublesome conditions are likely to occur. The U.S. Geological Survey has been conducting studies on continental drainage on a global basis. These studies and similar should provide some estimate of anticipated ocean loadings from terrestrial sources [49]. Simultaneously, the chemical industry predicts a doubling of chemical production in the U.S. in the next 15 yr. One can expect a continuing burden to the world oceans as other nations increase their chemical productivity. There are at present 135 maritime nations. Industrial and population trends can be used to anticipate those ocean regions which may experience exceptional stress from pollution. In addition, as we learn more about the existence and behavior of ocean currents, eutrophication foci may eventually be anticipated and pinpointed. For instance Sisler and Senftle [49] have suggested a mechanism involving the Earth's magnetic field to explain the abnormal concentration of any charged pollutant on or near the surface of the ocean. This would include a multiplicity of pollutants such as colloids, organic aggregates microorganisms and small plankton.

6. Summary and Conclusions

The previous sections have considered some large-scale time and space aspects of the impact of modern technology and population growth on the biosphere of the oceans

---

* While the Mediterranean Sea is now threatened with serious pollution, the situation was not apparent ten years ago. On the contrary, it was considered a nutrient-poor sea at that time [48].

and continents and possible consequences to the quality of the atmosphere. It seems obvious that man can shift natural equilibrium forces that make for a healthy balance in nature essential for life, in particular the composition of the atmosphere and the food potential of land and sea.

Thus far, a major disruption in the oceans and atmosphere is not apparent. We may, however, be looking at subtle changes in the environment, such as the accumulation of lead and other toxic chemicals, an increase in carbon dioxide and organic carbon compounds, atmospheric dust, and acid rainfall which may have far-reaching consequences in years to come. The equilibrium existing between physical, chemical, geological and biological forces is a subject for serious study on a global basis.

> "All things by almighty power
> Near and far
> Hiddenly connected are
> That thou canst not pick a flower
> Without disturbing of a star."
>
> Anon

References

1. Berkner, L. V. and Marshall, L. C.: 'On the Origin and Rise of Oxygen Concentration in the Earth's Atmosphere', *J. Atmos. Physics* **22**, 225–261 (1965).
2. Berkner, L. V. and Marshall, L. C.: 'Potential Degradation of Oxygen in the Earth's Atmosphere', Memo for file (1966).
3. Bryson, R. A. and Wendland, W. M.: 'Climatic Effects of Atmospheric Pollution', this volume, pp. 139–147.
4. Burlew, J. S. (ed.): 'Algal Culture – From Laboratory to Pilot Plant', Carnegie Institution of Washington Publication #6000, Washington, D.C. (1961).
5. Cole, LaMont C.: 'Can the World Be Saved', Paper presented at the 134th meeting of the American Association for the Advancement of Science (1967).
6. Commoner, B.: 'The Balance of Nature', Address to the Graduate School, U.S. Department of Agriculture, Washington, D.C. (1967).
7. Commoner, B.: 'Threats to the Integrity of the Nitrogen Cycle: Nitrogen Compounds in Soil, Water, Atmosphere and Precipitation', this volume, pp. 341–366.
8. Conference on Physical and Chemical Properties of Sea Water, Easton, Maryland, September 4–5, 1958; Washington National Academy of Sciences, National Resource Council (1959).
9. De Carlie, P. S.: Stanford Research Institute, personal communication.
10. *Encyclopedia of Oceanography, Encyclopedia of Earth Sciences Series*, Vol. 1 (ed. by R. W. Fairbridge), Reinhold Publishing Company, New York (1966).
11. *Equilibrium Concepts in Natural Water Systems. Advances in Chemistry Series* **67**. A Symposium Sponsored by Division of Water, Air and Waste Chemistry, 151st Meeting, American Chemical Society, Pittsburgh, Pennsylvania, March 23–24 (1966).
12. Felbeck, G. T.: 'Normal Alkanes in Much Soil Organic Matter Hydrogenolysis Products', *Trans. Comm. II and IV, Int. Soc. Soil. Sci.*, Aberdeen (1966).
13. Friedman, I.: U.S. Geological Survey, personal communication.
14. Gambell, A. W. and Fisher, D. W.: 'Chemical Composition of Rainfall in Eastern North Carolina and Southeastern Virginia', Geological Survey Water Supply Paper #1535-K (1966).
15. Gambell, A. W. and Fisher, D. W.: 'Chemistry of Atmospheric Precipitation', U.S. Geological Summary Report ACA-17-F (1966).
16. Glasstone, S.: 1958, *Sourcebook on Atomic Energy*, Van Nostrand, New York, 2nd Edition.
17. Goldberg, E. D.: 'The Chemical Invasion of the Oceans by Man', this volume, pp. 275–294.

18. *Handbook of Chemistry and Physics,* 47th edition, The Chemical Rubber Co. (1967).
19. Hires, R. I., Stroup, E. D., and Seitz, R. C.: 'Atlas of the Distribution of Dissolved Oxygen and pH in Chesapeake Bay 1949–1961', Chesapeake Bay Inst. Graphical Summary Report No. 3 (1963).
20. Johnson, F. S.: 'The Oxygen and Carbon Dioxide Balance in the Earth's Atmosphere', this volume, pp. 49–56.
21. Josephs, M. J. and Sanders, H. J.: 'Chemistry and the Environment', Am. Chem. Soc. Pub., Washington, D.C. (1967).
22. Keeling, C. D.: 'Carbon Dioxide from Fossil Fuel – Its Effect on the Natural Carbon Cycle and on the Global Climate', 49th Meeting, American Geophysical Union, Washington, D.C. (1968).
23. Keeling, C. D.: 'Carbon Dioxide in Surface Ocean Waters: Global Distribution', *J. Geophys. Res.* **73,** 4543–4553 (1968).
24. Keeling, C. D., Harris, T. B., and Wilkins, E. M.: 'Concentration of Atmospheric Carbon Dioxide at 500 and 700 Millibars', *J. Geophys. Res.* **73,** 4511–4528 (1968).
25. Keeling, C. D. and Waterman, L. S.: 'Carbon Dioxide in Surface Ocean Waters', *J. Geophys. Res.* **73,** 4529–4541 (1968).
26. Kuentzel, L. E.: 'Bacteria, Carbon Dioxide and Algal Blooms', *Proceedings of the 24th Annual Purdue Industrial Waste Conference,* Purdue University, Lafayette, Indiana (1969) (in press).
27. Mitchell, J. M., Jr.: 'A Reassessment of Atmospheric Pollution as a Cause of Long-Term Changes of Global Temperature', this volume, pp. 149–173.
28. Norton, R. B. and Warnock, J. M.: 'Seasonal Variation of Molecular Oxygen Near 100 Kilometers', *J. Geophys. Res.* **73,** 5798–5800 (1968).
29. Pales, J. C. and Keeling, C. D.: 'The Concentration of Atmospheric Carbon Dioxide in Hawaii', *J. Geophys. Res.* **70,** 6053–6076 (1965).
30. Pelczar, J. J., Jr. and Reid, R. D.: *Microbiology,* 2nd edition, McGraw-Hill, New York (1965).
31. Plass, G. N.: 'The Carbon Dioxide Theory of Climatic Changes', *Tellus* **8,** 140 (1956).
32. Plass, G. N.: *Proc. Conf. on Research in Climatology,* Scripps Inst. Oceanog. (1957).
33. Plass, G. N.: 'Carbon Dioxide and the Climate', *Amer. Sci.* **44,** 302 (1956).
34. Preul, H. C. and Schsoepfer, G. J.: 'Travel of Nitrogen in Soils', *J. Water Pollution Control Federation* **40,** 30–48 (1968).
35. Rabinowitch, E. I.: *Photosynthesis and Related Processes,* Interscience, New York 1945.
36. Rankama, K. and Sahama, T. G.: *Geochemistry,* University of Chicago Press, Chicago (1950).
37. Sisler, F. D.: 'Role of Earth Potentials in Organic Geochemical Processes Concerned with Petroleum', *Abh. Dent. Akad. Wiss. Berlin. Kl. Chem. Geol. Biol.* 199–204 (1966).
38. Sisler, F. D. and ZoBell, C. E.: 'Nitrogen Fixation by Sulfate-Reducing Bacteria Indicated by Nitrogen/Argon Ratios', *Science* **113,** 511–512 (1951).
39. Stewart, W. D. P.: 'Nitrogen-Fixing Plants', *Science* **158,** 1426–1432 (1967).
40. Sverdrup, H. U., Johnson, M. W., and Fleming, R. H.: *The Oceans,* Prentice-Hall, New York (1942).
41. Tatsumoto, M. and Patterson, C. C.: 'The Concentration of Common Lead in Sea Water' in J. Geiss and E. D. Goldberg (eds.), *Earth Science and Meteorities,* North-Holland Publishing Co., Amsterdam (1963).
42. Watt, B. K. and Merrill, A.: 'Composition of Foods', *Agri. Handbook,* No. 8, U.S. Department of Agriculture (1963).
43. Harris, R. C.: 'Ecological Implications of Mercury Pollution in Aquatic Systems', *Biol. Conservation* **3,** 279–283 (1971).
44. Harris, R. C., White, D., and Macfarlane, R.: 'Mercury Compounds Reduce Photosynthesis by Plankton', *Science* **170,** 736–737 (1970).
45. Kerr, P. C., *et al.*: 'The Interrelation of Carbon and Phosphorous in Regulating Heterotrophic and Autotrophic Populations in Aquatic Ecosystems', U.S. Dept. of Interior, Water Pollution Control Research Series, U.S. Govt. Printing Office, Washington, D.C. (1970).
46. Moore, S. A. and Harris, R.: 'Polychlorinated Biphenyls: Effects on Marine Phytoplankton Communities', *Nature* (1972) (in press).
47. Sisler, F. D.: 'Biochemical Fuel Cells', in D. J. D. Hockenhull (ed.), *Progress in Industrial Microbiology,* Vol. 9, J. and A. Churchill Co., London (1971).
48. Sisler, F. D. and Olson, F. C. W.: 'Improving the Fertility of the Mediterranean', *New Sci.* **16,** 696 (1962).

49. Sisler, F. D. and Senftle, F. E.: 'Possible Influence of the Earth's Magnetic Field on Geomicro-biological Processes in the Hydrosphere', in Carl Oppenheimer (ed.), *Symposium on Marine Microbiology*, Charles C. Thomas, Springfield, Ill. p. 159 (1963).
50. Turekian, K. K.: 'Rivers, Tributaries, and Estuaries', in Donald Hood (ed.), *Impingement of Man on the Oceans*, Hood, Wiley–Interscience, N.Y. (1971).

For Further Reading

1. Effects of population growth and technology on pollution of the sea:
   Panel Reports of the Commission on Marine Science, Engineering and Resources, Vol. 1: *Science and Environment*, U.S. Government Printing Office, Washington, D.C. 1969.
2. Significance of photosynthesis in oxygen and carbon dioxide equilibria:
   E. I. Rabinowitch, *Photosynthesis*, Vol. 1 and 2, Interscience Publishers, New York, 1956, 2088 pp.
3. Pollution from large-scale use of nitrogen fertilizers and other agricultural chemicals:
   U.S. Dept. of Agriculture, *Wastes in Relation to Agriculture and Forestry*, Pub. No. 1065, U.S. Government Printing Office, Washington, D. C., 1968.
4. Chemical and physical interactions between the earth, oceans and atmosphere:
   Special report from Chemical and Engineering News, American Chemical Society, *Chemistry and Environment*, A.C.S. Publications 1967.

THE DEPENDENCE OF ATMOSPHERIC TEMPERATURE
ON THE CONCENTRATION OF CARBON DIOXIDE

SYUKURO MANABE

Geophysical Fluid Dynamics Laboratory/ESSA, Princeton, N.J., U.S.A.

Abstract. Numerical computations using a radiative, convective equilibrium model of the atmosphere predict an increase of 0.8°C in temperature of the Earth's surface by the end of this century, based on the anticipated increase in CO_2.

1. Introduction

It seems to be certain that the concentration of carbon dioxide in the atmosphere is indeed increasing with time. I shall discuss how the world temperature may be affected by such an increase.

As you know, carbon dioxide is nearly transparent to visible light but it is a strong absorber of infrared radiation particularly in the wavelength from 12 to 18 μ; consequently, an increase of atmospheric carbon dioxide could act much like a glass in a greenhouse to raise the temperature of the lower atmosphere.

The dependence of world temperature upon the concentration of carbon dioxide has been evaluated by various authors, e.g., Plass [7], Kaplan [1], Kondratiev and Niilisk [2], and Möller [6]. The latter's conclusion is quite different from those of preceding authors. One shortcoming of these studies is that their estimates were obtained from the computation of the heat balance of the Earth's surface instead of the atmosphere as a whole. Here, I would like to discuss the possibility of estimating this dependence by using the mathematical model of the whole atmosphere.

2. Calculation Procedure

Two of the most fundamental processes controlling the thermal structure of the atmosphere are radiative transfer, and moist or dry convection. Recently, we have been successful in obtaining the state of radiative, convective equilibrium of the atmosphere from the numerical integration of the model with both of these two processes [4, 5]. By comparing the states of equilibrium, which were obtained for various CO_2 concentrations, it was possible to estimate the dependence of atmospheric temperature upon CO_2 concentration.

Although the major constituents of the earth's atmosphere are nitrogen and oxygen, they hardly absorb the atmospheric radiation. However, minor constituents such as water vapor, carbon dioxide and ozone, have strong absorption bands and affect the field of both solar and terrestial radiation. In our computation, the radiative effects of these gaseous absorbers as well as that of clouds are calculated by using the equation of radiative transfer.

S. Fred Singer (ed.), The Changing Global Environment, 73–77. All rights reserved.
Copyright © 1975 by D. Reidel Publishing Company, Dordrecht–Holland.

The state of radiative convective equilibrium was approached asymptotically by the numerical time integration of the model starting from the initial condition of an isothermal atmosphere. In order to simulate the macroscopic behavior of moist convection, we introduced a very simple concept of a so-called 'convective adjustment'. Whenever the vertical temperature gradient exceeds the neutral gradient for moist convection, it was assumed that the neutral lapse rate* is restored instantaneously by the effect of the free moist convection.

3. Results

Figure 1 shows the approach towards the state of equilibrium. Towards the end of this time integration, the magnitude of the net downward solar radiation is almost exactly equal to that of net upward terrestial radiation at the top of the atmosphere, i.e., the atmosphere is in complete thermal equilibrium as a whole. In Figure 2, the state of radiative, convective equilibrium for the hemispheric mean insolation is compared with the U.S. standard atmosphere. The agreement between the two distributions is excellent.

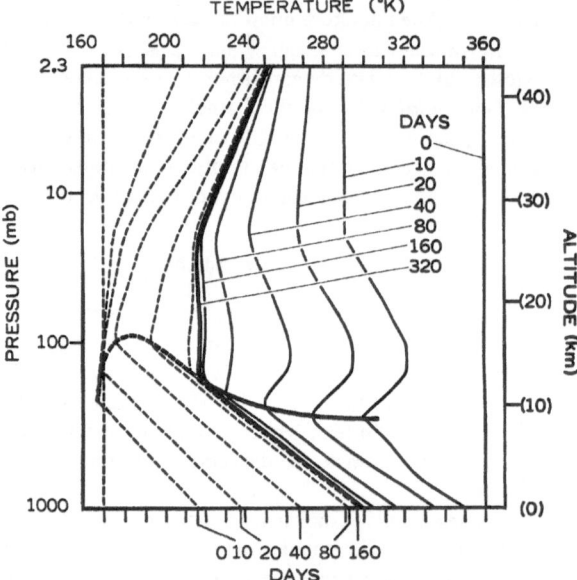

Fig. 1.   Approach towards the state of radiative, convective equilibrium. The solid and dashed lines show the approach from a warm and cold isothermal atmosphere. (By Manabe and Strickler [4]).

* The neutral lapse rate is assumed to be 6.5 °C km $^{-1}$ for this study.

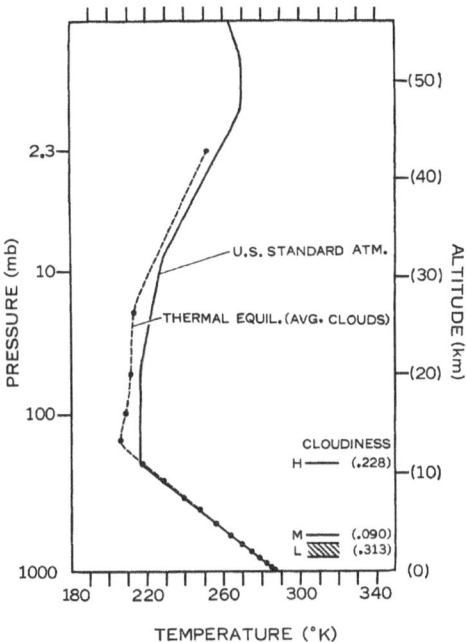

Fig. 2.   Dashed line shows the radiative, convective equilibrium of the atmosphere with cloudiness indicated in the right hand side of the figure. The solid line shows the U.S. Standard Atmosphere. (By Manabe and Strickler [4]).

Encouraged by this agreement, we decided to evaluate the dependence of the equilibrium temperature upon the concentration of carbon dioxide by using this model. In the equilibrium computation described so far, the state of radiative equilibrium has been obtained for the given distribution of absolute humidity. It is well known, however, that the warmer the atmosphere is, the more moisture it usually contains. As Möller [6] pointed out, the atmosphere tends to preserve the general level of relative humidity rather than that of absolute humidity through the process of condensation and evaporation. Owing to the dependence of the so-called greenhouse effect upon the concentration of water vapor, the equilibrium temperature of the atmosphere with a given distribution of relative humidity is almost twice as sensitive to the change of CO_2 concentration as that of the atmosphere with a given distribution of absolute humidity. Table I shows the results of our computation.

This table indicates that the doubling or halving the CO_2 concentration increases or decreases the surface temperature of the atmosphere by about 2.3 °C. Suppose the concentration of CO_2 increases by about 25 % from 1900 A.D. to 2000 A.D. as the U.N. Dept. of Social and Economic Affairs predicts [8], the resulting increase of surface temperature would be about 0.8 °C, which may have significant effect upon the

TABLE I

Change of equilibrium temperature of the Earth's surface in °C corresponding to various changes of CO_2 content of the atmosphere (Manabe and Wetherald [5]).

Change of CO_2 content (ppm)	Fixed absolute humidity		Fixed relative humidity	
	Average cloudiness	Clear	Average cloudiness	Clear
300→150	−1.25	−1.30	−2.28	−2.80
300→600	+1.33	+1.36	+2.36	2.92

climate of the earth's surface. Figure 3 shows how the vertical distribution of temperature depends upon the CO_2 content. It is interesting that in the stratosphere, the larger the CO_2 concentration is, the colder is the temperature.

4. Discussion

So far, we have discussed the dependence of equilibrium temperature upon the concentration of carbon dioxide. In order to discuss how the latitudinal distribution of

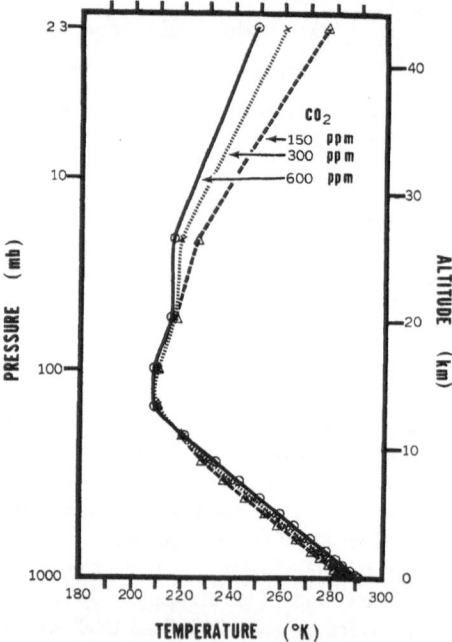

Fig. 3. Vertical distribution of temperature in radiative, convective equilibrium for various values of CO_2 content. (By Manabe and Wetherald [5]).

temperature is affected by the change of CO_2 concentration, it is necessary to construct the three-dimensional model of the atmosphere which involves not only the effect of radiation and convection but also the dynamics of the large scale eddies and the hydrologic cycle of water and snow.* Such a model has been constructed at the Geophysical Fluid Dynamics Laboratory of ESSA. We are preparing to carry out a series of numerical experiments for various CO_2 concentrations by using this model.

In order to discuss the very long range evolution of climate it is necessary to consider the exchange of CO_2 between the ocean and the atmosphere. Recently Manabe and Bryan [3] have attempted to construct a general circulation model of a joint ocean-atmosphere system, which could be used for such a purpose. One of the major difficulties in constructing such a model is our lack of knowledge about the distribution of the coefficient of vertical mixing by the small scale eddies in the ocean. I hope the improvement of the knowledge of the turbulent process in the ocean will enable us to study the probability of climatic instability by using such a joint model.

References

1. Kaplan, L. D.: 'The Influence of Carbon Dioxide Variations on the Atmospheric Heat Balance', *Tellus* **12**, 204–208 (1960).
2. Kondratiev, K. Y. and Niilisk, H. I.: 'On the Question of Carbon Dioxide Heat Radiation in the Atmosphere', *Geofis. Pura Appl.* **46**, 216–230 (1960).
3. Manabe, S. and Bryan, K.: 'Climate Calculation with a Combined Ocean-Atmosphere Model', *J. Atmos. Sci.* **26**, 786–789 (1969).
4. Manabe, S. and Strickler, R. F.: 'Thermal Equilibrium of the Atmosphere with a Convective Adjustment', *J. Atmos. Sci.* **21**, 361–385 (1964).
5. Manabe, S. and Wetherald, R. T.: 'Thermal Equilibrium of the Atmosphere with a Given Distribution of Relative Humidity', **24**, 241–259 (1967).
6. Möller, F.: 'On the Influence of Changes in the CO_2 Concentration in Air on the Radiation Balance of the Earth's Surface and on the Climate', *J. Geophys. Res.* **68**, 3877–3886 (1963).
7. Plass, G. N.: 'The Carbon Dioxide Theory of Climatic Change', *Tellus* **8**, 140–154 (1956).
8. 'World Energy Requirement in 1975 and 2000', in *Proceedings of the International Conference on the Peaceful Uses of Atomic Energy*, pp. 3–33, United Nations Department of Economic and Social Affairs (1956).

General Bibliography

For the more recent discussion of this subject by the author, see:
Manabe, S.: 'Estimate of Future Change of Climate Due to the Increase of Carbon Dioxide in the Air,' in W. H. Matthews, W. H. Kellog and G. D. Robinson (eds.), *Man's Impact on the Climate*, pp. 249–264, MIT Press, Boston, Mass. (1971).

Also see:
Study of Man's Impact on the Climate, in C. L. Wilson, *et al.* (eds.), *Inadvertent Climate Modification*, *SMIC*, 308 p., MIT Press, Boston, Mass. (1971).

In Part III of this book, discussion of the use of mathematical models of climate for the study of climatic change is made. On pages 238–240, the possible climatic change due to the change of carbon dioxide is discussed, based upon the results from a simple three-dimensional model of climate as well as other models.

* Since snow or ice has a large reflectivity for solar radiation, the incorporation of the snow hydrology into the model may significantly increase the sensitivity of the model climate to the change in the amount of atmospheric absorbers such as CO_2.

EXCHANGE OF CO_2 BETWEEN ATMOSPHERE AND SEA WATER: POSSIBLE ENZYMATIC CONTROL OF THE RATE*

RAINER BERGER and WILLARD F. LIBBY

Institute of Geophysics, University of California, Los Angeles, Calif., U.S.A.

Abstract. Surface and sub-surface ocean water differ in exchange characteristics with atmospheric CO_2. The possibility of control by an enzyme like carbonic anhydrase is discussed.

It has been discovered [1] that sea waters can differ markedly in their rates of equilibration with atmospheric CO_2. Surface waters on the average possess less than half the bomb C^{14} content of tropospheric air and are rising only in a matter of years [2, 3] to full equilibrium.

It has been well known for some time now [4] that in the analogous problem in mammals – the ready elimination of CO_2 from the blood for expiration in the lungs – that a special enzyme, carbonic anhydrase, is essential. So the question arises: Is there, perhaps, enzymatic control over the interchange of carbon dioxide between the atmosphere and the oceans?

Our procedure has been to take 50-gal seawater samples in polyethylene lined barrels to China Lake in order to avoid the Los Angeles smog and to vigorously aerate them with clean desert air (some 200 l h^{-1}) for periods of days and then to

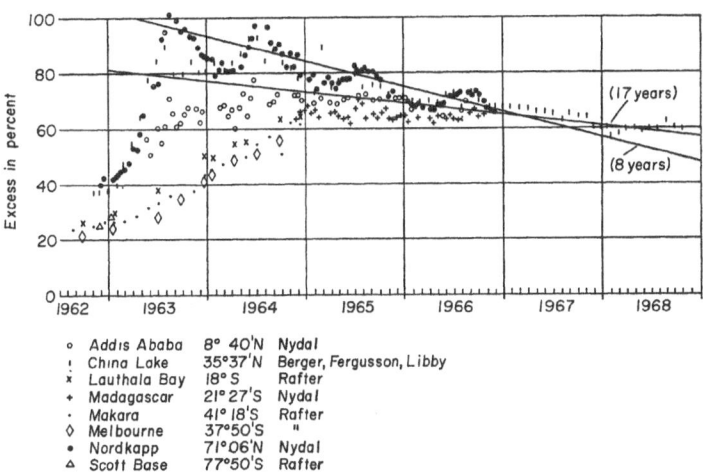

o	Addis Ababa	8° 40'N	Nydal
!	China Lake	35°37'N	Berger, Fergusson, Libby
x	Lauthala Bay	18°S	Rafter
+	Madagascar	21°27'S	Nydal
.	Makara	41°18'S	Rafter
◊	Melbourne	37°50'S	"
•	Nordkapp	71°06'N	Nydal
△	Scott Base	77°50'S	Rafter

Fig. 1. Radiocarbon in atmospheric CO_2.

* Publication No. 730 of the Institute of Geophysics, University of California.

test for the bomb C^{14} content. Figure 1 shows the course of the bomb C^{14} content of the surface Mojave Desert air at this location over the last seven years (analogous data from elsewhere in the world are included for comparison).

Qualitatively our results show that

(1) the exchange rate of surface waters (Santa Monica Beach) is very slow ($\sim 1/470$ day^{-1} unimolecular rate constant).

(2) carbonic anhydrase at 100 mg for 50 gal increases the rate very substantially (about a factor of 20) as do 10 mg quantities.

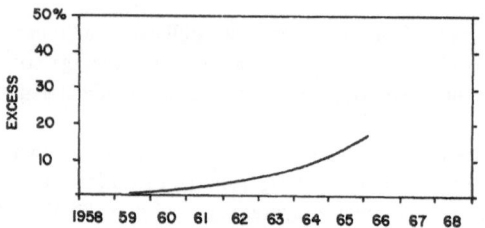

Fig. 2. Radiocarbon in ocean at 50° N.

TABLE I

CO_2 exchange rates for seawater

(50-gal samples in polyethylene lined barrels capped, trucked to China Lake, and aerated at 200 l hr^{-1} for indicated period with or without addition of the enzyme carbonic anhydrase.)

A. Surface waters (Santa Monica Beach at foot of Sunset Boulevard)

Sample No.	Date	Initial $\Delta C^{14}\%$	Treatment	Final ΔC^{14}	Exchange time (Days, cf. Table II)
1	12/22/65	14.9	17 days	17.3	460
2	1/8/66	15.1	68 days	22.7	480
3	4/12/66	15.4	100 mg CA +14 days	47	19
4	5/12/66	14.0	100 mg CA +3 days	29.6	10

B. Waters from 200 ft depth

5	3/9/67 200′ well at Naval Pt. Mugu	4.7	7.6 days	23.9	14
6	5/26/67 200′ Submarine USS Baya 33°20′N (Off Catalina Is.) 118°17′W	6.7	3.0 days	34.2	5
7	7/25/68 200′ Submarine USS Baya 31°40′N (Off Catalina Is. in 2200 fathoms total depth) 120°20′W	20.7	2.7 days	37.1	5

(3) the rate of water from 200 ft depth is rapid and shows about the same characteristics as surface water to which the carbonic anhydrase has been added.

The general level of bomb C^{14} in surface seawaters as reported by Münnich and Roether [3] in 1967 are given in Figure 2. Comparison of Figures 1 and 2 shows that after several years the sea still lags substantially behind tropospheric air.

In order to test our theory of a biochemically controlled rate analogous to that found in mammalian systems we have studied the waters off the Southern California Coast both at the surface (Samples 1–4) and at 200 feet depth (Samples 5–7). The data are assembled in Table 1.

It is clear that Santa Monica beach water lacks some quality which Pt. Mugu and Catalina waters from 200 ft depth possess. This quality is matched by 100 mgms of carbonic anhydrase added to 50 gal.

Since only limited localities have been studied no very general conclusions can be drawn but the results suggest that

(1) the quality may be due to an enzyme like carbonic anhydrase which might well be isolatable and may be derived from sea life known to produce it [5–8].

(2) there may be extensive areas of the sea devoid of the quality imparted by carbonic anhydrase and therefore slow to dissolve CO_2.

(3) that this quality may not long survive contact with air.

TABLE II

Exchange rate calculations

$$C^{14}O_{2a}+CO_{2s} = CO_{2a}+C^{14}O_{2s}$$

$$\frac{[C^{14}O_{2a}]}{[CO_{2a}]} = \gamma_a$$

$$\frac{[C^{14}O_{2s}]}{[CO_{2s}]} = \gamma_t \quad \text{Eq. 1}$$

$$[C^{14}O_{2s}] = \gamma_0 + (\gamma_a - \gamma_0)(1 - e^{-t/\tau})$$

$$[CO_{2s}] = \gamma_0 e^{-t/\tau} + \gamma_a(1 - e^{-t/\tau})$$

γ_t, γ_0 are measured, $(\gamma_a - \gamma_0)$ calculated from Figure 1 and τ is calculated from Equation (1).

$$\tau = \frac{t}{-\ln\left(1 - \frac{(\gamma_t - \gamma_0)}{(\gamma_a - \gamma_0)}\right)}$$

$$1 - \frac{(\gamma_t - \gamma_0)}{(\gamma_a - \gamma_0)} \exp(-t/\tau)$$

Sample	γ_0	γ_a	γ_t	t(days)	$\gamma_t - \gamma_0$	$\gamma_a - \gamma_0$	ln	τ(days)
1	14.9	80	17.3	17	2.4	65.1	−0.37	460
2	15.1	75	22.7	68	7.6	60.0	−0.14	480
3	15.4	75	47.0	14	31.6	60.0	−0.75	19
4	14.0	75	29.6	3	15.6	61.0	−0.31	10
5	−4.7	65	23.9	7.6	28.6	70.0	−0.53	14
6	6.7	65	34.2	3.0	27.5	58.0	−0.63	5
7	20.7	63	37.1	2.7	17.1	42.0	−0.53	5

The rates measured here for Santa Monica beach surface waters agree well enough with the work of others [3] for the sea as a whole so we can have some confidence in the tentative conclusions. However, additional work is necessary to isolate and identify the enzyme and to measure its oxidative stability and to assay the waters of the seas for it.

Acknowledgements

We thank Captains V. H. L. Duckett and W. A. White, officers and men of USS 'Baya', G. Plain, USNWC China Lake for assistance, and L. Provasoli, Haskins Laboratories, New York and J. D. Strickland, Scripps Institute of Oceanography for advice.

This research was supported in part by the National Science Foundation G-628.

References

1. Berger, R. and Libby, W. F.: UCLA Radiocarbon Dates VII, *Radiocarbon* 10, 149; 1969, *ibid.* 11, No. 1, 194 (1968).
2. Nydal, R.: *J. Geophys. Res.* 73, 3617 (1968).
3. Münnich, K. O. and Roether, W.: *Radioactive Dating and Methods of Low Level Counting*, International Atomic Energy Agency, Vienna 1967, p. 93–104 (1967).
4. White, A., Handler, P., and Smith, E. L.: *Principle of Biochemistry*, McGraw-Hill, p. 663 (1964).
5. Wilbur, K. M. and Owen, G.: in *Physiology of Mollusca* (ed. by K. M. Wilbur and C. M. Yonge), Academic Press, New York, p. 232 (for molluscs) (1964).
6. Nicol, J. A. C.: *The Biology of Marine Animals*, Sir Isaac Pitman & Sons, London, p. 191 (for fish) (1968).
7. Goreau, T.: *Endeavour* 20, 32 (for corals) (1961).
8. Isenberg, H. D., Lavine, L. S., and Weissfellner, H.: *J. Protozoology* 10, 477 (for coccoliths) (1963).

For Further Reading

1. M. N. Hill (ed.), *The Sea*, Vols. 1, 2 and 3, Interscience Publishers, 1962.
2. C. E. Junge, *Air Chemistry and Radioactivity*, Academic Press, New York, 1963.
3. Y. Bottinga and H. Craig, 'Oxygen Isotope Fractionation Between CO_2 and Water and the Isotopic Composition of Marine Atmospheric CO_2', *Earth and Planetary Science Letters* 5 (1969), 285.
4. P. Kilho Park, 'Oceanic CO_2 System: An Evaluation of Ten Methods of Investigation', *Limnology and Oceanography* 14 (1969), 179.

THE GLOBAL BALANCE OF CARBON MONOXIDE

LOUIS S. JAFFE

The George Washington University, Washington, D.C., U.S.A.

Abstract. Carbon monoxide, the most abundant air pollutant found in the community atmosphere, generally exceeds the mass of all other air pollutants combined (excluding CO_2). Total annual tonnages of CO from man-made sources exceeded 132 million metric tons (MT) in the United States and 359 MT (3.6×10^{14} g) globally in 1970. The largest single technological source of CO is the motor vehicle. This contribution is about 66% of the anthropogenic CO emissions in the U.S.A. and about 55% of man-made CO sources globally in 1970. Community concentrations are quite variable, unequally distributed, being related to proximity to anthropogenic sources, patterns of human activity, and meteorologic factors; they range from about 1 to > 70 ppm with brief, peak levels as high as 140 ppm.

While localized buildup of CO in cities may still represent a serious health hazard, man's CO production is relatively minor when considered on a global basis compared to the larger natural sources. These include the atmospheric oxidation of methane, the ocean, decay of chlorophyll in plants, and terpene oxidation from plant sources. Their total source strength is estimated as 3 to 14 times the man-made source, based on current calculations. Background concentrations, a product of both natural and anthropogenic sources, normally range from 0.04 to 0.20 ppm: with occasional concentrations as high as 0.80 ppm at ground level in remote clean air areas, the higher concentrations being found in the northern hemisphere.

1. Introduction

The review to follow presents (1) data on anthropogenic sources of CO in the United States and globally; (2) discussion of the characteristics of CO in community atmospheres; (3) a discussion of global background levels, distribution, and estimated mean lifetime; (4) natural sources of CO; (5) chemical reactions and fate of CO, including a review of several potential CO removal processes.

2. Anthropogenic Sources of Carbon Monoxide

Carbon monoxide is produced globally in large quantities by the incomplete combustion of carbonaceous materials used as fuels* for transportation and heating; it also is generated in industrial processing and refuse burning. Motor vehicle exhaust is by far the largest single technological source of CO emissions [1, 2, 3, 4, 5, 6].

2.1. UNITED STATES

More than 132 MT** of CO were emitted to the atmosphere in 1970 from these major man-derived technological sources in the United States alone (Table I). The

---

* Carbon dioxide (CO_2) is not normally regarded as an air pollutant *per se* in the common usage of the term; instead it is considered to be the normal end product of combustion of organic fuels and substances. About 15×10^9 tons of CO_2 are emitted yearly through the combustion of fossil fuels worldwide; about ½ is retained in the atmosphere. (See Singer, p. 35.)
** We denote 10^6 metric tons by MT.

S. Fred Singer (ed.), The Changing Global Environment, 83–110. All rights reserved.
Copyright © 1975 by D. Reidel Publishing Company, Dordrecht–Holland.

TABLE I

Estimated anthropogenic carbon monoxide emission sources – United States 1970 (10^6 metric tons)[a]

Source category	Emissions[a]	Percent of total[b]
Man made sources		
1. Fuel combustion in stationary sources	0.7	0.5
Steam and electrical	0.1	0.1
Industrial	0.1	0.1
Commercial and institutional	0.2	0.1
Residential	0.3	0.2
2. Transportation – mobile sources	100.6	75.8
Motor vehicles – gasoline	86.9	65.5
Motor vehicles – diesel	0.7	0.5
Railroads	0.1	0.1
Watercraft	1.5	1.1
Aircraft	2.7	2.0
Other non-highway use	8.6	6.5
3. Solid waste disposal	6.5	4.9
Municipal incineration	0.3	0.2
On-site incineration	0.4	0.3
Open-burning	4.1	3.1
Conical burning	1.8	1.4
4. Industrial process losses	10.3	7.8
5. Miscellaneous	14.4	10.9
Structural fires	0.2	0.2
Coal refuse burning	0.3	0.2
Agricultural burning	12.5	9.4
Prescribed burning	1.4	1.1
Total all categories (Anthropogenic sources)	132.6	100

[a] Totals in MT can be converted to short tons by multiplying by 1.1023×10^6.
[b] Totals may not add up due to rounding.

principal source, about 100 million tons or more than 75% from all technological sources in the United States, was the combustion of fossil fuels in mobile sources.

While the numbers of motor vehicles in use in the United States have been increasing yearly, nevertheless, the CO emissions from motor vehicles and other forms of transportation have begun to level off slightly within the past three years due to the introduction of motor vehicle pollution controls.

Urban CO levels are expected to decrease with the introduction of progressively more effective pollution control requirements; a 90% reduction in CO emissions is required for 1976 motor vehicles as compared with CO emissions from 1970 model vehicles. Inasmuch as these control devices are installed in all new vehicles sold domestically at the time of manufacture, there being no present national requirement for pollution controls in older vehicles already in use, and because it takes several years for a significant attrition or turnover in older motor vehicles in use, there will be a lag in decrease in CO emission from motor vehicles in the United States until several years after the introduction of the 1976 model. This anticipated decrease in CO emissions, however, will be counterbalanced, to some degree, by an increase in the total number of motor vehicles in use and by the deterioration of control devices, with use and age.

Agricultural burning and industrial processes are now the second and third largest technological sources of CO, respectively, while solid waste combustion is the fourth largest source of CO in the United States. The smallest general man made source of CO presently is fuel combustion from stationary sources.

2.2. GLOBAL ANTHROPOGENIC CO EMISSION SOURCES

Global CO emissions from combustion of fossil fuels are estimated to approximate 359 MT for the calender year 1970 [4]. (See Table II.)

Data for derivation of the 1970 global CO emissions from various sources are based

TABLE II

Estimated global anthropogenic CO sources – 1970

	Fuel consumption[a] world MT yr^{-1c}	CO emission[b] world MT yr^{-1c}
Mobile sources		
Motor vehicles	439	199
Gasoline		197
Diesel		2
Aircraft	84	5
(Aviation gasoline, jet fuel)		
Watercraft		18
Railroads		2
Other motor vehicles (non-highway)		26
(Construction equipment, farm		
tractors, utility engines, etc.)		
Stationary sources		
Coal and lignite	2983	4
Residual fuel oil	682	<1
Kerosene	69	<1
Distillate fuel oil	411	<1
Liquified petroleum gas (LPG)	34	<1
Industrial processes		41
(Petroleum refineries, steel mills, etc.)		
Solid waste disposal (urban and industrial)	1130	23
Miscellaneous		41
Agricultural burning,		
Coal bank refuse, etc.,		
Structural fires, etc.)		
Total anthropogenic CO		359

[a] Based on production data. Assumption is made that consumption equals production
[b] CO emission obtained by multiplying weight of fuel by appropriate 'CO factor'.
[c] To convert million metric tons (MT) to short tons multiply by 1.1023×10^6.

upon: (1) estimates of the comparative United States [4, 74] and world wide (less the United States) consumptions respectively of motor gasoline, aviation gasoline and jet fuel for aircraft, kerosene, and residual and distillate fuel oils, [75], (2) 1970 estimates of comparative consumptions of coal and other fuels based on the United Nations 1971 Statistical Yearbook [75] and the United States Department of Interior Bureau of Mines Minerals Yearbook, Vol. I, 1970 [77]; and (3) extrapolation of the 1970 United States estimated CO emissions [74], which are founded on use of weighted emissions factors for each source derived from data on current air pollution control practices in the U.S.A. [76], and substitution of other weighted emission factors for the remainder of the world for each source and/or fuel (based on the general lack of air pollution controls in most countries and the use of minimal air pollution abatement practices only in the more advanced countries).* The emission factors used for the global emission sources in Table II are listed in study by Jaffe [4] (Table 3). The global CO emission from watercraft include CO emissions from all types of vessels including international sea-going ships which carried 2833 million tons of cargo. International vessels were not included in the domestic United States CO emission inventory for watercraft. Additionally, ships are a more important form of commerce in foreign countries than in the U.S.A.

The global emission from industrial processes is extrapolated from the United States CO emission from industrial processes [74] based on a comparison of the 1970 relative gross national products expressed in equivalent United States dollars: World G.N.P. 3817×10^9 [132] versus United States G.N.P. 976.4×10^9 [133]. The estimate of 41 MT for global CO emissions from industrial processes obtained in this manner is a far more conservative estimate than that projected in the M.I.T. study which estimated a global CO emission from iron foundries alone of 15 MT (16.5 million short tons), Reference [134] Tables 5 and 6) based on 1968 data. Iron foundries contributed but 15.3% (based on 50% control) of the total CO emission from all industrial sources in the United States in 1970, the largest single source of CO emission from industrial sources in the United States being carbon black production furnaces, followed by emission from petroleum refineries [76]. Data on global agricultural burning is derived from Ref. 74 and 78 for the United States, and from an evaluation of agricultural waste removal practices in other countries, with the assistance of the Foreign Agricultural Service, U.S. Department of Agriculture. Data on miscellaneous combustion sources including coal bank refuse burning and structural fires are extrapolated from U.S.A. data [4].

There is a strong imbalance in CO emissions between the northern and southern hemispheres. On the basis of gasoline consumption, by far the largest anthropogenic source, 95% of the anthropogenic CO is emitted in the northern hemisphere. Although comparative hemispheric emission data for other anthropogenic sources are not available, it is likely that other combustion sources of CO are also greater in the northern hemisphere.

* Japan has since instituted a stringent emission control program which should significantly reduce CO and other air pollutant ambient levels by 1975.

3. Characteristics of Environmental CO in Community Atmospheres

3.1. TEMPORAL VARIATIONS

The concentration of CO in metropolitan areas varies widely with time and place and is dependent on human activity and meteorological factors [23–32]. Continuous air monitoring of CO at Continuous Air Monitoring Program (CAMP) stations in selected American cities and in special aerometric studies has revealed some distinct temporal patterns of variations in urban CO levels. There are distinct diurnal, weekly, and seasonal modes which correspond to the volume of motor vehicle traffic, traffic speed, and meteorological conditions [23–33, 79–83].

3.1.1. *Diurnal Patterns*

Community CO levels follow a regular diurnal pattern dependent primarily on human activity [23–27]. Ambient CO levels in communities correlate remarkably well with traffic volume and traffic speed, the highest levels being found most often in places where vehicular traffic is heaviest [24, 30–34] and vehicular speed is low [32–34, 79–83]. Similar findings have been reported in studies in large foreign urban com-

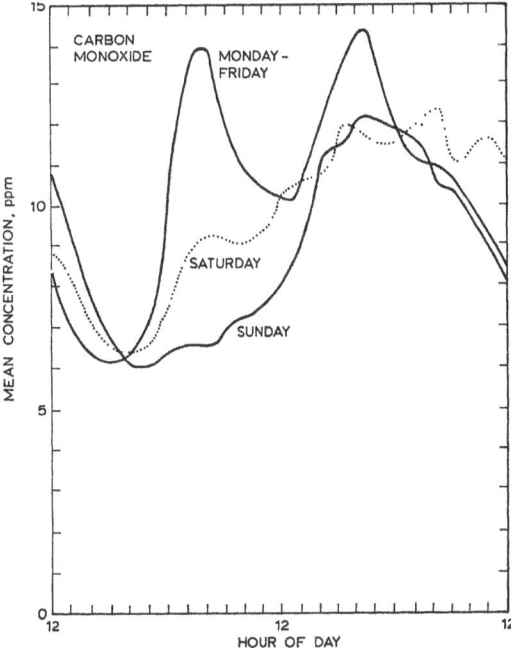

Fig. 1.   Diurnal variation of carbon monoxide levels on weekdays, Saturdays, and Sundays in Chicago, 1962–64. (From Ref. [25], Dept. of Commerce.)

munities such as London [28], Paris [29], and Frankfurt [31]. More recently such findings have also been observed in Mexico City as well [80].

While the exact shape of the CO curve is dependent on local traffic patterns, two daily peaks corresponding to the morning and evening traffic 'rush' hours occur in most communities (see Figure 1). The initial daily maxima are found between 7:00 and 9:00 a.m., coincident with heavy morning automobile traffic volume; the second peak is reached in the late afternoon and early evening [26, 27]. Within a community, there is little change in the time of occurrence of the daily morning maximum CO levels during the year [23, 26]. Although a late afternoon rise is evident in all seasons, a very pronounced evening 'rush hour' peak is found only in the winter CO curve. An exception to these general observations may be found in 'downtown' New York City, where there is a rapid rise in the morning CO levels corresponding to the morning

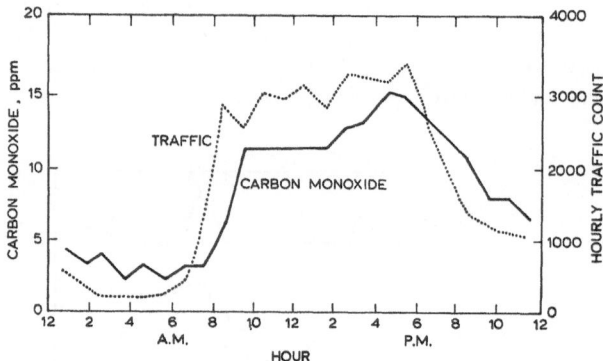

Fig. 2.   Hourly average carbon monoxide concentration and traffic count in mid-town Manhattan.
(From Johnson *et al.* [30].)

motor vehicle 'rush hour' traffic, with a uniformly high plateau lasting until afternoon; then a slower step rise in CO concentrations begins and builds to a single peak in the late afternoon [30] (see Figure 2). The shape of this curve is indicative of saturation levels of traffic [30].

3.1.2. *Weekly and Seasonal Patterns*

Peak concentrations are higher on weekdays than on Saturdays, which in turn, are higher than on Sundays and holidays, corresponding to the relative traffic volumes [24, 26] (see Figure 1).

Community CO levels disclose distinct seasonal patterns related mainly to seasonal meteorology. The mean ambient CO concentrations are generally highest in autumn, followed by the summer, spring, and winter, respectively [1–3]. The tendency toward increased atmospheric stability and low wind speeds in the fall and summer contribute substantially to the occurrence of high community CO levels (1, 26, 31, 79).

3.2. Meteorological Factors

The rates of emission and dispersion of CO determine the level at a given location. Both macrometeorological elements, suchs as atmospheric stability and wind speed, and micrometeorological elements, such as mechanical turbulence, play a role in the rate of dispersion of ambient CO [1, 27].

During prolonged periods of air stagnation, characterized by poor or inadequate diffusion, which occur periodically in most urban communities, the atmospheric levels of CO and other air pollutants will build up. For example, in the fall of 1964, during unusually prolonged and severe inversions, CO concentrations measured at air monitoring stations in Los Angeles and Sacramento, Calif. exceeded 30 ppm (35 mg m^{-3} for 8 hour periods [32]. Lawther *et al.* [28] reported a peak community CO level reading in downtown London at street level on a calm day of 235 ppm (270 mg m^{-3} in 1957.

3.3. Community Carbon Monoxide Concentrations

Data on community CO levels have been obtained from continuous air monitoring program (CAMP) stations located in 'downtown', off-street locations in a number of large U.S. cities; commuter traffic surveys; state, regional and municipal aerometric surveys; and special studies. These data indicate that a wide range of ambient CO levels exist within any community with significant localized differences dependent on proximity to traffic, traffic volume, type of traffic, and meteorological variables. These community CO levels range from 1 to >140 ppm, the latter being present only as brief peaks in heavy traffic.

Brice and Roesler [33], in a comparative study of CO levels in several U.S. cities participating in the aforementioned CAMP network, found that mean commuter traffic CO concentrations, based on 30-min integrated samples, were from 1.3 to 6.8 times greater than simultaneous mean aerometric levels measured at the corresponding CAMP station unit. Lynn *et al.* [34], in a study of commuter CO exposures in a number of U.S. cities, found that people in moving vehicles, particularly those in heavy traffic, are at times exposed to sustained levels of 50 or more ppm CO. Such mean CO levels, based on integrated 30-min aerometric samples taken in traffic, were found in the central business areas of 11 out of the 15 large U.S. cities surveyed. Very brief peak CO exposures, however, in all of these cities usually far exceeded 50 ppm and reached as high as 147 ppm in Los Angeles arterial traffic and 141 ppm in New York expressway traffic.

Larsen and Burke [35] using a mathematical model, recently have statistically analyzed extensive aerometric CO data from a variety of sampling sites in many U.S. cities. This analysis, based on maximum annual 8-hour averaging time concentrations at the most polluted 5% of the sites, revealed that the maximum annual 8-hour average CO concentrations in vehicles in downtown heavy traffic was about 115 ppm CO, about 75 ppm in vehicles in expressways and/or arterial routes, about 40 ppm in central commercial and mixed industrial areas, and about 23 ppm in residential areas.

Thus, the estimated CO concentrations in heavy traffic in city streets were almost 3 times the CO levels found in the central urban areas and 5 times the CO levels found in residential areas.

Carbon monoxide is also a serious atmospheric pollutant in urban areas of other industrialized countries [28, 29, 31, 38]. Lawther *et al.* [28], for example, reported average CO levels in central London ranging from 10 to 55 ppm over 10 min sampling periods with a mean concentration of 36 ppm for the 8-hour period from 11:00 a.m. to 7:00 p.m. and a maximum 10-min CO concentration of 155 ppm during the evening rush hour. Bank and McEachern [80], reported CO levels as high as 100–200 ppm in the central business area of Mexico City.

Mathematical urban diffusion models for CO have been developed by Stanford Research Institute [84], and others [85]. Extensive empirical studies and CO field measurements have also been obtained for determining the temporal-spatial variations and distribution of CO concentrations over urban areas in relation to traffic density, traffic speed, and meteorological factors [27, 31, 78, 82, 84, 85]. Additionally, noteworthy is an ongoing field CO measurement program being conducted by the State of California, Division of Highways, wherein a number of CO measurements are being taken simultaneously at various levels above and near traffic corridors in Los Angeles, Oakland, and other regions in California to determine the impact of traffic volume, traffic speed, and meteorological factors on CO concentrations and localized distribution for purposes of improved highway design.

3.4. SPECIAL AREA CARBON MONOXIDE CONCENTRATIONS

While significant concentrations of CO occur in community air of urban areas, particularly in traffic, even higher concentrations often exceeding 100 ppm for sustained periods, have been reported in underground garages and tunnels, and at loading platforms [36–38]. Waller *et al.* [37], for example, found the mean hourly CO levels in the Blackwall Tunnel in London in 1958–59 generally averaged > 100 ppm during rush hours with the mean hourly concentration reaching as high as 295 ppm on some days during the morning rush hours. Mean hourly concentrations, on the other hand, in the Sumner Tunnel in Boston usually were less than 100 ppm except for the evening rush hours when an hourly mean of 126 ppm was reached [36]. Most modern tunnels, however, now have CO alarm systems which automatically trigger auxiliary ventilating systems when the CO levels reach a prescribed CO concentration considered to be undesirable.

4. Global CO Background Levels, Distribution,
and Estimated Mean Lifetime

4.1. BACKGROUND LEVELS

Atmospheric CO was first identified by Migeotte [9] in 1949 as a trace constituent of the air on the basis of its presence in the solar spectrum. These findings indicated that CO is a world-wide constituent of the atmosphere.

The amount of CO measurable in relatively unpolluted ('clean') air is small. Junge reported that the background level of CO in the lower atmosphere is in the range of 0.01 to 0.2 ppm [48], based on a limited number of infrared solar spectra taken at Mt. Wilson, California; Ottawa, Canada; Jungfraujoch, Switzerland; and Columbus, Ohio, with an average concentration of about 0.1 ppm. Robbins *et al.* [50] determined that North Pacific marine air contained as little as 0.025 ppm while the non-urban air mass over continental California contained levels of 0.05 to 1.0 ppm CO. Studies in northern Alaska indicate a range of 0.055 to 0.260 ppm, averaging 90 ppb (0.09 ppm) [51]. Robbins *et al.* also measured typical background levels of CO ranging from 0.24 to 0.90 ppm at Camp Century, Greenland, and concluded that the variability of CO in unpolluted areas is a characteristic of the air mass in transit and reflects the prior history of the air mass.

4.2. DISTRIBUTION OF CARBON MONOXIDE IN THE ATMOSPHERE

Latitudinal variations of CO in remote areas have been obtained on cruise ships by Robinson and Robbins [86] in the USNS Eltanin and USNS Perseus ocean trips and by Junge [87, 88] on the Meteor expedition. The concentration in the northern hemisphere is variable, but averages nearly twice that of the southern hemisphere in the troposphere. Diurnal variations have been observed by Swinnerton [89] in remote areas in surface air. Robinson and Robbins [90] in the aforementioned surface measurements of CO, during 5 voyages in the North and South Pacific, found that atmospheric CO levels were the highest, 0.20 ppm, in the North Pacific, and the lowest, 0.04 ppm, in the South Pacific. They suggest that the maximum CO 'clean air' zone, 0.20 ppm at at 40 ° to 50 °N latitude, is related to the larger northern hemisphere anthropogenic CO pollution sources. Similar hemispheric differences of CO levels in clean surface air were observed by Seiler and Junge [87].

Measurements by Seiler and Junge [87], and Junge *et al.* [88] on commercial aircraft flights between Frankfurt and Tokyo and Frankfurt and Johannesburg indicate that CO is relatively well mixed vertically throughout the upper troposphere. The value of the mixing ratio in the upper troposphere immediately below the tropopopause in general ranged between 0.1 and 0.15 ppm, averaging 0.13 ppm in the high latitudes [87] and 0.12 ppm in the low latitudes [88], which is in good agreement with the mean value reported for clean surface air. They also found no marked differences in CO concentration between the two hemispheres in the subtropics at altitudes between 8 and 10 km, implying the presence of large natural sources.

Seiler and Junge [91] and Junge *et al.* [88] found that the mixing ratio decreased rapidly just above the tropopause in the lower stratosphere. It has been suggested by Seiler and Junge [87], and by Pressman and Warneck [92], that recombination with OH provides an effective sink in the stratosphere. Hesstvedt's calculations [93] indicate that a mixing ratio of 0.1 ppm is obtained in the troposphere 2 km below the tropopause, while 2 km above the tropopause, the mixing ratio has dropped to 0.02 ppm. There appears to be an efficient sink for CO in the lower stratosphere. (See section on Removal Processes for CO).

4.3. Estimated mean lifetime of atmospheric carbon monoxide

Carbon monoxide is a relatively long-lived substance in the lower atmosphere. The precise mean residence time, however, is not known with certainty. Estimates in the troposphere range from a lower limit of 0.1 year [52, 53] to about 5 yr [50]. Based on the studies of CO in ice samples and in aerometric monitoring at Camp Century, Robbins *et al.* [90] have tentatively concluded that the background levels of CO do not appear to be rising at the present time.

Bates and Witherspoon [10] first estimated the mean lifetime of CO in the atmosphere to be <4 yr, based on an estimated global anthropogenic production rate of CO of 8×10^{17} molecules cm^{-2} yr^{-1}, (i.e. 209 MT yr^{-1}), a small natural production, and a global average mixing ratio of 0.1 ppm. Robinson, *et al.* [3] estimated the annual global CO production rate in 1966 to be 231 MT, due almost entirely to human activity with only small natural sources present. With an average global mixing ratio of 0.10 ppm, they derived a mean CO lifetime of about 2.7 yr, in essential agreement with Bates and Witherspoon.

Weinstock [53] calculated the CO atmospheric lifetime without knowledge of the global source strength by a variation of the carbon dating technique, using radioactive ^{14}C in atmospheric CO as a tracer. The rate of ^{14}C production in the atmosphere by cosmic radiation is known more accurately than the total source strength of ^{12}CO. In addition, Pandow *et al.* [94] and McKay *et al.* [95] have demonstrated that in air 90% of the ^{14}C produced is initially fixed as ^{14}CO before it is further oxidized to CO_2. McKay, *et al.* [95], have also reported measurements of the specific activity of CO collected from air that give an indication of the atmospheric ^{14}CO content. From these data, Weinstock [53] deduced an atmospheric ^{14}CO residence time of $\tau \simeq 0.1$ yr.

Thus, if the sinks for both ^{14}CO and ^{12}CO are the same, Weinstock's value for the CO residence time implies the existence of much larger sinks and sources than heretofore recognized, and also that large natural sources of CO are involved. Weinstock indicates that his value is probably too low because his calculations of the CO oxidation rate in the troposphere were based on data pertaining to CO oxidation in both troposphere and the stratosphere. (A more exact calculation increases the lifetime of CO by roughly a factor of two to 0.2 yr) Junge, *et al.* [88], in a review of Weinstock's data, also confirm that a mean residence time of 0.2 yr is warranted. Wofsy *et al.* [96] more recently calculated the CO lifetime to be about 0.3 yr (in view of the calculated very large atmospheric CH_4 source strength and its subsequent oxidation to CO by OH radicals.)

Weinstock's calculations thus suggested that a large, relatively constant natural source of CO would resolve the anomaly: i.e. an increasing anthropogenic source function with no apparent increase in background concentrations. His data implied that there is a total source function which is at least ten times as large as the anthropogenic source function; this implies a sink ten times as large as that necessary to explain constant global background (CO) concentrations with fossil fuel consumption only.

5. Natural Sources of Atmospheric Carbon Monoxide

Until very recently, natural sources of CO had been considered to be quite negligible when compared to global anthropogenic sources. Early references had reported that some CO is produced in volcanic and marsh gases, as well as in natural gases found in coal mines [7]. Carbon monoxide had also been reported to be formed during electrical storms [8]. Bates and Witherspoon [10] reported that CO is also formed in the upper atmosphere by photodissociation of CO_2. (See also Figure 3.)

Lately, however, a number of large geophysical, geochemical and biological sources have been postulated. Although the precise production of natural CO from various sources is not entirely known, present evidence suggests that it is on the order of 3–14 times greater than anthropogenic sources, depending on the particular study cited. Stevens *et al.* [97, 101], Weinstock and Niki [98], McConnell *et al.* [99], and Wofsy *et al.* [96], among others, have discussed possible natural atmospheric CO sources.

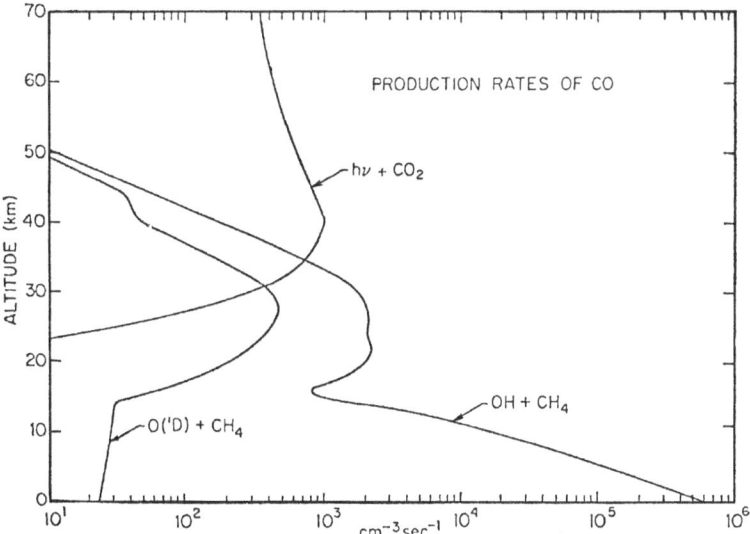

Fig. 3.   Rate of Production of CO from model calculations at standard summer atmosphere 30° N latitude (from Wofsy *et al.*).*

Stevens *et al.* [97, 101], in a comprehensive ongoing study, are isotopically analyz- ing the CO from the atmosphere, as well as that from natural sources, to ascertain the contribution of each to the atmosphere. CO emissions have distinctive isotopic com- positions characteristic of the source. By analyzing the isotopic composition of CO

* This model implies a surface source for CH_4 of about 2×10^{11} molecules an $^{-2}$s $^{-1}$, as well as uni- form mixing in the hyposphere.

LOUIS S. JAFFE

TABLE III

Identifiable species of atmospheric CO in rural illinois

Variety	^{18}O Enrichment[a] (%)	^{18}C Depletion[b] (%)	Principal occurrence	Source	Production rate in Northern Hemisphere
AGA[c]	2.46	2.74			
1	0.5	3.0	Principal species everywhere. Increased abundance in summer	Methane	
2	0.5	2.4	In varying amounts with 1. Increased concentration in winter and spring. Also in marine air of low northern latitudes	Probably methane	$> 3 \times 10^9$ ton y^{-1}
3	1.6–1.8	2.8	Lesser abundant heavy oxygen species during summer	Unknown	$\sim 5 \times 10^7$ ton month^{-1} during summer
4	2.5–3.3	2.2–2.6	Major species during autumn	Degradation of chlorophyll	2–5×10^8 tons during autumn
5	2.0–2.5	2.7	Major species during winter and early spring	Primarily anthropogenic	3–6×10^7 ton month^{-1} during winter

[a] With respect to the accepted oxygen isotopic standard, standard mean ocean water.
[b] With respect to the accepted carbon/isotopic standard, Peedee belomnite.
[c] AGA, average global automobile. (Source: Argonne National Laboratory, Stevens et al. (97, 101).

samples collected at many locations and at various times of the year Stevens et al. [97, 101], were able to trace the origin of several different species of CO. Then by comparing the average isotopic composition of atmospheric CO with that of a CO species whose production rate is known, they could then estimate the production rates of other species. For each isotopic determination, 87-liter samples were collected at 2 to 3 atmospheres pressure and stripped of moisture, carbon dioxide, and oxides of nitrogen. The remaining CO was then oxidized to CO_2, which is more easily purified, and the ^{18}O/^{16}O and ^{13}C/^{12}C ratios were determined with an isotope-ratio mass spectrometer. Calculations then provide the ^{18}O enrichment and the ^{13}C depletion of the CO sample with respect to accepted oxygen and carbon isotopic references (Table III).

There are five major varieties of atmospheric CO, two with light oxygen and three with heavy oxygen. (Table III) Isotopic analysis of atmospheric CO shows that it consists principally of light oxygen varieties. These two species (which are less ^{18}O enriched) are produced year round and are deduced to originate from the atmosphere itself. The remaining three species are heavy-oxygen varieties (more ^{18}O enrichment) and are minor constituents of atmospheric CO. Production of each of the latter is associated with a specific season of the year, and each may represent a mixture.

Measurements at a number of locations indicate that the constant concentration (0.10 to 0.15 ppm) of less enriched CO (varieties 1 and 2) occurs throughout the world. The production rate of the less enriched species – as well as that of the other species – were estimated by determining the average relative concentration, on a global basis, of automobile CO and using this figure as an internal standard for isotopic dilution calculations.

5.1. Geophysical and Geochemical Sources

In addition to the large geophysical and geochemical sources reported below, forest and prairie fires contribute a substantial amount of atmospheric CO emissions in the USA (see Table I) as well as globally. Carbon monoxide formed by electrical storms is clearly a natural source. Recently CO has been reported to be present in high concentrations in rainwater [108].

5.1.1. *Atmospheric Oxidation of Methane and Formaldehyde*

Atmospheric methane [99] and formaldehyde [102] have been lately suggested as natural sources of CO. McConnell *et al.* [99] have noted that the annual methane source is of the order of 900 MT (9×10^{14} g) and believe that the most likely product is CO, suggesting a possible mechanism involving OH radicals. They deduced that the oxidation of atmospheric CH_4 by OH radicals can provide over 90% of the CO produced in the troposphere. Methane is converted to CH_3 by reaction with OH or with O (^{1}D) (with photolysis playing a role at high altitudes). By a three-body reaction CH_3 is converted to CH_3O_2 which ultimately forms formaldehyde, H_2CO. Formaldehyde is photodissociated to form HCO which reacts with oxygen primarily to form CO. If all of the methane were converted to CO, this would produce about 1500 MT/yr^{-1}, more than four times as much as from anthropogenic sources. There are many sidepaths to this chemical mechanism and uncertainties in many of the rates. However, this does indicate a significant natural source for CO.

Calvert [102] also considered formaldehyde as a source of CO. The chemistry producing CO from formaldehyde is the same as that suggested by McConnell *et al.* [99]. A rough calculation of CHO radical concentration using $k(OH + H_2CO) = 7 \times 10^{-12}$ and $k(CHO + O_2) = 1 \times 10^{-13}$ and a H_2CO density of 5×10^{10} cm^{-3} with the OH and HO_2 densities given above, gives data to calculate a rate of CO formation of 1.3×10^6 molecules cm^{-3}s^{-1}. From this a lifetime of 0.4 yr is found for average solar conditions. Including this reaction system, it is seen that OH can participate in both the creation and loss of CO. The source function postulated is large enough to be the natural source postulated by Weinstock [98] and may be significantly larger than the man-made CO.

Weinstock and Niki [98], based on atmospheric models have recently calculated that the concentrations of hydroxyl radicals proposed as being present in the troposphere could account, through oxidation of methane, for a very large production of CO, ~ 5000 MT (5×10^{15} g yr^{-1}), amounting to a rate some 14 times greater than the production rate of CO from the combustion of fossil fuels. The average concentration of OH radicals in the troposphere required to achieve this is 2.3×10^6 molecules cm^{-3}, with a daytime concentration of twice that. Levy [103] and McConnell *et al.* [99] have deduced OH concentrations in the troposphere of this magnitude from purely photochemical considerations. Studies of atmospheric models of CO, CH_4 and CO_2 by Wofsy *et al.* [96] however, lead to a preliminary estimate that the atmospheric oxidation of methane to form CO is only 3 times greater than the global anthropogenic sources of CO (see Figure 3). This conclusion was based on a one-dimensional eddy diffusion model [96]; other studies by these investigators are in progress. Stevens *et al.* [97, 100, 101] in the aforementioned field studies suggest that methane may be the largest single source of natural CO and account for the production rate of the less enriched species of more than 3000 MT yr^{-1} in the Northern Hemisphere [100]. They estimate that all natural sources produce more than 3500 MT, or about ten times the amount produced by man [100]. (Table III). (See revised anthropogenic estimates, Table II). Growth of vegetation and decay of chlorophyll account for 100 MT a year [104]. The remaining 400 MT come from the ocean and from a variety of other, not yet identified natural sources [100]. Photochemical reactions involving naturally emitted terpenes from plants are estimated to contribute about 12 million tons of CO annually on a global basis [5].

5.1.2. *Oceanic Sources*

Oceans have been found to be a source of CO. A number of studies by Swinnerton and co-workers [22, 105, 106] with measurements of CO in the Atlantic and Pacific ocean waters of the northern hemisphere have yielded data that the CO content of the surface waters ranged from 7 to about 90 times (average 28 times) the calculated atmospheric equilibrium CO. Seiler and Junge [87] independently confirm the earlier work by Swinnerton [22] that the surface waters of the Atlantic have much larger CO concentrations (10–40 times greater) than the atmospheric equilibrium CO. Estimates by Junge *et al.* [88], based on the selected value of 15 g O_2 cm^{-2} yr^{-1} atm^{-1} for the exchange coefficient, indicate that the oceans in the Northern hemisphere produce about 75 MT CO yr^{-1} (or about 21 % of the 359 MT currently estimated global man-made CO production [4]). Linnenbom *et al.* [106] have revised their previous estimates of the oceanic CO contribution to the atmosphere, based on an average CO concentration of 10^{-5} ml/L, with a new estimate of about 90 MT of CO flux for the *northern hemisphere*, and 120 MT as the total global oceanic CO contribution to the atmosphere (or 61 % of the current estimate for global anthropogenic CO sources) assuming that the CO sea water saturation in the southern ocean is similar to that found in sea water measured in the northern hemisphere ocean waters.

The sources of the excess CO found in sea water have not been entirely determined. Swinnerton *et al.* [107] considered photochemical reactions on dissolved organic matter in surface ocean waters to be a large potential source. Rainwater, which empties into the ocean, is supersaturated with CO measuring up to 200-fold super-saturation relative to the partial pressure of the gas in the atmosphere [108]. Carbon monoxide is also produced by various biological organisms found in sea water including marine algae and by colonies of marine hydrozoan jelly-fish known as sipho-nophores [18, 19].

While kelps and other marine algae contribute only a small portion of the CO content of sea water, the carbon monoxide generated by the millions of siphonophores found in the oceans of the world contributes substantially to the relatively high concentrations of CO found in sea water. Diurnal variations of CO in surface sea water have been reported by Swinnerton *et al.* [22, 105, 108] to indicate the system is bio-logically dependent.

5.1.3. *Charged-Particle Deposition Phenomena*

Green *et al.* [110] in a recent study of the production of CO by charged-particle deposition mechanisms have determined that several electrical discharge-type natural phenomena produce CO in the atmosphere. The most significant of these natural charged-particle deposition-mechanisms of phenomena are: (1) lightning in the tropos-phere, (2) coronal discharges within and between clouds, and (3) the impact of photoelectrons in the ionosphere produced by the extreme ultraviolet radiation of the Sun. The report of high concentrations of CO in raindrops [108] lends credence to the lightning and cloud corona mechanism of CO formation. The total global CO pro-duced by these and minor electrical discharge atmospheric phenomena have been estimated by Green *et al.* [110] to be small when compared to global anthropogenic CO sources.

5.1.4. *Atmospheric Photochemical Production of CO*

Another natural source of CO is the result of photochemical degradation of various reactive organic compounds involved in the formation of photochemical smog [11, 12]. Photochemical reactions involving naturally emitted terpenes from plants are oxidized by O_3 or NO_2 to produce CO in the atmosphere. Went [113] estimated that about 200 MT of volatile organics of plant origin are dispersed globally each year. Went later raised his estimates to 1000 MT yr^{-1}. Robinson and Moser [5] estimated that the photochemical oxidation of terpenes from plant sources could yield about 12 MT of CO based on Went's original study and about 60 MT tons based on Went's revised data.

5.2. BIOLOGICAL SOURCES

5.2.1. *Production of CO by Vegetation*

Small quantities of CO are formed by vegetation during seed germination and seedling growth of higher plants [13, 14], and by certain marine brown algae or kelps [15, 16]. Float cells of the kelp *Nereocyctis* have been found to have CO concentrations of up to 800 ppm [3]. Microorganisms have been shown to produce CO from plant flavon-oids [17].

5.2.2. *Oxidation of Chlorophyll and Tetrapyrole Compounds (Bilin Biosynthesis)*

Laboratory studies by Crespi *et al.* [104] using pure cultures of blue-green algae indicate that blue-green algae have two characteristic photosynthetic pigments, phyco-cyanin and phycoerythrin, whose prosthetic groups, phycocyanobilin and phycoery-throbilin, result from the oxidation of a tetrapyrole macronucleus, such as chlorophyll. They consider that the chemistry of this process is probably similar to the conversion of the heme of hemoglobin in animals to the bile pigments, a process which produces one molecule of CO for each molecule of heme transformed to bile pigment [115, 116, 117]. The macrocycle ring is opened by oxidation of the carbon atom at the α-methine bridge position to yield a linear tetrapyrole and a molecule of CO [115, 117]. The chemical relationship between chlorophyll and algae bilins is very similar to that of heme and bile pigment, and each molecule of tetraphyrole converted to bilin would produce one molecule of CO. Thus, as the algae grow they evolve CO during the synthesis of the bile pigment [118].

5.2.3. *Catabolism of Chlorophyll*

The degradation of chlorophyll in dying or senescent plant material is another way for plants to produce CO, for the degradation of chlorophyll may well follow a course analogous to the degradation of heme [104]. All photosynthetic organisms that fix CO_2 and produce O_2 contain chlorophyll. Crespi *et al.* estimate that the annual degradation of chlorophyll could yield 60 MT of CO globally. (Together with the bilin biosynthesis, plants may be the source of 100 MT of CO per year globally.) Stevens *et al.* have observed CO emissions from trees in their field studies using radioactive isotopes [97]. They estimate that in the northern hemisphere an autumnal burst of CO from chlorophyll destroyed by trees and other plants corresponds to an emission of 500 MT tons of CO over a 1.5 month period, a similar order of magnitude to Crespi's data [104].

5.2.4. *Production of CO by Marine Invertebrates*

As indicated above, CO is produced by colonies of marine hydrozoan jelly-fish known as siphonophores [18, 19] and by the float cells of *Physalia physalis* (Portuguese Man-of-War) [20] a surface dwelling siphonophore found in bays and along shore lines of the oceans. These invertebrate colonies are widespread and make up a large portion of the plankton in the warmer oceans of the world and contribute substantially to the high concentrations of CO found in sea water.

Another biological source of CO is endogenous CO produced in man and animals as a by-product of heme catabolism [21].

To sum up, natural sources contribute a major share of atmospheric CO present in the troposphere and stratosphere, several times greater than all anthropogenic sources combined. Present estimates indicate that the natural CO from all sources annually, about 3500 MT in the Northern Hemisphere, constitutes about 90% of the global CO. Anthropogenic sources in the Northern Hemisphere are estimates to be about 341 MT.

6. Fate of Atmospheric Carbon Monoxide; Removal Processes

6.1. HORIZONTAL TRANSPORT

Kwok *et al.* [128] have included CO convective transport in an atmospheric circulation model for a four-week simulation of dispersion from the North American Continent, Europe and Asia. It took about two days for the American sources to reach the north

Atlantic, and after six days discernible contours linking the two sources were evident. This may be regarded as a limiting case, since horizontal eddy diffusion and CO sinks were neglected, and any natural source was neglected as well. The study does show the obvious rapid convective mixing which occurs in the Northern Hemisphere, and the very slow interchange between hemispheres.

6.2. CHEMICAL REACTIONS IN THE ATMOSPHERE

In the absence of scavenging processes, the currently estimated emissions of CO from both technological and natural sources, on the order of about 3900 MT in the Northern hemisphere alone, would be sufficient to raise the atmospheric background by 0.69 ppm yr^{-1}.

Based on the increasing mass of anthropogenic CO released globally and the newly uncovered far larger natural global sources of CO, it is apparent that the background CO levels would have surpassed 1 ppm in <2 yr were there no sink mechanisms, yet the background levels have remarkably not increased essentially for a century or more [90]. Obviously, there is a global balance and one or more effective scavenging or sink mechanisms for removal of atmospheric CO exist.

The rate of chemical oxidation of CO in the dense lower atmosphere had been heretofore considered to be very slow and unimportant [10, 12]. It was believed to be essentially chemically inert in the lower atmosphere, apparently not reacting with other constituents of urban air to a significant degree. Consideration, however, had been given to the possibility that some very rapid reactions might occur between CO and certain intermediates of photochemical smog reactions. One possible intermediate was the hydroxyl (OH) radical [11], which can be produced by the photolysis of aldehydes to produce perhydroxyl, and then be reduced to the hydroxyl radical. While the possible importance of radical reactions in the troposphere had long been considered, it had been believed that radicals and metastable species were not present in sufficient concentrations to be important in the normal troposphere [12].

In 1969, Weinstock suggested that OH radicals could be an effective mechanism for the removal of CO in the troposphere. He indicated that the concentration of OH required to maintain CO at a concentration of 0.1 ppm is about 7×10^4 molecule cm^{-3}, which could only be realized by a regenerative mechanism for OH because of the large amount of CO involved [53].

Levy [103] developed a simplified steady state photochemical model of the normal unpolluted troposphere, and demonstrated that OH radicals could be formed in sufficient concentrations to provide the necessary sink for CO. A chain reaction is involved:

The driving photochemical reaction for this system is

$$O_2 + hv(2900 \text{ Å} < \lambda < 3400 \text{ Å}) \rightarrow O(^1D) + O_2. \tag{1}$$

The principal reactions of metastable atomic oxygen are quenching

$$O(^1D) + M \rightarrow O(^3P) + M \tag{2}$$

and the formation of hydroxyl radicals

$$O(^1D) + H_2O \rightarrow 2OH. \tag{3}$$

A short chain reaction is then initiated by hydroxyl radical attack of ozone

$$OH + O_3 \rightarrow HO_2 + O_2 \tag{4}$$

and of carbon monoxide

$$OH + CO \rightarrow CO_2 + H \tag{5}$$

followed immediately by

$$H + O_2 + M \rightarrow HO_2 + M. \tag{6}$$

The chain reaction is completed by the oxidation of nitric oxide, which reforms the hydroxyl radical

$$HO_2 + NO \rightarrow NO_2 + OH. \tag{7}$$

A longer chain starts with hydroxyl radical attack of methane

$$OH + CH_4 \rightarrow CH_3 + H_2O \tag{8}$$

followed immediately by

$$CH_3 + O_2 + M \rightarrow CH_3O_2 + M. \tag{9}$$

The methylperoxyl radical oxidizes nitric oxide to give a methoxy radical

$$CH_3O_2 + NO \rightarrow NO_2 + CH_3O \tag{10}$$

which reacts with molecular oxygen to form formaldehyde and a hydroperoxyl radical

$$CH_3O + O_2 \rightarrow H_2C{=}O + HO_2. \tag{11}$$

The chain is then completed by reaction [7]. The important loss reactions for the radicals are

$$OH + OH \rightarrow H_2O + O \tag{12}$$

$$HO_2 + OH \rightarrow H_2O + O_2 \tag{13}$$

and

$$HO_2 + HO_2 \rightarrow H_2O_2 + O_2. \tag{14}$$

These chain reactions, which rapidly interconvert hydroxyl and hydroperoxyl radicals, may provide the dominant mechanism for removing atmospheric carbon monoxide and methane and for producing formaldehyde in the normal atmosphere. Westburg and Cohen [111] developed a photochemical model of a smog chamber that contained a chain reaction for removing CO.

Recent studies by McConnell *et al.* [99] and by Weinstock and Niki [98] in addition to Levy [103] have indicated that OH radicals may be present in the troposphere and stratosphere in sufficient concentration to oxidize CH_4 to CO. OH radicals may also be present in sufficient strength to oxidize CO to CO_2. Hence, the sink and source of CO are proportional to OH, and to a fair approximation the importance of CH_4 oxidation is independent of the average value of OH.

The chemistry of these chain reactions is quite complex and is based largely on theoretical models. Conflicting reports of many investigators were presented at a recent Symposium on Sources, Sinks, and Concentrations of Carbon Monoxide and Methane in the Earth's Environment [96, 119–127, 129]. The resolution of sink mechanisms for CO must await actual field measurements, underway or being planned.

6.3. VERTICAL TRANSPORT TO THE STRATOSPHERE (STRATOSPHERIC SINK)

Pressman and Warneck [92] have analyzed the stratospheric sink mechanism and indicate that the effectiveness of the stratosphere as a chemical sink for CO depends to a large extent upon the rate at which CO can be transported into the stratosphere. The exchange of air between the troposphere and the stratosphere is limited by the tropopause layer which acts as a barrier to convective transport. Data indicate that mixing in the troposphere in a given hemisphere is relatively fast and occurs within a few weeks; the exchange between the northern and southern hemispheres is slower, requiring about 1 yr; the storage time of air in the lower stratosphere, below 20 km, is about 1–2 yr; mixing within a hemisphere between troposphere and stratosphere is of the same order of magnitude. According to Equation (5) CO is oxidized to CO_2 in the stratosphere by the presence of OH radicals, but not regenerated further. The presence of OH radicals in the stratosphere results from the interaction of photochemically generated, excited oxygen atoms with water vapor. While a variety of species can react with CO, OH appears to be the most efficient. Two theoretical models were studied. Based on these studies and on the constraints including the absolute value of the atmospheric CO concentration, Pressman and Warneck conclude that the stratosphere consumes 11 % of the total CO inventory of the troposphere per year [92].

6.4. BIOLOGICAL REMOVAL (TERRESTRIAL AND MARINE BIOSPHERE SINK)

Another possible removal mechanism of atmospheric CO is the presence, in significant numbers, of microorganisms and plants that can metabolize CO.

(a) *Soil bacteria*. The Earth's surface is a possible agent in the removal of CO from the atmosphere. Carbon monoxide in contact with the soil may be oxidized to CO_2 or converted to methane (CH_4) by common specific anaerobic methane-producing soil microorganisms, *Methanosarcina Barkerii* and *Methanobacterium formicum*, in the presence of moisture.

This action has been demonstrated in the laboratory by Schnellen [56], who showed that pure cultures of these bacteria utilize CO as a source of carbon and convert CO into methane. Schnellen found that *Ms. Barkerii* is capable of effecting a considerable conversion of CO to CH_4 according to the equation:

$$4CO + 2H_2O \rightarrow CH_4 + 3CO_2. \tag{15}$$

Stephenson [57], however, indicates that CO, in the absence of H_2, reacts with water in these bacteria in two stages as follows:

$$4CO + 4H_2O \rightarrow 4CO_2 + 4H_2. \tag{16}$$

and

$$CO_2 + 4H_2 \rightarrow CH_4 + 2H_2O. \tag{17}$$

In the presence of H_2, these bacteria convert CO directly into methane and water:

$$CO + 3H_2 \rightarrow CH_4 + H_2O. \tag{18}$$

An aerobic soil bacterium, *Bacillus oligocarbophilus (Carboxydomonas oligocarbo-phila)*, found in and isolated from arable soil, has also been demonstrated to utilize CO as a source of carbon by oxidizing CO to CO_2. The organism, when cultivated on simple organic media free from other carbon sources, oxidizes CO to CO_2, providing a source of energy [58, 59]. Another bacterium, *Closteridium welchii*, when grown in the presence of CO, has been reported to produce lactic acid as a fermentation product [60].

Yagi [112] reported the enzymic conversion *in vitro* of CO to CO_2 by cell-free extracts of the sulfate reducing bacteria, *Desulfovibrio desulfuricans*, in the presence of sulfite as an oxidant. The bacterial extracts were prepared by subjecting the cell suspension to sonic disintegration. The enzyme conversion of CO to CO_2 by the bacterial extract was demonstrated by means of ^{14}CO used as a tracer. This study corroborates the potentiality of soil bacteria as a sink for atmospheric CO.

Recently, preliminary studies have been made by investigators at Stanford Research Institute (Inman *et al.* [130], Ingersoll *et al.* [131]) on the relative uptake of CO by various soils. A potting soil mixture depleted CO in a test atmosphere containing 120 ppm CO to near 0 ppm within 3 hr. Maximum activity occurred at 30 °C. Steam sterilization of the soil, the addition of antibiotics, highly saline (about 10%) solutions, and anaerobic conditions, all prevented CO uptake. Sterilized soil inoculated with nonsterile soil acquired activity with time. Samples of various natural soils differed in their ability to remove CO from the air. Acidic soils with a high content of organic matter were generally the most active. These investigators concluded that soil microorganisms, particularly 16 species of soil fungi, effectively absorbed carbon monoxide. Soils under cultivation were consistently lower in CO activity than the same soils under natural vegetation. These investigators concluded that the soils in the United States have the potential capacity for serving as a major sink for all of the anthropogenic CO emissions in the United States. Data for the field studies have been corrected for the influence of environmental variable based on laboratory studies; the potential uptake capacity of the soils of the coterminous United States was estimated to be 505×10^6 tons yr^{-1} [131].

Seiler *et al.* [129] in ongoing studies of CO uptake by various soils observed that the uptake of atmospheric CO is influenced considerably by the temperature of the soils. Precise conclusions, however, must await the publication of this study.

(b) Absorption or Retention by Vegetation. The process of plant respiration may also serve as a potential CO removal process, but this is not firmly established. Plants

are known to be scavengers for a wide variety of atmospheric materials. Although CO is not toxic to vegetation at concentrations found in ambient air, nevertheless, a number of reactions have been noted in plants after exposure to relatively high concentrations of CO: Ducet and Rosenberg [61] report the inhibition of respiration processes by CO, Burris [62] cites the inhibition of nitrogen fixation by CO, and Carr [63] has observed a variety of visible changes when plants are exposed to CO. Evidence of the fact that CO is a phytotoxicant at high concentrations reinforces the speculation that CO entering a plant during the respiration process will undergo subsequent reaction within the plant.

6.5. BIOCHEMICAL REMOVAL (BIOCHEMICAL SINK)

A potential biochemical removal process for CO is the binding of CO to the porphyrin-type compounds that are widely distributed in plants and animals. In particular, the heme compounds, such as hemoglobin found in man and animals, which are analogous to porphyrin compounds found in plants, are known to bind CO. It must be noted, however, that practically all of the CO absorbed by these heme compounds is eventually discharged from the blood of man and animals and only a small fraction is retained [64]. Nevertheless, this type of process in vegetation may have important potential for scavenging atmospheric CO. Permanent removal from the environment, however, would depend on whether CO subsequently entered into some reaction process to form CO_2 when the porphyrin compound is degraded.

6.6. ADSORPTION ON SURFACES

The catalytic adsorption of CO on hot metallic surfaces is well known. Adsorption of CO on copper surfaces, on charcoal, on Pyrex glass, and on quartz have been reported in laboratory studies at temperatures of 300 °C and above [68–72]. Kummler et al. [52] indicate that the gas-phase reaction of CO with nitrous oxide (N_2O), which is considered to be too slow to be of importance in the atmosphere, is, nevertheless, catalyzed in the presence of certain surfaces such as charcoal, carbon black, and glass in laboratory exposures of 300 °C and above, wherein the CO is oxidized to CO_2.

$$CO + N_2O \xrightarrow[\text{surface}]{} CO_2 + N_2. \tag{19}$$

These investigators have extrapolated the reported reaction rates to lower temperatures (300 K $\cong$ 27 °C) such as found in the lower atmosphere, and consider that such a reaction of the two gases in the presence of such surfaces is feasible in the lower atmosphere [52]. This conclusion is supported in part by laboratory studies wherein Gardner and Petrucci [72] measured and observed the chemisorption of CO on metallic films such as copper, cobalt, and nickel oxides at room temperature by means of infrared spectroscopy.

The necessary data for evaluation of the catalytic efficiency of such commonly found surfaces as metals, soils and atmospheric particulates in the adsorption of CO at realistic ambient temperatures found in the lower atmosphere, however, are present unavailable. Hence, the possibility of such commonly found surfaces serving as a significant sink for atmospheric CO by adsorption is unclear and presently uncertain.

6.7. Oceanic sink

There is no evidence at present that the oceans are a sink for CO, since no process or reaction has been discovered that would remove CO from the atmosphere [22, 89, 106]. As mentioned previously, Swinnerton *et al.* found that the CO concentration of the surface waters at all sampling points was much greater than that in equilibrium with the CO measured above the water. These findings [22, 89, 106] indicate that the ocean in the areas studied is not a sink for atmospheric CO, but, indeed, serves as an additional natural source.

7. Summary

While localized buildup of CO in cities may still represent a serious health hazard, man's CO production is relatively insignificant when considered on a global basis. Natural CO sources, heretofore considered to be negligible, have been determined to far exceed anthropogenic CO sources in the atmosphere as a whole. The largest single natural source is believed to be the oxidation of atmosphere methane by OH radicals. The background levels of CO, based upon analyses of ice samples in the Arctic and Antarctic, appear not to have increased essentially over a period of many centuries. To maintain this global balance of CO, in consideration of the continuing large global increases of man-made CO sources due to technology and combustion, one or more major global scavenging processes must be operating.

References

1. Jaffe, L. S.: 'Ambient Carbon Monoxide and its Fate in the Atmosphere', *J. Air Pollution Control Assoc.* **18** (8), 534–540 (1968).
2. Air Quality Criteria for Carbon Monoxide: National Air Pollution Control Administration Publication No. AP–62. Environmental Health Service, Public Health Service. U.S. Department of Health, Education and Welfare, March 1970, Washington, D.C. (1970).
3. Robinson, E. and Robbins, R. E.: 'Sources, Abundance and Fate of Gaseous Atmosphere Pollutants', Stanford Research Institute Project Pr–6755. Menlo Park, Calif. (1968).
4. Jaffe, L. S.: 'Carbon Monoxide in the Biosphere: Sources Distribution, and Concentrations', in *Symposium on Sources, Sinks and Concentrations of Carbon Monoxide and Methane in the Earth's Environment*, St. Petersburg Beach, Florida, August 15–17, 1972, *J. Geophys. Res.* **78** (24), 5293–5305 (1973).
5. Robinson, E. and Moser, C. E.: 'Global Gaseous Emissions and Removal Mechanisms', Second International Clean Air Congress, December, 1970, Washington, D.C., in H. M. Englund and W. T. Berry (eds.), *Proceedings of the Second International Clean Air Congress*, Academic Press, New York (1971).
6. Jaffe, L. S.: 'Sources, Characteristics and Fate of Atmospheric Carbon Monoxide', Paper presented at the Conference on Biological Effects of Carbon Monoxide, January 1970, New York Academy of Sciences (1970).
7. Flury, F. and Zernik, F.: *Schädliche Gase, Dampfe, Nebel, Rauch und Staubarten*, Julius Springer, Berlin (1931).
8. White, J. J.: 'Carbon Monoxide and its Relation to Aircraft'. *U.S. Naval Med. Bull.* **30**, 151 (1932).

9. Migeotte, M. V. and Neven, L.: 'The Fundamental Band of Carbon Monoxide at 4.7 μ in the Solar Spectrum', *Phys. Rev.* **75**, 1108–1109 (1949).

10. Bates, D. R. and Witherspoon, A. E.: 'The Photochemistry of some Minor Constituents of the Earth's Atmosphere', *Monthly Notices Roy. Astron. Soc.* **112**, 101–124 (1952).

11. Leighton, P. A.: 'Photochemistry of Air Pollution, IX', in *Phys. Chem.*, A Series of Monographs, Academic Press, New York (1961).

12. Altshuller, A. P. and Bufalini, J. J.: 'Photochemical Aspects of Air Pollution: a Review', *Photochem. Photobiol.* **4**, 97–146 (1965).

13. Wilks, S. S.: 'Carbon Monoxide in Green Plants', *Science* **129**, 964–966 (1965).

14. Siegel, S. M., Renwick, G., and Rosen, L. A.: 'Formation of Carbon Monoxide during Seed Germination and Seedling Growth', *Science* **137**, 683–684 (1962).

15. Loewus, M. W. and Delwiche, C. C.: 'Carbon Monoxide Production by Algae', *Plant Physiol.* **38** (4), 371–374 (1963).

16. Chapman, D. J. and Tocher, R. D.: 'Occurrence and Production of Carbon Monoxide in some Known Algae', *Can. J. Botany* **44**, 1438–1442 (1966).

17. Westlake, D. W., Roxburgh, J. M., and Talbot, G.: 'Microbial Production of Carbon Monoxide from Flavonoids', *Nature* **189**, 510–511 (1961).

18. Barham, E. G.: 'Siphonophores and the Deep Scattering Layer', *Science* **140**, 826–828 (1963).

19. Barham, E. G. and Wilton, J. W.: 'Carbon Monoxide Production by a Bathypelagic Siphonophore', *Science* **144**, 860–862 (1964).

20. Wittenberg, J. B.: 'The Source of Carbon Monoxide in the Float of *Physalia Physalis*, the Portuguese Man-of-War', *J. Exp. Biol.* **37**, 698–705 (1960).

21. Coburn, R. F., Blakemore, W. S., and Forster, R. E.: 'Endogenous Carbon Monoxide Production in Man', *J. Clin. Invest.* **42**, 1172–1178 (1963).

22. Swinnerton, J. W., Linnenbom, V. J., and Lamontague, R. A.: 'The Ocean: A Natural Source of Carbon Monoxide', *Science* **167**, 984–987 (1970).

23. Dickinson, J. E.: *Air Quality of Los Angeles County*. Technical Progress Report, Vol. II, Los Angeles Air Pollution Control District, Los Angeles, Calif. (1961).

24. Brief, R. S., Jones, A. R., and Yoder, J. S.: 'Lead, Carbon Monoxide, and Traffic, a Correlation Study', *J. Air Pollution Control Assoc.* **10**, 384–388 (1960).

25. Department of Commerce: The Automobile and Air Pollution: A Program for Progress, Part II. Subpanel reports to the panel on electrically powered vehicles. Commerce Technical Advisory Board. U.S. Government Printing Office, Washington, D.C. (1967).

26. Department of Health, Education and Welfare: Continuous Air Monitoring Projects. 1962–1967 Summary of monthly means and maximums. National Air Pollution Control Administration Publication No. APTD 69-1. Public Health Service, Arlington, Va. (1966).

27. McCormick, R. A. and Xintaras, C.: 'Variation of Carbon Monoxide Concentrations as Related to Sampling Interval, Traffic and Meteorological Factors', *J. Appl. Meteorol.* **1** (2), 237–243 (1962).

28. Lawther, P. J., Commins, B. T., and Henderson, M.: 'Carbon Monoxide in Town Air: an Interim Report', *Ann. Occupational Hyg.* **5**, 241–248 (1962).

29. Chovin, P.: 'Carbon Monoxide: Analyses of Exhaust Gas Investigations in Paris', *Envir. Res.* **1**, 198–216 (1967).

30. Johnson, K. L., Dworetsky, L. H., and Heller, A. N.: 'Carbon Monoxide and Air Pollution from Automobile Emissions in New York City', *Science* **160** (3823), 67–68 (1968).

31. Georgii, H. W. and Weber, E.: 'Untersuchung der Kohlenoxyd-emission in einer Grosstadt' [Investigations of Carbon Monoxide Emissions in a Large City], *Intern. J. Air Water Pollution* **6**, 179–195 (1962).

32. Tebbens, B. D.: 'Gaseous Pollutants in the Air', in A. C. Stern (ed.), *Air Pollution*, Vol. I Second Edition, Academic Press, New York, Ch. 2, pp. 31–32 (1968).

33. Brice, R. M. and Roesler, J. F.: 'The Exposure to Carbon Monoxide of Occupants of Vehicles Moving in Heavy Traffic', *J. Air Pollution Control Assoc.* **16**, 597–600 (1966).

34. Lynn, D. A., Tabor, E., Ott, W., and Smith, R.: 'Present and Future Commuter Exposures to Carbon Monoxide'. Paper Nos. 67-5, presented at the 60th Annual Meeting, Air Pollution Control Association, Cleveland, Ohio (1967).

35. Larsen, R. I. and Burke, H.: 'Ambient Carbon Monoxide Exposures'. Paper 69-167, presented at the 62nd Annual Meeting Air Pollution Control Association, June 1969, New York (1969).

36. Conlee, C. J., Kenline, P. A., Cummins, R. L., and Konopinski, V. J.: 'Motor Vehicle Exhaust at Three Selected Sites', *Arch. Environ. Health* **14**, 429–446 (1967).
37. Waller, R. E., Commins, B. T., and Lawther, P. J.: 'Air Pollution in Road Tunnels', *Brit. J. Ind. Med.* **18**, 250–259 (1961).
38. Trompeo, G., Turletti, G., and Giarrusso, O. T.: 'Concentrations of CO in Underground Garages', *Rass. Med. Ind.* **33**, 392–393 (1964).
39. Fischer, E. R. and McCarthy, M. Jr.: 'Study of Reaction of Electronically Excited Oxygen Molecules with Carbon Monoxide', *J. Chem. Phys.* **45**, 781–784 (1966).
40. Grave, W. M. and Long, F. J.: 'Kinetics and Mechanisms of the two Opposing Reactions of the Equilibrium: $CO + H_2O = CO_2 + H_2$', *J. Am. Chem. Soc.* **76**, 2602–2607 (1954).
41. The Merck Index of Chemicals and Drugs, Seventh Edition, *Carbon Monoxide*, (ed. by P. G. Stecher), Merck and Company, Inc., Rahway, N. J., p. 212 (1960).
42. Garvin, D.: 'The Oxidation of Carbon Monoxide in the Presence of Ozone', *J. Am. Chem. Soc.* **76**, 1523–1527 (1954).
43. Harteck, P. and Donders, S.: 'Reactions of Carbon Monoxide and Ozone', *J. Phys. Chem.* **26**, 1734–1737 (1957).
44. Zatsiorskii, M., Kondrateev, V., and Solnishkova, S.: 'Izluchenie plameni $CO + O_3$ i mekhanism etoi reaktsii' [Radiation of the flame of $CO + O_3$ and the mechanism of this reaction], *Zn. Fiz. Kim.* **14**, 1521–1527 (1940).
45. Brown, F. B. and Crist, R. H.: 'Further Studies on the Oxidation of Nitric Oxides: The Rate of Reaction between Carbon Monoxide and Nitrogen Dioxide', *J. Chem. Phys.* **9**, 840–846 (1941).
46. Penndorf, R.: 'The Vertical Distribution of Atomic Oxygen in the Upper Atmosphere', *J. Geophys. Res.* **54**, 1–38 (1949).
47. Heller, A. N. and Walters, D. F.: 'Impact of Changing Patterns of Energy Use on Community Air Quality', *J. Air Pollution Control Assoc.* **15**, 423–428 (1965).
48. Junge, C. E.: *Air Chemistry and Radioactivity*, Academic Press, New York (1963).
49. Locke, J. L. and Herzberg, L.: 'The Absorption due to Carbon Monoxide in the Infrared Solar Spectrum', *Can. J. Phys.* **31**, 504–516 (1953).
50. Robbins, R. C., Borg, K. M., and Robinson, E.: 'Carbon Monoxide in the Atmosphere', *J. Air Pollution Control Assoc.* **18**, 106–110 (1968).
51. Cavanagh, L. E., Schadt, C. F., and Robinson, E.: 'Atmospheric Hydrocarbons and Carbon Monoxide Measured at Point Barrow, Alaska', *Environ. Sci. Technol.* **3**, 251–257 (1969).
52. Kummler, R. H., Grenda, R. N., Baurer, T., Bortner, M. H., Davis, J. H., and MacDowall, J.: 'Satellite Solution of the Carbon Monoxide Sink Anomaly', (abstract) *Eos Trans. AGU* **50**, 174 (1969).
53. Weinstock, B.: 'Carbon Monoxide: Residence Time in the Atmosphere', *Science* **166**, 224–225 (1969).
54. Haagen-Smit, A. J. and Wayne, L. G.: 'Atmospheric Reactions and Scavenging Processes', Chapter 6, in A. C. Stern (ed.), *Air Pollution*, Vol. I, 2nd ed., Academic Press, New York, p. 181 (1968).
55. Harteck, P. and Reeves, R. R., Jr.: Some Specific Photochemical Reactions in the Atmosphere', Paper No. 20, presented at the Symposium on the Chemistry of the Natural Atmosphere. American Chemical Society, 154th Annual Meeting, Chicago, Ill. (1968).
56. Schnellen, C. G.: *Onderzoekingen over de methaangisting*, Doctoral thesis, Technische Wetenschap, Delft, Rotterdam, The Netherlands (1947).
57. Stephenson, M.: *Bacterial Metabolism*, 3rd ed., Longmans, Green and Company, New York City (1949).
58. Kaserer, H.: 'Die Oxydation des Wasserstoffes durch Microorganismen', *Zentr. Bakteriol., Parasitenk.*, Abt. **11** (16), 681–696 (1906).
59. Rabinovitch, E. I.: *Photosynthesis and Related Processes*, Interscience Publishers, New York (1945).
60. Waksman, S. A.: *Principles of Soil Microbiology*, Williams and Wilkins Company, Baltimore (1929).
61. Ducet, G. and Rosenberg, A. J.: 'Leaf Respiration', *Ann. Rev. Plant Physiol.* **13**, 171–200 (1962).
62. Burris, R. H.: 'Biological Nitrogen Fixation', *Ann. Rev. Plant Physiol.* **17**, 155–184 (1966).
63. Carr, D. J.: 'Chemical Influences of the Environment', in *Encycl. Plant Physiol.* **16**, 773–775 (1961).

64. Tobias, C. A., Lawrence, J. H., Roughton, F. J. W., Root, W. S., and Gregerson, M. I.: 'The Elimination of Carbon Monoxide from the Human Body with Possible Conversion of CO to CO_2', *Am. J. Physiol.* **145**, 253–263 (1945).

65. Douglas, E.: 'Carbon Monoxide Solubilities in Sea Water', *J. Phys. Chem.* **71**, 1931–1933 (1967).

66. Swinnerton, J. W., Linnenbom, V. J., and Cheek, C. H.: 'Distribution of Methane and Carbon Monoxide between the Atmosphere and Natural Waters', *Environ. Sci. Technol.* **3** (9), 836–838 (1969).

67. Skirrow, G.: 'The Dissolved Gases – Carbon Dioxide', in J. P. Riley and G. S. Skirrow (eds.), *Chemical Oceanography*, vol. I, Academic Press, New York, pp. 312–317, Ch. 7 (1965).

68. Madley, D. G. and Strickland-Constable, R. F.: 'The Kinetics of the Oxidation of Charcoal with Nitrous Oxide', *Trans. Faraday Soc.* **49**, 1312–1324 (1953).

69. Smith, R. N. and Mooi, J.: 'The Catalytic Oxidation of Carbon Monoxide by Nitrous Oxyde on Carbon Surfaces', *J. Phys. Chem.* **59**, 814–819 (1955).

70. Strickland-Constable, R. F.: 'Part played by Surfaces Oxides in the Oxidation of Carbon', *Trans. Faraday Soc.* **34**, 1074–1080 (1938).

71. Krouse, A.: 'Mechanism of Catalytic Oxidation of CO with N_2O', *Bull. Acad. Polon. Sci., Ser. Sci. Chem.* **9**, 5 (1961).

72. Gardner, R. A. and Petrucci, R. H.: 'The Chemisorption of Carbon Monoxide on Metals', *J. Am. Chem. Soc.* **82**, 5051–5053 (1960).

73. Bisselle, C., Goldstein, S., Lilienthal, M., and Pikul, R.: *Environmental Trends: Radiation, Air Pollution, Oil Spills*, Report No. MTR-6013, The Mitre Corporation (1971).

74. United States Environmental Protection Agency: Office of Air Programs Data File of Nationwide Emissions. Environmental Protection Agency, Research Triangle Park, North Carolina 27711, July 1972 (to be published in Nationwide Inventory of Air Pollution Emissions AP-073) (Revised) (In Preparation).

75. United Nations Statistical Yearbook, 1971, 23rd Edition, United Nations, New York, N.Y. (1972).

76. United States Environmental Protection Agency: *Compilation of Air Pollutant Emission Factors*, Publication No. AP-42 (Revised). U.S. Environmental Protection Agency, Office of Air Programs, Research Triangle Park, North Carolina (1972).

77. United States Department of Interior: Bureau of Mines 1970 Minerals Yearbook, Vol. I, Washington, D.C. (1973).

78. Wadleigh, C. H.: *Wastes in Relation to Agriculture and Forestry*, Miscellaneous Publication 1065, United States Department of Agriculture, Washington, D.C. (1968).

79. Georgii, H. W., Busch, E., and Weber, E.: *Investigation of the Temporal and Spatial Distribution of the Emission Concentrations of Carbon Monoxide in Frankfurt/Main*, Report No. 11 of the Institute for Meteorology and Geophysics, University of Frankfort/Main (Transaction No. 0477, NAPCA) (1967).

80. Bank, M. and McEachern, D. M.: *A Carbon Monoxide Profile of a Main Traffic/Artery in Mexico City*, Paper No. 70–13, presented at the 63rd Annual Meeting of the Air Pollution Control Association, June 14–18, 1970, St. Louis, Mo.

81. Colucci, J. M. and Begeman, C.: 'Carbon Monoxide in Detroit, New York, and Los Angeles Air, *Envir. Sci. Technol.* **3**, 14–47 (1969).

82. Dworetsky, L.: *Final Report on the Study of Air Pollution Aspects of Various Roadway Configurations*, Urban Expressway Air Pollution Study for the New York City Department of Air Resources, General Electric Co., Reentry and Environmental Systems Div., Philadelphia, Pa. (1971).

83. Wolf, P. C.: 'Carbon Monoxide Measurement and Monitoring in Urban Air', *Envir. Sci. Technol.* **5**, 212–218 (1971).

84. Ott, W. and Eliassen, R.: *An Urban Survey Technique for Measuring the Spatial Variation of Carbon Monoxide Concentrations in Cities*, Paper No. 72–17, presented at the 65th Annual Meeting of the Air Pollution Control Association, June 1972, Miami Beach, Fla.

85. Johnson, W. B., Ludwig, F. L., Dabberdt, W. S., and Allen, R. J.: *An Urban Diffusion Simulation Model for Carbon Monoxide*, Proceedings of the 1972 Summer Computer Simulation Conference, June 14–16, 1972, San Diego, Calif.

86. Robinson, E. and Robbins, R. C.: 'Atmospheric Background Concentrations of Carbon Monoxide', *Ann. N.Y. Acad. Sci.* **174**, 89–95 (1970).

87. Seiler, W. and June, C.: 'Carbon Monoxide in the Atmosphere', *J. Geophys. Res.* **75**, 2217–2226 (1970).
88. Junge, C., Seiler, W., and Warneck, P.: 'The Atmospheric ^{12}CO and ^{14}CO Budget, *J. Geophys. Res.* **76**, 2866–2879 (1971).
89. Swinnerton, J. W., Linnenbom, V., and Lamontagne, R.: 'Distribution of Carbon Monoxide Between the Atmosphere and the Ocean', *Ann. N.Y. Acad. Sci.* **174**, 96–101 (1970).
90. Robbins, R. E., Cavanagh, L. A., Salzs, L. J., and Robinson, E.: 'The Analysis of Ancient Atmospheres', in *Symposium on Sources, Sinks and Concentrations of Carbon Monoxide and Methane in the Earth's Environment*, St. Petersburg Beach, Florida, August 15–17, 1972, *J. Geophys. Res.* **78** (24), 5341–5344 (1973).
91. Seiler, W. and Junge, C.: 'Decrease in the Mixing Ratio above the Polar Tropopause', *Tellus* **21** (3), 447–449 (1969).
92. Pressman, J. and Warneck, P.: 'The Stratosphere as a Chemical Sink for Carbon Monoxide, *J. Atmos. Sci.* **27**, 155–163 (1970).
93. Hesstvedt, E.: 'Vertical Distribution of CO near the Tropopause', *Nature* **225**, 50 (1970).
94. Pandow, M., McKay, C., and Wolfgang, R.: 'The Reaction of Atomic Carbon with Oxygen Significance for the Natural Radio Carbon Cycle', *J. Inorg. Nucl. Chem.* **14**, 153–158 (1960).
95. McKay, C., Pandow, M., and Wolfgang, R.: 'On the Chemistry of Natural Radiocarbon', *J. Geophys. Res.* **68**, 3929–3931 (1963); *Phys. Res.* **68**, 3929–3931 (1963).
96. Wofsy, S. C., McConnell, J. C., and McElroy, M. B.: 'Atmospheric CH_4, Co, and CO_2', in *Symposium on Sources, Sinks, and Concentrations of Carbon Monoxide and Methane in the Earth's Environment*, St. Petersburg Beach, Florida, August 15–17, 1972, *J. Geophys. Res.* **77**, 4477–4493 (1972).
97. Stevens, C. M., Krout, L., Walling, D., Venters, A., Engelkemeir, A., and Ross, L. E.: 'The Isotopic Composition of Atmospheric Carbon Monoxide', in Symposium on *Sources, Sinks, and Concentrations of Carbon Monoxide and Methane in the Earth's Environment*, St. Petersburg Beach, Florida, August 15–17, 1972, *Earth Plan. Sci. Letters* **16**, 147–165 (1972).
98. Weinstock, B. and Niki, R.: 'Carbon Monoxide in Nature', *Science* **176**, 290–292 (1972).
99. McConnell, J. C., McElroy, M. B., and Wofsy, S. C.: 'Natural Sources of Atmospheric CO', *Nature* **233**, 187–188 (1971).
100. Maugh, T. H.: 'Carbon Monoxide: Natural Sources Dwarf Man's Output', *Science* **177**, 338–339 (1972).
101. Stevens, C. E.: *Fate of Carbon Monoxide in the Atmosphere*, APRAC Status Report, Presented at the Automotive Air Pollution Research Symposium, APRAC, Chicago, Ill. (1971).
102. Calvert, J. G., Kerr, J. A., Demerjian, K. L., and McQuigg, R. D.: 'Photolysis of Formaldehyde as a Hydrogen Atom Source in the Lower Atmosphere', *Science* **175**, 751–75 (1972).
103. Levy, H.: 'Normal Atmosphere: Large Radical and Formaldehyde Concentrations', *Science* **173**, 141–143 (1971).
104. Crespi, H. L., Huff, D., DaBoll, H. F., and Katz, J. J.: *Carbon Monoxide Emissions by Fresh-Water Algae*, Argonne National Laboratory, Argonne, Ill., 60439, Final Report to the Coordinating Research Council (1972).
105. Wilson, D. F., Swinnerton, J. W., and Lamontagne, R. A.: 'Production of Carbon Monoxide and Gaseous Hydrocarbons in Seawater: Relation to Dissolved Organic Carbon', *Science* **168**, 1577–1579 (1970).
106. Linnenbom, V. J., Swinnerton, J. W., and Lamontagne, R. A.: 'The Ocean as a Source for Atmospheric Carbon Monoxide', in *Symposium on Sources, Sinks, and Concentrations of Carbon Monoxide and Methane in the Earth's Environment*, St. Petersubrg Beach, Florida, August 15–17, 1972, *J. Geophys. Res.* **78** (24), 5333–5340 (1973).
107. Swinnerton, J. W., Linnenbom, V., and Lamontagne, R.: 'Distribution of Carbon Monoxide Between the Atmosphere and the Ocean', *Ann. N.Y. Acad. Sci.* **174**, 96–101 (1970).
108. Swinnerton, J. W., Lamontagne, R. A., and Linnenbom, V. J.: 'Carbon Monoxide in Rainwater', *Science* **172**, 943–945 (1971).
109. Lamontagne, R. A., Swinnerton, J. W., and Linnenbom, V. J.: 'Nonequilibrium of Carbon Monoxide and Methane at the Air-Sea Interface', *J. Geophys. Res.* **76** (21), 5117–5121 (1971).
110. Green, A. E. S., Sawada, T., Edgar, B. C., and Uman, M. A.: 'Production of CO by Charged-Particle Deposition Mechanisms', in *Symposium on Sources, Sinks, and Concentrations of Carbon*

Monoxide and Methane in the Earth's Environment, St. Petersburg Beach, Florida, August 15–17, 1972, *J. Geophys. Res.* **78** (24), 5284–5291 (1973).

111. Westburg, K. and Cohen, H.: Aerospace Corporation Report ATR-70-(8107)-1, Aerospace Corporation, El Segundo, California (1969).

112. Yagi, T.: 'Enzymic Oxidation of Carbon Monoxide', *Biochem. Biophys Acta* **30**, 194–195 (1958).

113. Went, F. W.: 'Organic Matter in the Atmosphere and its Possible Relation to Petroleum Formation', *Proc. Nat. Acad. Sci.* **46**, 212–221 (1960).

114. Went, F. W.: 'On the Nature of Aitken Condensation Nuclei', *Tellus* **18**, 549–556 (1966).

115. Sjostrand, T.: 'The Formation of Carbon Monoxide by *in vitro* Decomposition of Hemoglobin in Bile Pigments', *Acta Physiol. Scand.* **26**, 328–333 (1952).

116. Sjostrand, T.: 'The Formation of Carbon Monoxide by Decomposition of Hemoglobin *in vivo*'. *Acta Physiol. Scand.* **26**, 338–344 (1952).

117. White, P.: 'Carbon Monoxide Production and Heme Catabolism', in *Biological Effects of Carbon Monoxide*; *Ann. New York Acad. Sci.* **174**, 23–31 (1970).

118. Troxler, R. F., Brown, A., Lester, A., and White, P.: 'Bile Pigment in Plants', *Science* **167**, 192–193 (1970).

119. Nicolet, M. and Peetermans, W.: 'On the Vertical Distribution of Carbon Monoxide and Methane in the Stratosphere,' in *Symposium on Sources, Sinks, and Concentrations of Carbon Monoxide and Methane in the Earth's Environment*, St. Petersburg Beach, Florida, August 15–17, 1972.

120. Shimazaki, T. and Cadle, R. D.: 'Theoretical Model of Vertical Distributions of CO and CH_4 in the Mesosphere and Upper Stratosphere', in *Symposium on Sources, Sinks, and Concentrations of Carbon Monoxide and Methane in the Earth's Environment*, St. Petersburg Beach, Florida, August 15–17, 1972, *J. Geophys. Res.* **78** (24), 5352–5361 (1973).

121. Kummler, R. H. and Baurer, T.: 'A Temporal Model of Tropospheric Carbon Hydrogen Chemistry', in *Symposium on Sources, Sinks, and Concentrations of Carbon Monoxide and Methane in the Earth's Environment*, St. Petersburg Beach, Florida, August 15–17, 1972, *J. Geophys. Res.* **78** (24), 5306–5316 (1973).

122. Davis, D. D., Wong, W., Payne, W. A., and Stief, L. J.: 'A Kinetic Study to Determine the Importance of HO_2 in Atmospheric Chemical Dynamics: Reaction with CO_2', in *Symposium on Sources, Sinks, and Concentrations of Carbon Monoxide and Methane in the Earth's Environment*, St. Petersburg Beach, Florida, August 15–17, 1972, *Science* **179**, 280 (1973).

123. Westenburg, A. A. and de Haaf, N.: 'The Role of $CO-HO_2$ Reaction in the Atmospheric CO Sink', in *Symposium on Source, Sinks, and Concentrations of Carbon Monoxide and Methane in the Earth's Environment*, St. Petersburg Beach, Florida, August 15–17, 1972, *J. Phys. Chem.* **76**, 1580 (1972).

124. Levy, H. II: 'The Tropospheric Budgets for Methane, Carbon Monoxide, and Related Species', in *Symposium on Sources, Sinks, and Concentrations of Carbon Monoxide and Methane in the Earth's Environment*, St. Petersburg Beach, Florida, August 15–17, 1972, *J. Geophys. Res.* **78** (24), 5325–5332 (1973).

125. Goldman, A., Murcray, D. G., Murcray, F. H., Williams, W. J., Brooks, J. N., and Bradford, C. M.: 'Vertical Distribution of CO on the Atmosphere', in *Symposium on Sources, Sinks, and Concentrations of Carbon Monoxide and Methane in the Earth's Environment*, St. Petersburg Beach, Florida, August 15–17, 1972, *J. Geophys. Res.* **78** (24), 5273–5283 (1973).

126. Weinstock, B. and Chang, T. H.: 'Carbon Monoxide Balance', in *Symposium on Sources, Sinks, and Concentrations of Carbon Monoxide and Methane in the Earth's Environment*, St. Petersburg Beach, Florida, August 15–17, 1972, *Tellus* **26** (1974).

127. Whitten, R. C., Sims, J. S., and Turco, R. P.: 'A Model of Carbon Compounds in the Stratosphere and Mesopsphere', in *Symposium on Sources, Sinks, and Concentrations of Carbon Monoxide and Methane in the Earth's Environment*, St. Petersburg Beach, Florida, August 15–17, 1972, *J. Geophys. Res.* **78** (24), 5362–5374 (1973).

128. Kwok, L., Langlois, W., and Ellefsen, R.: *IBM J. Res. Dev.* (1971).

129. Seiler, W.: 'Carbon Monoxide in the Atmosphere,' in *Symposium on Sources, Sinks, and Concentrations of Carbon Monoxide and Methane in the Earth's Environment*, St. Petersburg Beach, Florida, August 15–17, 1972, *Tellus* **26** (1974).

130. Inman, R. E., Ingersoll, R. B., and Levy, E. A.: 'Soil: A Natural Sink for Carbon Monoxide', *Science* **172**, 1229–1231 (1972).

131. Ingersoll, R. B., Inman, R. E., and Fisher, W. R.: 'Soils Potential as a Sink for Atmospheric Carbon Monoxide', *Tellus* **26** (1974).
132. Block, H. and Kornei, L.: *The World's Product at the Turn of the Decade: Research Study*, United States Department of State, Washington, D.C. (1972).
133. United States Department of State: World Data Handbook. Department of State Publication 8655, General Foreign Policy Series 264, Washington, D.C. (1972).
134. *Man's Impact on the Global Environment*. Report of the Study of Critical Environmental Problems (SCEP). Sponsored by the Massachusetts Institute of Technology, The MIT Press, Cambridge, Mass. (1970).
135. Black, R. J., Muhick, J., Klaee, A. J., Hickman, H. L., Jr., and Vaughn, R. D.: *The National Solid Wastes Survey: An Interim Report*. United States Department of Health, Education and Welfare, Cincinnati, Ohio (1968).

For Further Reading

1. A. C. Stern (ed.), *Air Pollution*, Second Edition, Vols. I and II, Academic Press, New York, 1968.
2. Cleaning our Environment: The Chemical Basis for Action, 1969. A Report by the Subcommittee on Environmental Improvement. Committee on Chemistry and Public Affairs, American Chemical Society, Washington, D.C.
3. Restoring the Quality of Our Environment, November 1965. Report of the Environmental Pollution Panel, President's Advisory Committee. The White House, Washington, D.C.
4. W. Ott, J. F. Clarke and G. Ozolins, Calculating Future Carbon Monoxide Emissions and Concentrations from Urban Traffic Data, Public Health Service Publication No. 999-AP-41, Department of Health, Education and Welfare, Washington, D.C., 1967.

GASEOUS ATMOSPHERIC POLLUTANTS FROM URBAN
AND NATURAL SOURCES

ELMER ROBINSON

Washington State University, Pullman, Wash., U.S.A.

and

ROBERT C. ROBBINS

Stanford Research Institute, Menlo Park, Calif., U.S.A.

Abstract. Major aspects of the circulation through the atmospheric environment of a number of gaseous pollutants have been estimated, including source magnitudes, residual atmospheric concentrations, and scavenging processes. The compounds considered include the major sulfur and nitrogen pollutants as well as CO. One-third of the sulfur reaching the atmosphere comes from pollutant sources, mainly as SO_2. Within the atmosphere there is a net transfer of sulfur from land to ocean areas. In the global atmospheric nitrogen cycle, pollutant emissions of NO_2 play only a minor role, and the atmospheric nitrogen cycle is apparently dominated by natural emissions.

1. Introduction

The atmosphere is a complex chemical system in which the emissions from urban pollution sources mix with emanations from the natural environment. By considering both the pollutant and the natural sources it is possible to improve our understanding of the impact of air pollutants on the atmospheric environment.

This discussion will cover a number of the common atmospheric pollutants: SO_2, NO_2, and CO and their counterparts from the natural environment. We will look at these compounds on an integrated global basis, and will consider the sources of the pollutants, their atmospheric concentrations, their reactions, and available scavenging mechanisms.

This discussion is a brief summary of a portion of the research carried out for the American Petroleum Institute at Stanford Research Institute. Substantiating calculations and additional discussion for much of the material presented in this paper are given in the basic research report published by the API [28].

2. SO_2 and the Sulfur Cycle

The sulfur compounds in the atmosphere come from both the natural environment and from air pollution emissions. Natural sulfur compound emissions are SO_4 aerosols* produced in sea spray, H_2S from the decomposition of organic matter in swamp areas, bogs, and tidal flats, and probably organic sulfur compounds such as dimethyl sulfide from a large number of biological processes in the soil and vegetation [20].

---

* For convenience the notation SO_4 will be used in this presentation with the understanding that this is the sulfate ion and that it is actually present as a compound such as $(NH_4)_2SO_4$ or H_2SO_4.

S. Fred Singer (ed.), The Changing Global Environment, 111–123. All rights reserved.

Areas of volcanic activity are also a minor source of H_2S. The emissions of SO_2 come almost exclusively from pollution sources. Some H_2S is also of industrial origin.

2.1. SO_2 SOURCES

Annual worldwide pollution emissions of SO_2 have been estimated to be 147×10^6 tons. Of this total, 70% is estimated to result from coal combustion and 16% from the combustion of petroleum products, mainly residual fuel oil. As Table I shows, the remaining tonnage is accounted for by petroleum refining and nonferrous smelting. These estimates are based on 1965 world data.

TABLE I

Estimated global emissions of sulfur compounds

Compound	Source	Estimated emissions (tons yr^{-1})	Emissions as sulfur (tons yr^{-1})
SO_2	Coal combustion	102×10^6	51×10^6
	Petroleum refining	6×10^6	3×10^6
	Petroleum combustion	23×10^6	11×10^6
	Smelting operations	16×10^6	8×10^6
H_2S	Industrial emissions	3×10^6	3×10^6
H_2S or org.S	Marine emissions	30×10^6	30×10^6
	Terrestrial emissions	70×10^6	70×10^6
SO_4	Marine emissions	130×10^6	44×10^6
		Total emissions	220×10^6

Previous estimates of world SO_2 emissions have been made by Katz [16]. For the years of 1937 and 1940, total emissions were about 69×10^6 tons and 78×10^6 tons, respectively. Thus, SO_2 emissions have roughly doubled in the period between 1940 and 1965. A more recent estimate of SO_2 emissions made by Peterson and Junge [24] indicates pollutant emissions of 166×10^6 tons for 1970. At least part of the change is due to the use of newer emission factors as well as to expanded pollutant emissions related to population growth. The result, however, is quite compatible with prior estimates.

The estimates for SO_2 emissions have been combined with estimates of H_2S and SO_4 in Table I to show the estimated world natural and pollution emissions of sulfur compounds. The total emission, expressed as sulfur, is 220×10^6 tons. This estimate is not precise, because about 50% of the total comes from estimates of H_2S emissions from land and ocean areas. The natural sulfur emissions are expressed as either H_2S or organic S; both are probably involved, but the ratio cannot be determined at this time.

2.2. ATMOSPHERIC REACTIONS AND SCAVENGING PROCESSES

There has been considerable interest in SO_2 scavenging reactions for many years, and much research has centered around the need for specific catalysts to promote SO_2

oxidation in liquid droplets. However, a more realistic process for foggy atmospheres involves ammonia. Junge and Ryan [15] found that SO_2 had low solubility in water droplets of low pH, but that ammonia promoted the solubility of SO_2 by neutralizing the acid in the droplets formed by the absorbed SO_2. Extrapolation of laboratory experiments to realistic atmospheric conditions indicates that SO_2 lifetimes in foggy conditions might be as short as one hour. The fact that ammonium sulfate is commonly identified in atmospheric particulate samples lends support to this scavenging-reaction.

Rainout processes within clouds and washout resulting from falling rain may also be quite effective in scavenging SO_2 as Bielke and Georgii [1] have shown.

Sulfur dioxide oxidation is not confined to fog or rain conditions. While direct oxidation by molecular oxygen has been shown to be insignificant, photochemical oxidation of SO_2 in mixtures with NO_2 and hydrocarbons is probably one of the more significant scavenging systems for SO_2. The resultant aerosol formed by this system is H_2SO_4. The reaction can proceed with very low concentrations of the constituents [26].

An integrated system to explain the reaction of SO_2 in the atmosphere is not now available. Under daytime, low humidity conditions the photochemical processes that form H_2SO_4 or sulfate aerosols seem to be most important. At night and with high humidity, fog, or rain, absorption into water drops with subsequent oxidation to SO_4 is probably the most important process.

Sulfur dioxide is also scavenged from the atmosphere by vegetation. In vegetation scavenging it is possible to consider the rate of deposition by using Chamberlain's (1960) concept of a deposition velocity. On the basis of chamber studies of SO_2 intake by vegetation [17], the calculated deposition velocity for SO_2 is about 1 cm s^{-1}. For a concentration of 1 ppb this deposition velocity predicts an SO_2 deposition rate of 2.5 μg m^{-2} day^{-1}.

Once the sulfur is in aerosol form as SO_4, precipitation scavenging by clouds and rain are effective removal processes. Particles are also removed from the atmosphere as dry fallout. For precipitation processes the rate of scavenging is dependent upon intensity of precipitation activity and on the size of the aerosol.

2.3. SO_2 BACKGROUND CONCENTRATIONS

Sulfur as SO_2 and SO_4 has been measured in polluted atmospheres for many years, and voluminous statistics are available. However, in our analysis of the total cycle of sulfur in the environment concentration data are needed for the clean ambient atmosphere. These data are very sparse.

Vertical profiles taken over Nebraska indicate values of less than 0.3 ppb SO_2 in the upper portions of the troposphere [9]. These are in line with the few other available measurements of surface SO_2 in very remote places, such as the 0.3 ppb found in Hawaii and 1 ppb on the southeast coast of Florida by Junge [14]. Similar values, of 0.3–1 ppb SO_2, were recently found by Cadle et al. [2] in Antarctica and by Lodge and Pate [19] in the Panama Canal Zone. Over wide areas of the Central Atlantic,

TABLE II

Average tropospheric concentrations of
sulfur compounds

Compound	Average concentration	Average concentration as sulfur
SO_2	0.2 ppb	0.28 μg m^{-3}
H_2S	0.2 ppb	0.28 μg m^{-3}
SO_4	2 μg m^{-3}	0.7 μg m^{-3}

Kühme [18] found no SO_2 above the limit of detection, which was about 0.3 ppb. From these values we have tentatively concluded that the average tropospheric SO_2 concentration on a global basis is about 0.2 ppb.

Table II summarizes our present estimates of background concentrations for SO_2 and the other important atmospheric sulfur compounds.

2.4. THE ENVIRONMENTAL SULFUR CYCLE

Our present calculations, along with additional data from the literature, permit us to estimate the circulation of sulfur in various compound forms through our environment. Figure 1 shows our estimate of this sulfur circulation. Some of the values used in this calculation are reasonably well known, i.e., pollutant emissions and total depositions, but some data must be considered very speculative and have been adjusted reasonably to balance the cycle, e.g., land and sea emissions of H_2S and organic S.

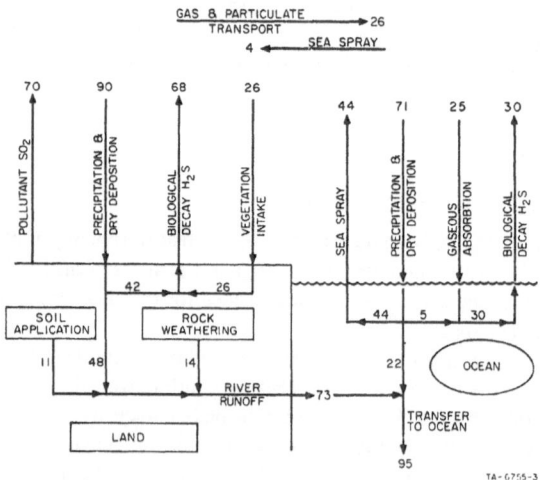

Fig. 1. Environmental sulfur circulation. Units: 10^6 tons yr^{-1} sulfur.

To understand this circulation better, we can examine its various components. The sulfur annually discharged to the sea by the world's rivers is 73×10^6 tons; this results from sulfur accumulated from weathering rocks, 14×10^6 tons; sulfur applied to the soils as fertilizer, etc., 11×10^6 tons; and sulfur deposited on the soil by precipitation and dry deposition, 48×10^6 tons. These amounts were estimated by Eriksson [7].

The atmosphere-land portion of the cycle contains 70×10^6 tons of pollutant sulfur as SO_2 and H_2S emitted to the atmosphere; 90×10^6 tons of sulfur, mostly as SO_4, deposited from the atmosphere to the land; a loss of 68×10^6 tons of gaseous sulfur from decaying vegetation, and an intake of sulfur by vegetation from the atmosphere of 26×10^6 tons. The 90×10^6 tons deposited includes 80%, or 70×10^6 tons, in rain and the remainder as dry deposition [14]. As indicated, 48×10^6 tons of this deposited sulfur is carried off by rivers, and 42×10^6 tons is absorbed by vegetation and then released again. The intake of sulfur by vegetation is estimated to be 26×10^6 tons based on calculations using a deposition velocity of 1 cm s^{-1} and an ambient concentration of 0.4 ppb (0.5 μg m^{-3}). This is twice the average tropospheric concentrations listed in Table II, which are based on the argument that ground level concentrations over land would be higher than the average for the whole troposphere. The 68×10^6 tons estimated for the emission of sulfur gases from vegetation decay results from a summation of the atmospheric vegetation intake, 26×10^6 tons, and the excess of deposition over river carryoff, 42×10^6 tons. This assumes that there is no net accumulation in the surface soils, which seems reasonable. This value is close to Eriksson's [7] estimate of 77×10^6 tons for H_2S emissions from land areas; however, no data are available with which to check this value. Land areas also gain 4×10^6 tons of sulfur from sea spray [6].

The net result of this land circulation is an excess of 26×10^6 tons of sulfur which must be deposited in the ocean if there is to be no net accumulation in the atmosphere.

Deposition of sulfate in the ocean in rain and as dust is 71×10^6 tons [14]. The ocean also absorbs 25×10^6 tons of gaseous sulfur calculated on the basis of an SO_2 concentration of 0.2 ppb and a deposition velocity of 0.9 cm s^{-1} [6, 7]. The ocean surface is a source of 44×10^6 tons of sulfur in sea spray [7] and 30×10^6 tons of sulfur as H_2S or biological organic S. There is a tropospheric transfer of 4×10^6 tons of sulfur from the ocean to the land. The emission of 30×10^6 tons is obtained on the basis of what is needed to balance the 100×10^6 tons of gaseous and solid pickup by the ocean and the transfer from sea to land. There are no data that would provide a check as to whether or not this is reasonable. It is a significantly smaller value than the approximately 200×10^6 tons estimated by Eriksson [7] and Junge [14] for similar calculations.

The end result of this cycle is an accumulation of sulfur in the oceans of 95×10^6 tons, which is the sum of pollutant emissions, sulfur applied to the soil, and rock weathering.

This compilation of facts and discussion about sulfur in our environment points up a number of interesting things especially relative to pollutant and natural sources of sulfur. With regard to estimates of gaseous sulfur emissions, our evaluation of available data indicates that natural emissions of sulfur, as organic S or H_2S, are about

30% greater than are the estimated industrial emissions of SO_2 and H_2S, i.e., 100×10^6 tons as S from natural sources compared to 76×10^6 tons from SO_2. With regard to sulfur pollutants, the most significant fact is that SO_2 is the only significant pollutant and the transformation of SO_2 to SO_4 in the form of H_2SO_4 occurs in a matter of days, perhaps about four. Most of the emitted SO_2 becomes SO_4 in the atmosphere as a result of several possible photochemical or physical reactions. This rapid reaction rate plus ready absorption of SO_2 by vegetation contributes to a rapid decrease in concentration outside emission source areas. In the ambient troposphere the majority of the sulfur is present as SO_4.

It is unlikely that we can adequately evaluate the circulation of sulfur and the relative importance of the various sulfur compound sources until considerably more data are gathered over the oceans and remote land areas of the world.

3. NO$_2$ and the Nitrogen Cycle

The total circulation of nitrogen compounds through the Earth's invironment is relatively little known. However, since there are numerous compounds involved, the nitrogen circulation is obviously a complex one. Of the seven significant atmospheric forms of nitrogen, the only significant pollutants emitted by man's activities are NO and NO_2.

3.1. SOURCES OF NO$_2$

There are indications that significant amounts of NO_2 can result from NO formed by bacterial action in the reduction of nitrogen compounds under anaerobic conditions. For example, Junge [14] points out that the formation of NO from HNO_3 occurs in acid soils.

The fixation of nitrogen by lightning has been investigated in a number of studies, but most writers conclude that lightning is unimportant in the fixation of nitrogen [8]. It also seems unlikely that stratospheric processes are significant in the lower atmospheric nitrogen cycle.

The pollution sources of NO and NO_2, usually considered together and expressed as NO_2, are primarily those combustion processes in which the temperatures are high enough to fix the nitrogen in the air and in which the combustion gases are quenched rapidly enough to reduce the subsequent decomposition. Worldwide NO_2 emissions can be roughly estimated on the basis of estimates of fuel combustion processes using available NO_2 production ratios [21].

Table III summarizes our estimates of the emission sources of NO_2 and other important nitrogen compounds. The data on pollutant emissions are based on average source emissions. The natural emissions are primarily based on the cycle of nitrogen compounds that has been derived in this study and on the amounts that would be needed to balance the cycle. Probably of most interest is the balance between pollutant and natural emissions of NO_2. It is our estimate that these are in the ratio of about 1 to 15, the natural emissions being much greater than the pollutant emissions.

TABLE III

Estimated annual global emissions of nitrogen compounds

Compound	Source	Source magnitude (tons yr^{-1})	Estimated emissions (tons yr^{-1})	Emissions as nitrogen (tons yr^{-1})
NO$_2$	Coal combustion	3074×10^6	26.9×10^6	8.2×10^6
	Petroleum refining	$11\,317 \times 10^6$ (bbl)	0.7×10^6	0.2×10^6
	Gasoline combustion	379×10^6	7.5×10^6	2.3×10^6
	Other oil combustion	894×10^6	14.1×10^6	4.3×10^6
	Natural gas combustion	20.56×10^{12} (ft^3)	2.1×10^6	0.6×10^6
	Other combustion	1290×10^6	1.6×10^6	0.5×10^6
Total NO$_2$			52.9×10^6	16.1×10^6
NH$_3$	Combustion		4.2×10^6	3.5×10^6
NO$_2$	Biological action		768×10^6	234×10^6
NH$_3$	Biological action		1160×10^6	960×10^6
N$_2$O	Biological action		592×10^6	376×10^6

3.2. ATMOSPHERIC NO$_2$ CONCENTRATIONS

Data on NO$_2$ concentrations outside city areas are widely scattered, but show definitely that NO$_2$ is present in trace amounts in remote locations. Analyses by Lodge and Pate [19] in Panama indicate a dry season average for NO$_2$ of 0.9 ppb and a rainy season value of 3.6 ppb. A few samples of NO have indicated concentrations up to 6 ppb. Junge [13] has reported NO$_2$ concentrations in Florida averaging 0.9 ppb and Hawaiian data averaging about 1.3 ppb.

In Ireland O'Connor [23] reported average NO$_2$ concentrations for North Atlantic air to be 0.3 ppb in April and 0.2 ppb in September and October. More recently NO$_2$ and NO data from Pike's Peak, Colorado, and from a remote area of North Carolina have been reported [10, 27]. The Colorado data indicate an average NO$_2$ concentration of 4.1 ppb and 2.7 ppb for NO. In the Appalachian area of North Carolina, average concentrations were 4.6 ppb for NO$_2$.

From these scattered data we have concluded that in continental areas average levels of NO$_2$ are about 4 ppb. Ocean areas and land areas north of 65° N and south of 65° S appear to be characterized by lower concentrations, perhaps 0.5 ppb, or less.

Table IV summarizes our estimates of atmospheric concentrations of nitrogen compounds, and the mass of nitrogen present in these compounds throughout the atmosphere. In this tabulation N$_2$O accounts for more than 97% of the tonnage, calculated on the basis of N content. Of the 3% of the materials which show a certain degree of reactivity and change, 80% of the N is present as NH$_3$.

3.3. ATMOSPHERIC SCAVENGING REACTIONS FOR NO$_2$

NO and NO$_2$ are important as pollutants mainly because of their participation in photochemical reactions, which constitute a major scavenging process.

The reaction of NO with O$_3$ in the atmosphere to form NO$_2$ is rapid, and since there

TABLE IV

Background concentrations of
nitrogen compounds

Compound	Ambient concentration	Atmospheric mass as N
N_2O	0.25 ppm	1500×10^6 tons
NO/NO_2		
Land	4 ppb (as No_2)	12×10^6
Ocean	1 ppb	3×10^6
NH_3	6 ppb	30×10^6
NO_3	$0.2 \ \mu g \ m^{-3}$	0.2×10^6
NH_4	$1.0 \ \mu g \ m^{-3}$	4.1×10^6

is always some background O_3 from stratospheric transport, it is generally assumed that NO_2 rather than NO is the predominant of the two species found in clean atmospheres. However, the previously quoted Panama data of Lodge and Pate [19] and the Colorado data of Hamilton *et al.* [10] indicate that this assumption for the ambient atmosphere may have to be altered.

The O_3 oxidation of NO_2 to N_2O_5 is about 500 times slower than the O_3 oxidation of NO; nevertheless it is rapid enough to limit the half-life of 1 ppb NO_2 in the atmosphere in the presence of 5 ppb of O_3 to about two weeks. However, the residence time of NO_2 based on our atmospheric nitrogen cycle is only three days. This short time can probably be explained by the following scavenging reaction of NO_2:

$$3NO_{2(v)} + H_2O_{(v)} \leftrightarrows 2HNO_{3(v)} + NO_{(v)}$$

The equilibrium constant is 0.004 atm^{-1} [22] at 25 °C, but with the extreme excess of water vapor found in the atmosphere 10% of the NO_2 is converted to HNO_3 at equilibrium. This HNO_3 vapor is rapidly removed by reaction with atmospheric ammonia and absorption into hygroscopic particles. At relative humidities higher than 98%, condensation of dilute nitric acid droplets will occur at 10 °C and a HNO_3 vapor concentration of 0.1 ppb. All of the HNO_3 eventually becomes nitrate salt aerosol.

3.4. NO_2 AND THE GENERAL NITROGEN CYCLE

This analysis of atmospheric nitrogen compounds provides us with some rough estimates of global sources and sinks. We have combined these source and sink data into a simplified nitrogen circulation as shown in Figure 2 [31]. The basic features of this circulation are as follows:

(a) The N_2O cycle is independent of the rest of the system, except that it draws on soil nitrogen for formation. One sink for N_2O is above 30 km and results in the production of 5×10^6 tons y^{-1} of NO.

(b) Deposition and destruction of N_2O by soil bacteria is the major sink.

(c) Nitrogen appears to enter the soil regime through fixation in biological processes. Hutchinson [12] calculated an annual nitrogen fixation rate of 130×10^6 tons.

(d) NH_4 particulate deposition, 269×10^6 tons, is determined from NH_4 content of rainfall, 213×10^6, plus dry deposition of 25%. The source of the NH_4 is gaseous NH_3 released from the biosphere.

(e) NH_3 gaseous deposition, 910×10^6 tons, was obtained by analog with SO_2 using a deposition velocity, v_g, of 1 cm/sec and a concentration of 6 ppb. For lack of data no differentiation between land and ocean areas has been attempted.

(f) NH_3 gaseous emissions were calculated to balance gaseous NH_3 deposition and particulate depositions of NH_4. This resulting emission rate, 1160×10^6 tons, indicates a residence time of about two weeks.

(g) NO_3 particulate deposition, 462×10^6 tons, is determined by the NO_3 content of rainfall, 366×10^6 tons, plus 25% for dry deposition. The major source for the nitrogen in this NO_3 deposition is NO_2 from the biosphere.

(h) NO_2 pollution emissions are 50×10^6 tons. The source of this nitrogen is the atmosphere.

(i) NO from the biosphere, 501×10^6 tons, provides sufficient NO_2 after oxidation to balance an atmospheric residence time of about seven days (768×10^6 tons of NO_2).

(j) NO_2 gaseous deposition was obtained using a deposition velocity, v_g of 0.5 cm s^{-1} and a global average concentration of 2 ppb.

(k) Tonnages of nitrogen added to the soil annually as fertilizer, 20×10^6 tons (U.S. Statistical Abstracts, 1967), and the nitrogen carried to the ocean by rivers, 13×10^6 tons [12], are included.

This circulation is very much simplified and should be considered only as our initial approximation. For example, no division is made for variations in conditions over land and ocean areas or for differences due to latitude, except as these changes affected

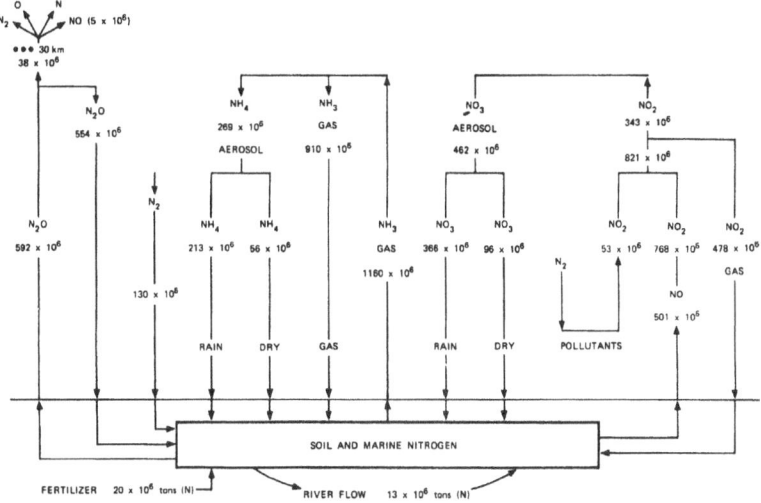

Fig. 2.   Simplified nitrogen compound circulation. Units: tons yr^{-1}.

the assessment of nitrate and ammonium deposition. A very important qualification of the results of this cycle has been pointed out by Hidy and Brock [11] and several other authors, namely that the nitrate to sulfate deposition ratio from our sulfur and nitrogen cycles (Figures 1 and 2) is significantly different from some of the ratios obtained in other ways. Figures 1 and 2 show annual global depositions of about 400×10^6 tons of sulfate and 460×10^6 tons of nitrate. The indicated nitrate/sulfate ratio is therefore 1.15. Hidy and Brock [11], as well as Peterson and Junge [24], point out that particulate filter samples from nonurban areas in the U.S. when analyzed for sulfate and nitrate indicate a nitrate/sulfate ratio of about 0.2, i.e. a five-fold excess of sulfate over nitrate. There is some indication that the filter collection of nitrate particles may not be a completely effective system in the results of the airborne filter data of Cadle *et al.* [3] where filters not readily wet by liquid droplets such as nitric acid showed significantly lower nitrate concentrations than parallel filters of different composition. Thus, the 1 to 5 nitrate/sulfate ratio based on filter samples may be low. Comparative precipitation data for the U.S. given by Junge [14] shows that the relatively low pollution areas west of the Rockies have a ratio of about 1 to 2.

As noted previously, Figure 2 does not attempt to delineate separate cycles for land and ocean areas, although lower deposition rates were used for ocean areas to get the indicated nitrate deposition rates. If instead of the uniform global nitrate deposition assumed for Figure 2, it were assumed that the land/ocean nitrate deposition ratio were 4:1 as suggested for atmospheric NO_2 background concentrations, the nitrate deposition rate would decrease by about one-half. On a global basis this might be more in line with present data on sulfate. A reduction of 50 % in the nitrate deposition would be about 230×10^6 tons and would be balanced by reducing the natural NO emission by about 110×10^6 tons. The calculated NO emission from the biosphere would then be about 390×10^6 tons instead of the 501×10^6 tons indicated in Figure 2. Whether such a calculated result would be more nearly correct than what is shown in Figure 2 is questionable. Certainly more definitive concentration data in both remote land and ocean areas would be useful in making a clearer assessment of the situation.

While this nitrogen compound cycle obviously leaves much to be desired, it still seems clear that natural emission processes are responsible for the major share of the atmospheric nitrogen compounds.

4. Atmospheric Carbon Monoxide

Carbon monoxide has been considered an important toxic atmospheric pollutant for many years because of its prevalence in automobile exhaust and in the effluents from combustion.

A comprehensive discussion of CO in the atmosphere is presented by Jaffe*; however, the CO cycle is still open to much speculation and we would like to point out

* See p. 83–110.

some of the general features of the global CO cycle for which an atmospheric model must account.

Background measurements of CO in the atmosphere by Cavanagh *et al.* [4], Robinson and Robbins [30, 32], Seiler and Junge [33], and Swinnerton, *et al.* [34, 35] show that CO concentrations in the northern hemisphere are considerably higher than in the southern hemisphere. In the north and south Pacific the ratio of northern to southern hemisphere average concentrations was 0.14 ppm to 0.06 ppm, or about 2:1. Aircraft sampling by Seiler and Junge [33] shows that CO concentrations measured in the troposphere are significantly higher than observed in the lower stratosphere. The observed concentrations were about 0.1 ppm below the tropopause and 0.03 to 0.01 ppm above the tropopause. On the basis of these and other observed CO concentrations, the pattern of CO in the Earth's atmosphere, away from the direct influence of pollutant sources, is beginning to assume some reasonable structure. Higher concentrations are indicated for the northern hemisphere compared to the southern, as would be expected on the basis that about 90 % of the world's fuel combustion takes place in the northern hemisphere. Highest background levels, about 0.2 ppm in the remote Pacific Ocean area, seem to occur between 40° N and 50° N which also generally agrees with the latitudinal distribution of combustion. Mid-latitude, northern hemisphere concentrations over the Atlantic Ocean seem to be somewhat higher than the data for the Pacific. This could well reflect the emissions from the United States and Canada.

At the equator in both the Pacific and Atlantic CO concentrations of about 0.09 to 0.10 ppm are shown by current data. In the southern hemisphere data are sparse but in the Pacific a minimum concentration of about 0.04 ppm at about 50° S is indicated. One set of samples indicated increasing concentrations of CO southward from 65° S toward McMurdo in Antarctica [29].

At present our understanding of the cycle of CO through the environment is very imperfect, but it is to be hoped that the various research programs now being conducted will lead to an understanding of this important problem.

5. Summary

This analysis of pollutant emissions has attempted to relate the gaseous emanations from both man-made and natural sources to the circulation through the atmosphere of several common pollutants. In some instances natural sources on a global basis are of greater magnitude than are the urban pollution sources. Some of the most important of the natural sources are in the nitrogen cycle, where emissions of both NO and NH_3 are predominantly from natural sources in the biosphere. In the circulation of sulfur compounds, the emission of H_2S and/or organic sulfur from biological processes is estimated to be a major source of sulfur which eventually is detected as a sulfate aerosol. In the CO cycle there are strong indications that natural sources are present – in the biosphere, in the ocean, and perhaps from atmospheric chemical reactions.

The fact that trace amounts of a wide variety of compounds emanate from the natural environment is important for several reasons. The presence of these natural sources emphasizes the fact that there are processes present in the atmosphere which can scavenge the common pollutants emitted from urban sources. In the case of the more reactive gases – SO_2, NO_2, etc. – these scavenging processes act relatively rapidly. The scavenging cycle is apparently as short as a day or so for SO_2.

The importance of the identification of organic sulfur emissions from vegetation [20] should not be underestimated. Rasmussen [25] points out that the several organic sulfur compounds emitted by vegetation are strongly reactive photochemically in very dilute concentrations simulating possible atmospheric emission concentrations. He also points out the wide probable range of natural sources of organic sulfur compounds. This could correlate with the apparent ubiquitous presence of sulfate aerosols in the atmosphere.

These scavenging cycles have to be recognized if we are going to achieve an understanding of our environment. However, the fact that these cycles exist cannot be used as an excuse for permitting any urban area to reach levels of pollution that are detrimental in any way to its residents. None of the scavenging cycles that have been described are effective enough within the time-frame of a few hours for them to have any significant effect within a given source area. Thus, even with these natural processes some form of control of pollutant sources must be used to protect a local area. However, we can apparently rest somewhat more easily with regard to the possibility of the accumulation of many pollutants on a global scale. The natural scavenging processes are effective when the time scale is increased.

There are many aspects of the cycling of trace materials through the atmosphere that we do not understand. As our understanding increases we can more adequately use the atmosphere as a renewable resource and also conserve its properties as one of our most important environmental factors.

Acknowledgments

This study is part of a larger research program on gaseous contaminants in the atmosphere supported by the American Petroleum Institute. The authors are pleased to acknowledge this support.

References

1. Bielke, S. and Georgii, H.-W.: *Tellus* 20, 435–41 (1967).
2. Cadle, R. D., Fischer, W. H., Frank, E. R., and Lodge, J. P.: *J. Atm. Sci.* 25, 100 (1968).
3. Cadle, R. D., Lazrus, A. L., Pollock, W. H., and Shedlovsky, J. P.: *The Chemical Composition of Aerosol Particles in the Tropical Stratosphere*, paper presented at International Conference on Tropical Meteorology, *Am. Meteorol. Soc.*, Honolulu, Hawaii (1970).
4. Cavanagh, L. A., Schadt, C. F., and Robinson, E.: *Environ. Sci. Technol.* 3, 251–257 (1969).
5. Chamberlain, A. C.: *Int. J. Air. Poll.* 3, 63 (1960).
6. Eriksson, E.: Part I, *Tellus* 11, 375–403 (1959).
7. Eriksson, E.: Part II, *Tellus* 12, 63–109 (1960).
8. Georgii, H.-W.: *J. Geophys. Res.* 68, 3963 (1963).
9. Georgii, H.-W.: personal communication to C. E. Junge (1967).

10. Hamılton, H. L., Worth, J. J. B., and Ripperton, L. A.: *An Atmospheric Physics and Chemistry Study on Pikes Peak in Support of Pulmonary Edema Research*, Research Triangle Institute, North Carolina, for Army Research Office, Contract No. DA-HC19-67-C-0029 (1968).
11. Hidy, G. M. and Brock, J. R.: in H. M. Englund and W. T. Berry (eds.), *Proceedings of the Second International Clean Air Congress*, pp. 1088–1096, Academic Press, New York (1971).
12. Hutchinson, G. E.: in G. P. Kuiper (ed.), *The Earth as a Planet*, University of Chicago Press, Chicago (1954).
13. Junge, C. E.: *Tellus* **8**, 127 (1956).
14. Junge, C. E.: *Air Chemistry and Radioactivity*, Academic Press, New York, p. 123 (1963).
15. Junge, C. E. and Ryan, T.: *Quart. J. Roy. Meteorol. Soc.* **84**, 46–55 (1958).
16. Katz, M.: in P. L. Magill, F. R. Holden and C. Ackley (eds.), *Air Pollution Handbook*, McGraw-Hıll Book Company, Inc., New York (1958).
17. Katz, M. and Ledingham, G. A.: National Research Council of Canada, in *Effect of Sulfur Dioxide on Vegetation*, NCR No. 815, Ottawa (1939).
18. Kühme, H.: private communication to C. E. Junge (1967).
19. Lodge, J. P. and Pate, J. B.: *Science* **153**, 408 (1966).
20. Lovelock, J. E., Maggs, R. J., and Rasmussen, R. A.: *Nature* **237**, 452–453 (1972).
21. Mayer, M.: *A Compilation of Air Pollution Emission Factor*, U.S. Public Health Service, Division of Air Pollution, Cincinnati, Ohio (1965).
22. McHenry, L. R. J.: *The Vapor Phase Reaction between Nitrogen Oxides and Water*, M.S.thesis in chemical engineering, University of Illinois (1953).
23. O'Connor, T. C.: *Atmospheric Condensation Nuclei and Trace Gases*, Final Report, Dept. of Physics, University College, Galway, Ireland, Contract No. DA-91-591-EUC-2126 (1962).
24. Peterson, J. T. and Junge, C. E.: in W. H. Matthews, W. H. Kellogg and G. D. Robinson (eds.), *Man's Impact on the Climate*, p. 317, The MIT Press, Cambridge, Mass. (1971).
25. Rasmussen, R. A.: private communication (1972).
26. Renzetti, N. A. and Doyle, G. J.: *Int. J. Air Poll.* **2**, 327 (1960).
27. Ripperton, L. A., Worth, J. J. B., and Kornreich, L.: *Nitrogen Dioxide and Nitric Oxide in Non-Urban Air*, Paper No. 68-122, 61st Annual Meeting Air Pollution Control Assocation, June (1968).
28. Robinson, E. and Robbins, R. C.: *Sources, Abundance, and Fate of Gaseous Atmospheric Pollutants*, Final Report SRI Project PR-6755, for Amerıcan Petroleum Institute, New York, February and Supplemental Report (1968).
29. Robinson, E. and Robbins, R. C.: *Antarctic Journal of the U.S.* **194**, Sept.-Oct. (1968).
30. Robinson, E. and Robbins, R. C.: *J. Geophys. Res.* **74**, 1968–1974 (1969).
31. Robinson, E. and Robbins, R. C.: *J. Air Pollution Control Assoc.* **20**, 303–306 (1970).
32. Robinson, E. and Robbins, R. C.: *Ann. New York Acad. Sci.* **174**, 89–95 (1970).
33. Seiler, W. and Junge, C. E.: *Tellus* **21**, 447–449 (1969).
34. Swinnerton, J. W., Linnenbom, V. J., and Check, C. H.: *Environ. Sci. Techn.* **3**, 836 (1969).
35. Swinnerton, J. W., Linnenbom, V. J., and Lamontagne, R.: *Science* **167**, 934 (1970).

Suggested Further Reading

1. C. E. Junge, *Air Chemistry and Radioactivity*, Academic Press, New York, 1963.
2. P. A. Leıghton, *Photochemistry of Air Pollution*, Academic Press, New York, 1961.
3. A. C. Stern (ed.), *Air Pollution*, 2nd edition, Academic Press, New York, 1968, Vol. I, Chapters 2, 6 and 11.

GLOBAL STRATOSPHERIC EFFECTS
OF SUPERSONIC TRANSPORTS

S. FRED SINGER

University of Virginia, Charlottesville, Va., U.S.A.

Abstract. SST's emit pollutants, such as particulates, water vapor and nitrogen oxides. Since their residence time in the stable stratospheric layers is measured in years, appreciable concentrations can build up, comparable to naturally occurring concentrations. Effects of these pollutants are judged to be small as well as reversible. The major unresolved problem relates to whether appreciable amounts of stratospheric ozone are destroyed. A research program to settle outstanding questions is now underway.

The prospect of a large commercial fleet of supersonic transport planes (SST's) has precipitated considerable debate concerning the potential environmental effects of the exhaust emissions, first on the stratosphere, as the proposed cruise mode altitude is 20 km, and then on the global climate over which the stratosphere exercises considerable control. The SST 'issue' is not only important on its own scientific merits. It may well be the archetype of many future controversies as the impact of modern technology becomes more and more global.

The stratosphere, that layer of the atmosphere between roughly 12 and 50 km, is

TABLE I

Statistics of emissions from one GE-4 engine, cruise mode[a]

Constituent	lb h^{-1}
Ingested air and consumed fuel	
air	1 380 000
fuel	33 000
Unused air	
N_2	1 030 000
O_2	208 000
Ar	19 300
Combustion products	
CO_2	103 500
H_2O	41 400
CO	1 400
NO	1 400
SO_2	33
Soot (Particles)	5
Unused fuel	
Hydrocarbons	16.5

[a] From SCEP (1970).

S. Fred Singer (ed.), The Changing Global Environment, 125–134. All rights reserved.
Copyright © 1975 by D. Reidel Publishing Company, Dordrecht–Holland.

characterized by a sharp temperature inversion and very weak vertical mixing (Figure 1). Especially because of this, any gases or particles injected into it have a potentially long residence – on the order of years (SCEP, 1970). Thus the concern seems legitimate that the injection of pollutants into this layer may build up to concentrations rivaling in importance the natural constituents. Of particular importance are the effects the projected concentrations may have on the radiative balance and photochemical reactions, and in turn, how these changes will affect the global climate.

The majority of the atmospheric ozone is produced and resides in the stratosphere. Ozone exerts a well-known and important shielding effect on the solar radiation below a wavelength of about 3400 Å; a general decrease in the total ozone column by as little as 1% could have detectable effects on the biosphere, including man. Many substances, especially H_2O and various oxides of nitrogen, are known to react with ozone, and therefore have the potential to alter the ozone column. A review of such processes will be of major concern here, as will the direct effects of these and other contaminants on the general radiative equilibrium.

Table I lists the concentrations of the various exhaust pollutants of the SST engines as estimated by the manufacturer, and while these now appear to need revision (SMIC 1971) they give an idea of what must be considered.

1. Carbon Dioxide

Carbon dioxide (CO_2) currently has an average atmospheric mixing ratio of about 324 ppm with an annual growth rate of 0.3% or 1.0 ppm. This rapidly increasing mixing ratio is primarily due to the combustion of large quantities of fossil fuels. Approximately 50% of the CO_2 thus released remains active in the atmosphere and biosphere while the remainder is semi-permanently removed in an average time of 5 yr by various sinks of which the oceans are most important. Although the CO_2 emissions of the SST exhaust could by themselves cause a slight radiative cooling of the stratosphere, and 'greenhouse' heating of the lower atmosphere, the SST has only negligible effects with respect to the much larger source of CO_2 pollutants. The effects of CO_2 from other sources are discussed elsewhere [18, 19, 20, and 23].

2. Particles

The SST will increase the stratospheric particulate loading through direct injections of soot and through photochemical production of particles from SO_2, NO_x, and hydrocarbons. The effect of particles injected into the stratosphere has been studied by observing changes following volcanic eruptions. After the Agung eruption in 1963, the temperature of the equatorial stratosphere was observed to rise 6 or 7 °C above its previous ambient [11]. Coupled with this increased dust loading (from this and other sources) was a decrease in direct solar radiation (SMIC, 1971) and a global temperature decline. Mitchell [18] in fact relates the 0.3 °C global temperature decline since 1940 to the radiative effects of particles injected into the stratosphere by volcanoes, begin-

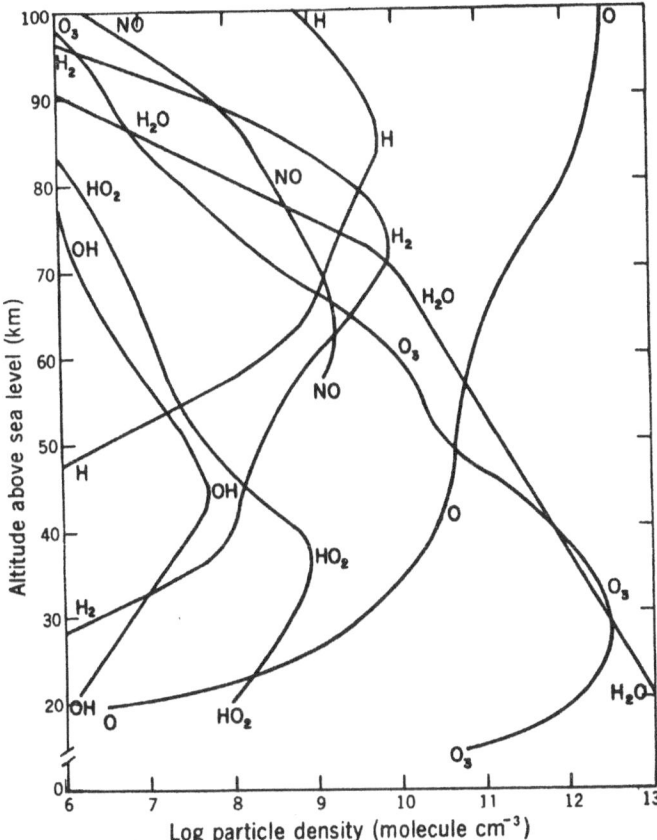

Fig. 1. Daytime equilibrium profiles of the various stratospheric constitutents, constructed from experimental data and theoretical calculations for a hydrogen-oxygen atmosphere. For comparison, the O_2 and total particle densities at 20, 30, and 45 km are 17.6, 16.9, 15.9 and 18.3, 17.6, 16.6 respectively. From Cadle and Allen [8].

ning in 1947 with the eruption of Hekla in Iceland and continuing with present eruptions.

Direct samples of stratospheric particles revealed a maximum concentration occurring at 18 to 20 km with the majority of particles being sulfates [6]. The origin of these particles is believed to be the oxidation and condensation in the stratosphere of H_2S and SO_2 which has been transported up through the tropopause. Of particular importance to the present discussion is the formation of particles from SO_2. It is now believed that SO_2 (a pollutant which comes largely from combustion of fossil fuels, smelting, and petroleum refining [22], but also possibly from natural sources) upon entering the stratosphere through the equatorial tropopause is quickly oxidized by O_3

and O to SO_3, which is then almost immediately hydrolyzed to H_2SO_4. During their 1 to 2 yr residence in the stratosphere, these initially submicron-sized droplets of H_2SO_4 grow to larger size through coagulation.

Calculations by SCEP (1970) indicate that a fleet of 500 SST's using fuel containing 0.05% sulfur by mass would on a worldwide average inject particles with a concentration equaling or exceeding the background pre-Agung concentrations.*

On a local basis, in high traffic areas, the particulate loading will rival the post-Agung values. As the local effects of the SST are predominantly on the northern stratosphere, the stratospheric adjustments cannot be exactly compared with those following Agung. It is conceivable, however, that there may be a local temperature rise in the stratosphere. On a global scale, however, the increased particulate loading due to the SST alone, being an order of magnitude less than the post-Agung loading, should produce no significant radiative modification.

The SST then is just one of many sources of particles and attention must be directed toward all, as their combined effects could be important [18].

The increase in atmospheric turbidity due to other human-related activities is accelerating [21] and may be more serious than the effects of the SST.

3. Water Vapor

Water is a natural constituent of the stratosphere with an average mixing ratio of 3 ppm. The sources of this natural equilibrium concentration have been studied and are believed understood. (See Table II.)

An estimate of the net flux of H_2O into the stratosphere resulting from Hadley cell circulation has been put at 2.2×10^{14} g yr^{-1} by Weickmann and Van Valin [2]. (This assumes a cell rising at 0.02 cm s^{-1} with 3 ppm water entering the stratosphere at the tropical tropopause; it is a conservatively low limit since higher vertical velocities may prevail.**) The value compares well with that published by Newell [11] which was used in the SCEP study. But this publication did not consider two other important sources which add several times as much water vapor.

Cloud tops, especially during severe local thunderstorms, are often observed to penetrate the tropopause and enter the lower stratosphere. Due to the difference in water vapor content and temperature between the cloud and the stratosphere and a strong wind shear, considerable quantities of water are delivered to the stratosphere. Using an estimate of 100 such storms per day, each losing 2×10^{10} g of water to the stratosphere, Weickmann and Van Valin [2] estimate that 8×10^{14} g of water are annually injected into the stratosphere by this source.

Methane is released into the atmosphere at a rate of about 1.6×10^{15} g yr^{-1}. As the average mixing ratio of methane is only about 1 ppm it appears that there must exist several strong sinks, aerobic bacteria probably being the largest. In the atmosphere

* A reduction in the permissible sulfur content of SST fuel would decrease the effects.
** Lateef [16] quotes a velocity of 0.04 cm s $^{-1}$, and Reed and Vlcek [17] quote a velocity of 0.05 cm s $^{-1}$.

TABLE II

Water vapor injection into the stratosphere

	$g \ yr^{-1}$
Hadley cell circulation[a]	2.2×10^{14}
Severe local storms[a]	8.0×10^{14}
Methane oxidation[b]	1.1×10^{14}
Methane oxidation (human related)[b]	5.5×10^{13}
SST Exhaust[c]	7.8×10^{13}

[a] Weickmann and Van Valın (1972)

[b] Singer (1971)

[c] SCEP (1970). Assuming 500 SST's flying 7 h day^{-1}

up to about 10 km the methane mixing ratio remains rather constant, with the major sink being oxidation by OH to carbon monoxide [5]. Above 10 km the mixing ratio falls off rapidly. This is due primarily to oxidation of methane to carbon dioxide and water, 1 gm of CH_4 producing 2.25 gm of H_2O. Calculations of the strength of the methane sink in the stratosphere put the rate of loss of methane at 3×10^{-13} g cm^{-2} s^{-1}, resulting in an annual injection of water into the stratosphere of 1.1×10^{14} g [4]; he estimates that a substantial fraction (perhaps as much as 50 %) of the methane could be produced by human-related activities; thus this source may be of the same magnitude as the SST source.*

It is estimated that a fleet of 500 SST's would annually inject about 7.8×10^{13} g of water into the stratosphere (SCEP, 1970). This might increase the worldwide equilibrium mixing ratio of water by about 0.2 ppm, or by 2 ppm in the heavily traveled air corridors. Three possible consequences must be considered. (i) The increased water vapor would slightly lower the temperature of the stratosphere, through radiative cooling, and might raise the ground level temperature by up to 0.1 °C through the 'greenhouse effect' [10]. (ii) The Earth's radiation balance is highly dependent on the albedo, with the Earth's cloud cover an important highly variable constituent. Clouds, specifically the nacreous or mother-of-pearl variety are occasionally observed in the high northern and southern latitudes during their respective winters, indicating that the necessary conditions for cloudiness can exist in the stratosphere, i.e. -87°C and a mixing ratio of 3 ppm at 30 mb pressure (SCEP, 1970). The increased water vapor due to the SST would result in conditions more favorable to cloud formation. The immediate result of increased cloudiness might even be a general temperature rise of the lower atmosphere as these high clouds have a small reflectivity but a low emission temperature [19, 20]. As the atmospheric dynamics of cloud formation is only vaguely known, the ultimate effects of an increased cloud cover is uncertain. It should, however, be noted that the high-altitude cirrus (contrails) in the high troposphere from conventional jet aircraft have effects [21] similar to those that might result from the

* Of course the water vapor so produced will have the same effect on ozone as H_2O from other sources; in addition, during the oxidation process of CH_4 atomic oxygen is consumed, causing a reduction in ozone production.

SST. Even if the SST fleet never materializes, these competitive effects must be considered. (iii) Water, finally, can affect the ozone balance of the stratosphere through the action of the water-derived radicals OH and HOO. SCEP (1970) considered the effects of water vapor on the ozone shield and found a significant decrease only above 50 km where the ozone is already quite rarefied. The total ozone column, even for atmospheres with as high as 20 ppm water was not appreciably altered. They, therefore, concluded that the increased water vapor content would not cause significant difficulties.

4. Oxides of Nitrogen

It is well known that photochemical reactions near sea level involving NO_x and hydrocarbons produce smog. In the stratosphere these reactions would be of only minor importance. However, the possibility that NO_x could affect the stratospheric ozone balance is sufficiently plausible to deserve a detailed review.

In their initial study, SCEP rejected as insignificant the effects of the various oxides of nitrogen (NO_x) on stratospheric chemistry. The increase in the equilibrium mixing ratio of NO_x, resulting from a fleet of 500 SST's flying 7 h a day, was projected by SCEP (1970) to be 6.8 ppb as a global average, or ten times this value in areas of high traffic.

The natural abundance of the various nitrogen oxides in the stratosphere are not well known. An upper limit for the nitrogen dioxide (NO_2) concentration at 15 km has been reported to be 30 ppb [9], and nitric acid (HNO_3) has been measured at 16 km [12] as 8 ppb. Nitric oxide (NO) concentration profiles have thus far been unavailable. The sources of stratospheric NO have, however, been identified: (i) oxidation of nitrous oxide which has diffused upward through the tropopause following production in the biosphere; (ii) downward migration from a mesosphere source where it has known concentrations of 790 ppb above 74 km and 300 ppb at 60 km [15]. The best estimates for the stratospheric NO concentration, based on model calculations, seems to be on the order of a few ppb.

A detailed study [3] of the kinetics of ozone formation and destruction in an atmosphere containing nitrogen oxides and water vapor (or more specifically the water-derived radicals hydroxyl (OH) and perhydroxyl (HO_2)) concluded that the SST exhaust would significantly diminish the ozone column. The reactions considered in the model atmosphere are listed in Table III. Of these, reactions a, b, c, e, f, g, h, x, y, and aa were considered most important, on the basis of reasonable estimates for the rate constants.

The basic mechanism is the simple catalytic pair

$$\begin{array}{l} NO + O_3 \rightarrow NO_2 + O_2 \\ \underline{NO_2 + O \rightarrow NO + O_2} \\ O + O_3 \rightarrow O_2 + O_2 \end{array} \tag{1}$$

in which ozone is destroyed with no net change in the NO_x concentration; thus the cycle repeats, with NO_x as a catalyst continually destroying ozone. To estimate the

TABLE III

Reactions used in the Johnston model stratosphere, together with photo-chemical (j) and kinetic (k) rate constants [3]

$O_2 + h\nu$ (below 242 nm) $\rightarrow O + O$, $j_a[O_2]$	(a)
$O + O_2 + M \rightarrow O_3 + M$, $k_b[O][O_2][M]$	(b)
$O_3 + h\nu$ (190 to 350 nm, 450 to 650 nm) $\rightarrow O + O_2$, $j_c[O_3]$	(c)
$O_3 + h\nu$ (above 313 nm) $\rightarrow O_2 + O$ (^3P), $j_{c'}[O_3]$	(c')
$O_3 + h\nu$ (below 313 nm) $\rightarrow O_2 + O$ (^1D), $j_{c''}[O_3]$	(c'')
$O + O + M \rightarrow O_2 + M$, $k_d[O]^2[M]$	(d)
$O + O_3 \rightarrow 2O_2$, $k_e[O][O_3]$	(e)
$NO + O_3 \rightarrow NO_2 + O_2$, $k_f[NO][O_3]$	(f)
$NO_2 + O \rightarrow NO + O_2$, $k_g[NO_2][O]$	(g)
$NO_2 + h\nu$ (260 to 400 nm) $\rightarrow NO + O$, $j_h[NO_2]$	(h)
$2NO + O_2 \rightarrow 2NO_2$, $k_i[NO]^2[O_2]$	(i)
$NO + O + M \rightarrow NO_2 + M$, $k_j[NO][O][M]$	(j)
$NO_2 + O_3 \rightarrow NO_3 + O_2$, $k_k[NO_2][O_3]$	(k)
$NO_3 + h\nu$ (visible) $\rightarrow NO + O_2$, $j_l[NO_3]$	(l)
$NO_2 + NO_3 + M \rightarrow N_2O_5 + M$, $k_m[NO_2][NO_3][M]$	(m)
$N_2O_5 + M \rightarrow NO_2 + NO_3 + M$, $k_n[M][N_2O_5]$	(n)
$N_2O_5 + h\nu \rightarrow 2NO_2 + O$, $j_p[N_2O_5]$	(p)
$N_2O_5 + O \rightarrow 2NO_2 + O_2$, $j_q[N_2O_5][O]$	(q)
$N_2O_5 + H_2O \rightarrow 2HNO_3$, $k_r[N_2O_5][H_2O]$	(r)
$HO + NO_2 + M \rightarrow HNO_3 + M$, $k_s[HO][NO_2][M]$	(s)
$HNO_3 + h\nu$ (below 300 nm) $\rightarrow HO + NO_2$, $j_t[HNO_3]$	(t)
$HO + HNO_3 \rightarrow H_2O + NO_3$, $k_u[HO][HNO_3]$	(u)
$O(^1D) + M \rightarrow O(^3P) + M$, $k_v[M][O(^1D)]$	(v)
$O(^1D) + H_2O \rightarrow 2HO$, $k_w[H_2O][O(^1D)]$	(w)
$HO + O_3 \rightarrow HOO + O_2$, $k_x[HO][O_3]$	(x)
$O + HOO \rightarrow HO + O_2$, $k_y[O][HOO]$	(y)
$O + HO \rightarrow H + O_2$, $k_z[O][HO]$	(z)
$H + O_2 + M \rightarrow HOO + M$, $k_{aa}[H][O_2][M]$	(aa)
$HOO + HO \rightarrow H_2O + O_2$, $k_{bb}[HOO][HO]$	(bb)
$HOO + NO \rightarrow HNO_3$, $k_{cc}[HOO][NO]$	(cc)
$HOO + NO \rightarrow HO + NO_3$, $k_{dd}[HOO][NO]$	(dd)

effects of this reaction on the ozone column, a rate equation for odd atomic oxygen is derived from the reactions of Table III and the assumption of a steady state for NO_2, H, and HOO:

$$\frac{d[O] + [O_3]}{dt} = 2j_a[O_2] - 2k_e[O][O_3] - 2k_g[O][NO_2] - 2k_y[O][HOO] \quad (2)^*$$

In comparing the last two terms in Equation (2) which are ozone destruction terms, it is permissible to neglect the last term and therefore the effects of H_2O relative to NO_x because under assumed stratospheric conditions the NO_x effect turns out to be 80 to 800 times greater than that of H_2O at the 20 km level [3].

Further, Equation (2) allows comparison of the oxygen-NO_x atmosphere and the basic oxygen (Chapman model) atmosphere as a determinant of ozone equilibrium.

* j_a is a photo chemical rate constant, k_e, k_g are kinetic rate constants, and [O], [O_3], [O_2], [HOO], and [NO_2] are concentrations.

This is done by defining a catalytic destruction ratio ϱ, where

$$\varrho = 1 + k_g\,[NO_2]/k_e[O_3] \tag{3}$$

Equation (2) then becomes

$$\frac{d[O] + [O_3]}{dt} = 2j_a[O_2] - 2k_e\varrho[O][O_3] \tag{4}$$

From Equation (3) and the values of k_g and k_e, it is seen that if $[NO_x]/[O_3]$ is as little as 10^{-3}, then ϱ becomes $(1 + 4.6)$, and the equilibrium ozone concentration will be decreased by about a factor $\varrho^{1/2}$, or 2.3.

The effects of NO_x are of course felt only in the limited section of the stratosphere to which the exhaust emissions of the SST have diffused. To determine the effects that the estimated injections of NO_x would have on the total ozone column, Johnston integrated the stratosphere from top to bottom under various assumed amounts of vertical mixing after emission of pollutants at cruise altitude. The minimum reduction was found to be 3% and the maximum was found to be 50%. As it was estimated that even a 1% reduction in the total ozone column would result in 10000 additional cases of skin cancer each year in the United States [7], even a 3% reduction might be considered unacceptable.

Ashby et al. [1] have also made a study of the effects of water vapor and NO_x on the stratospheric ozone balance. They devised a computer model for the stratosphere to predict 16 concentration profiles,* the most important of which was ozone. Basically, their model solves the differential equation:

$$\frac{\partial n_i}{\partial t} = Q_i[n_j] - L_i\,(n_i, n_j) - \frac{\partial}{\partial z}\left\{-D\left[\frac{\partial n_i}{\partial z} + \left(\frac{1}{T}\frac{\partial T}{\partial z} + \frac{1}{H}\right)n_i\right]\right\} \tag{5}$$

for each of the 16 atmospheric constituents (n_i), where Q and L are production and loss rates, D is a vertical diffusion coefficient, T is the temperature, and H is the scale height. The solution was not averaged over the day and night as many previous workers had done; instead they calculated numerically in short time steps, over a 48 h period.

The model was first tested by solving for the 16 ambient profiles. Having been successful at this they next studied effects of H_2O and NO_x in quantities comparable to those injected by an SST fleet. They conclude that under the worst conditions the total ozone column would be diminished by only 0.5%. They attribute their difference from Johnston to the removal of NO at night, ultimately turning into inactive HNO_3. Thus the NO concentration is effectively buffered.

At present it is not entirely clear which of these models is correct. Johnston [24] has pointed to several serious flaws in the Ashby model. Foley and Ruderman [25], however, have developed an independent approach to test the effects of NO_x on the ozone balance. They point out that a large amount of NO_x is produced in an atmospheric nuclear explosion, and most of it eventually is transported to the stratosphere.

* $[O_2]$, $[O_3]$, $[H_2O]$, $[N_2O]$, $[HNO_3]$, $[NO_2]$, $[NO]$, $[N_2O_5]$, $[H_2O_2]$, $[HO_2]$, $[O(^3P)]$, $[OH]$, $[NO_3]$, $[N]$, $[O(^1D)]$, $[H]$

There have been several periods of active above-ground nuclear testing, the latest occurring between October 1961 and December 1962 during which the U.S. and USSR detonated about 340 megatons. Foley and Ruderman conclude that about 6×10^{34} molecules of NO were delivered to the stratosphere, which is about 3 times the amount which 500 SST's flying 7 h a day would inject during the same period.

A worldwide network of ozone monitoring stations has been in operation since the early 1960's. Although there have been periodic fluctuations in the ozone column, there has been no long-term decrease. Also, there have been no short-term decreases after major detonations which could be correlated with nuclear explosions. Foley and Ruderman therefore conclude that NO_x injections have not appreciably diminished the worldwide ozone content [25].

5. Conclusions

We can draw several conclusions about the effects of SST exhausts on worldwide ozone:

(i) The possible effects of NO_x deserve further experimental and theoretical study. In particular, the theory must consider the full dynamic situuation, including circulation and time variations, rather than averages.

(ii) Effects, if they occur, are not irreversible.

(iii) Natural injections of various pollutants, as well as other human activities, must be considered alongside any SST injection of pollutants.*

An attempt to resolve many of the existing problems of the SST's environmental effects is currently underway by the Climatic Impact Assessment Program (CIAP) of the U.S. Department of Transportation which has been mandated to 'assess, by 1974, the impact of climatic changes resulting from propulsion effluents of vehicles in the stratosphere, as projected to 1990'. To accomplish this, CIAP has directly funded over 50 research groups, elicited support from many other governmental agencies, and has gained considerable international participation.

The final CIAP report will be a series of six monographs. The first of these, 'The Natural Stratosphere of 1974', will gather existing information on stratospheric composition, supplement it with new information, including measurement (ground based, in situ, satellite, and aircraft) of the present stratospheric constituents, and suitably precise laboratory determinations of the various important chemical parameters, in order that a model of the existing stratosphere can be constructed, both to gain a more complete understanding of the stratospheric processes and to serve as a base to gauge future perturbations.

The other five monographs, 'The Engine Emissions in the Stratosphere of 1990 A.D.', 'The Perturbed Stratosphere of 1990 A.D.', 'The Perturbed Troposphere of 1990 and 2020 A.D.', 'The Biological Effects of the Tropospheric Changes', and 'The

---

* We have become aware only recently (see p. 12) of the importance of free chlorine in reducing the concentration of stratospheric ozone. Freons from aerosol spray cans have been identified as a source, but little is known about the importance of natural sources: salt spray from the ocean, volcanic injections, and extraterrestrial sources.

Social and Cost Measures of the Biological Changes', will use this information and projections of engine emissions and fleet size, to determine the environmental effects of the SST.

References

1. Ashby, R. W., Shimazaki, T., and Weinman, J. A.: 'Effect of Water Vapor and Oxides of Nitrogen on the Composition of the Stratosphere', *Bull. Am. Met. Soc.* **53**, 219 (1972).
2. Weickmann, H. K. and Van Valin, C. C.: 'The Sources and Sinks of Water Vapor in the Upper Atmosphere', *Bull. Am. Met. Soc.* **53**, 219 (1972).
3. Johnston, H.: 'Reduction of Stratospheric Ozone by Nitrogen Oxide Catalysts from Supersonic Transport Exhaust', *Science* **173**, 517 (1971).
4. Singer, S. F.: 'Stratospheric Water Vapour Increase Due to Human Activities', *Nature* **233**, 543 (1971).
5. Weinstock, B. and Niki, H.: 'Carbon Monoxide Balance in Nature', *Science* **176**, 290 (1972).
6. Junge, C. E. and Manson, J. E.: 'Stratospheric Aerosol Studies', *J. Geophys. Res.* **66**, 2163 (1961).
7. McDonald, J. E.: 'Congressional Record', S3904 (1971).
8. Cadle, R. D. and Allen, E. R.: 'Atmospheric Photochemistry', *Science* **167**, 243 (1970).
9. Ackerman, M. and Frimout, D.: 'Measure de l'absorption stratospherique du rayonnement solaire de 3.05 a 3.70 μ', *Bull. Acad. Roy. Belg., Cl. Sc.* **55**, 948 (1969).
10. Manabe, S. and Wetherald, R. T.: 'Thermal Equilibrium of the Atmosphere with a Given Distribution of Relative Humidity', *J. Atmos. Sci.* **24**, 241 (1967).
11. Newell, R. E.: 'Modification of Stratospheric Properties by Trace Constituent Changes', *Nature* **227**, 697 (1970).
12. Fried, P. M. and Weinman, J. A.: 'Vertical Distribution of Nitric Acid Vapor in the Stratosphere', *Bull. Am. Met. Soc.* **51**, 1006 (1970).
13. Newell, R. E.: 'Water Vapor in the Stratosphere by the Supersonic Transporter', *Nature* **226**, 70 (1970).
14. Newell, R. E., Kidson, J. W., and Vincent, D. G.: 'Annual and Biennial Modulations in the Tropical Hadley Cell Circulation', *Nature* **222**, 76 (1969).
15. Rearce, J. B.: *J. Geophys. Res.* **74**, 853 (1969).
16. Lateef, M. A.: *Monthly Weather Rev.* **96**, 286 (1968).
17. Reed, R. J. and Vlcek, C. L.: *J. Atmos. Sci.* **26**, 163 (1969).
18. Mitchell, J. M., Jr.: 'Reassessment of Atmospheric Pollution as a Cause of Long-Term Changes of Global Temperature', this volume, pp. 149–173.
19. Manabe, S.: 'The Dependence of Temperature on the Concentration of Carbon Dioxide', this volume, pp. 73–77.
20. Manabe, S.: 'Cloudiness and the Radiative, Convective Equilibrium', this volume, pp. 175–176.
21. Bryson, R. and Wendland, W. M.: 'Climate Effects of Atmospheric Pollution', this volume, pp. 139–147.
22. Robinson, E. and Robbins, R. C.: 'Gaseous Atmospheric Pollutants From Urban and Natural Sources', this volume, pp. 111–123.
23. Singer, S. F.: 'Environmental Effects of Energy Production', this volume, pp. 25–44
24. Johnston, H.: Private communication to Dr David Elliott (1972).
25. Foley, H. M. and Ruderman, M. A.: 'Stratospheric Nitric Oxide Production from Past Nuclear Explosions and its Relevance to Projected SST Pollution', *J. Geophys. Res.* **78**, 4441 (1973).

General References

1. *Inadvertent Climate Modification – Report of the Study of Man's Impact on Climate* (SMIC), MIT Press, Cambridge, Mass. 1971.
2. *Man's Impact on the Global Environment – Report of the Study of Critical Environmental Problems* (SCEP), MIT Press, Cambridge, Mass. 1970.
3. Nicolet, M.: 'Aeronomic Chemistry of the Stratosphere', *Planet. Space Sci.* **20**, 1671, 1972.
4. Johnston, H.: 'Pollution of the Stratosphere', *Env. Cons.* **1**, 163 (1974).

PART III

EFFECTS OF ATMOSPHERIC POLLUTION
ON CLIMATE

INTRODUCTION

Human activities are not only increasing the content of carbon dioxide in the atmosphere (see Part I), but also the particle content: dust, smoke, aerosols, even water droplets and ice particles in the high stratosphere, and rocket exhausts in the mesosphere above the stratosphere.

The effects are not at all well understood; even the immediate effects are difficult to predict and our confidence in predicting long-term effects is not high. Yet, not only local weather patterns, but indeed the world's climate, are involved.

Bryson and Wendland attempt to delineate the current thinking and review available scientific studies. They opt for a downward trend in planetary temperatures, over the long run, because of the increased albedo effects of atmospheric particulate material. This trend may accelerate, they believe, if high altitude planes produce a significant increase in the global cirrus cover through contrails.

Mitchell uses a different, but complementary approach, but with differing results. He concludes that the carbon dioxide increase is more effective in raising planetary temperatures than is the human-derived particulate loading in reducing temperatures. Natural dust loads dominate at present; in the future, however, anthropogenic dust effects could eventually have an important influence on climate.

On the other hand, Ellsaesser concludes, after critical examination of various data sources, that the human-related particle loading is only 13 % of the natural loading, on a worldwide scale, and that the anthropogenic contribution is growing slowly – if at all.

Manabe examines some of the fundamental assumptions in a critical manner. The induced temperature changes depend on the optical properties of the particles. Under certain circumstances, the increased particulate loading could even raise planetary temperatures.

Schaefer describes more local effects on the weather produced by air pollution. He is especially concerned with the sources of small particles that are introduced into the atmosphere; they can provide condensation nuclei for water vapor or lead to the formation of ice crystals, sometimes over very wide areas.

Many of the foregoing discussions are tied together in the review by Landsberg which deals with climatic changes on global, regional and local scales. It concentrates on the last five centuries and on future changes produced by CO_2 and dust. The review deals in detail with rural and urban effects and with air pollution.

CLIMATIC EFFECTS OF ATMOSPHERIC POLLUTION*

REID A. BRYSON and WAYNE M. WENDLAND

Dept. of Geography and Meteorology, University of Wisconsin, Madison, Wis., U.S.A.

Abstract. The trend of world temperature in this century appears to be directly related to the trends of atmospheric carbon dioxide content and atmospheric turbidity (dustiness). Both are believed by various scholars to be related to human activities. Since 1940, the effect of the rapid rise of atmospheric turbidity appears to have exceeded the effect of rising carbon dioxide, resulting in a rapid downward trend of temperature. There is no indication that these trends will be reversed, and there is some reason to believe that man-made pollution will have an increased effect in the future.

1. World Distribution of Atmospheric Particulates

Most northern hemisphere meteorologists live in North America, Europe, The Far East, and India. Among these, interest in air pollution is largely directed toward the immediate problems of industrial pollution in cities, and indeed this is the most apparent problem. These are also the regions of most dense observation. However, large sections of the world are characterized by at least high seasonal levels of dust and smoke which are general in distribution rather than concentrated in the cities. Surveys of the literature and personal reconnaissance indicate that this area includes much of Africa, Arabia, southern Asia (especially northwestern India and West Pakistan), China, and Brazil. Of these areas Brazil, Southeast Asia, and Central Africa have a blue haze, probably smoke from agricultural burning, and the rest a brown haze of mineral aerosols (Figure 1). Little attention has been paid to the climatic effect of these millions of square miles of 'dust cloud' cover, the ultimate fate of the dust when advected out of the region, or the general quantity of material resident in the atmosphere. In some of these dusty regions it would appear that the high pollution level is sufficiently constant that it does not attract the attention of the resident meteorologists as a current problem.

The Harmattan haze of Africa, the summer dust storms of India, and the dust flowing out of China on the winter monsoon are well known in descriptive terms, but there is little quantitative information on the densities and depths of the dust clouds. Field observations in India indicate that common densities are 600–800 μgm m^{-3} up to a height of 3000 to 9000 m [1]. Comparison of visibilities suggests that over much of the brown dust areas, densities similar to those over India are common. Densities are not known in the blue haze areas, though the low general visibilities of less than three miles often observed during the dry seasons and observed height of the smoky layer reaching in excess of 5000 m indicates that the densities are not negligible.

---

* This work was sponsored by National Science Foundation Grant GP-5572X1.

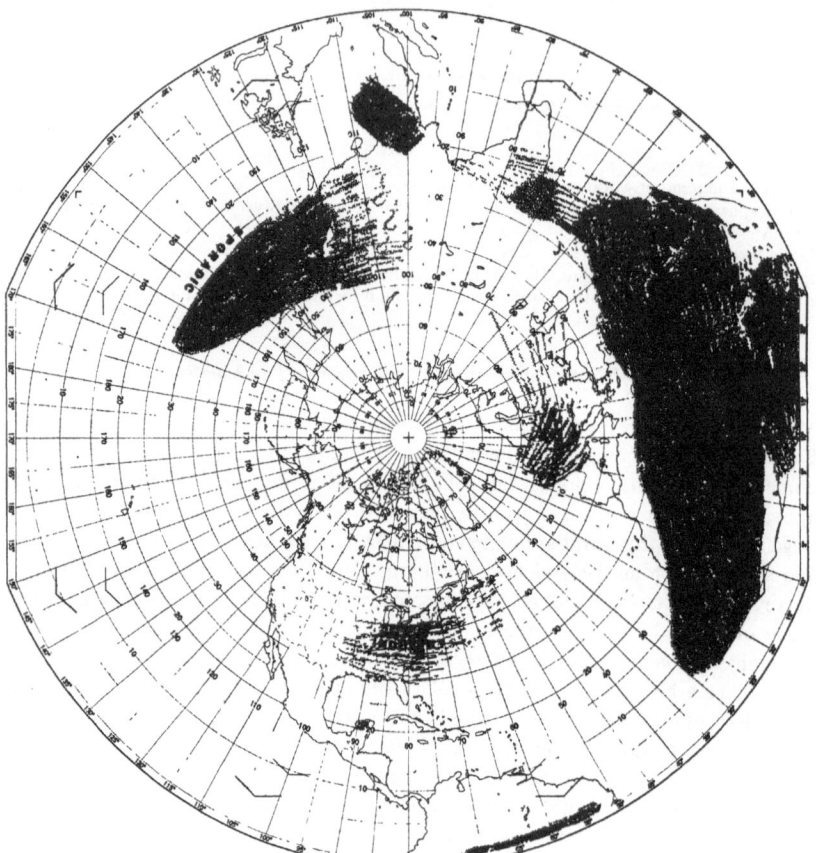

Fig. 1.   Schematic distribution of non-aqueous aerosols in the northern hemisphere.

2. Sources

It is generally assumed that the brown dust consists of clay and silt sized particles deflated from the desert surfaces, i.e. is natural [1], while the blue haze is due to slash-and-burn agriculture. There are indications, however, that even in the brown dust areas the rate of deflation and the quantity of dust resident in the atmosphere is strongly influenced by human activites. With atmospheric mixing, this implies that the total worldwide particulate load, adding the usual particulate pollution of the industrialized nations, is subject to human influence of considerable magnitude. Qualitative evidences of this human contribution are to be found in the denser dust streams crossing the Atlantic to the Caribbean after the North African tank battles of World War II had disturbed the 'desert pavement'.

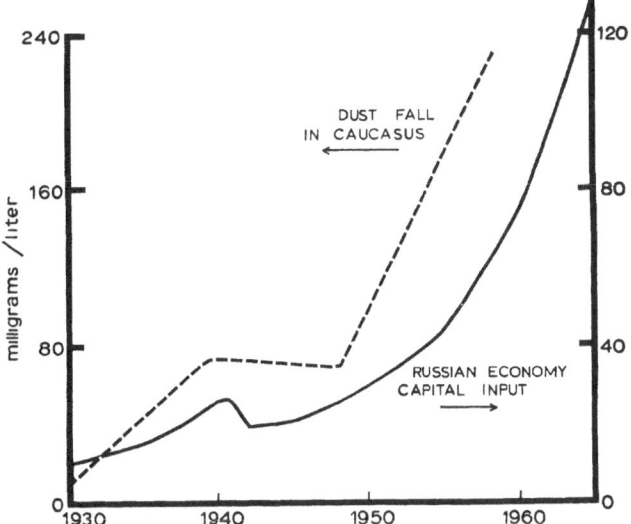

Fig. 2.   Dust fall in the high Caucasus [3], and capital input into the Russian economy [28].

Since DDT is a very, very small fraction of the pollution put into the environment by man, the fact that as much as 14 parts per billion of the dust that falls on Barbados is DDT (DDE) clearly indicates a considerable total human contribution to the dust fall [2]. Even clearer evidence is seen in the parallelism of dust-fall in the Caucasus measured by Davitaia [3] and the capital input to the Soviet economy (Figure 2). Davitaia has expressed the belief that the trend of his dust-fall measurements is due to mechanization and industrialization of eastern Europe [4].

3. Trends

It is from the consideration of such trends of dustiness or turbidity that we may investigate the relation of man, dust, and climate. There are far too few studies of long term trends, but most studies have shown a rapid increase in the period since the 1930's quite in addition to those contributions due to volcanic activity. McCormick and Ludwig [5] have shown that the turbidity of the atmosphere over Washington, D.C. and Davos, Switzerland increased 57% (1905–64) and 88% (1920–58) respectively. Davitaia's study of dust-fall on the high snowfields of the Caucasus indicates very little variation from 1790 to 1930, then a catastrophic rise of 19-fold to 1963 only leveling off during World War II (Figure 3). Data from eight middle- and high-latitude Asiatic stations presented by Pivovarova [6] support the recent upward trend in atmospheric turbidity.

Similar studies of trends over the continent are very sensitive to changes of wind

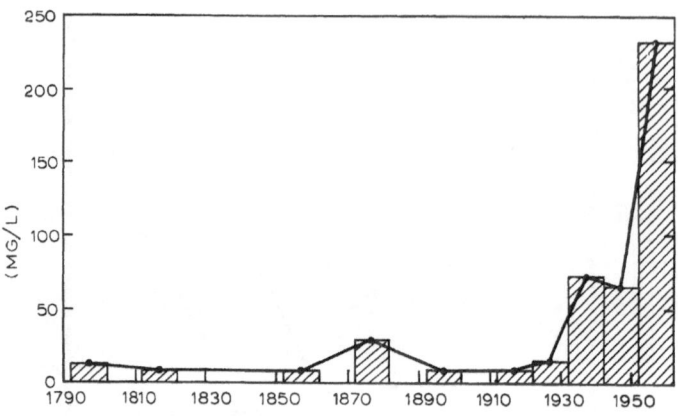

Fig. 3.   Dust fall in the high Caucasus [3].

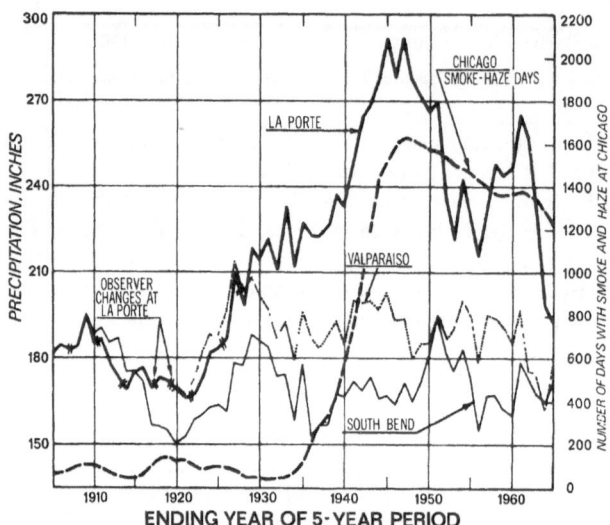

Fig. 4.   Number of smoke-haze days per five years at Chicago. Values plotted at end
of five year period. [7]

direction. The number of smoky days in Chicago rose from about 20 yr^{-1} in the decades prior to 1930 to a high of about 320 yr^{-1} in 1948 [7] (Figure 5). It has decreased somewhat in recent years as the frequency of east winds decreased. It is likely that many industrial areas show the same trend as Chicago.

4. Effects

The local climatic effects of particulate air pollution are better known than the regional and global effects, but far too little is known at any scale. Particulate effluents from industry and energy production apparently seed clouds [7, 8]. Over a city the smoke pall changes the character of the radiation falling on the city [9], especially reducing the ultraviolet penetration, rearranging the long wave radiation balance and providing condensation nuclei which contribute to greater precipitation over a city and a higher frequency of fog [10].

The regional effect of large quantities of suspended dust has been studied in North-west India [1, 11]. It has been found that the presence of the dense dust in that area affects radiative transfer through the atmosphere such that diabatic cooling of the mid-troposphere by infrared divergence is increased 30 to 50%. Das [12] and Naga-tani [13] have shown that the effect of increased diabatic cooling is to increase the mean subsidence rate over the desert. This in turn increases the aridity and enhances deflation of more dust from the desert surface. Since a thick grass cover develops naturally inside animal exclosures on the Rajasthan Desert, it is evident that much of the desert itself may be man-made. Indeed, there is some evidence that it was made by the Indus civilization [14]. Thus we must consider the possibility that the 'natural' dust load of the atmosphere over the Indian desert is largely man-made, or at least the result of animal domestication.

Vukovich and Chow [15] have shown in their numerical simulation of the atmos-phere that a distributed heat source can stabilize the long waves in the upper air-streams. The results of the Indian dust study suggest that the effect of the dust-enhanced diabatic cooling is of such a magnitude that it should be taken into account in such calculations. Yet *there are no systematic observations of dust densities and distribution.*

There have been many papers written about the climatic effects of sunspots and the rising carbon dioxide content of the atmosphere, and a few on the global climatic effect of dust other than volcanic [16, 17, 18].

The theory is not complete but there are enough data to at least get a statistical hint of the order of magnitude of the effects. The well-known equation for the radiative equilibrium of the Earth is

$$S\pi R^2(1-\alpha) = 4\pi R^2 I_t$$

where S = intensity of solar beam at top of the atmosphere, i.e. the 'solar constant'; R = radius of the Earth; α = global albedo; I_t = outward radiation intensity at the top of the atmosphere.

The usual values of S and α agree quite well with satellite measurements of I_t [19]. We can approximate the surface temperature through

$$I_t = I_0 - \Delta I = \varepsilon\sigma T_0^4 - \Delta I$$

where I_0 = the emitted radiation of the Earth's surface; T_0 = 'mean' surface temperature of the Earth; $\varepsilon\sigma =$ the effective emissivity of the surface times the Stefan-Boltz-mann constant; and $\Delta I =$ the difference between the upward radiation from the sur-face of the Earth and the outward radiation at the top of the atmosphere, i.e. the 'greenhouse effect'.

In other words, the temperature of the surface of the Earth is a function of green-house effect (in part, CO_2 concentration), the Earth-atmosphere albedo (in part, atmospheric dust concentration), and the solar constant. For the present discussion, we assume the latter is indeed constant. However, *climatic change* cannot be modelled with the above equations, because they represent equilibrium between input and out-put of heat, and do not include heat storage changes.

The temporal trend of mean temperature for the northern hemisphere is shown as the dash line in Figure 5a [20]. However, because of the ocean's great heat capacity, the Earth's temperature responds slowly to changes in the causal mechanism(s), *i.e.*, the Earth's temperature integrates and therefore lags any imbalance imposed on thermal equilibrium. Taking this storage effect into account, the change in magnitude of the 'cause' can be calculated and is shown as the solid line in Figure 5a. If the above hypothesis of cause-response is correct, the magnitudes of the causal mechanism implied by the solid line of Figure 5a should be primarily explained by changes in CO_2 and dust concentrations.

Using the calculations of Manabe and Wetherald [21], we may draw the curve of temperature effect due to increasing CO_2, shown as the solid line of Figure 5b. The dash line is the variation of northern hemisphere mean annual temperature adjusted for lag, and reproduced from Figure 5a. Interestingly, the effect of changing CO_2 concentration explains only a small percent of the temperature variance.

Figure 5c presents estimates of the dust loading of the atmosphere over the last century. The solid line represents the exponential march of the 'dust-veil index' of Lamb [22]. He compiled a rather thorough listing of volcanic eruptions from which he calculated the index. The exponential plot is used because attenuation is essentially an exponential response. The magnitude of this curve purports to express stratospheric dust loading. Lacking better data, we suggest that the dust observed in the Caucasus by Davataia [3] may be a rough approximation to the loading of the troposphere, and this in turn is mostly anthropogenic.

The total attenuation of solar radiation due to the tropospheric and stratospheric dust loading is shown as the solid line in Figure 5d. The change in the Earth's mean annual temperature adjusted for lag and CO_2 is repeated here as the dash line, and it is seen that the causal mechanism of the above hypothesis explains much of the tem-perature variation of the last century. The warming attributed to increasing concen-tration of CO_2 has apparently been more than equalled by the cooling attributed to the increased dust loading of the atmosphere.

It has been argued that the carbon dioxide increase was the work of man also, in the burning of fossil fuels, but as Deevey [25] has pointed out, the radiocarbon evidence suggests that the increase does not appear to be of fossil carbon, thus is more likely to be due to increased oxidation of plant materials such as soil humus and bogs. Yet

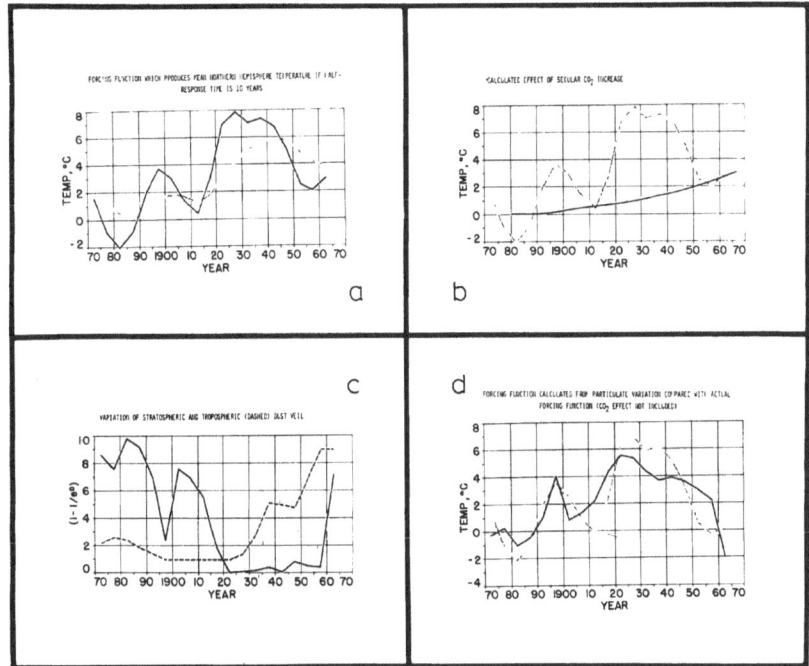

Fig. 5. (a): Temporal trend of mean temperature of the northern hemisphere (dash), and reconstructed temporal change of causal mechanism of temperature change (solid). (b): Temperature change due to CO_2 concentration alone (solid), and change of causal mechanism of temperature change reproduced from (a). (c): Change of tropospheric dust loading (dash) and change of stratospheric dust loading (solid). (d): Temporal change in causal mechanism corrected for CO_2 concentration (dash) and calculated temperature change due to total atmospheric dust loading (solid).
After Bryson [17].

even here one must recognize the role of increased cultivation, forest clearing, and bog draining. In either case man is inadvertently modifying the climate.

There is another effect of increased turbidity that is more important locally than the mean world temperature change, and that is the effect on the meridional radiation gradient and thus on the thermal Rossby number and the circulation pattern. Increased turbidity should reduce the meridional radiation gradient and thus weaken the westerlies. Lamb [26] has shown that such weakening has been characteristic of the 1960's. This change of gradient can have locally profound climatic effects.

In the preceding paragraphs we have not considered, for lack of enough information, other possible widescale climatic modifications due to human activity -- such as worldwide cloud seeding by automobile exhaust [8], the effect of jet contrails on cloud cover, or the effect of atmospheric turbidity in reducing the ultraviolet intensity in the Vitamin D band as hinted at by the work of Sastri and Das [27].

It is possible, however, to make a crude estimate of the effect of contrails. Taking 3000 as the number of jet aircraft in the air, averaging 500 mi h^{-1}, 50% making contrails, which last an average of 2 h and spread to a width of $\frac{1}{2}$ mile we have

$$3000 \times 500 \times 0.5 \times 2 \times 0.5 \text{ mi}^2 \text{ of contrails.}$$

Dividing by the area of the region in which most of these aircraft are operating we find a 5–10% increase in cirrus in the North American-Atlantic-Europe area or about a twentieth of this for the world. This is not negligible.

5. The Future

It appears that population growth, mechanization, and industrialization have now made man the equivalent of other natural processes in his effect on climate. The industrial revolution is still underway in large parts of the world, and if we can attribute either the carbon dioxide increase and/or the recent increase of atmospheric turbidity to human activites it appears that there is little that any one nation can do to reverse the trend. Nevertheless, if the analysis of this paper is correct or even nearly correct, it behooves us to study the problem much more intensively than we have. We would be pleased to be proven wrong. It is too important a problem to entrust to a half-dozen part-time investigators.

References

1. Peterson, J. T. and Bryson, R. A.: *Proc. of First Nat'l Conf. Wea. Mod.*, Albany, N.Y. (1968).
2. Risebrough, R. W., Huggett, R .J., Griffen, J. J., and Goldberg, E. D.: *Science* 159, 3820 (1968).
3. Davitaia, F. F.: *Trans. Soviet Acad. Sci., Geogr. Ser.* No. 2 (1965).
4. Davitaia, F. F.: Personal Communication, Georgian Acad. of Science, Tbilisi, U.S.S.R. (1968). He also reported a similar trend of dust concentration from the Altai Mts.
5. McCormick, R. A. and Ludwig, J. H.: *Science* 156, 1358 (1967).
6. Pivovarova, Z. I.: World Meteor. Organ. Tech. Note No. 104, p. 181 (1970).
7. Changnon, Jr., S. A.: *Bull. Amer. Met. Soc.* 49, 1 (1968).
8. Schaefer, V. J.: *Proc. of First Nat'l Conf. Wea. Mod.*, Albany, N.Y. (1968).
9. Lettau, H. and Lettau, K.: *Tellus* 21, 208 (1969).
10. Landsberg, H. E.: *Symposium – Air Over Cities*, Sanitary Eng. Center Tech. Rep. A62-5, Cincinnati, Ohio (1962).
11. Bryson, R. A. and Baerreis, D. A.: *Bull. Amer. Met. Soc.* 48, 3 (1967).
12. Das, P. K.: *Tellus* 14, 2 (1962).
13. Nagatani, R. M.: M.S. thesis, Dept. of Meteor., Univ. of Wis., Madison, unpublished (1968).
14. Singh, G.: personal communication, Birbal Sahni Institute of Paleobotany, Lucknow, India (1968).
15. Vukovich, F. M. and Chow, C. F.: Research Triangle Inst., Research Triangle Park, N.C., Final Rep. Cont. AF19(628)-5834 (1968).
16. Machta, L.: *Bull. Amer. Met. Soc.* 53, 402 (1972).
17. Bryson, R. A.: in N. Polunin (ed.), *The Environmental Future*, pp. 133–177 (1972).
18. Hamilton, W. L. and Seliga, T. A.: *Nature* 235, 320 (1972).
19. Vonder Haar, T.: Ph.D. thesis, Meteor. Dept., Univ of Wis., Microfilm Library, Ann Arbor, Mich. (1968).
20. Reitan, C. H.: Ph. D. thesis, Dept. of Meteor., Univ. of Wis.-Madison, Microfilm Library, Ann Arbor, Mich. (1971).
21. Manabe, S. and Wetherald, R. T.: *J. Atmos. Sci.* 24, 241 (1967).

22. Lamb, H. H.: *Phil. Trans. Royal Soc. London A* **266**, 425 (1970).
23. Ångstrom, A.: *Tellus* **14**, 435, (1962).
24. Mitchell, Jr., J. M.: *Ann. N.Y. Acad. Sci.* **95**, 1 (1961).
25. Deevey, Jr., E. S.: *Sci. Amer.* **209**, 10 (1958).
26. Lamb, H. H.: *Geogr. J.* **132**, 2 (1966).
27. Sastri, V. D. P. and Das, S. R.: *J. Opt. Soc. Amer.* **58**, 3 (1968).
28. Powell, R. P.: *Sci. Amer.* **219**, 6 (1968).

For Further Reading

1. 'All Other Factors Being Constant...', *Weatherwise* **21**, No. 2 (April, 1968), 56–61.
2. L. J. Battan, *The Unclean Sky*, Doubleday & Company, New York, 1966.
3. R. A. Bryson and J. E. Kutzbach, Air Pollution, Assn. Amer. Geogr. Resource Paper No. 2, Washington, D.C., 1968.
4. R. G. Ridker, *Economic Costs of Air Pollution*, Frederick A. Praeger, New York, 1967.
5. A. C. Stern (ed.), *Air Pollution*, Vols. 1 and 2, Academic Press, New York, 1962.
6. W. Bach, *Atmospheric Pollution*, McGraw-Hill, New York, 1972.

A REASSESSMENT OF ATMOSPHERIC POLLUTION AS A CAUSE OF LONG-TERM CHANGES OF GLOBAL TEMPERATURE

J. MURRAY MITCHELL, JR.

National Oceanic and Atmospheric Administration, Silver Spring, Md. 20910, U.S.A.

Abstract. Two globally extensive forms of atmospheric pollution (carbon dioxide and particulate loading) are each considered from the viewpoint of long-term changes in their total abundance, and the impact of such changes on the equilibrium temperature of the Earth.

A comparison of the observed levels of atmospheric CO_2 since 1958 with estimates of the fossil CO_2 input to the atmosphere from human activities indicates that between 50 and 75% of the latter has remained in the atmosphere. The present-day CO_2 excess (referred to 1850) is estimated at 11%; the excess is conservatively projected to increase to 15% by 1980, 22% by 1990, and 32% by 2000 A.D. Changes of mean atmospheric temperature due to CO_2, calculated by Manabe *et al.* [20] as about 0.3 °C per 10% change of CO_2, are sufficient to account for only about one-third of the observed warming of the Earth between 1880 and 1940, but would appear capable of contributing a further warming of about 0.6 °C between the present time and the end of the century.

The total global atmospheric loading by small particles is estimated at about 4×10^7 tons at present, of which about 1×10^7 tons is derived directly or indirectly from human activities. If the anthropogenic fraction should grow in the future at about 4% yr^{-1}, the total loading would increase to a level at the end of this century about double that of the 19th century, and 60% above present-day levels. At present, the total anthropogenic loading is estimated to exceed the average stratospheric loading by volcanic dust during the past 120 yr, and to equal about one-fifth of the stratospheric loading following the 1883 eruption of Krakatoa. The impact of anthropogenic particle loading changes on mean temperature cannot be reliably determined from present information although a cooling effect is likely.

Of the two forms of pollution, it appears that the carbon dioxide increase is more influential in raising planetary temperatures than the anthropogenic particle increase is in lowering planetary temperatures. (If, however, both the CO_2 and particle inputs to the atmosphere should grow at equal rates in the future, the relative importance of particle effects will increase and could eventually become dominant.) In balance, the net thermal impact of all global-scale pollution (including thermal pollution) is likely to be one of warming, perhaps increasingly so after 2000 A.D. It is concluded that the cooling of climate since 1940, apparently still in progress, is a natural phenomenon plausibly related to an enhanced stratospheric loading by volcanic dust in the period. Natural variations of climate have been larger than those probably induced by human activities during the past century, but the rapidity with which human impacts on atmospheric quality threaten to grow in the future has disturbing climatic overtones that demand better understanding.

1. Introduction

In recent years there has been growing speculation that air pollution is responsible for world-wide disturbances of weather and climate. There are two basic lines of reasoning that have fed such speculation.

One line of reasoning is of a circumstantial nature. It begins with the realization that, according to various meteorological evidence, the large-scale climate of the Earth has fluctuated during the past century. This fluctuation happens to have occurred at a time when modern industrial man was first emerging as a powerful manipulator of his physical environment. Inasmuch as the fundamental cause or causes of this cli-

S. Fred Singer (ed.), The Changing Global Environment, 149–173. All rights reserved.

150 J. MURRAY MITCHELL JR.

matic fluctuation are not yet identified, it is argued that – unless or until demonstrated otherwise – one or more inadvertent byproducts of man's (historically unprecedented) activities cannot be ruled out as possible contributory causes.

The other line of reasoning is more direct, and in some respects more compelling. This one takes off from the fact that locally heavy concentrations of air pollution, such as we find in cities and heavy-industry areas, have undeniable effects on local weather and climate. Among these effects are reduced solar radiation and visibility, increased fog and low cloudiness, and perhaps anomalous temperature conditions as well [1]. Such effects, incidentally, are consistent with theoretical principles of atmospheric physics, and are predictable to some degree. Now since certain types of air pollutants have atmospheric residence times measured in weeks or more [2], these pollutants are obviously capable of being spread by atmospheric circulations, in diluted concentrations, over very wide geographical areas. It presumably follows, then, that certain meteorological effects of these pollutants should extend over equally wide geographical areas, and, in the case of at least one pollutant (carbon dioxide, whose atmospheric residence time is measured in years rather than months), the effects should be strictly global. From a qualitative viewpoint, such reasoning is obviously sound. It is left for us first to establish the prevailing concentration of each geographically extensive form of air pollution, together with its changes over the years, and second to determine as reliably as possible the quantitative meteorological or climatological effects of each such pollution form.

Before we comment further on the idea that man is responsible in any way for altering world climate, it would seem prudent for us to keep in mind that world climate has unquestionably varied throughout the earth's history, and sometimes dramatically so. It has, in fact, varied significantly during earlier centuries of the Christian Era [3], long before man's capacity for influencing his physical environment on a large scale had developed. The obvious implication of this is that world climate is clearly capable of variation through *natural* causative agencies, whatever those agencies may specifically be. Thus, the climatic fluctuation of the past century could also be attributable entirely to natural causes, as for example some type of non-linear thermodynamic interaction between the atmosphere and the oceans, or a more remote environmental disturbance that can be expected to influence climate (for example variable radiation from the sun, or episodes of unusually great volcanic activity) [4].

On the other hand, if man is not yet in nature's league as a potent climate-regulating force, he is almost certainly destined to become such a force in the rather near future. Of special concern in this connection is the dangerous circumstance that man may well arrive at that point *inadvertently* before he arrives there deliberately, and that he will find himself unequipped to arrest or reverse undesirable climatic developments that he may have set in motion unwittingly. Fom any point of view, therefore, it is a matter of some urgency that we identify the causes of modern-day climatic instability, and make an accurate determination of man's impact on world climate both present and future.

2. Scope of This Discussion

There are many different aspects of the problem of large-scale air pollution/climate relationships that are in need of much more intensive investigation. Some involve the possible influence of pollutants on cloudiness and precipitation, along such lines as those Schaefer has considered in this volume. Others involve the planetary heat budget and the equilibrium temperature of the earth's surface, such as Bryson and Manabe have considered.

My remarks will be confined to two specific forms of pollution, and their apparently competing impacts on the thermal climate of the Earth. These are (1) carbon dioxide and (2) atmospheric particles (aerosol). Both have already been introduced by other contributors to this volume, so in my remarks I will obviously be treading over some of the same ground. I do so, without apology, for two reasons. First, it has recently become possible to diagnose the carbon dioxide factor with considerably more insight than ever before, and so a fresh look at the CO_2 situation seems particularly appropriate. Second, with regard to atmospheric particle loading, I feel that this factor requires re-examination in three fundamental respects: (1) the extent to which anthropogenic sources of particles are contributing to the present total atmospheric loading; (2) the relative importance of natural and anthropogenic factors in accounting for variations of total atmospheric loading; and (3) uncertainties as to the thermal impact of particle loading changes.

Let us first turn briefly to the evidence for changes of the average temperature of the Earth, which are to be interpreted as one (and, practically, very important) manifestation of the variations of the planetary heat budget during the past century.

3. Recent Planetary Temperature Trends and Their Significance

By analysis of climatological data for stations distributed as uniformly as possible over the Earth's surface, it can be established that the mean temperature of the whole planetary atmosphere, at least in its surface layers, has fluctuated systematically during the past century [5]. The data reflect a net world-wide warming of about 0.6 °C (1.0 °F) between the 1880s and the 1940s, followed by a net cooling of about 0.2 °C (0.3 °F) between the 1940s and 1959, the most recent year of data in the original analysis. Additional data for 1960–68 as analyzed by Reitan [6], were later incorporated into the analysis in order to estimate the further movement of world mean temperature after 1960 (see Figure 1). On this tentative basis, it appears that the cooling trend which first set in during the 1940s has continued essentially up to the present time, and that the net temperature drop in the last quarter-century has now accumulated to almost 0.3° C (0.5 °F). To date, then, roughly half of the warming that occurred during earlier decades of the century has been erased by subsequent cooling (Figure 1). One cannot say offhand whether or how long this cooling will continue in the future.

The observed fluctuation in world mean temperature is believed to indicate a systematic change in the planetary heat budget during the past century. Although the

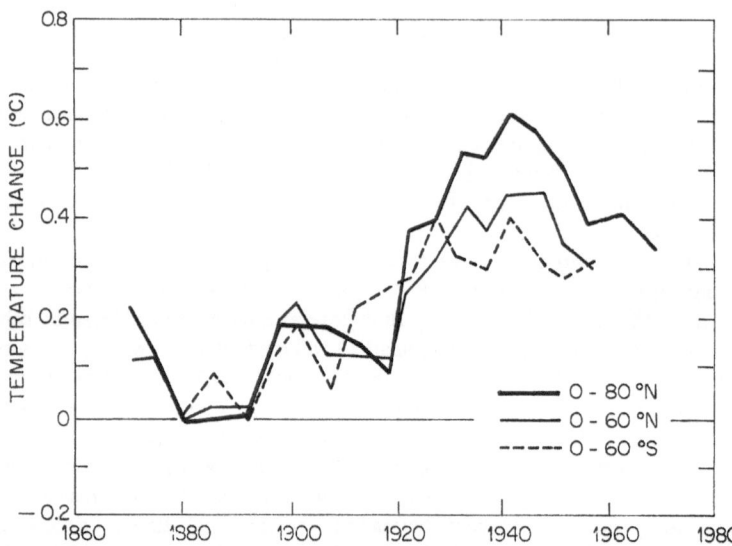

Fig. 1. Changes of annual mean temperature of the Earth during the past century, shown separately for three ranges of latitude. Data to 1960 after Mitchell [5]; updating of curve for 0–80°N after Reitan [6].

magnitude of the fluctuation may appear extremely small, in comparison to the familiar seasonal temperature changes for example, it is symptomatic of larger changes in certain geographical areas – most notably in the Arctic – that are evidently associated with concurrent changes in the pattern of atmospheric circulation [3, 5]. Moreover, the magnitude of the fluctuation, of the order of 0.6°C, is an appreciable fraction of the planetary temperature difference of about 6°C that is believed to have distinguished glacial from interglacial conditions during the Pleistocene Ice Age [7]. Indeed, the temperature changes of the past century have been paralleled by well-documented and widespread changes in alpine glacier movement, Arctic pack-ice cover, world sea levels, and floral and faunal limits. In such respects, the temperature fluctuation reflects a climatic disturbance of no inconsiderable practical significance. Our ignorance of the future course of that disturbance is a strong motivation for us to establish its causes.

4. The Secular Increase of Atmospheric Carbon Dioxide

4.1. DISCUSSION OF THE EVIDENCE FOR INCREASING CO_2

Since late in the 19th century, man has released carbon dioxide to the atmosphere at an ever-accelerating rate through the combustion of fossil fuels (primarily lignite, coal, petroleum and natural gas). The present rate of CO_2 production by man is estimated at 16×10^9 metric tons yr^{-1} [8, 9]. Until recently, however, it was not clear how much of this gas has accumulated in the atmosphere, and how much has been removed by a net transfer to the oceans and to the terrestrial biomass.

Early efforts had been made by Callendar and others [10] to estimate the actual CO_2 accumulation in the atmosphere by means of a statistical comparison of historical measurements of atmospheric concentrations available during the past 100 yr. These had suggested a systematic increase from a representative 19th century base level of about 290 parts per million (ppm) to about 330 ppm by 1960, an increase of 14% in the period. If these figures are correct, they imply that nearly 75% of all CO_2 released by man has remained in the atmosphere. But according to Keeling, a proper consideration of the effect of differences of instrumentation reveals that these figures are probably too large [11].

By 1965, Keeling et al. [12] had found it possible to refine such estimates on the basis of a CO_2 monitoring program begun seven years earlier in cooperation with the U.S. Weather Bureau (now NOAA). In that program, which is still continuing, precision gas-analyzer measurements have been made at Mauna Loa Observatory in Hawaii and at the Pole Station in Antarctica. Inasmuch as these measuring sites are at high altitudes, and as the measurements are made in such a manner as to permit detection and removal of the disturbing effects of local upwind sources, monthly average CO_2 'background' concentrations can be evaluated at these stations which rarely deviate by more than 1 ppm from their true ambient values. These ambient values, in turn, are expected to remain within 2 or 3 ppm of the worldwide annual average atmospheric concentration (after adjustment for local seasonal effects) since the rate of mixing of the gas by planetary-scale air motions is rapid in comparison to the mean residence time of CO_2 molecules in the atmosphere, of the order of 5 yr [13].

It is therefore highly significant that Keeling and his colleagues were able to detect a systematic increase in CO_2 concentrations, in both the Mauna Loa and Pole Station records, which averaged about 0.7 ppm yr^{-1} between 1958 and 1963 [12]. Since then, evidence of a continuation of the CO_2 increase has accumulated from various sources [14], as shown in Figure 2. This evidence establishes the increase as a global scale phenomenon, and reveals that while the rate of increase may have varied from year to year (being slower in the mid-1960s than at other times) it has averaged 0.8 ppm yr^{-1} between 1958 and 1971. Since 1968, the increase appears to have accelerated to at least 1.0 ppm yr^{-1}.

On the reasonable premise that the observed increase is attributable to the accumulation of fossil CO_2 in the atmosphere, it is of interest to compare the observed increase with the cumulative input of CO_2 from fossil fuel combustion and other man-made sources (principally limestone kilning) since 1860. Early estimates of the fossil CO_2 input to the atmosphere, e.g. those of Revelle and Suess [15], have now been supplemented by newer (and somewhat lower) estimates which I adopt here [16]. These are indicated by the upper curve in Figure 3, in terms of the percentage increase of atmospheric CO_2 above an assumed pre-industrial CO_2 content of 290 ppm to be expected if all fossil CO_2 were to accumulate in the atmosphere with zero loss to the oceans or to the terrestrial biosphere. The observed increase between 1958 and 1972

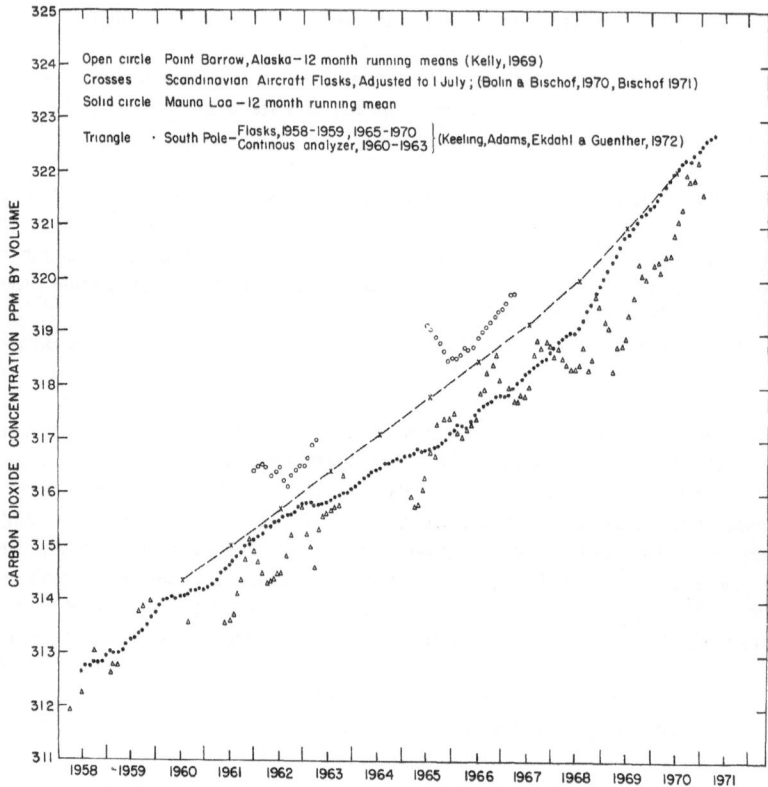

Fig. 2.   Changes of atmospheric carbon dioxide concentration, 1958 to 1971, at various locations.
For references cited in the figure, see [14]. After Machta and Telegadas [9].

is also shown in the figure as three parallel curve segments, one for each of three
assumed pre-industrial CO_2 contents (lower curve for 295 ppm, middle curve for
290 ppm, and upper curve for 285 ppm) the correct value of which is not known but
likely to have been within this range [10]. The inflections in the observed trends are
intended to represent the smoothed changes in rate of increase evident in the Mauna
Loa record, and may or may not be of global significance.

Comparison of the cumulative fossil CO_2 input curve with the observed CO_2
concentration changes, shown together in Figure 3, indicates that between 50 and
75 % of the input to date has remained in the atmosphere (depending on the value
chosen for the pre-industrial CO_2 level in the atmosphere). It should be noted that the
proportion remaining in the atmosphere, when calculated by comparing the incre-
mental fossil input *since* 1958 with the observed atmospheric CO_2 increase since 1958,
has varied considerably from year to year but has averaged only about 50 % between

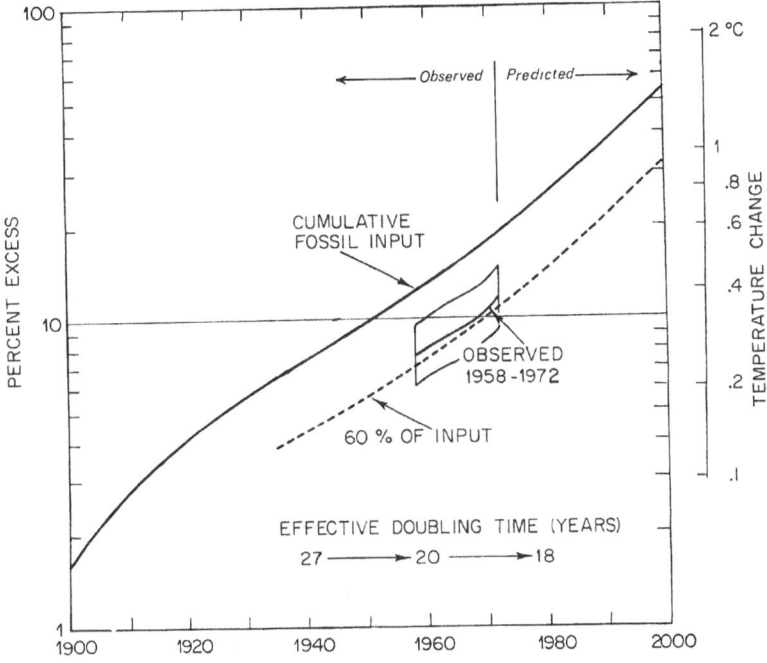

Fig. 3. Growth of atmospheric carbon dioxide, 1900 to 2000 A.D., expressed as percent excess above 19th century base level. Cumulative fossil input curve represents potential atmospheric growth if all fossil CO_2 were to remain in the atmosphere without loss to the oceans or biosphere. Wavy line segments, labelled 'observed 1958–1972', are observed CO_2 growth at Mauna Loa for three assumed 19th century base levels between 285 and 295 ppm (see text). Dashed curve indicates tentative long-term increase of actual atmospheric CO_2 content. Projections after 1970 are based on assumption that fossil CO_2 input will increase 4 % yr^{-1} to 1980 and 3.5 % yr^{-1} thereafter. Temperature scale at right after Manabe and Wetherald [20].

1958 and 1972 [17]. These two results are difficult to reconcile unless one or more of the following circumstances apply: (1) the fossil input to the atmosphere has been consistently underestimated by 10 to 15%; (2) the pre-industrial atmospheric concentration of CO_2 deduced by Callendar [10] and others was underestimated by several ppm; or (3) the average fraction of input taken up by the oceans and/or the terrestrial biosphere has been substantially greater in the years since 1958 than earlier in the century. Further study of the problem is required to resolve the matter, but tentatively it appears that the last mentioned of these explanations is the most likely.

What, then, can be said about the future growth of atmospheric CO_2 levels? Clearly, the answer to this question depends on what can be said in answer to two other questions. First, at what rate will future fossil CO_2 inputs to the atmosphere increase in future years? Second, what fraction of these future inputs will be withdrawn by absorption in the oceans and by photosynthetic fixation in long-lived elements of the terrestrial biomass? In the period 1950 to present, the worldwide total CO_2 input to the atmosphere is estimated to have grown by at least $4.7\% \, yr^{-1}$ [16, 17]. For various reasons this rate can be anticipated to slow somewhat in the future. Machta *et al.* [8, 9] have suggested a future growth rate of 4.0% between now (1972) and 1980, and 3.5% between 1980 and 2000 A.D. Adopting these rates, the future growth of accumulated fossil input to the end of the century would be as shown by the upper curve in Figure 3. By assuming that the fraction of total CO_2 input that remains in the atmosphere will average 60% in the future, the further growth of atmospheric CO_2 levels then becomes as shown by the dashed line in Figure 3. Although for the purposes of this paper I shall adopt this estimate of future CO_2 growth, it should be emphasized that for various reasons the actual future growth may turn out to be greater than indicated here. First, it is more likely than not that the assumed future growth rates of fossil CO_2 input are a bit low. Second, it appears from evidence just coming to light at this writing [11] that the observed atmospheric CO_2 levels (e.g. those summarized in Figure 2) have been consistently underestimated by a few ppm, which suggests that the fraction of fossil input remaining in the atmosphere has been even higher than indicated in Figure 3. Third, theoretical studies [18] imply that the decreased pH of the mixed ocean layer that will follow from increasing atmospheric CO_2 levels may have the effect of lowering the capacity of the oceans to absorb as much of the fossil CO_2 input to the atmosphere as formerly, and so the fraction of future fossil CO_2 input which remains airborne may substantially increase in the future. Fourth, to the extent that future increases of atmospheric CO_2 will be successful in raising atmospheric temperatures above their present levels (see below), the upper layers of the oceans are likely to participate in the temperature increase in which case even more CO_2 will be shifted from the oceans to the atmosphere. Thus, a very real possibility exists that future atmospheric CO_2 levels may increase between now and the end of the century by a considerably greater amount than indicated by the dashed line in Figure 3. It cannot be entirely precluded, in fact, that if circumstances favor a large enough CO_2 shift from the upper ocean layers to the atmosphere in future decades, the atmospheric CO_2 levels might climb *above* the curve of cumulative fossil input shown in Figure 3, either before or after the end of the century.

To summarize, the following conclusions with respect to the CO_2 content of the atmosphere are to be emphasized:

(1) Measurements made since 1958 in various parts of the world are consistent in indicating that the average CO_2 concentration of the atmosphere has been increasing in the past 15 yr at an average rate of 0.8 ppm $(0.25\%) \, yr^{-1}$. This rate of increase is probably accelerating, and may now (1972) be as much as 1.0 ppm (0.3%) yr.

(2) The average planetary atmospheric CO_2 concentration is now (1972) about 324 ppm, or 11% above the apparent 19th-century, pre-industrial concentration of about 290 ppm. This is an appreciably smaller increase over the past century than that estimated by Callendar [10], but it is a much larger increase than that inferred from the rate of radiocarbon dilution measured in tree wood (the Suess effect) [19].

(3) The observed long-term increase of atmospheric CO_2 levels is almost certainly

attributable to fossil fuel combustion (plus a small contribution by limestone kilning), and of all the fossil CO_2 that has been injected into the atmosphere from this source somewhere between 50 and 75% appears to have remained in the atmosphere.

(4) Assuming that the total input of fossil CO_2 to the atmosphere will grow at 4% yr^{-1} to 1980 and at 3.5% yr^{-1} thereafter, and assuming that an average of 60% of all future input will remain in the atmosphere, the atmospheric CO_2 concentration – compared with a 19th century base level of 290 ppm – will have grown 45 ppm (15%) by 1980, 65 ppm (22%) by 1990, and about 95 ppm (32%) by 2000 A.D. This, however, is to be regarded as a conservative estimate of future growth since a number of circumstances neglected here may turn out to require a considerable upward revision of these projected increases.

4.2. CO_2: SIGNIFICANCE TO THE THERMAL STATE OF THE ATMOSPHERE

Undoubtedly the most reliable determination yet available of the effect of changes of atmospheric CO_2 on the equilibrium temperature distribution in the atmosphere is that based on the numerical investigations of Manabe *et al.* [20]. According to these authors, increases of CO_2 result in a warming of the entire lower atmosphere, the amount of warming being dependent in part on whether the atmosphere is held at a fixed absolute humidity or at a fixed relative humidity during the CO_2 change. For the assumption of fixed *absolute* humidity, together with conditions of surface albedo, cloudiness, solar radiation, and other parameters chosen as typical of the mid-latitudes in an equinoxial season, the temperature effect is such that a 10% increase of CO_2 concentration (from 300 to 330 ppm) would lead to a warming of about 0.2 °C (0.3°F). For the assumption of fixed *relative* humidity, all other conditions remaining the same, the temperature effect of a 10% CO_2 increase is nearly doubled to 0.3 °C (0.5°F). Manabe and Wetherald [20] reason that the latter assumptions are probably more realistic. Moreover, recent studies [21] of the thermal effects of CO_2 increases in a three-dimensional general circulation model (in which large-scale atmospheric dynamics, seasonal changes of solar radiation, and air-sea heat exchange processes are all taken into account) yield average atmospheric temperature changes that compare very closely in magnitude with those predicted by the simpler models [20]. It is important to keep in mind, however, that in none of these models has any allowance been made for such potentially important thermal feedback effects as cloud-cover changes. Therefore, the thermal effects of CO_2 changes in the real atmosphere are very likely to be different (i.e., smaller) than those predicted by these models, but as yet the difficulty of modeling cloud interaction effects leaves us with no reliable means of verifying this.

At least tentatively, these modeling experiments suggest that it is possible to ascribe *not more* than about one third of the observed world-wide warming trend between 1880 and 1940 to the secular increase of CO_2. We are apparently required to look elsewhere to account for the remaining two thirds of this warming. On the other hand, since the theoretical relationship between CO_2 concentration and equilibrium temperature is nearly linear in the limited range of CO_2 variations involved, we should note that the rate of warming due to CO_2 was probably greater since 1940 than it was at any earlier time. In fact, more than half of the net warming to date that is attributable to CO_2 since the 19th century (0.4 °C in 1972) is expected to have occurred since 1940. This, of course, was during the time when world-mean temperatures are estimated to have *declined* by about 0.3 °C. Thus it appears that the climatic cooling mechanism coming into operation during the past quarter century has been roughly three times more influential on temperature than the CO_2 effect in the same period.

In summary, I suggest that, according to the best information now available, probably not more than one third of the planetary temperature disturbance of the past century is attributable to variations of atmospheric CO_2. Other mechanisms are evidently required to account for part of the warming observed between 1880 and 1940, as well as for the cooling observed since 1940 which has occurred in spite of further warming contributions by CO_2 in that period. The temperature contribution of CO_2 changes anticipated in the future, neglecting all other mechanisms of climatic change, will consist of a further warming (above present day temperature levels) of order 0.1 °C (0.2 °F) by 1980, 0.3 °C (0.5 °F) by 1990, and 0.6°C (1.0°F) by 2000 A.D. It is therefore possible that, if other causative factors in climatic change do not also become more important in the future than in the past, carbon dioxide will win out by the end of this century as the dominant factor in determining the future course of planetary temperature.

5. Changes of Atmospheric Particle Loading

5.1. Comparison of Human and Natural Contributions to Atmospheric Particle Loading

In the past several years a number of authors have cited various kinds of evidence pointing to a long-term growth of the particle content of the atmosphere [22]. It has generally been supposed that this is a real global-scale phenomenon, and that it is attributable to human activities. More than one author has, in fact, reasoned that increases of particle loading by man have been the cause of the latter-day cooling of global climate already noted, and that further increases in the future to be expected from anthropogenic sources could, within a few decades, trigger the onset of a man-made ice age.

No doubt the particle content of the atmosphere has a non-trivial impact on climate. If the particle content is changing systematically over the years – through either natural or human agencies – the equilibrium temperature of the Earth (among other things) is likely to respond to the change to one extent or another. However, it is to be recognized that in most hitherto published analyses of the problem of atmospheric particle loading and its impact on global climate, certain assumptions have been made in common which preordain the more or less drastic climatic impacts inferred from such analyses but which on closer scrutiny appear to require re-examination. To put the situation in perspective, let us consider in turn three questions:

First, what does the observational evidence of long-term increases of atmospheric particle loading tell us about the magnitude and probable geographical scale of such increases? Second, to what extent are human activities directly or indirectly contributing to (observed) changes of atmospheric particle loading, and how do these anthropogenic contributions compare with the particle loading contributed by natural geophysical phenomena? Third, what do we know about the optical properties of atmospheric particles, and about the interaction of particles with the energy budget of the Earth-atmosphere system, which determine the thermal impact of atmospheric particle loading changes?

With regard to the magnitude and geographical scale of long-term changes of particle loading, it should be mentioned in passing that significant trends observed in and around urban/industrial centers have been documented [23] which in some locations appear actually to have reversed direction in recent years with the enforcement of local air pollution control ordinances. Such statistics as these, however, tell us very little about regional or global-scale trends of 'background' atmospheric particle loading, which are of primary concern to global-scale climate and which involve smaller particles (in the 0.1 to 5 μm range) than those to which urban/industrial air quality measurements primarily refer.

Evidence for systematic changes of large-scale background particle loading has come from a variety of indirect sources. This evidence derives in part from dustfall measurements in alpine glaciers [24], changes of electrical conductivity of the atmosphere over the North Atlantic Ocean [25], atmospheric turbidity measurements at mountain locations such as Mauna Loa (Hawaii) and Davos (Switzerland) [26], and starlight extinction data available from astronomical observatories [27]. For the most part these point to a real increase of total atmospheric particle loading during the past century, but leave in doubt both the magnitude of the increase and the extent to which it is a regional rather than a global-scale phenomenon. Recent analyses of the Mauna Loa solar radiation record [26] and electrical conductivity measurements over the oceans [28] strongly imply that, except in plumes downwind from the east coasts of North America and Asia (and around the Indian subcontinent), the increases are mainly continental in scope and do not extend to the bulk of the world's oceanic areas [28]. Thus, allusions one can find in the literature to 'global' trends of atmospheric particle loading (or of turbidity) are likely to be an exaggeration in terms. At the same time, however, the geographical scale of the problem is probably sufficient to have at least an indirect bearing on global climate, a point to which I shall return toward the end of the paper.

With regard to the contribution of human activities to the total particle loading of the global atmosphere, I shall now consider this question from two points of view. First, I propose to synthesize from various sources a (necessarily rough) estimate of the total flux of particulate material into the atmosphere, including the anthropogenic flux circum 1970; and to infer from a consideration of applicable particle residence times in the atmosphere the mean atmospheric loading that is consistent with this estimated total flux. Second, I propose to estimate the long-term growth rate of atmospheric loading from both direct and indirect anthropogenic sources, and to compare the changes of total loading from these sources with the changes of loading by a natural source of particles about which some quantitative information is available (volcanic dust that reaches the stratosphere).

Recently, a number of estimates have been made of the total mass of particles entering the atmosphere from all sources, including particles that form in situ (as sulfates and nitrates) by chemical reactions involving gaseous emissions [29]. In most respects these estimates are all in rather good agreement, but in some a considerable difference of opinion remains that I have attempted to resolve through personal arbitration with some of the authors. In particular, the important contribution of smoke particles from African grass fires, set by man during the dry season, is believed by Flohn [29] to be grossly underestimated in most assays of this kind. After making due allowances for this, I show the results of my synthesis in Table I.

According to Table I, the worldwide aggregate source strength of atmospheric particles is at present nearly 3×10^9 metric tons yr^{-1}, of which about half (1.6×10^9

TABLE I

Estimated global production of particulate matter[a] (Millions of metric tons per year)

Circum 1970 data modified after SCEP[b]	A. All particle sizes			B. Particles $<5\mu m$ diameter		
	Natural	Man-made or Man-triggered	Total	Natural	Man-made or Man-triggered	Total
Direct particle production						
Sea salt	1000	—	1000	500	—	500
Wind-blown dust	200	300	500	100	150	250
Volcanic tephra	250	—	250	25	—	25
Forest fires/Slash-burn agriculture	35[c]	100[d]	135	5[e]	60[d]	65
Cosmic dust	10	—	10	<1	—	<1
Industrial processes	—	45	45	—	12	12
Power and heating plants	—	35	35	—	10	10
Other human activities	—	30	30	—	8	8
Sub totals	1495	510	2005	630	240	870
Percent	75%	25%	100%	72%	28%	100%
Particles formed from gases						
Sulfates (conversion of SO_x)	420	220	640	335	200	535
Nitrates	75	40	115	60	35	95
Hydrocarbons	75	15	90	75	15	90
Sub totals	570	275	845	470	250	720
Percent	67%	33%	100%	65%	35%	100%
Totals	2065	785	2850	1100	490	1590
Percent	72%	28%	100%	69%	31%	100%

[a] Ref. [29].
[b] Adjusted toward estimates of H. Flohn (SMIC and personal communication) for slash/burn agriculture.
[c] Forest fires caused by man assumed to be offset by control of natural fires.
[d] Not included in SCEP.

tons yr^{-1}) consists of particles smaller than 5 μm diameter. This 'small-particle' fraction is of special interest for two reasons. First, as noted in the next section, particles between about 0.1 and 5 μm are of unique climatic significance (the mass of particles smaller than 0.1 μm is negligible). Second, the particles in this size range, unlike the larger particles, have sufficiently long residence times in the atmosphere that large-scale wind systems can carry them over intercontinental distances before they are altogether depleted by rainout and other removal mechanisms [2, 30]. In general, these residence times range from a few days for particles entering and remaining in the lower troposphere, to a week or two for those reaching the upper troposphere, and to at least a year for those injected well into the stratosphere [2, 30, 31]. *Thus, all further references to particles in this paper should be understood to refer to small particles only (those less than 5 μm in diameter).*

Of the total particle source strength, about 20% is estimated to derive from direct inputs by human activities (of order 3×10^8 tons yr^{-1}). Another 10% is estimated to derive from natural sources that arise *indirectly* from human activites, such as grass and forest fires set by man, slash-burn agricultural practices, wind-blown dust associated with man's disturbance of natural ground cover, and similar inputs (of order 1.5×10^8 tons yr^{-1}).

Adopting a mean residence time of nine days (0.025 yr) as a weighted average for all small particles (with the exception of the one percent or less of the total particle mass that reaches the stratosphere, to be considered later), we can readily obtain a crude estimate of the total particle loading of the global *troposphere* by multiplying this residence time by the total source strength of small particles, from Table I. In this way we find that the total loading is now (circum 1970) of the order 4×10^7 tons. In the same way we find that, of this total loading, roughly 1×10^7 tons is being contributed by human activities (20 to 30% of the total). This leaves about 3×10^7 tons contributed by natural phenomena, plus the stratospheric loading which will be considered later. While all these numbers seem to be reasonable, it cannot be emphasized too strongly that they are extremely crude estimates and may easily be in error by as much as 50%.

I now consider the question of how the contribution of human activities to the total particle loading of the atmosphere is likely to have changed since the nineteenth century, and how it might be expected to change in the future. For lack of a better approach, I shall assume that the total flux of particles into the atmosphere by human activities grows with time in a fixed proportion to the increase of fossil fuel combustion over the years, as estimated in turn by Keeling [16] and as projected into the future by Machta *et al.* [8, 9]. In making this assumption I neglect the fact that advances in combustion technology have resulted in a trend to higher combustion efficiency and to the increasing use of so-called 'clean fuels' such as natural gas. I also neglect the impact of smoke abatement ordinances and other emission controls that will tend to become increasingly effective in the future. On the other hand, I think it likely that future increases of fossil fuel consumption will continue to serve as a reasonable index of future increases of antropogenic particle production from sources other than combustion products per se, which according to Table I add up to a major fraction of the total anthropogenic production today. From information about past and anti-cipated future growth rates of fossil CO_2 flux to the atmosphere, discussed earlier, I therefore adopt a mean growth rate of 4% yr^{-1} as a rough index of the long-term increase of total particle flux into the atmosphere from anthropogenic sources between 1940 and 2000 A.D. While this may indeed lead to an overestimate of the particle source strengths toward the end of the century, it will at least give us an idea of what to expect in the future *if present economic trends continue and if effective air pollution curbs fail to take hold by* 2000 *A.D.*

In line with the foregoing assumptions, it is instructive to consider how the total particle loading of the atmosphere would have varied since 1900, and would continue to vary in the future, if, as I have suggested, the anthropogenic contribution to the total loading appreciates at a rate of approximately 4% yr^{-1} and if the loading from natural particle sources remains constant. This situation leads us to expect a doubling and redoubling of total loading at the sharply super-exponential rates shown in Figure 4, where the curve labelled $f = 0.2$ refers to the man-made fraction of total loading and that labelled $f = 0.3$ refers to the man-made plus man-triggered fractions of total loading (from Table I). It is rather obvious from Figure 4 that, if the anthropo-genic contributions to total loading are anywhere near the fractions estimated in this paper, the changes of total loading to date from human activities would rather easily

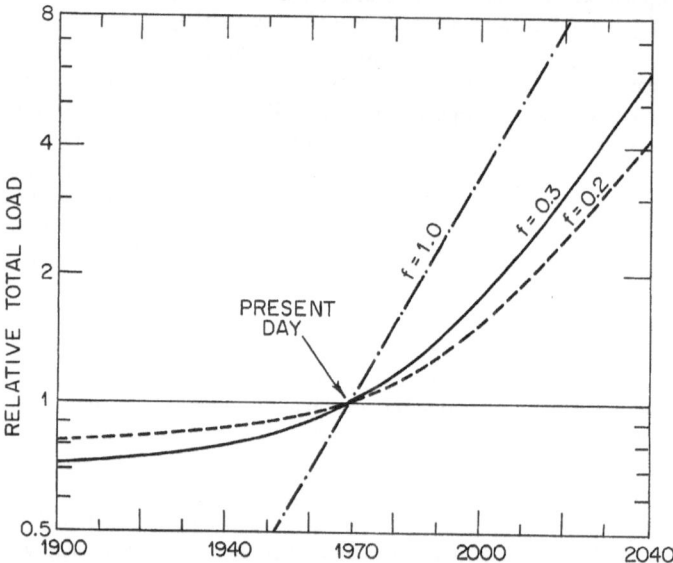

Fig. 4. Relative growth of total atmospheric loading by particles, assuming that the anthropo genic contribution to total loading increases at a uniform rate of 4% yr^{-1}. Parameter f represents the fraction of total loading from anthropogenic sources in 1970, assumed equal to 0.2 for man-made sources and 0.3 for man-made plus man-triggered sources. Curve for $f = 1$ is added for comparative purposes only.

have escaped detection until very recent years, and any climatic impacts of these changes would be unlikely to have become noticeable until after 1940. By the same token, future increases of total loading would soon accelerate to a readily observable extent, and any climatic impacts attributable to changes of atmospheric particle content would be likely to become rapidly more significant in future decades.

As a prelude to any consideration of the possible climatic impact of the long-term growth of particle loading (as estimated in Figure 4), it is appropriate to consider the variability of total atmospheric loading that is attributable to natural phenomena. There can scarcely be any doubt that *all* sources of atmospheric particles identified in Table I are inherently variable in magnitude from year to year. In large part these variations have a meteorological origin, especially in the case of sea salt, wind blown dust, and smoke from grass or forest fires. To some extent, variations may also arise from extra-meteorological sources as in the case of volcanic dust. Unfortunately, no reliable information exists as to variations of total atmospheric loading from all natural sources combined, which is one reason that the estimation of the anthropogenic changes from long-term dustfall and turbidity observations has proven so difficult. On the other hand, large and relatively rapid changes of loading by volcanic dust, injected into the stratosphere by explosive volcanic eruptions, are readily detectable in available turbidity observations throughout the world. Rough estimates of the mass of volcanic dust in the stratosphere can be derived either from studies of the unique optical effects of dust in the upper atmosphere [32, 33] or from inferences drawn through available information about the eruptions themselves [32, 34].

Previously I attempted to construct a chronology of stratospheric dust loading by volcanic activity between 1850 and 1969 [34]. The chronology was developed on the basis of information about forty explosive eruptions in the period listed by Lamb [32], together with an order of magnitude estimate of the tonnage of fine dust injected into the stratosphere by each eruption, and an assumed residence time of 14 months for this dust. (For further details concerning the method of construction of this chronology, the reader is referred to my earlier paper [34].) After being updated through 1972, the volcanic chronology is reproduced here in Figure 5.

For comparison with the volcanic dust chronology since 1850, two other curves have been added to Figure 5. One (the upper horizontal line) represents the total particle loading of the troposphere contributed by natural phenomena, estimated earlier as about 3×10^7 tons, which for lack of better information I take to be constant in the period. The other (heavy rising curve in the lower part of the figure) represents the long-term growth of total atmospheric particle loading contributed by human activities, where it is assumed that the particle loading from anthropogenic sources has been appreciating at the same relative rate as the CO_2 source strength to a present-day total of about 1×10^7 tons.

Some interesting conclusions follow from Figure 5. To the extent that the various contributions to total atmospheric particle loading have been realistically estimated in the figure, it follows that:

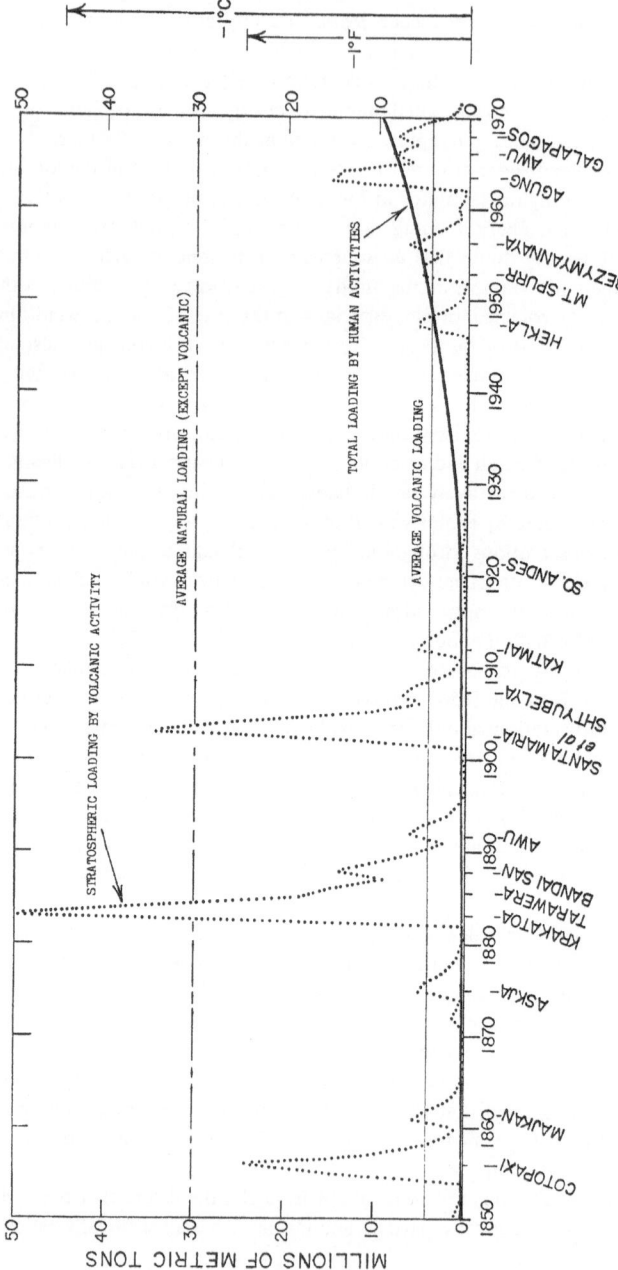

Fig. 5. Estimated chronology of annual average global atmospheric particle loading, 1850 to 1972, by volcanic activity (stratospheric loading only, dotted curve), by human activities (heavy solid curve), and by all natural sources other than volcanic (assumed constant at 30 million tons, dash-dotted line). Estimated calibration of volcanic loading curve in terms of planetary temperature influence is shown at right.

(1) On certain relatively infrequent occasions during the past century volcanic activity has been responsible for temporary increases of total loading to levels roughly double the average loading. The largest increase (since 1850) followed the eruption of Krakatoa in 1883, and the most recent increase of note followed the eruption of Gunung Agung (Bali) in 1963.

(2) At the present time, the total particle loading from anthropogenic sources is evidently greater than the *average* stratospheric particle loading by volcanic dust during the past century.

(3) Since 1940, after which the recent world-wide cooling trend set in, roughly comparable increases of atmospheric particle loading have been contributed by quickening volcanic activity and by human activites. The volcanic activity increase, however, has been much more irregular than the anthropogenic increase since 1940. At present (1972) the volcanic loading has temporarily fallen to negligible levels, notwithstanding two recent eruptions (those of Alaid, Kamchatka, in 1972 and Helgafjell, Iceland, in progress at this writing) which have contributed very little additional dust to the stratosphere.

In connection with the changes of total atmospheric particle loading in the most recent years, it is noteworthy that two recent re-evaluations of the Mauna Loa record of solar radiation since 1958, which in its original form had seemed to confirm a systematic degradation of atmospheric transparency in the period, now shed a very different light on the matter [9, 35]. The results of both of these new studies, despite the altogether different methods of analysis on which each was based, are consistent in showing a clearly detectable and sudden loss of atmospheric transparency after the 1963 Gunung Agung eruption followed by a gradual recovery that brought the transmission back to pre-Agung levels by 1971 (see Figure 6). The recovery since 1963 has been much slower than a consideration of stratospheric residence times of volcanic

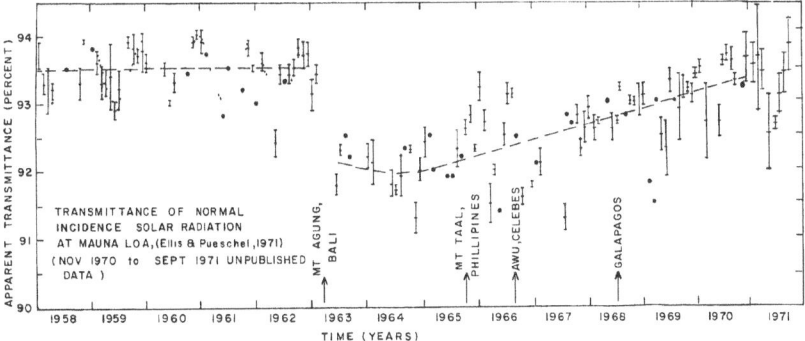

Fig. 6.   Changes of apparent atmospheric transparency, 1958 to 1971, at Mauna Loa Observatory. Data shown as monthly means with associated sampling variability (vertical bars). Important volcanic eruptions are indicated. After Ellis and Pueschel [26] and Machta [13].

dust would lead one to expect, and may be a result either of an underestimate of the residence time appropriate to Gunung Agung or of the addition of dust from later eruptions identified in Figure 6. In any event, the Mauna Loa record is consistent with the changes of global volcanic dust loading in the same period as indicated in Figure 5, but interestingly enough it shows no additional trend that might be associated with the anthropogenic increase of atmospheric loading also indicated in Figure 5. The failure of the Mauna Loa record to reflect the anthropogenic increase can be interpreted in three ways: (1) The anthropogenic increase is undetected at Mauna Loa because it is mostly confined to layers of the atmosphere beneath the 3500 m altitude of the Mauna Loa station; (2) the sources of the anthropogenic increase are geographically so remote that, in line with the findings of Cobb [28], most particles do not remain airborne long enough to reach Mauna Loa; and/or (3) the global particle loading from anthropogenic sources has not increased as rapidly since 1958 as I show in Figure 5. In abeyance of further studies to clarify the situation, I tentatively suggest that the second of these interpretations is the most reasonable one.

To summarize, the following conclusions with respect to the particle loading of the atmosphere are to be emphasized:

(1) The total atmospheric loading by particles smaller than 5 μm in diameter, of primary concern to thermal conditions of global climate, is estimated to aggregate to roughly 4×10^7 tons at the present time.

(2) Of this loading, about 90% on the average is confined to the troposphere. The remaining fraction, about 10% on the average, is contained in the stratosphere and consists primarily of volcanic dust together with sulfate particles in the Junge layer [2] (which may also derive largely from volcanic activity). The stratospheric loading may vary by as much as two orders of magnitude from year to year depending on the timing of important volcanic injections.

(3) Of the total loading, roughly 1×10^7 tons (25%) represents the loading maintained at the present time from sources directly or indirectly associated with human activities. This anthropogenic loading is assumed to have grown by about 4% yr^{-1} since 1940 (and at variable rates before 1940), in which case it would have tripled between the early 1940s and the present.

(4) If the anthropogenic particle loading continues to grow at 4% yr^{-1} in the future, the *total* atmospheric loading from all sources will increase above its present (circum 1970) level about 10% by 1980, about 30% by 1990, and about 60% by 2000 A.D. Relative to the mean total loading in the 19th century (assumed to have been 3×10^7 tons), a doubling to 6×10^7 tons could reasonably be expected by the end of the century unless effective measures can be instituted before then to reduce both particulate and SO_x emissions to the global atmosphere.

5.2. ESTIMATION OF THE THERMAL EFFECT OF VARIABLE ATMOSPHERIC PARTICULATE LOADING DUE TO MAN'S ACTIVITIES

It is now well established from both theory and observation that atmospheric particles of the order of 0.1 to about 5 μm in diameter are of special concern to the heat balance of the Earth, since these are relatively abundant in the atmosphere and are highly effective in scattering, absorbing, and attenuating solar radiation (see, e.g. [1, 29, 36]). On the other hand, particles in this size range have a relatively small impact on terrestrial long-wave radiation (e.g. [36, 37]) – small enough to be neglected for purposes of this discussion. Larger sized particles that might interfere more strongly with long-wave radiation, much less abundant other than locally in duststorms or near urban/industrial centers, can also be neglected for purposes of this discussion.

In the meteorological literature concerned with the climatic impact of atmospheric particles, it has commonly been assumed in the past that the interaction of the particles with solar radiation (and with the thermal balance of the atmosphere) is confined more or less entirely to pure scattering effects [22]. Recently, however, it has become apparent that the absorption efficiencies of at least some kinds of atmospheric particles are sufficiently large [37, 38] that the role of absorption cannot be neglected either for the purpose of interpreting the source of solar radiation attenuation or for that of estimating the net impact of particles on atmospheric temperature [39, 40, 41]. Unfortunately the relative efficiencies of absorption and backscattering of solar radiation by particle populations commonly found in the atmosphere are neither easily measured by direct instrumental means nor very reliably inferred from theory [39]. In applicable Mie scattering theory, several physical properties of the particles must be known which are likewise difficult to measure, including the complex index of refraction (both real and imaginary parts), the particle size spectrum, and the particle shape (if other than spherical).

The two kinds of particle effects on solar radiation (backscatter and absorption) have competing impacts on atmospheric temperature. Both effects deny a (normally small) part of the incoming solar radiation from reaching the vicinity of the Earth's surface. The part denied to the surface by backscatter represents energy reflected back by the particles toward space and therefore irretrievably lost as a source of heating of the Earth-atmosphere system. The part denied to the surface by absorption, on the other hand, represents energy absorbed by the particle-bearing layers of the atmosphere, and thereby heats the atmosphere instead of being *available* to heat the Earth's surface. Thus, the total energy *available* for heating the Earth-atmosphere system tends always to be reduced by the presence of atmospheric particles, provided only that the particles are capable of some backscattering (as all real particles are). Moreover, the amount by which the total solar energy available for heating is reduced by the particles is smaller if the particles are capable of absorbing as well as backscattering solar radiation.

This picture of particle effects on the net solar heating of the Earth-atmosphere system is incomplete without due consideration of certain properties of the Earth's surface, namely the surface reflectance (or albedo) and the surface water content. These two properties are of paramount importance in determining the extent to which the *available* solar energy denied to the surface by particle backscatter and absorption *would actually have been used* for heating in the absence of the particles, and *where* in the atmosphere such heating would have occurred in the absence of the particles. These matters have been explored quantitatively in two recent papers [40], and lead us to the following principal conclusions:

(1) With regard to stratospheric particles, as for example volcanic dust veils, a net cooling of the Earth's surface and lower atmosphere is to be expected following an increased loading of such particles. This is the case regardless of the extent to which the particles may be absorbers as well as scatterers of solar radiation. The amount of cooling to be expected, as a function of total stratospheric particle loading, has been empirically estimated (by analyses of global average temperature changes following major volcanic eruptions) to be as shown by the temperature scale on the right margin of Figure 5 [5, 6, 34].

(2) With regard to particles introduced into the troposphere, which includes nearly all anthropogenic particles except those from aircraft, the fact that these particles are typically concentrated near the Earth's surface results in a situation that contrasts sharply with that of stratospheric particles. In this case, *particles cannot be assured of having a cooling effect on climate unless it can be established that the particles are considerably less efficient as absorbers of solar radiation than as backscatterers of solar radiation.* Except in certain cases, as for example sulfate particles and wind-blown dust consisting largely of quartz, most particles in the atmosphere are not known to have sufficiently small absorption efficiencies to be assured of having a cooling effect on climate, and may possibly result in a warming instead.

(3) To the extent that atmospheric particles attenuate solar radiation at the Earth's surface (whether through absorption or backscatter), the particles are very likely to reduce the rate of evaporation of water from the surface and for this reason lead to a decrease of average cloudiness and precipitation over the Earth as a whole. The decrease of cloudiness, in turn, could suffice to admit more solar radiation to the surface in which case any initial cooling effect of the addition of particles to the atmosphere would tend to be offset by the warming effect of the radiation increase. It is unlikely, however, that the cloudiness change would exactly offset the initial thermal reaction to particles, and do so in the same geographical areas, so this mechanism cannot be relied upon to argue that global temperature may be wholly insensitive to particle loading changes. Especially if the particle loading is non-uniform over the Earth, as suggested earlier in this paper, geographical inequalities in the radiative effects of particles can perhaps induce large-scale changes of atmospheric circulation, and associated *regional* changes of temperature that are quite unlike the planetary average temperature change. Such possibilities deserve careful study be means of suitable general circulation modeling experiments that should become feasible in the next few years.

The key point to be emphasized here is that the net thermal impact of long-term increases of anthropogenic particle loading cannot be assessed with any reliability from currently available information. While this thermal impact is likely to be in the direction of cooling, it cannot be ruled out that the net impact is either very small or actually one of warming instead. The possibility of a warming effect of particle increases is especially favored when the added particles are concentrated near the

Earth's surface, and when the region of the atmosphere being considered is the lower troposphere over the oceans or over vegetated land areas (e.g. forests, and farmland during the growing season) [40].

Of a number of existing theoretical studies of the dependence of global mean temperature on particle loading increases expected from human activities, probably the best known and most widely cited study is that of Rasool and Schneider [42]. Those authors assumed an idealized atmosphere in which all the added particles are confined to the lowest kilometer of the atmosphere, the thermal lapse rate is held fixed at all levels, the particles are assumed to have relatively small absorbing efficiencies with respect to solar radiation, and no account is taken of evaporation or cloudiness effects. Under these modeling conditions the results they obtain as to the magnitude of the global temperature response to particle loading increases are best viewed as *an upper limit to the expected cooling effect*, especially the temperature response in the near-surface region of the atmosphere sampled by conventional climatological data (such as those used to construct Figure 1). Thus we may combine Rasool and Schneider's analysis with the estimate of anthropogenic loading changes as shown in Figure 4, to draw the following conclusion: Relative to mean global temperature levels prevailing in the 19th century, the long-term increase of total atmospheric particle loading attributable to human activities would be expected to have resulted in a net temperature change of unknown direction or magnitude but bounded on the cooling side by magnitudes of $-0.1\,°C$ by 1930, $-0.2\,°C$ by the early 1940's, $-0.3\,°C$ by 1960, and about $-0.4\,°C$ at the present time (circum 1970). This lower bound of temperature change will decrease in the future to the order of $-0.5\,°C$ by 1980 and to the order of $-1.0\,°C$ by the year 2000 A.D. if the anthropogenic particle loading continues to appreciate at $4\%\ \mathrm{yr}^{-1}$. It is interesting to note that these lower bounds as theoretically derived by Rasool and Schneider are in reasonably good agreement with the cooling effect of equal particle loading changes by volcanic dust in the stratosphere, as calibrated by empirical means [34] and shown by the temperature scale applying to volcanic dust in Figure 5.

6. Conclusions

I conclude by considering in combination the thermal effects of the apparent growth of both the CO_2 and the particle content of the global atmosphere, making suitable provisions for the uncertainties involved in establishing the particle effects. As a point of departure in this synthesis, I adopt the long-term growth of CO_2 content and the calculated temperature response to this growth after Manabe and Wetherald [20], both shown in Figure 3. I also adopt the long-term growth of total particle loading of the atmosphere inferred from Figures 4 and 5, together with the calculated temperature response after Rasool and Schneider [42] which I have argued should be interpreted as an estimate of the *maximum* cooling effect of the particle loading changes from anthropogenic sources. This results in the two curves of temperature change (normalized to 19th century levels) shown in Figure 7. The curve of the observed temperature

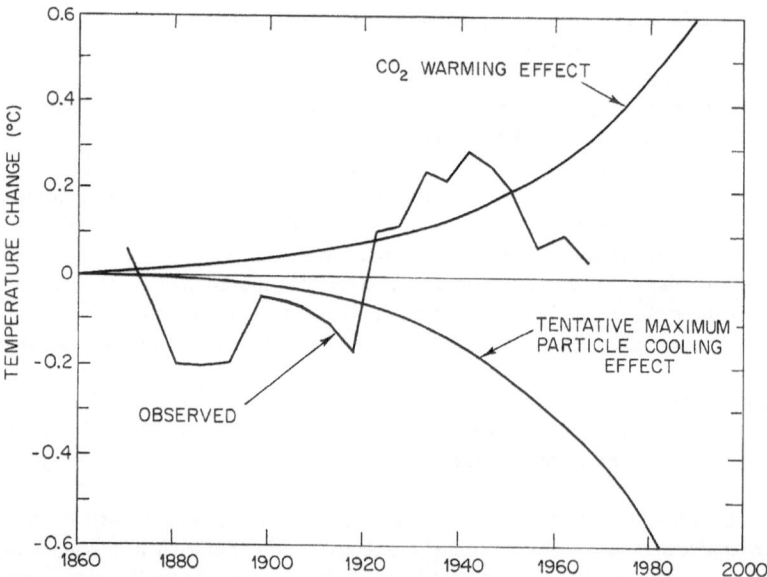

Fig. 7.   Trends of mean temperature, 1860 to 2000 A.D. Upper smooth curve represents estimated warming contribution by carbon dioxide, assuming growth of CO_2 as shown by dashed curve in Figure 3 and temperature response as calculated by Manabe and Wetherald [20]. Lower smooth curve represents probable maximum cooling contribution by total atmospheric particle loading changes, assuming growth of particles as shown by curve for $f = 0.3$ in Figure 4 and temperature response as calculated by Rasool and Schneider [42]. Broken curve is the observed temperature change in the Northern Hemisphere, equivalent to the curve for 0–80°N in Figure 1 corrected for local warming effects of urban growth around some of the stations whose data were used in constructing the curve.

change for the Northern Hemisphere, taken from Figure 1 after correction for what is believed to be the magnitude of local urban growth effects on the observations, is also shown in Figure 7.

It is seen in Figure 7 that until recent years neither pollution effect was apparently sufficient to account for an important part of the observed fluctuation of temperature during the past century. Beginning in the 1930's and 1940's, either pollution effect by itself might have sufficed to make an appreciable contribution to the observed temperature change, but if the two effects (CO_2 and particles) have been additive it is interesting to note that the *joint* temperature effect of both pollutants would still have been negligible. By the 1950's and 1960's, it is seen that either pollution effect alone had begun to induce temperature changes that are comparable in magnitude to the observed change, but still the joint temperature effect of both pollutants (i.e. the average of the two effects) remained very small in the period.

The tendency for a balancing of each pollution effect by the other is indicated in

Figure 7 to extend right up to the present time. However, even if the two pollutant effects on temperature are accurately described by the curves in Figure 7, it is clear that there is little room for complacency about the longer-range future. Should further inputs of both CO_2 and particles to the atmosphere grow in the future at similar per annum rates, as tentatively projected in this paper, then the particle effect on temperature must sooner or later dominate over the CO_2 effect and the fortuitously close balance between the two effects which may have prevailed in the past cannot continue to prevail much longer. Indeed, already by 1980 it appears from Figure 7 that the maximum particle cooling effect on the atmosphere will exceed the CO_2 warming effect by about 0.1 °C, and that this gap may widen to 0.2 °C or more by the end of the century. More important, however, is the fact that the temperature effects of the two pollutants, as well as the future growth of atmospheric loading by these pollutants, have been only grossly estimated in this paper. Therefore, the results shown in Figure 7 are at best crude approximations to reality, and the tendency for the two pollution effects to grow in the nearly equal and opposite directions shown should not be taken too literally.

It should be emphasized once more that the particle cooling effect evaluated by Rasool and Schneider may best be viewed as an upper limit to the actual cooling possibly caused by real particles in the real atmosphere. It is possible, as I noted earlier, that the net effect of particles on the temperature of the lower atmosphere may be one of warming rather than cooling. Thus, the combined effect of both CO_2 and particles on average temperature is quite likely to be one of warming.

Eventually the input of CO_2 into the atmosphere will diminish as the reserves of fossil fuels are drawn down and other sources of energy (e.g. nuclear and solar energy) take their place. Presumably this turning point will be reached some time after 2000 A.D. Thereafter, the CO_2 content of the atmosphere should begin a decline and its climatic warming effect go into reverse. By that time, however, direct thermal pollution of the atmosphere will very possibly have grown to such an extent that our planetary climate will tend to be maintained at relatively warmer levels than to-day's.

The cooling trend of global climate, which began in the 1940's and which is presumably continuing today at least in the higher latitudes of the Northern Hemisphere, appears from this analysis probably to have a natural origin. The cause of the cooling is not unlikely to be related to the acceleration of volcanic activity during the past quarter century, which has maintained a relatively high stratospheric dust loading following a period of 30 yr (1915 to 1945) in which the loading from volcanic sources had been well below average. Owing to possible thermal lag effects of the oceans [43], the cooling may persist for a time even if new explosive volcanic eruptions do not occur in the near future to renew the stratospheric dust reservoir. Whether attributable to volcanic activity or otherwise, it is to be expected that natural climatic fluctuations will continue to occur in future decades – just as they have always occurred in the past – and add considerable 'noise' to the future climatic changes wrought by human activities.

J. MURRAY MITCHELL JR.

References

1. U.S. Public Health Service: *Air Over Cities*, SEC Tech. Report A62-5, Cincinnati (1962).
   U.S. Public Health Service: *Air Quality Criteria for Particulate Matter*, NAPCA Publ. No. AP-49, U.S. Government Printing Office (1969).
   Robinson, E.: in A. C. Stern (ed.), *Air Pollution*, Academic Press, N.Y., Vol. I, pp. 349–400 (1968).
2. Junge, C. E.: *Air Chemistry and Radioactivity*, Academic Press, N.Y., p. 228 (1963).
   Robinson, E. and Robbins, R. C.: *Sources, Abundance and Fate of Gaseous Atmospheric Pollutants*, American Petroleum Institute, N.Y. (1968).
3. Lamb, H. H. and Johnson, A. I.: *Geogr. Ann.* **41**, 94–134 (1959); *ibid.* 43, 363–400 (1961).
   Lamb, H. H.: in H. Flohn (ed.), *World Survey of Climatology*, Vol. 2, Elsevier Publishing Company, Amsterdam, pp. 173–249 (1969).
   Lamb, H. H.: *Climate, Present, Past and Future*, Vol. 1, Methuen, London/Barnes and Noble, N.Y. (1972).
4. Mitchell, J. M., Jr.: in H. E. Wright and D. G. Frey (eds.), *The Quaternary of the United States*, Princeton University Press, pp. 881–901 (1965).
   Mitchell, J. M., Jr.: in R. W. Fairbridge (ed.), *Encyclopedia of Atmospheric Sciences and Astrogeology*, Reinhold Publishing Co., pp. 211–213 (1967).
5. Mitchell, J. M., Jr.: *Ann. N.Y. Acad Sci.* **95**, 235–250 (1961).
   Mitchell, J. M., Jr.: in *Changes of Climate, Arid Zone Research*, **XX**, UNESCO, Paris, pp. 161–181 (1963).
   Wilson, C. L. *et al.* (eds.), *Inadvertent Climate Modification* (SMIC Report), MIT Press, Cambridge, Mass., pp. 40–45 (1971).
6. Reitan, C. H.: *An Assessment of the Role of Volcanic Dust in Determining Modern Changes in the Temperature of the Northern Hemisphere*, Ph.D. Thesis, Univ. of Wis.-Madison, Xerox University Microfilms, Ann Arbor, Mich. (1971).
7. Emiliani, C.: in K. K. Turekian (ed.), *The Late Cenozoic Glacial Ages*, Yale University Press, New Haven, pp. 183–197 (1971).
8. Machta, L.: in D. Dyrssen and D. Jagner (ed.), 'The Changing Chemistry of the Oceans', *Nobel Symp.* 20, Almqvist and Wiksell, Stockholm, John Wiley and Sons, N.Y., pp. 121–145 (1972).
9. Machta, L. and Telegadas, K.: in W. N. Hess (ed.), *Weather Modification*, John Wiley and Sons, N.Y. (in press) (1973).
10. Callendar, G. S.: *Tellus* **10**, 243–248 (1958); C. E. Junge in [2].
11. Keeling, C. D.: personal communication.
12. Pales, J. C. and Keeling, C. D.: *J. Geophys. Res.* **70**, 6053–6075 (1965).
    Brown, C. W. and Keeling, C. D.: *J. Geophys. Res.* **70**, 6077–6085 (1965).
13. Machta, L.: personal communication.
14. Kelly, J. J., Jr.: *An Analysis of Carbon Dioxide in the Arctic Atmosphere near Barrow, Alaska, 1961 to 1967*, Scientific Report, Dept. of Atmospheric Sciences, University of Washington (1969).
    Bolin, B. and Bischof, W.: *Tellus* **22**, 431–442 (1970).
    Bischof, W.: *Summary of Recent Carbon-Dioxide Measurements in the Atmosphere*, Summary Letter 11/1813, Institute of Meteorology, University of Stockholm (1971).
    Keeling, C. D., Adams, J. A., Ekdahl, C. A., and Guenther, P. R.: *Atmospheric Carbon Dioxide Variations at the South Pole, 1957–1970* (in preparation) (1972).
15. Revelle, R. and Suess, H. E.: *Tellus* **9**, 18–27 (1957).
16. Keeling, C. D.: *Industrial Production of Carbon Dioxide from Fossil Fuels and Limestone* (to appear in *Tellus*) (1973).
17. Rotty, R. M.: personal communication.
18. Pytkowicz, R. M.: *Fossil Fuel Burning and Carbon Dioxide – A Pessimistic View* (to appear in *Comments on Earth Sciences: Geophysics*, Gordon and Breach, N.Y.) (1973).
    Keeling, C. D., Bacastow, R., and Ekdahl, C. A.: *Diminishing Role of the Oceans in Industrial CO$_2$ Uptake During the Next Century* (in preparation) (1973).
19. Bolin, B. and Eriksson, E.: in B. Bolin (ed.), *The Atmosphere and Sea in Motion*, Rockefeller/Oxford Press, pp. 130–142 (1959); C. E. Junge in [2].
20. Manabe, S. and Strickler, R. F.: *J. Atmos. Sci.* **21**, 361–385 (1964).
    Manabe, S. and Wetherald, R. T.: *J. Atmos. Sci.* **24**, 241–259 (1967).
21. Wetherald, R. T.: personal communication.

22. Rasool, S. I. and Schneider, S. H.: *Science* **173**, 138–141 (1971).
    Barrett, E. W.: *Solar Energy* **13**, 323–337 (1971).
    Bryson, R. A. and Wendland, W. M.: in S. F. Singer (ed.), *Global Effects of Environmental Pollution*, Springer-Verlag N.Y./D. Reidel Publ. Co., Dordrecht, pp. 130–138 (1970).
    Budyko, M. I.: *Tellus* **21**, 611–619 (1969).
    McCormick, R. A. and Ludwig, J. H.: *Science* **156**, 1358–1359 (1967).
23. Peterson, J. T.: *The Climate of Cities, A Survey of Recent Literature*, National Air Pollution Control Admin., Publ. No. AP-59 (1969).
    Ludwig, J. H., Morgan, G. B., and McMullen, T. B.: *Eos*. **51**, 468–475 (1970).
24. Davitaia, F. F.: *Trans. Soviet Acad. Sci.*, Geogr. Ser., No. 2, pp. 3–33 (1965).
25. Cobb, W. E. and Wells, H. E.: *J. Atmos. Sci.* **27**, 814–819 (1970).
26. McCormick, R. A. and Ludwig, J. H.: *Science* **156**, 1358–1359 (1967).
    Ellis, H. T. and Pueschel, R. F.: *Science* **172**, 845–846 (1971).
    Pueschel, R. F., Machta, L., Cotton, G. F., Flowers, E. C., and Peterson, J. T.: *Nature* **240**, 545–547 (1972).
    Peterson, J. T. and Bryson, R. A.: *Science* **162**, 120–122 (1968).
27. Hodge, P. W., Laulainen, N., and Charlson, R. J.: *Science* **178**, 1123–1124 (1972).
28. Cobb, W. E.: *J. Atmos. Sci.* **30**, 101–106 (1973).
29. Wilson, C. L. *et al.* (eds.), *Man's Impact on the Global Environment* (SCEP Report), MIT Press, Cambridge, Mass. (1970).
    Wilson, C. L. *et al.* (eds.), *Inadvertent Climate Modification* (SMIC Report), MIT Press, Cambridge, Mass. (1971).
    Matthews, W. H., Kellogg, W. W., and Robinson, G. D. (eds.), *Man's Impact on the Climate*, MIT Press, Cambridge, Mass. (1971).
    Flohn, H.: personal communication; Machta, L.: personal communication; see also [9].
30. Englemann, R. J. and Slin, W. G. N.: (eds.), *Precipitation Scavenging (1970)*, U.S. Atomic Energy Comm., Oak Ridge, Tenn./National Technical Information Center, Springfield, Va.
    Poet, S. E., Moore, H. E., and Martell, E. A.: *J. Geophys. Res.* **77**, 6515–6527 (1972).
31. Dyer, A. J. and Hicks, B. B.: *Quart. J. Roy. Meteorol. Soc.* **94**, 545–554 (1968); C. E. Junge in [2].
32. Lamb, H. H.: *Phyl. Trans. Roy. Soc. London*, A **266**, 425–533 (1970).
33. Deirmendjian, D.: *Global Turbidity Studies. I. Volcanic Dust Effects – A Critical Survey*, Report R-886-ARPA, Rand Corp., Santa Monica (1971).
    Cronin, J. F.: *Science* **172**, 847–850 (1971).
    Volz, F. E.: *Appl. Optics* **8**, 2505–2517 (1969).
34. Mitchell, J. M., Jr.: in S. F. Singer (ed.), *Global Effects of Environmental Pollution*, Springer-Verlag N.Y./D. Reidel Publ. Co., Dordrecht, pp. 139–155 (1970).
35. Ellis, H. T. and Pueschel, R. F.: *Science* **172**, 845–846 (1971).
36. Bullrich, K.: 1964, in H. E. Landsberg (ed.), *Advances of Geophysics*, Academic Press, N.Y., pp. 99–260 (1964).
    Deirmendjian, D.: *Electromagnetic Scattering on Spherical Polydispersons*, American Elsevier, N.Y. (1969).
    Van de Hulst, H. C.: *Light Scattering by Small Particles*, John Wiley and Sons, N.Y. (1957).
    Plass, G. N.: *Appl. Optics* **5**, 279–285 (1966).
    Braslau, N. and Dave, J. V.: *Effect of Aerosols on the Transfer of Solar Energy through Realistic Model Atmospheres* (to appear in *J. Appl. Meteorol.*), (1973).
37. Robinson, G. D.: *Long-Term Effects of Air Pollution – A Survey*, Center for the Environment and Man, Inc., Hartford, pp. 21–24 (1970).
38. Eiden, R.: *Appl. Optics* **5**, 569–575 (1966); *ibid.* **10**, 749–754.
    Fischer, K.: *Beitr. Physik Atmosphäre* **43**, 244–254 (1970).
39. Ensor, D. S., Porch, W. M., Pilat, M. J., and Charlson, R. J.: *J. Appl. Meteorol.* **10**, 1303–1306 (1971).
40. Mitchell, J. M., Jr.: *J. Appl. Meteorol.* **10**, 703–714 (1971).
    Mitchell, J. M., Jr.: in *Proc. Internat. Symposium on Physical and Dynamical Climatology*, Leningrad, 1971, World Meteorological Organization, Geneva (in press) (1973).
41. Yamamoto, G. and Tanaka, M.: *J. Atmos. Sci* **29**, 1405–1412 (1972).
42. Rasool, S. I. and Schneider, S. H.: *Science* **173**, 138–141 (1971).
43. Dwyer, H. A. and Petersen, T.: *J. Appl. Meteorol.* **12**, 36–42 (1973).
    Bryson, R. A. and Wendland, W. M.: this volume, pp. 139–147.

CLOUDINESS AND THE RADIATIVE, CONVECTIVE
EQUILIBRIUM

SYUKURO MANABE

Geophysical Fluid Dynamics Laboratory/ESSA, Princeton, N.J., U.S.A.

Abstract. The dependence of the temperature of the Earth's surface upon the cloud cover at various altitudes is estimated. The effect of contrail on the surface temperature is discussed.

Professor Bryson (pp. 139–147) suggests that contrails may have a significant effect on the climate of the Earth's surface. Professor Schaefer (pp. 177–196) speculates that the increase of the atmospheric turbidity caused by human activity could affect the number of ice crystal nuclei and, accordingly, the cloudiness. With these thoughts in mind, I would like to discuss how clouds affect the temperature of the Earth's surface.

As you know, clouds have the following two radiative effects:

(1) They reflect solar radiation.

(2) They decrease the outgoing terrestrial radiation at the top of the atmosphere (because the temperature of cloud top is usually colder than that of the Earth's surface).

The first effect lowers while the second effect raises the temperature of the Earth's surface.

In order to evaluate the effect of clouds on the temperature of the Earth's surface quantitatively, Manabe and Wetherald [3] performed a series of computations of the radiative, convective equilibrium of the atmosphere with various distributions of cloudiness and with a given distribution of relative humidity. For our study, the reflectivity of solar radiation was assumed to be 20% for cirrus clouds, 48% for middle clouds (As), and 69% for low clouds based upon the study of Haurwitz [2]. For the computation of the flux of terrestrial radiation, low and middle clouds were assumed to be completely black, while cirrus clouds were assumed to be half black (according to Kuhn's recent measurements which show that cirrus are far from black [1]). Since I described how we obtained the radiative, convective equilibrium earlier (pp. 73–77), I shall not explain it any further.

Figure 1 shows the dependence of the equilibrium temperature of the Earth's surface upon cloud cover. According to this figure, an increase of low clouds lowers the temperature of the Earth's surface markedly because they have a large reflectivity and a relatively warm emission temperature. On the other hand, an increase of high clouds could raise the surface temperature because they have a low reflectivity and a low emission temperature. It is noteworthy that the increasing of low clouds by 3% has a comparable effect with the halving of the CO_2 content in lowering the equilibrium temperature of the Earth's surface. (Refer to the discussion on p. 157 concerning the effect of CO_2-content on the climate.)

According to the estimate of Professor Bryson, contrails cover about 0.8% of the

S. Fred Singer (ed.), The Changing Global Environment, 175–176. All rights reserved.
Copyright © 1975 by D. Reidel Publishing Company, Dordrecht-Holland.

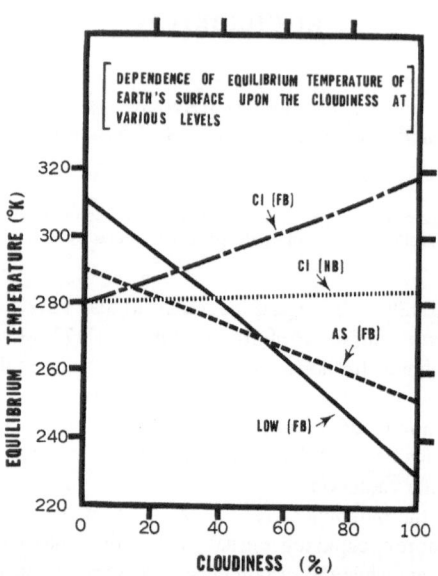

Fig. 1.   Radiative, convective equilibrium temperature at the Earth's surface as a function of cloudiness. (Cirrus, altostratus, low cloud.) FB and HB refer to full black and half black for terrestrial radiation respectively. (By Manabe and Wetherald [3].)

sky. If contrails have the same optical characteristics as Mr. Wetherald and I have assumed for cirrus, the contrail covering 1% of the sky would have a negligible effect upon the temperature of the Earth's surface. But if they are full black for the terrestrial radiation, 1% sky coverage of contrails would raise the equilibrium temperature by about 0.3°C and have a marginally significant effect. Our results are also highly dependent upon the reflectivity of solar radiation assumed for the computation. Therefore, it is necessary to establish the optical characteristics of ice clouds such as cirrus and contrails in order to determine whether or not the radiative effects of contrails are significant in modifying the climate of the earth.

References

1. Kuhn, P. M.: private communication.
2. Haurwitz, B.: 'Insolation in Relation to Cloud Type', *J. Meteor.* **5**, 110–113 (1948).
3. Manabe, S. and Wetherald, R. T.: 'Thermal Equilibrium of the Atmosphere with a Given Distribution of Relative Humidity', *J. Atmos. Sci.* **24**, 241–259 (1967).

General Bibliography

For the recent discussion of this subject, based upon the observation of cirrus cloud emissivities, see:
Cox, S. K.: 'Cirrus Clouds and the Climate', *J. Atmos. Sci.* **28**, 1513–1515 (1971).

THE INADVERTENT MODIFICATION OF THE ATMOSPHERE
BY AIR POLLUTION

VINCENT J. SCHAEFER

Atmospheric Sciences Research Center, State University of New York at Albany, N.Y., U.S.A.

Abstract. Recent measurements indicate that the particulate concentration near urban areas has increased an order of magnitude in the past 10 yr. Much of this increase can be attributed to increased automobile usage, incineration, and electricity generation. Airborne particles are active agents in cloud formation and precipitation. It is strongly indicated that particulate pollution from various sources is modifying the cloud patterns over large areas of the globe. In particular, lead iodide formed by reaction of lead from automobile exhaust with airborne iodine has been shown to be a major source of nuclei for ice crystal formation. Modification of clouds also changes the precipitation patterns and types, and the general weather of an area.

1. Introduction

The subject of my concern at the present time as it relates to global pollution is an aspect of the problem which may affect our welfare sooner than the more obvious deterioration of our environment as it relates to health, esthetics or property damage.

While it is quite possible that a disaster directly related to the build-up of particulate matter and gases in the atmosphere may occur within the next decade unless some major improvements are achieved toward reducing the rapid increase in air pollution now prevalent, I believe that only those already in poor health will be the immediate victims. After all, if one considers that the cigarette smoker insults his lungs with a concentration of particulate matter of at least 10 million particles per cm^3, – which is more than ten thousand times greater than the concentration of particles in country air and more than a hundred times worse than the air of a badly polluted city – one can realize that the human body can stand a considerable amount of physiological abuse. This is not to say we should push our bodies to their limit of compensatory abilities.

However, the phenomenon which concerns me most at the present time is the distinct probability that air pollution has already begun to inadvertently modify the atmosphere in which we live and the climate on which our complex civilization is based.

2. Recent Modification of Our Air Environment

There has been a very noticeable increase in air pollution during the past ten years over and downwind of the several large metropolitan areas of the U.S.A. such as the Northwest – Vancouver-Seattle-Tacoma-Portland; the West Coast from San Francisco-Sacramento-Fresno-Los Angeles; the Front Range of the Rockies from Boulder-Denver-Colorado Springs-Pueblo; the Midwest – Omaha-Kansas City-St.

S. Fred Singer (ed.), The Changing Global Environment, 177–196. All rights reserved.

Louis-Memphis; the Great Lakes area of Chicago-Detroit-Cleveland-Buffalo; and the Northeast – Washington-Philadelphia-New York-Boston. The worst accumulation of particulate matter occurs at the top of the inversion which commonly intensifies at night at levels ranging from 1000 to 4000 ft or so above the ground. This dense concentration of air-suspended particles is most apparent to air travellers. Thus, it has not as yet disturbed the general public except during periods of stagnant weather systems when the concentration of heavily polluted air extends downward and engulfs them on the highways, at their homes and in their working areas.

Until recently there is little question that except in very exceptional cases, natural processes dominated the mesoscale weather systems by initiating the precipitation mechanism. The effluent from the larger cities was quickly diluted by the surrounding 'country air' so that at a distance a few miles downwind of a city, little evidence of air pollution could be detected.

The recent spread of urban developments due to better roads and the massive proliferation of people and automobiles has led to a nationwide network of county, state, and interstate highways. This interconnection of thousands of smaller towns with large cities and the phenomenal increase in auto, truck and air traffic has caused a massive reduction in the regions which have 'country' type air. This increase in massive air contamination is of fairly recent origin. It is not easy to document this fact in the detail I would prefer since we have not had reliable automatic recording equipment for measuring Aitken, cloud and ice nuclei until the last few years. However, using simpler devices with which we made measurements at a number of scattered locations during the past 10 yr, we have in the past year used the same techniques to make comparative observations. The measurements indicate an increase in airborne particulates at these sites of a least an order of magnitude during this 10 yr period. At Yellowstone Park in the wintertime, which has the cleanest air we have found in the continental United States, the background levels of Aitken nuclei have increased from less than 100 to the 800–1000 per milliliter range within a 5-yr span. At Flagstaff, Arizona where in 1962 the background levels ranged from 100–300 the concentration now lies between 800–3000. At Schenectady, New York, the average concentration of these nuclei has risen from less than 1000 to more than 5000 with values occasionally exceeding 50000 ml^{-1}.

While it is difficult to ascribe these increases to any one cause, it is obvious that the increased demand for electric power, the large increase in garbage and trash incineration, and the automobile, are likely to represent the major sources of increased pollution, especially since many industrial plants have been forced to reduce their pollution due to more rigorous regulations.

Just as it is not easy to place the blame for increased air pollution specifically on the power plants, incinerators and automobiles, it is equally difficult to demonstrate clean-cut or unequivocal atmospheric modification to these sources. I am confident that in time there will be ample proof of these effects which are now inadvertently modifying the atmosphere.

The presence of high concentrations of visible as well as invisible particulates above and downwind of our cities produces a heat island effect as real as a sun-drenched Arizona desert or a semitropical island in the Carribean.

Those cities like Boston, New York and Philadelphia which are not affected by geographic barriers as are Los Angeles, Salt Lake City, or Denver are able to get rid of much of their effluent whenever the wind blows. Their plumes of airborne dirt extend as visible streamers for many miles downwind of the source areas. In the case of the metropolitan New York City-northeastern New Jersey complex, these plumes will be found in the upper Hudson Valley, in southeastern New England or over the Atlantic Ocean.

Commerical airline pilots flying the Atlantic are often able to pick up these pollution plumes hundreds of miles at sea. Hogan recently obtained data which provide a quantitative measurement of the New York effluents near the surface of the Atlantic between the U.S.A. and Europe. This same paper [1] amply demonstrates a similar zone of air pollutants being exuded over the seas surrounding Europe, the British Isles, and the east and west coast of the United States.

3. Properties of Maritime and 'Country' Air

We have known for 20 yr that maritime air is characterized by low levels of both cloud droplet and Aitken nuclei. Vonnegut showed [2] by a very simple experimental device that about 50 effective nuclei at lower water saturation droplet formation existed on the upwind coast of Puerto Rico where the trade wind clouds are seen. We were all much surprised when we established the nature of trade wind clouds during our research flights near Puerto Rico in 1948 [3]. Following these activities, I pointed out [4] the large difference noticeable even then between the 'raininess' of the clouds upwind of the island and those which formed over the land after entraining the polluted air from San Juan, the sugar fields and refineries, the cement mills and the myriads of charcoal pits which dotted the island, each sending out its plume of bluish smoke. In our studies in the vicinity of Puerto Rico we observed that in many instances trade wind clouds would start raining by the time the clouds had a vertical thickness of not more than a mile while those over or immediately downwind of the island often reached three times that thickness without raining.

During a continent-wide flight over a large area of Africa I found [5] an even more spectacular effect of inadvertent cloud seeding. As a result of the massive bush and forest burning initiated by the inhabitants preceding the onset of the rainy season, huge cumulus clouds, some of them reaching a height of more than 35000 ft (vertical thickness 4–5 miles) were observed which were not producing any rain. Instead the clouds grew so high that very extensive ice crystal plumes hundreds of miles long extended downwind of the convective clouds. No evidence of glaciation was observed in the side turrets of the clouds indicating a deficiency of ice nuclei at temperatures warmer than the homogeneous nucleation temperature of $-40\,^{\circ}$C. Thus it appeared that the precipitation process was being controlled almost entirely by coalescence and that so many cloud droplet nuclei were being entrained into the clouds from the fires

VINCENT J. SCHAEFER

below, that the coalescence process was impaired so that no rain developed. If ice nuclei were present, they were probably deactivated by the high concentration of smoke particles and gases flowing into the base of the clouds. Similar effects have been observed on a smaller scale in the Hawaiian Islands. During the trade wind cloud regime, clouds which form over sugar cane fields when they are burned prior to harvest are actually larger than the surrounding clouds but they have never been observed to rain even though smaller ones nearby produce showers. Warner more recently has documented such observations [6].

A further observation of secular change in the microphysics of clouds has been observed in the vicinity of large cities during airplane flights through convective clouds. The observations I have noted in particular were made in commercial twin engine planes over the past 10 yr. Of recent years it has been noticed that such clouds often have so many cloud droplets in them that visibility is restricted so much that the engine is hardly visible. In my earlier observations I can never recall being in clouds so opaque that the wing tips could not be seen. Several of my colleagues have reported similar experiences.

Perhaps the most impressive field evidence of inadvertent weather modification is the overseeding of supercooled clouds which is readily observed over and downwind of our northern cities in the wintertime.

4. Ice Crystals from Polluted Air

Although I have been observing such phenomena for more than 10 yr, the effect was brought to my attention in a vivid way during a flight from Albany to Buffalo on December 20, 1965. After flying above a fairly thin deck of supercooled stratus clouds downwind of the Adirondack Mountains, I noted a massive area of ice crystals above and downwind of Rochester, New York. The crystals were so dense that the reflection from the undersun* was dazzling as illustrated in Figure 1. Since that time I have observed similar high concentrations of crystals at low level above and downwind of most of the large northeastern cities such as New York, Albany, Utica, Syracuse, and Buffalo as well as Detroit, Chicago, Sacramento and Los Angeles. In all instances the ice crystals were observed at low level (below 5000 ft above the ground in most instances), and extending for at least 50 miles downwind of the city sources and without cirrus clouds above the areas affected. In a few instances when the plane passed through the crystal area, I observed the particles to be like snow dust, though in a number of instances after landing I observed very symmetrical though tiny hexagonal crystals drifting down from the sky.

---

* The undersun is an optical phenomenon caused by the specular reflection of the Sun from the surfaces of myriads of hexagonal plate ice crystals. In order to produce an undersun, it is necessary for the crystals to consist of smooth-surfaced plates which float with their long axes horizontal to the ground. They thus act as many tiny mirrors. If the crystals were not hexagonal plates but rather prisms, the optical effects would include under parhelia and other reflections which are well known and have been related to crystal types during our winter studies at Yellowstone Park.

Fig. 1.   Undersun photographed downwind of Rochester, New York December 20, 1965.

5. Misty Rain and Dust-Like Snow

For the past several years I have also been observing a number of strange snow and rain storms in the Capital District area in the east central part of New York State. These storms consist of extremely small precipitation particles. When in the form of snow, the particles are like dust having cross sections ranging from 0.02 cm (200 μ) to 0.06 cm (600 μ). When in the form of droplets, they often are even smaller in diameter, at times being so tiny that they drift rather than fall toward the Earth. When collected on clean plastic sheets, the precipitation is found to consist of badly polluted water.*

It is a well-known fact that precipitation 'cleanses the air'. In the past much of this cleansing action has been ascribed to the sweeping up of suspended aerosols by rain and snow. Little attention has been given to the possibility that submicroscopic particulates from man-made pollution may in fact be initiating and controlling pre-

---

* During the period November 10, 1969 to May 25, 1970, I observed and measured twenty of these unusual precipitation types. The mist or fine particle snow forms in clouds having a vertical thickness of several thousand feet, with little convection but long persistence. They range from layers of stratus to solid decks of strato cumulus. The visibility is restricted and the fine particle concentration tends to be 50 000 cc $^{-1}$ or higher.

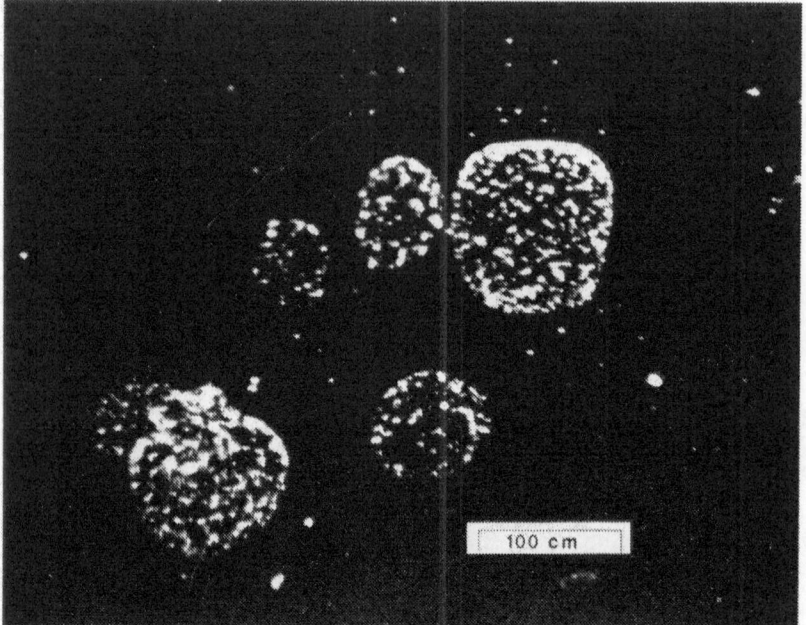

Fig. 2.   Residue remaining on glass slide after polluted, misty rain droplets have evaporated.

cipitation in a *primary manner* rather than being involved in the secondary process wherein precipitation elements coming from 'natural' mechanisms serve to remove the particles by diffusion, collision and similar scavenging processes.

My first evidence that there might be substances in urban air which would react with other chemicals was encountered while studying ice nucleation effects at the General Electric Research Laboratory in 1946 [7]. At that time I found that laboratory air contained aerosols which would react with iodine vapor to form very effective ice nuclei but that when the air was free of particulate matter, no further ice particles would form.

6. Potential Ice Nuclei from Auto Exhaust

In 1966 I published a paper [8] which suggested that air pollution in the form of automobile exhaust could account for the high concentration of ice crystals which I have observed downwind of the larger cities in the United States and in any area where a considerable number of automobiles are used. My laboratory studies have shown that submicroscopic particles of lead compounds produced from the combustion of leaded gasoline can be found at concentrations exceeding 1000 cm^{-3} in auto exhaust. These were measured by exposing auto exhaust samples to a trace of iodine vapor before or after putting the samples into a cold chamber operating at $-20\,^{\circ}$C. Presumably this reaction with iodine formed lead iodide which is an effective subli- mation nucleus for ice crystal formation. Evidence that the active ingredient in auto exhaust consists of submicroscopic particles of lead was determined by comparing its temperature ice nucleation activity pattern with that of lead oxide smoke produced by electrically sparking lead electrodes which was also reacted with a trace of iodine vapor. Figure 3 is a photomicrograph of replicated ice crystals formed on submicro- scopic lead compounds from auto exhaust reacted with iodine vapor. One of the problems related to the evidence that leaded gasoline is reponsible for the ice crystals observed in laboratory and field experiments concerned with auto exhaust is the source of the iodine needed to produce the lead iodide reaction. All evidence thus far en- countered shows that only a few hundred molecules of iodine are required to produce a nucleating zone for ice crystal formation. The amount of iodine reported in oceanic [9] air (the order of 0.5 U G. m^{-3}) is orders of magnitude greater than would be required to activate such particles.

I have recently completed further studies in Arizona, New York and France [10, 11] and have found that wood smoke and other organic sources add iodine to the air

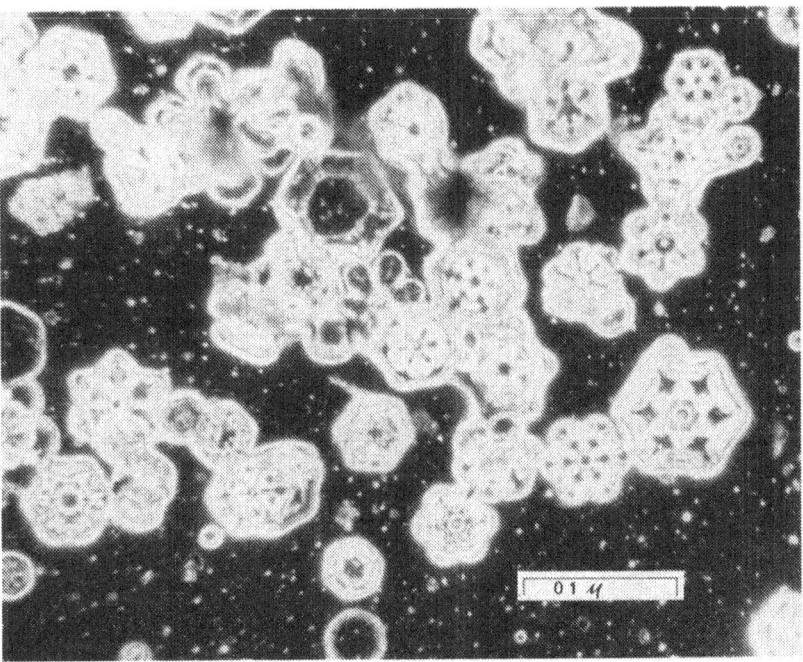

Fig. 3.   Photomicrograph of replicated ice crystals formed at $-16\,^{\circ}$C in a supercooled cloud on nuclei from auto exhaust exposed to trace of iodine vapor.

which could react with the auto exhaust submicroscopic lead compounds which are always present in urban pollution. Hogan has recently showed [12] that similar reactions will proceed from the vapor phase.

Admittedly we are dealing with chemical reactions in the realm of surface and even 'point' chemistry as Langmuir termed such molecule by molecule reactions. This is an area of particulate research for which there is very little experimental data or practical experience. The size of the primary lead particles from auto exhaust which are 0.008–0.010 μ diameter are far too small for analysis by any currently available chemical reaction techniques. All of my laboratory experiments indicate that the submicroscopic particles in auto exhaust which react with iodine vapor act only as nuclei for ice formation from the vapor phase. No evidence has been found that they act as freezing nuclei.

7. The Effect of Large Concentrations of Ice Crystals

The presence of high concentrations of tiny ice crystals in air colder than 0 °C over thousands of cubic miles raises interesting aspects of the dynamics of weather systems. Such crystals continually modify small supercooled clouds soon after they form. The net result is a reduction in the number of local rain or snow showers and the production of extensive sheets of 'false' cirrus. Bryson has pointed out [13] that cirrus sheets and

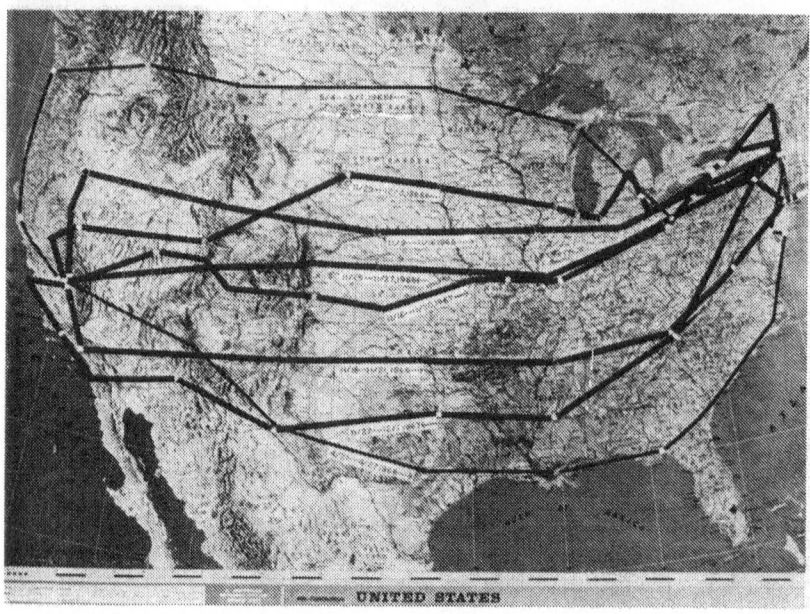

Fig. 4.   Flight routes of Project Air Sample 1–4 1966–1968.

even the presence on a large scale of airborne dust exerts a measurable decrease of insolation. If a much larger supply of moist air moves into such a region, the entrainment of high concentrations of crystals by more vigorous supercooled clouds may trigger the formation of a massive storm through the release of the latent heat of sublimation. Langmuir described [14] such a storm system which he believed was initiated and then intensified when dry ice in successive seeding operations was put into the lower level of a rapidly developing storm.

8. Findings of Project Air Sample

In order to determine whether or not polluted air above cities contained particles which would react with free iodine molecules, ten transcontinental flights have been made under our auspices during the Fall of 1966 and 1967, the Spring and Fall of 1968, and the Spring of 1969. The flight routes are shown in Figure 4. A Piper Aztec aircraft was fitted with instruments which could measure in a semi-quantitative manner the concentration of atmospheric particulates which would become ice nuclei by the reaction with iodine, and which would also measure natural nuclei for ice crystal formation. The iodine reactions were conducted in a cold chamber at $-20\,°C$. The

Fig. 5.   Experimental Laboratory for Cloud Physics and aircraft used for air pollution studies.

determination of naturally occurring nuclei was done at $-22\,°C$. In addition, measurements were also made of Aitken nuclei (a measure of polluted air) and cloud nuclei. This last measurement which is made at very low water saturation is also a measure of the degree of air pollution since values above about 50 cloud nuclei and 500 Aitken nuclei cm^{-3} is indicative of some degree of air pollution. The flight samples were made mostly just below the top of the haze layer which ranged from 1500 to 5000 ft above the ground throughout the flights. Figure 5 illustrates the aircraft used in Project Air Sample flights and the A.S.R.C. Experimental Laboratory. Of the 266 measurements in November 1966, 31 were made on the ground. All of these showed excessive pollution levels. Great care was exercised in making these observations to avoid contamination from the engine exhaust of the aircraft being used for the measurements.

At several locations observations were made above as well as within the upper part of the haze layer. In every instance the air above the visible top of the haze layer was low in lead particles while that just below the top or farther down showed very high concentrations.

All other locations where counts of the ice nuclei were low involved regions free of pollution sources. Of the 266 observations 108 or 40% of the measurements were in areas such as upwind of cities (9); above large lakes (8); above haze layers (22); and

Fig. 6.   Example of pollution which occurs along Lake Erie at Buffalo, New York, November 20, 1967.

woods and farms (33). The 60% remaining had values of potential ice nuclei of 100 1^{-1} or more. Some 115, all of them above or downwind of cities, had values in excess of 200 1^{-1} which I consider would lead to definite overseeding of the atmosphere with ice particles if suitable moisture was available. Values of 100 1^{-1} or more occurred at 101 of the stations. If concentrations of ice crystals that high occurred, the cloud would resemble a stable ice fog such as occurs at Fairbanks, Alaska or the Old Faithful area of Yellowstone Park [15] when the temperatures are colder than $-40\,^\circ$C. With crystal concentrations of this magnitude, the particles grow very slowly if at all and thus remain floating in the air for extended periods. This then reduces the incoming solar radiation to a noticeable degree. If such areas are extensive, they cannot help but cause changes in the weather patterns of the affected areas.

Similar findings characterized our second, third and fourth round-trip transcontinental flights covering more than 25000 additional miles and consisting of over 1500 more observations. In practically every instance where polluted air was present, high values in potential ice nuclei (using the iodine reaction) was found. The only exceptions were instances where the plumes of steel mills, forest fires and other highly concentrated effluents were measured in areas where auto exhausts could contribute very little if anything to the sampled air.

Fig. 7.   Snow showers falling from thin clouds about 80 miles downwind of Buffalo, New York, November 20, 1967.

Figure 6 illustrates the type of pollution which still occurs along Lake Erie at Buffalo and Figure 7 a zone of snow falling from low clouds about 80 miles downwind of Buffalo at a location where the iodine-activated nuclei had dropped from 5000 l^{-1} measured at Buffalo to 500 near Cayuga Lake and the Aitken count from 25000 ml^{-1} to 4500 as measured at 3000 ft above the terrain.

Fig. 8.   Example of severe pollution over New York City, November 23, 1966.

Figure 8 illustrates the fantastic amount of smoke which shrouded the metropolitan New York area on November 23, 1966 when one of the first dangerous smog alerts was sounded by New York City health officials. Just prior to obtaining this picture the airplane was flown up through the smog. At 600 ft the cloud nucleus count was 2000 ml^{-1}, the Aitken count 25000 and the ice nuclei measured were 0 for the natural background and 50000 to 100000 l^{-1} for ice nuclei activated with iodine vapor.

Figure 9 shows the conditions at Albany, New York, on the previous day. At 1200 ft the cloud droplet nuclei numbered 900 ml^{-1}, the Aitken count was 4000, the natural ice nucleus background was 0 but the concentration of ice nuclei activated with iodine was 50000 l^{-1}. These are concentrations which are commonly observed in the air below the top of the inversion over and downwind of all large cities. Although in many instances the smoke concentrations in such areas are not as spectacular as

Fig. 9.   Example of heavy air pollution at Albany, New York, November 22, 1966.

shown in Figure 7 and 8 since a large portion of the smog particles tend to be sub-microscopic, such areas are generally characterized by a veil of brownish gray haze.

9. Flight Observations of Inadvertent Seeding

It is quite feasible to detect and observe the massive systems of ice crystal nuclei which produce inadvertent effects on cloud and weather systems due to man's activities. This is accomplished most easily by riding on the sunny side of a jet aircraft.

I observed and photographed three such systems in 1967 during a flight from Buffalo, New York to Denver, Colorado by way of Chicago, returning directly from Denver to New York City.

9.1. ICE CRYSTALS RELATED TO POLLUTED AIR

On Wednesday, December 6, 1967, I left Buffalo at 1035 by Boeing 727 landing at Detroit and Chicago. Upon take-off I noted a heavy pollution pall over Buffalo extending westward to the horizon. Just west of Buffalo we climbed above stratiform clouds estimated to be at 15000 ft or lower which consisted of very high concentrations of ice crystals as established by an undersun. This extensive zone of ice crystals was observed all the way to Detroit and was associated with visibly polluted air. We flew at 20000 feet where the temperature was $-20\,°C$. Enroute from Detroit to Chicago I found the same condition to exist from the 1108 take-off until 1132 at which time only suspercooled clouds were visible. At the same time all evidence of polluted air disappeared, visibility between cloud decks was unlimited and no further trace of ice crystals could be seen as we landed at Chicago. The air pollution from Chicago was being carried to the southeast over Indiana about thirty miles south of our jet route.

9.2. OTHER SOURCES OF ICE NUCLEI

While I believe that the major source of high concentrations of nuclei for ice crystal formation in polluted air comes from lead particles which are converted to lead iodide, there certainly are other more localized sources. One of the most effective of these anthropogenic sources is the steel mills such as at Gary, Indiana, Buffalo, New York and Pittsburgh, Pennsylvania. Figure 10 is a photo obtained February 9, 1972 of the plumes from the steel mill complex at Gary, Indiana. This picture was obtained under fairly cold conditions, with a northwesterly air flow dominating the area. The moisture producing a supercooled cloud above the stacks of the main plume shifted to ice

Fig. 10.   Formation of large concentrations of snow particles in plume from steel mill complex at Gary, Indiana, February 9, 1972.

crystals within a distance of less than a mile, although a smaller plume adjacent to the main complex remained unaffected.

9.3. THE OCCURRENCE OF FINE PARTICLE PRECIPITATION IN POLLUTED AIR

An extremely interesting precipitation pattern sometimes occurs in moist air that is extensively polluted. I have observed this type of rain in some detail since the occurrence of a devastating ice storm which affected eastern New York from the early morning of December 4 to late at night of December 5, 1964. More than an inch of ice developed on all surfaces during the 48-hour period.

The ice storms of our region commonly occur in conjunction with an occluded front, with the moist air riding up over a wedge of colder air. Such a storm commonly produces sleet, which occurs when rain aloft falls into the colder air below and freezes into sleet. This then shifts to rain which coats trees, roads and other cold objects with a glaze of ice. In most instances the cold air in finally modified or displaced toward the end of the storm, at which time the glaze melts as the warmer air arrives at ground level. In the storm cited, a fine misty rain heralded the beginning of the storm about 0100 December 4, 1964, and continued falling throughout the period. The mist fell for days, finally ending with everything coated by several inches of clear ice. Great damage occurred. During the second night. I heard branches, power poles and lines falling at the rate of ten or more a minute. When the misty rain stopped falling, the cold air remained in the region for nearly a week, aggravating the problem of restoring power, telephone and other utilities.

During the past three years I have continued my observations of low level ice crystals in polluted air above and downwind of cities. Concentrations two to three orders of magnitude higher than in pollution-free air have been seen in scores of observations.

A typical example was observed and photographed on November 21, 1972 during a flight in a small twin-engine plane from the Schenectady County Airport to Wellsville, New York, a distance of about 200 miles. The flight route took us across the Helderbergs, the Schoharie Hills, the sparsely settled valleys south of the Finger Lakes, to our destination which is in the foothills of the Allegheny Mountains.

Shortly after passing the Helderberg escarpment, we encountered a thin but solid deck of clouds which, except for three holes about ten miles in diameter, persisted under us almost to our destination. There were no clouds above us, and the higher air was extremely dry, as was evident by the very short contrails produced by a score or so of jet aircraft which were observed.

The ground temperature was about $-10\,^\circ$C, which cooled to $-14\,^\circ$C at cloud top. The tops of the thin stratocumulus were limited by a strong inversion at 6500 ft, above which the temperature increased to $-8\,^\circ$C. This inversion was at least 1500 ft thick.

As we approached the eastern edge of the cloud shield, an extensive zone of ice crystals was encountered. This region of ice particles did not appear as virga, but rather as a stable area of very small crystals. As we climbed above this region, I observed a brilliant undersun similar to that already described.

As we moved over the cloud deck in a southwesterly course, the undersun disappeared and a large diameter glory was observed around our plane's shadow. At each of the three extensive 'holes' observed enroute to Wellsville, a similar sequence was observed. At the northeastern edge of the hole, a brilliant undersun appeared in the ice crystal area, with the glory no longer visible in the opposite direction. At the same time, the area showed considerable pollution, with a bluish haze extending for several hundred feet above the adjacent cloud tops. As we encountered the southwestern edge of the hole, a glory again appeared across the cloud tops, while the undersun disappeared at the edge of the hole.

Upon reaching the vicinity of Wellsville, the cloud shield ended. As we descended for a landing, I observed the same temperature profile as during our climb from Schenectady, with the inversion lid at the cloud top of 6500 ft and the temperature dropping from $-8°C$ at 7000 ft to $-14°C$ at cloud top, and then warming progressively to $-10°C$ at the surface of the Wellsville airport.

The undersun observed under such conditions (see e.g. Figure 1 and footnote) can only occur if the crystals are very small and flat, consisting of hexagonal plates or stellar forms with concentrations approaching $1000\ l^{-1}$. In most instances they are so small (200–1000 μ in their largest axis), that their falling velocity ranges from 10–30 cm s^{-1}. Thus they often evaporate before reaching the ground. If they reach the Earth's surface, their crystalline form is similar to that illustrated in Figure 3. Fig. 11 illustrates a collection of such crystals observed on January 7, 1966. They often are so small that the amount of snow which accumulates on the ground is hardly noticeable and can best be described as snow dust.

During the course of several hundred seeding experiments conducted in the wintertime at Yellowstone Park, many occurrences of undersuns have been observed in conjunction with the evaluation in seeding effectiveness of various types of seeding agents. To duplicate the brilliance of the undersuns commonly observed in the wintertime at low level in polluted air, it is necessary to have concentrations of at least 200 crystals per liter of air. In some instances the concentrations I have observed have certainly exceeded this amount by at least an order of magnitude. Thus the numbers of ice crystals observed downwind of cities are two to four orders of magnitude higher than is found in clean country air.

9.4. Ice crystals produced by dust from plowed land

Upon take-off at Chicago in a Boeing 707 at 1320 CST, the air was clear, several decks of stratiform clouds were visible with no evidence of ice crystals. Heading west I saw no ice crystals until 1424. Just previous to that time a peculiar zone of dusty air could be seen ahead of us extending toward the southwest. Within a few minutes a brilliant undersun could be seen which persisted for the next half hour. When we finally emerged from the affected zone over northeastern Colorado it was quite obvious that the 300 miles long zone of ice crystals was due to very extensive dust storms caused by 50–100 mph katabatic ground winds pouring down out of the Front Range of the Rockies and blowing top soil from the extensive wheat fields extending from north-

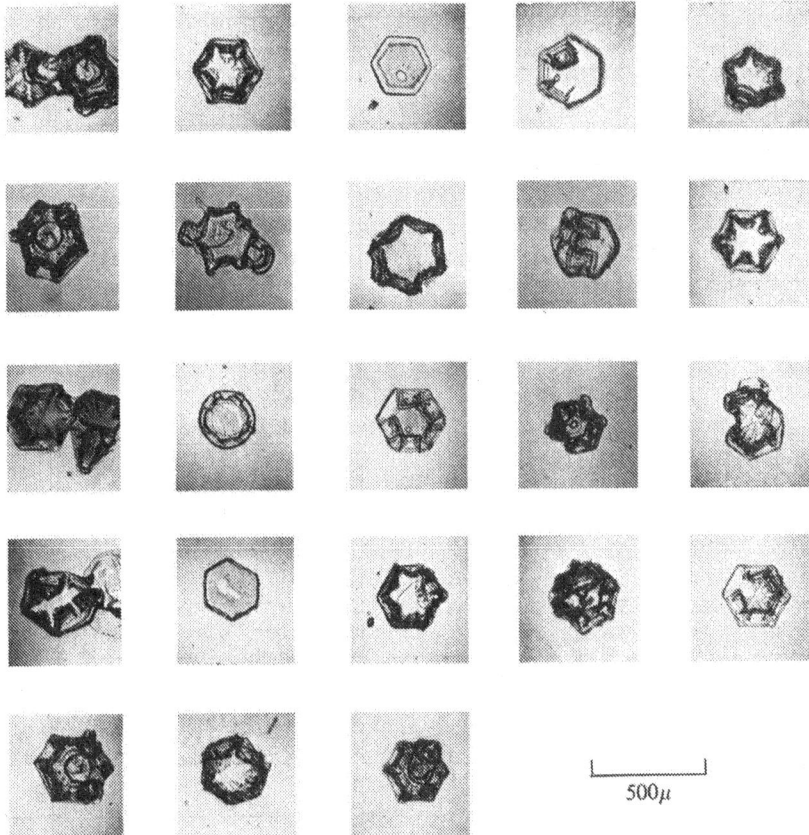

Fig. 11.   Ice crystals collected west of Schenectady, New York on January 7, 1969 during a fall of dust-sized snow.

eastern Colorado to the region about 50 miles east of Pikes Peak. The low level dust was rising only from tilled land; the grassy areas such as the Pawnee Grass Lands were unaffected.

A similar massive dust storm which produced very extensive cloud seeding was observed by me on the afternoon of April 12, 1967 between Amarillo, Texas and Denver, Colorado. This affected region was so extensive and had such a profound effect on the Great Plains and midwest weather systems that I was able to identify it and see its effects over western Illinois two days later.

On the return flight from Denver on December 8, a third source of inadvertent weather modification was observed. Take-off in a DC9 occurred at 1206 MST on a non-stop jet flight to New York City. Very fine snow was falling at the ground upon take-off. Four minutes afterward we climbed above an extensive area of ice crystals. A bright undersun became visible and was seen continuously all the way from the Denver area to the Atlantic Ocean east of New Jersey. Jet contrails appeared to be the source of these crystals throughout the entire flight which was conducted at 37000 ft. More than a dozen different planes were seen coming from the east within the flight corridor we were using, most of them several thousand feet below us. From time to time we were close to contrails being made by planes at our level but ahead of us.

The most striking effect observed as the sharp line of demarcation between the area affected by contrail seeding along our flight corridor and an extensive area of high altocumulus cloud (or cirrocumulus) which paralleled our zone at its southern extremity. This region of non-modified clouds was estimated to be about 10 to 20 miles away and extended over large regions of the country. I expect that an effect such as was observed could be seen on satellite cloud photographs.

Perhaps the most disturbing feature about inadvertent weather modification is that in a subtle manner it seems to be changing the nature of clouds over increasingly large areas of the globe. Much of our current consideration of cloud seeding assumes the ubiquity of supercooled clouds and the effectiveness of a seeding material for triggering the instability of such systems.

If pollution sources lead to increased dustiness from ill-used land, more cloud nuclei from burning trash and many more ice nuclei from the lead-permeated exhaust of internal combustion engines, not only will we lose the possible advantage we now have of extracting some additional water from our sky rivers, but we might even be confronted with a drastic change in our climatological patterns.

Interesting climatological evidence of inadvertent weather modification was found by Chagnon [16] to exist in the area downwind of the Chicago, Illinois-Gary, Indiana complex of extensive urban, highway and steel mill concentrations.

A very noticeable increase in precipitation and storminess is evident in the records of the past three decades. The LaPorte, Indiana region whose record is cited as evidence of this effect is downwind of the heavy pollution source mentioned above as well as its close proximity to a very moist air source in the form of Lake Michigan. It is a common observation to see a lake effect street of cumulus clouds extending in the convergence zone south and southeast of Lake Michigan. The combination of very moist air and an abundance of ice nuclei are apparently in very favorable juxtaposition for an optimum reaction to occur. The LaPorte anomaly was first observed by a local weather observer which was then evaluated by Chagnon. He found that there has been a notable increase in precipitation starting about 1925 with definite increases since that time also of the number of rainy days, thunderstroms and hail storms. There has been a 31% increase in precipitation, 38% of thunderstorms and 240% increase of hail incidences. The increases show a marked correlation with the production of steel.

Since these data were obtained entirely from an evaluation of the climatological records, it is of great importance that careful 'on-the-spot' field observations should be made in the LaPorte area to establish the atmospheric dynamics which are responsible for the apparent change in the precipitation pattern of that area. It is particularly important that the concentration of particulate matter by correlated with storm patterns. The weather systems at the mesocale level should especially be studied to determine whether the area receiving increased precipitation is in the center or edge of the city-industrial plume effluent and the properties of the moist air moving in from Lake Michigan.

10. Experimental Production of Large Areas of Ice Crystals

During the past 10 winters field operations have been developed by our Yellowstone Expeditions in which we have established certain relationships of ice crystal concentrations in the free atmosphere. The early morning inversions of the Old Faithful Geyser Basin in the wintertime often have liquid water contents ranging from 0.5 to 1 G m^{-3}. This rich supply of moisture is contained within a strong ground-based inversion having a vertical thickness of about 100 m. At a distance of 2000 m from a point source of seeding, ice crystal concentrations up to 10 000 l^{-1} have been measured. Such crystals at -12 °C are in the form of hexagonal plates with cross sections of from 10 to 1000 μ, with the size depending on concentration and moisture supply. Those of 200 μ occur typically at a concentration of 200 l^{-1} with a fall velocity of 10 cm s^{-1}. The brilliance of the undersun and related optical phenomena indicates that the number of crystals observed in areas caused by air pollution, jet contrails or dust storms often have concentrations as high or higher than observed in our experiments. Thus at Yellowstone we have an ideal outdoor laboratory to study some of the factors which must be better understood if we are to work out the physical interactions resulting from the inadvertent modification of the atmosphere.

11. The Need and Opportunity to Study These Phenomena

The effects cited are but a few examples of many which I have observed and photographed during the past few years. It is the rule rather than the exception that such massive zones of ice crystals can be observed over large areas of the country which can be related to man-caused modification.

Such occurrences must be exercising a detectable effect on the weather systems of the northern hemisphere. I feel that nowhere near enough effort has been directed toward the establishment of an organized and continuing study to determine the effect of such inadvertent seeding mechanisms on the synoptic weather patterns of our country. Such studies should have a major place in the World Weather Watch and the Global Atmospheric Research Project. I strongly recommend that the part played by atmospheric particulates should become an important research feature of this program.

There is a critical need for knowledgeable field scientists having an extremely broad scientific background who can work effectively in the real atmosphere under all types of conditions and extract quantitative and meaningful data from such systems.

Our Universities must place far more emphasis on this type of training than is being done at present. The eventual understanding of these complex interrelationships do depend on computers, electron microscopes, mass spectrometers and other costly instruments and equipment. However, the real atmosphere is the thing that must be understood and it is not enough to rely on data obtained by automatic instruments and uninformed field men as is too often the case. It is not easy to conduct efficient field operations. We must approach nature to an ever increasing degree but this confrontation must involve 'intelligent eyes', an understanding of the physics, chemistry and electricity of the reactions which can occur and a zeal to understand the things which combine to produce atmospheric phenomena.

References

1. Hogan, A.: 'Experiments with Aitken Counters in Maritime Atmosphere', *J. Rech. Atmos.* 3, 53 (1968).
2. Vonnegut, B.: 'Continuous Recording Condensation Nuclei Meter', *Proc. First Natl. Air Pollution Symposium*, Pasadena, Cal. 1, 36 (1950).
3. Schaefer, V. J.: 'Final Report Project Cirrus Part 1 Laboratory, Field and Flight Experiments', Report No. RL-785 General Electric Research Laboratory, Schenectady, N.Y. (1953).
4. Schaefer, V. J.: 'Artificially Induced Precipitation and its Potentialities', in W. L. Thomas (ed.), *Man's Role in Changing the Face of the Earth*, University of Chicago Press (1956).
5. Schaefer, V. J.: 'Cloud Explorations over Africa', *Trans. N.Y. Acad. Sci.* 20, 535 (1958).
6. Warner, J.: 'A Reduction in Rainfall Associated with Smoke from Sugar Cane Fires – An Inadvertent Weather Modification', *J. Appl. Met.* 7, 247–251 (1968).
7. Schaefer, V. J.: 'The Production of Clouds Containing Supercooled Water Droplets or Ice Crystals Under Laboratory Conditions', *Bull. Am. Met. Soc.* 29, 175 (1948).
8. Schaefer, V. J.: 'Ice Nuclei from Automobile Exhaust and Iodine Vapor', *Science* 154, 1555 (1966).
9. Junge, C. E.: 'Air Chemistry and Radioactivity', Academic Press, New York (1963).
10. Schaefer, V. J.: 'Ice Nuclei from Auto Exhaust and Organic Vapors', *J. Appl. Met.* 7, 113 (1968).
11. Schaefer, V. J.: 'The Effect of a Trace of Iodine on Ice Nucleation Measurements', *J. Rech. Atmos.* 3, 181 (1968).
12. Hogan, A.: 'Ice Nuclei from Direct Reaction of Iodine Vapor with Vapors from Leaded Gasoline', *Science* 158, 800 (1967).
13. Bryson, R. A.: 'Is Man Changing the Climate of the Earth?', *Saturday Review* p. 52, April 1, 1967.
14. Langmuir, I.: 'Results of the Seeding of Cumulus Clouds in New Mexico,' *The Collected Works of Irving Langmuir*, Pergamon Press, New York, Vol. II, pp. 245–262 (1962).
15. Schaefer, V. J.: 'Condensed Water in the Free Atmosphere in Air Colder than $-40°C$', *J. Appl. Met.* 1, 481 (1962).
16. Chagnon, S. A.: 'La Porte Weather Anomaly, Fact or Fiction', *Bull. Amer. Met. Soc.* 49, 4 (1968).

MAN-MADE CLIMATIC CHANGES

H. E. LANDSBERG

University of Maryland, College Park, Md., U.S.A.

Abstract. The climatic elements, such as temperature and precipitation, fluctuate naturally on a global, regional, and local scale. These fluctuations are of shorter or longer duration. Their causes are only partially known. They constitute a 'noise' pattern against which man-made changes, if any, have to be evaluated. Anthropogenic changes on a global scale are projected by some to cause rise in atmospheric carbon dioxide from cumbustion of fossil fuels, increase in atmospheric dust content from human activities, and clouds produced by aircraft exhaust. All of these interfere with the radiation balance but there is no present indication that they have up to now caused any significant changes. This may not hold in the future when it is foreseen that CO_2, e.g., may readily double in a few decades. A number of primitive models have been designed to estimate effects on global temperature. None of them indicate changes of more than $2\,°C$, a moderate amount gaged by natural changes even since the last Pleistocene glaciation.

Regionally notable changes have been observed because of deforestation and various agricultural practices. These include on the smallest scale human manipulations, such as orchard heating, irrigation, shelter belts, and soil management. The greatest inadvertent changes have taken place in urbanized areas, where islands of higher temperature and lower wind speed have appeared. These, together with air pollution, have altered cloud and precipitation amounts and patterns. The measurable climatic changes in urban areas are growing from microclimatic to mesoclimatic size as metropolitan regions grow. It is concluded that while man's influence on climate is as yet restricted to local environments close monitoring of the global atmosphere at all levels is indicated to watch for indications of possible larger-scale effects.

1. Introduction

His environment has become one of man's major concerns in recent years. Has he produced by his activities major and perhaps irretrievable changes? An attempt was made in 1970 to take a look at this important problem for the atmosphere which governs our climatic environment [1]. The importance of the theme has attracted widespread interest in the scientific community. Hence there is a continuous flow of contributions to this theme. In a year's time many new and important papers have been published. They throw new light on the subject in some respects but have also deepened controversies in others. It seems, therefore, appropriate to revise, and somewhat expand, the earlier review.

Let us define climate as the total ensemble of weather conditions at a point or over a smaller or larger area. Weather is proverbially fickle. Climate has a connotation of greater stability but nevertheless it fluctuates locally and globally in irregular pulsations. Man, in his evolution and history, has been governed by climate. He is a creature of a tropical environment but his cultural adaptability has permitted his adjustment to other environments at a rate that is many orders of magnitude faster than his genetic responses [2]. He has developed technologies that enable him to survive now in all parts of the globe. It has been alleged that this very technology has had effects on climate, a question that will be examined in detail below. There have also been many

S. Fred Singer (ed.), The Changing Global Environment, 197–234. All rights reserved.
Copyright © 1975 by D. Reidel Publishing Company, Dordrecht-Holland.

proposals to change local and global climates to suit man's needs. Some small scale experiments have, indeed, been carried out but the larger schemes have not been tried because their feasibility has not been demonstrated and costs are exorbitant. This is in a sense fortunate because at the level of our present knowledge, it is impossible to predict whether or not beneficial effects in one place might not be vitiated by detrimental results elsewhere. Undoubtedly, even if the gross results were a general improvement in anthropocentric terms, delicate ecological balances in the remainder of the biosphere would be upset.

The nagging question "Has man inadvertently changed the global climate or is he about to do so?" has not been answered. There has been widespread discussion about it. Much of this has been characterized by misunderstandings, exaggerations, and distortion. The daily press, the popular magazines, and mass meetings are not the right forum to ascertain facts and forge the knowledge necessary for intelligent action. According to some of the soothsayers mankind is threatened by acute oxygen deprivation, imminent heat death, and a new glacial period. Obviously, these prophecies, if they come to pass at all, are unlikely to happen all at the same time. Atmospheric processes of this type are very slow and there will be ample time to sort fact from fiction.

A number of eminently qualified individuals and groups have recently addressed themselves to the problem [3, 4, 5]. Their reviews, in depth or in breadth, all come to the rather similar conclusion that there are great uncertainties in this field. These can only be removed by a substantial research effort which initially will entail a comprehensive surveillance program of a wide variety of atmospheric elements many of which have not been adequately monitored in the past.

2. Natural Climatic Fluctuations

It is necessary at the outset to differentiate natural climatic changes from those ascribable to man.

The Earth's atmosphere has, since the formation of the planet, undergone a slow but continuous evolution. Because of differences in the absorptive properties of various constituents, present in different proportions during the course of this evolution, the energy balance near the surface has undergone parallel changes. The greatest event in this process has been the emergence of substantial amounts of oxygen. The prevalent opinion is that this was photosynthetically produced by plants [6, 7], although there are some divergent views [8]. For our purpose here it is important to state that the oxygen supply in the atmosphere is essentially constant. Man has had no effect on it with definite evidence that the proportion has stayed constant since the beginning of the century [9]. In Broecker's words: "Molecular oxygen is one resource that is virtually unlimited" [10]. It is well to recall here also the replenishment rate by the plants' photosynthetic processes. One hectare of young, vigorous forest will produce 10 tons of oxygen and consume 30 tons of carbon dioxide.

Climatically important is the photochemical development of ozone in the upper atmosphere. There it forms an absorbing layer for short-wave ultraviolet radiation from the Sun, a fact which is significant for the forms of organic life now in existence. Ozone also creates a warm stratum aloft that has important energetic implications for atmospheric circulation. However, water vapor and carbon dioxide (CO_2), with major absorption bands in the infrared, have a more profound influence on the atmospheric energy balance. They absorb a substantial fraction of the dark radiation emitted by the Earth's surface. Even more important for the Earth's energy budget is atmospheric water condensed or sublimated into clouds. These reflect substantial portions of the incoming short-wave radiation from the Sun and play a determining role for the planetary albedo. At night, clouds intercept outgoing radiation from the Earth's surface and radiate it back.

Over the past two decades Budyko [11] has gradually evolved models of global climate, using an energy balance approach. Among other relevant factors, these models incorporate incoming solar radiation, albedo, and outgoing radiation. They neglect as yet non-linear effects which might affect the element usually striven for, namely the atmospheric temperature near the surface. Such effects are particularly introduced by the ocean-atmosphere interactions. These are not likely to change the basic results but they can introduce substantial time lags and rhythmic oscillations. Somewhat altered models have been advocated by Sellers [12, 13, 14]. Budyko's calculations show that a 1.5 to 2% decrease in the incoming radiation or a 5 to 10% increase in albedo could bring about a major global glaciation [15, 16]. He frankly admits that the theory of climate is so complex that one can only speculate. Yet, in his opinion, the balances are so delicate that man's pollution activities, continued for another century, might initiate a major climatic fluctuation.

Long before man became an influence on the globe climate showed vigorous fluctuations. These have been documented geologically and biologically for eons of Earth's history. Here reference to pertinent literature must suffice [17, 18, 19, 20]. As regards causes in the face of the undisputed fact that major changes in climate have occurred no consensus exists, even if we disregard in this context such major geological upheavals as continental break-up and drift, varying size of ocean surface, and position of continents with respect to the poles. A major theory has always been that radiation that reaches the earth from the Sun is not constant (although this input has paradoxically been designated as the 'solar constant'). This fluctuation of input can be caused by a change in solar energy output. This hypothesis has been advocated by the astrophysicist Öpik in a series of contributions [see, e.g. 21]. Our observations of the solar constant are still very deficient. Extrapolations of observation at higher levels in the atmosphere are afflicted by uncertainties of the same order of magnitude as the value of the fluctuations that are being searched for. Observations from high-altitude balloons by Kondratyev and co-workers suggest a higher output at sunspot numbers between 80 and 100 than for higher or lower activity on the Sun [22]. It remains a task for space exploration to verify this. Theoretical calculations show that a 1% change in the solar constant could lead to a global temperature change of 1.5°C.

Another hypothesis attributes the changes of radiation received at the upper bound of the atmosphere to variations of the position of the Earth in space with respect to the Sun. The basic mathematical model for this was formulated by the astronomer Milankovitch [23]. It included the periodic fluctuation of the Earth's axis, its precession, and the eccentricity of its orbit. From these elements he calculated an insolation curve back into time and the corresponding temperature of the Earth. He tried to correlate minima with the Pleistocene glaciations. Considerable support for these views has been obtained by isotope investigations, especially using the $^{18}O/^{16}O$ ratios in marine shells and glacier cores. Lower ^{18}O amounts correspond to lower temperatures [24, 25, 26, 27, 28, 29]. These studies show that in moderate latitudes of the northern hemisphere temperatures were during glaciations about 7°C lower than at present and in equatorial regions about 3–5°C. In the last 400000 yr there were about 7 or 8 major glacial periods [30]. Budyko [11] and others [31] raise some doubts that Milankovitch's theory can explain glaciations but admit that it explains some smaller temperature fluctuations. Sellers made some new calculations for one of the elements, obliquity, and finds that sea level temperatures for the extremes of 21°55′ and 24°24′ change only from 14.1°C to 13.5°C, with the current value of 23°27′ yielding 13.7°C [32]. This is a rather small range.

For the last few thousand years there is other evidence in the ^{18}O content of glacier water. This is inversely related to a solar activity index, which can be inferred from auroral frequencies that can be traced back to 522 B.C. [33, 34, 35]. Again, low values of ^{18}O reflect low temperatures at the time when the precipitation forming the glacier fell.

There have been many attempts to establish solar variations in the instrumentally observed meteorological data. A large number of statistical studies have found traces of sunspot cycles, their multiples and subharmonics [36]. But the effects seem to be neither universal nor completely proven. If the solar component is present, it is not sufficiently prominent to constitute more than one of the components of climate that make up the observed noise pattern.

In the context of this discussion we will disregard all the major geological forces that act as terrestrial influences on climate (orogenesis, ocean-floor spreading, polar wandering), except one, namely volcanic eruptions. This will be referred to again below. Other important natural changes include extent of snow and ice covers, cloudiness, and vegetation which all affect the terrestrial albedo [37]. These factors interact through very complex feedback mechanisms with the radiative balance, temperature, and precipitation. These have not yet been completely disentangled. Hence, the absence of an adequate theory of natural climatic changes considerably complicates the analysis of possible man-made changes.

3. The Climatic Seesaw

Instrumental records of climate have become available only in the last three centuries, with a very limited coverage of the globe in the first two of these. There are some supplementary natural sources which permit one to get qualitative assessments for earlier periods such as pollen associations [2] and somewhat more quantitatively, tree rings [38, 39]. In historical times there are diaries, chronicles listing crop conditions, and records of freezing of waterways and major storm catastrophes. Even though some of this is very tenuous evidence, careful analysis can reveal at the least the general climatic character of a decade or half century [40, 41]. These studies indicate, for example, that the 10th century was a warm period, that the first half of the 15th century was decidedly cool but that by the end of the 16th century a recovery had set in.

The period of precise knowledge starts at the end of the 17th century when instruments to measure atmospheric elements became available [42]. However, most of our longest uninterrupted series of observations do not start until the middle of the 18th century. From that time onward a considerable number of observations are available, at least for the land areas of the northern hemisphere. These observations give a reasonably objective view of the climatic fluctuations for the last two and a half centuries. This is, of course, the interval in which man and his activities have multiplied rapidly. Most of the long series of data cover Europe [42, 43, 44, 45, 46, 47, 48, 49] but some cover the Far East [50, 51, 52] and North America [53].

Recently a time series of temperature and precipitation values from all available early data sources, has been reconstructed for the eastern seaboard of the United States. In this series Philadelphia has been used as index location, not only because it has a considerable amount of early observations but also because of its central location with respect to other observational material in the 18th and early 19th century [54]. There are still some minor gaps for years in which the observational base was inadequate. These curves are very characteristic of the type found at other locations. They reflect the restlessness of the atmosphere very well. Some analysts have simply considered the variations to be quasi-random. Without going into the details of the statistical characteristics of these series, let me here only state that they do not reflect any pronounced one-sided trends. However, there are definite long or short intervals in which considerable one-sided departures from the mean are notable. On corresponding curves representing Western and Central Europe as well as the Pacific area of Japan we find fairly similar trends to Eastern North America. The late 18th century was fairly warm, most of the 19th century cool and wet, and the first half of the 20th century showed a decided temperature rise. [54, 55, 56] These trends appear to be fairly uniform over the Northern hemisphere. The temperature range has been about 1 to 2 °C. The past two decades have shown cooling trends in the moderate and higher latitudes of the northern hemisphere but it is too early to ascertain if this has been a hemispheric or global event. It is worthwhile noting that Japan apparently still has shown temperature rises to about 1965 at a rate of 0.2 °C per decade in areas where city growth is not a factor [57]. An independent check on temperature trends is afforded by ground water temperature measurements. These reflect mean annual temperature fairly well. In the United States, in rural areas since 1790 a rise of 1.5 to 2 °C has been documented [58].

In the precipitation patterns, except for the pluvial tendency of last century, 'noise' masks all trends. For shorter intervals droughts alternate with high precipitation. Sometimes these aberrations from the mean can be quite spectacular. An example is the seasonal snowfall on Mount Washington, New Hampshire. There snowfall increased from an average of 4.5 m in the winters of 1933/34 to 1949/50 to a seasonal average of 6 m in the interval 1951/52 to 1966/67 [5]. Laymen are apt to ascribe to such observations validity far beyond the local area and extrapolate a single point to represent hemispheric or global conditions. This has to be guarded against. Even indices that integrate various climate influences cannot be made the basis for categorical statements. Glacier conditions are typical for this type of index. For example, the glaciers on the west coast of Greenland have been repeatedly surveyed since 1850. In consonance with temperature trends in lower latitudes, they showed their farthest advance in the 7th decade of the 19th century and have been retreating ever since [60]. This fits the temperature patterns to the 1950 turning point but deviates thereafter. Some glaciers in other parts of the world have shown advances; however, the tendency is as yet not universal. The question whether these changes reflect (1) relatively short-term temperature fluctuations, (2) alterations in the alimenting precipitation, (3) a change in radiation balance, or (4) a combination of these factors remains unanswered.

Many of the fluctuations with short duration are likely to be only an expression of atmospheric interaction with the oceans. External or terrestrial impulses may affect the energy budget and cause an initial change in atmospheric circulation. This, in turn, with notable time lag affects the oceans and through feedback mechanisms induces pulsations in the atmosphere [61, 62, 63]. The oceans are essentially the atmosphere's memory. They have a very large thermal inertia and both horizontal and vertical motions are slow. Namias in a series of very cogent investigations [64, 65, 66, 67, 68, 69, 70] has shown the intimate relationship of some climatic anomalies to atmosphere-ocean interaction. He concluded, for example, that drought conditions on the eastern seaboard of the United States in the 1960s were directly related to the prevailing wind system and the sea surface temperatures in the vicinity but that the real dominant factor was a wind-system change in the North Pacific. Such teleconnections complicate interpretations of local or even regional climatic data tremendously. The world-wide effect of changes in the Pacific wind patterns, with accelerations and decelerations, is directly tied to large-scale breaks in the regime of sea-surface temperatures. These occur in irregular sequences of about 5 yr and result in seemingly small temperature changes of $0.5\,^\circ$C over the whole North Pacific. But, according to Namias, these relatively innocuous changes cause differences of 8×10^{18} g in the annual amount of water evaporated from the surface of the ocean. The consequences for world-wide cloud and rain formation are obvious. Namias even thinks that repetitive conditions of this type can "lead to climatic fluctuations of a much longer time scale than a decade." He further states: "It may be short-sighted to invoke extra-terrestrial or man-made activity to explain these fluctuations." It is against this background that we must weigh the evidence that man has indeed wrought climatic changes.

4. Carbon Dioxide

The fact that the atmospheric gases play an important role in the energy budget of the Earth was recognized early. First Fourier and then Pouillet and Tyndall expressed the idea that the gases acted as a 'greenhouse'. Later it was recognized that the spectrally selective absorptivity of gases is the important factor. Carbon dioxide (CO_2) and water vapor are effective absorbers in the infrared. Their role in the atmospheric heat budget is the interception of long-wave radiation from the Earth. This has been mislabelled as 'greenhouse effect'. Although glass in a greenhouse is opaque for long-wave radiation and helps in trapping absorbed short-wave radiation or heat generated inside, the seclusion of the space from convective and advective air flow is an essential part of the functioning of a greenhouse. In the free atmosphere such flow is, of course, always present. Hence, it seems highly desirable to eliminate the term 'greenhouse effect' from the vocabulary of treatises dealing with climatic change.

After discovery of prominent absorption bands of CO_2 in the infrared its variable content in the atmosphere appeared as convenient explanation for climatic changes in geological history. Arrhenius, in 1908, [71] made the first quantitative estimates of the magnitude of the effect, which he mainly attributed to fluctuating volcanic activity, although he also mentioned the burning of coal as a minor source of CO_2. The possibility that CO_2 resulting from man's activities could be an important factor in the Earth's heat balance was not seriously considered until Callendar, in 1938, [72] showed evidence of a gradual increase in CO_2 concentration in the Earth's atmosphere. But it was Plass [73] who initiated the modern debate on the subject based on his detailed study of the CO_2 absorption spectrum. An important question in this context is: How much has CO_2 increased as a result of the burning of fossil fuels? It is quite difficult to ascertain even the mean amount of CO_2 in the surface layers, especially near vegetation. There are large diurnal and annual variations, in which not only the photosynthetic processes play a role but soil breathing is also prominently involved. Various authors have reported concentrations ranging from 210 to 500 parts per million (ppm). The daily amplitudes during the growing season are about 70 ppm [74, 75]. Nearly all early measurements were made in environments where such fluctuations took place. This, together with the lack of precision of the measurements, means that our baseline – atmospheric CO_2 concentration prior to the spectacular rise in fossil fuel consumption of this century – is very shaky. Bray [76] made a careful statistical reassessment of all CO_2 measurements since 1857. He concludes that the increase from the quarter century 1857/81 to the quarter century 1932/56 was most probably 25 ppm. But he, too, stressed the difficulty of extrapolating from samples in 'micro-atmospheres' to the general atmosphere. The end of that interval coincided with the International Geophysical Year during which sampling in polar regions, on high mountains, and in oceans with precision equipment started. Continuous data from these spots have since been supplemented by samples of the free atmosphere collected by various vehicles. These base our estimates on a somewhat firmer but not yet altogether secure basis [77, 78, 79]. Bolin and Bischof [80] estimate the 1968 mean value of CO_2 in the atmosphere at 320 ppm. Akiyama [81] finds in the same year over the Pacific Ocean near Japan and over the East China Sea values between 318 and 332 ppm. The difficulty of sampling were also pointed out by Junge and Czeplak [82], to which Georgii added the fact that CO_2 content changes when crossing the tropopause and may not have a really constant mixing ratio [83]. All this underlines the SCEP [Wilson and Matthews, 84] recommendation for better monitoring – a bid that applies as well too all other atmospheric constituents and admixtures.

Casting some of the doubts aside, it seems reasonable to assume the following values: Anthropogenic CO_2, averaged for the whole atmosphere 1.6 ppm yr^{-1} [82]; present rate of increase in the atmosphere 0.7 ppm yr^{-1}. [85]; past increase since 1860 about 10%. None of this is spectacular but various bold extrapolations have been made into the 21st century. Present estimates for the value of CO_2 in the atmosphere by the year 2000 range from 340 to 400 ppm [80, 82, 86], roughly a 10 to 30% increase over present values. Much of this depends on the sinks for CO_2 which at present are not completely known. The photosynthetic process in plants keeps an approximate equilibrium between O_2 and CO_2 in the atmosphere. It is estimated that 150×10^9 tons of CO_2 yr^{-1} are used in photosynthesis [87]. Unless the total volume of plant material increases, an equivalent amount is returned to the atmosphere by decay [88]. This volume is one of the unknowns in the estimates of the CO_2 balance. There are good prospects that satellite surveillance will in the future be able to provide information on plant material on land and in the oceans for a bulk estimate and possible changes with time. However, the oceans are the major terrestrial sink for CO_2. The equilibrium with the bicarbonates dissolved in seawater determines the amount of CO_2 in the atmosphere. In this exchange of CO_2 between atmosphere and ocean, the temperature of the surface water enters as a factor. More CO_2 is absorbed at lower temperatures than at higher ones. We have already noted the fluctuating nature of sea-surface temperatures, variations that are principally governed by the wind conditions. Also the interchange of the cold waters of the ocean depths and the warm surface waters through upwelling and downward mixing in the very irregular spurts is a governing element in the CO_2 exchange [89, 90]. Effects at the sea-surface such as films can play an important role. Also the possibility that enzymes facilitate the absorption of CO_2 has to be taken into account [91, 92]. Thus it is obvious that long-range estimates of how much CO_2 will disappear in the oceanic sink are very tenuous. The bold extrapolators simply assume a constant rate of removal. Even if we accept the values projected for future increases in atmospheric CO_2 the question of what this will do to global temperatures cannot be unambiguously answered. The answers depend on other variables, such as the water vapor and clouds present in the atmosphere. Calculations have been made based on various assumptions. Most reliance has been placed on the model of Manabe and Wetherald [93]. This model predicts, given no change from present cloudiness and average humidity, an increase of 2 °C in global temperature at the surface for a doubling of CO_2 from 300 to 600 ppm. At the same time the lower stratosphere would cool by 15°C. This is essentially the result that was accepted by SCEP [84] and others. Recent work by Manabe [94] based on a radiative convective equilibrium model yields a numerical value of a global temperature increase of 0.8°C at the surface for a 33 % increase in CO_2 from 300 to 400 ppm. Rasool and Schneider [95] have re-evaluated the absorption aspects of CO_2. They come to the conclusion that the CO_2 effect is self-limiting. Even if CO_2 should increase by a factor of 10, surface temperatures will, according to their calculation, increase only by a factor of 2.5 °C. Doubling of atmospheric CO_2 would cause a rise of 0.7 °C. This is a little less than half of Manabe and Wetherald's value. Such doubling is nowhere in sight even by the most pessimistic estimates for the next century. Mitchell, in his assessments of the possible contributions of CO_2 to the rise of temperature in the early part of the century, places it at a small fraction of a degree C [96].

All serious studies of the CO_2 effect have pointed out the role of water vapor and cloudiness. Because of the obvious feedback mechanisms between temperature changes and these elements via evaporation and condensation processes there are continuously compensating mechanisms at work. And, quite independently, the water vapor is a far more effective barrier to terrestrial heat loss than CO_2. This was pointed out by Möller [97], who stated that, as far as surface temperature is concerned, the anticipated CO_2 increase can be compensated for by a 1 % increase in cloudiness or a 3 % decrease in water vapor. We are far from being able to observe these highly variable constituents to that level of precision. Others have contributed to the same theme [98, 99] and we may quote here Gates [100], who stated that the "CO_2 hypothesis should not be considered likely to be the primary reason for major climatic changes", and Johnson [101], who expressed it as follows, "the risk of serious perturbation (of climate) appears small...". Monteith [86], in addition, cautioned against taking too much stock in the projections into the future on the basis of the highly uncertain assumptions about photosynthetic CO_2 used by plants. Also our estimates of CO_2 production by natural causes are very inaccurate. These sources are principally organic decay and volcanic eruptions.

It seems quite safe to state that any rise of CO_2 in the atmosphere during the last century, whether man-made or natural, has had minimal impact on global temperature and any such change has simply been submerged in the climatic noise. Equally, any CO_2 production by man's combustion of fossil fuels is highly unlikely to induce any appreciable climatic changes. This is not to say that continuous surveillance of CO_2 should be abandoned. Indeed, this gas should be included with all the others in the comprehensive monitoring program that has been initiated by the World Meteorological Organization.

5. Dust

The influence on climate of suspended dust in the atmosphere was first recognized in relation to volcanic eruptions. Observations of solar radiation at the Earth's surface following the spectacular eruption of Krakatoa in 1883 showed measurable attenuation. The particles stayed in the atmosphere for about 5 yr [102]. There was also some suspicion that summers in the northern hemisphere were cooler after that eruption.

The inadequacy and unevenness of the temperature observations make this conclusion doubtful. The main exponent of the hypothesis that volcanic dust controls terrestrial climatic fluctuations was W. J. Humphreys [103]. The effect is twofold: dust scatters some of the incoming solar radiation back, thus increasing the Earth's albedo; and it absorbs some radiation before it reaches the surface, thus redistributing the energy available to the atmosphere. In the major volcanic eruptions the dust gets into the stratosphere where its residence time may reach several years in the absence of scavenging processes. A re-evaluation of some early radiation records from the first decades of the 20th century shows an effect of the great 1912 Katmai eruption [104]. The subsequent decades were volcanically quiescent.

The interest in volcanic dust was suddenly revived by the eruption of Mt. Agung in March of 1963, injecting a large amount into the stratosphere. There were not only spectacular sunsets [105] but also there appears to have been a cooling trend of surface temperatures since [96, 106]. In contrast, the stratosphere has shown warming in the tropical latitudes, which has been associated with the eruption [107, 108]. The interpretation in terms of magnitude is somewhat ambiguous because of the quasi-biennial oscillation in the tropical stratosphere.

The Mount Agung eruption was followed in 1960s by others. Material from at least three reached stratospheric levels: Mount Taal, 1965; Mount Mayon, 1968; and Fernandina, 1968. Some dust from recent Icelandic eruptions, such as Hecla, 1970, also seem to have penetrated the tropopause. Actually this series was preceded in 1956 by a dust injection from Bezymianny in Kamchatka [109]. The existence of all this volcanic dust in several layers in the lower stratosphere and their effect has been documented in a series of papers by Volz [110, 111, 112], who also noted that turbidity effects from the Mt. Agung eruption were measurable for two years in the northern hemisphere. Unz estimated that maximal concentrations were as much as 50 particles cm^{-3} [113]. It should also be noted here that these volcanic eruptions do not only eject dust but also gases, including SO_2 and CO_2. They may also cast large quantities of water vapor aloft, especially in underwater explosions, some of which may penetrate the stratosphere. Chemical interaction among these constituents and other atmospheric aerosols may account for the ammonium sulfate layer first discovered by Junge in the stratosphere.

The climatic consequences of volcanic eruptions have been examined again in recent years. Mitchell attributes to them a major role in several cooling episodes during the past century and a half, including the most recent one [96]. Lamb has just published a monograph [114], in which he subjects the whole problem to a thorough and rather exhaustive examination, using his wealth of accumulated data and experience in the field of climatic fluctuations. He developed a 'dust veil index' which is based on the depletion of solar radiation. He shows that the lowering of surface temperature in middle latitudes is a function of this index, the areal extent of the veil and its duration. A large dust veil results in the weakening of the general circulation and the zonal westerlies. This results in more meridional motions which cause more polar outbreaks in the northern hemisphere. However, the effect of volcanic dust is in reality more suspected than established without doubt. Only the optical effects are proven; the climatic consequences have yet to be fully verified [303].

As yet man cannot compete in dust production with major volcanic eruptions. He is making a good try but presently he is an order of magnitude behind on a global scale. Another principal difference is the fact that most of his solid products stay near the ground. In the lower layers of the atmosphere they are intercepted by vegetation, or are eliminated by fallout and washout. The larger particles seem to have an average residence time of about 10 days. Nature produces occasionally a great deal of dust from blowing sand in deserts. These dust masses have been seen travelling hundreds of miles on satellite pictures. Volz [115] has documented optically Sahara dust as far west as the Caribbean. Man also contributes by poor agricultural practices to soil blowing. But only rarely does this dust travel long distances by global standards.

Yet there are particulates, both man-made and natural, that seem to be increasing. These particles generally have diameters of less than 10^{-4} cm and are now usually designated as Aitken nuclei. Measurements made in the early 1930 gave values of around 100 of these cm^{-3} as global background, a value which seems to have doubled in recent years [116, 117, 118, 119, 120, 121]. This is at least partially backed up by measurements of the atmospheric electric conductivity over the North Atlantic Ocean where a reduction of 20% indicates a doubling of the atmospheric aerosol in the last half century. However, no such change was noted over the South Pacific [122]. There are, of course, many sources of Aitken nuclei in nature, especially salts from ocean spray. But man-made combustion processes are a prodigious source of such nuclei and it is hence plausible that, with a constant natural background, some increase due to man's activities has taken place. This still leaves us to answer the question: What is the effect of the man-made aerosol? There is general agreement that it depletes the direct solar radiation and increases the scattered radiation from the sky. Measurements of direct solar radiation in many places show a gradual increase in turbidity [104, 123, 124, 125, 126, 127]. Many of the measurements were taken in or near settlements and in industrial regions and over a few decades indicate increases from 10 to 100% in turbidity. Even the antarctic data show an increase but local contamination cannot be excluded. Indeed Porch et al. [128] report that it is still possible to observe at remote stations occasionally a ratio of a transmission coefficient of the actual atmosphere to the ideal with only Rayleigh scattering near unity. There had been reports that seemed to support a gradual increase in turbidity at Mauna Loa which is remote and usually above the trade inversion that would confine most aerosols produced near the surface [129]. A careful re-evaluation of the Mauna Loa data show that direct radiation there has not decreased. From 1958 to 1963 there was only a small annual variation. Then the Agung eruption decreased transmissivity but by 1970 the atmosphere at that Hawaiian station reached about its prior transparency [130].

Even though the turbidity increases ascribable to man seem to be at most regional, it seems to be useful to pursue the effects of an attenuation of the solar radiation by atmospheric aerosols. This weakening is in part, at least, caused by backscattering of solar radiation to space. This is equivalent to an increase in the Earth's albedo, which is being interpreted as a heat loss resulting in lowered temperatures. Bryson, with an important reservation, advocated this hypothesis. He stated [131]: "all other factors being constant, an increase in atmospheric turbidity will make the Earth cooler by scattering away more incoming sunlight. A decrease of dust should make it warmer." This model is far too simplified and the reservation of "all other factors being constant" essentially admits that in reality the other factors vary too. Barrett [132] has pursued this theme further. He developed a model in which he calculated for various latitudes the depletion of energy received at the surface because of dust. For geometrical reasons this is more pronounced at higher than lower latitudes. He deduces that an order-of-magnitude increase in the atmosphere will redistribute the energy balance sufficiently to cause changes in the general circulation.

But these approaches are far too categorical and simple. In reality the optical and consequent thermal effects of aerosols depend on the size spectrum, their height in the atmosphere, and the absorptivity of the particles. These properties have been studied in detail by a number of authors [133, 134, 135, 136, 137]. It is quite clear that most man-made particulates stay in lower atmospheric layers. Temperature inversions attend to that. The tropopause is a very effective barrier and has often been cited as a dust horizon. There is no evidence that there have been any substantial man-made dust injections into the stratosphere since the treaty banning nuclear tests in the atmosphere became effective and was adhered to by the major nuclear powers. The optical analyses show, first of all, that backscatter of particles is outweighed at least 9 to 1 by forward scattering. There is also substantial absorption of energy by the dust. This applies to a limited extent also to the outgoing terrestrial radiation although that is mainly governed by water vapor [137]. The process is very wave-length dependent [22] and the discrepancies between observed and calculated values of aerosol absorption are large. The most recent assessment by Mitchell [138] indicates that a reasonable model shows a surface temperature loss from stratospheric aerosols, and a tropospheric aerosol – except in deserts – will cause warming. This is quite at variance with the popular conception. Another point deserves mentioning and that deals with the annual variation of turbidity, which shows a maximum in summer. This is not the season of greatest fuel use. It coincides with the time when forest areas are often covered by a blue haze. It is caused by photochemical reactions among hydrocarbons emitted by vegetation. These seem to be highly effective Aitken nuclei, a view first advanced by Went [139] and also discussed in other work [140].

If one examines assertions that since 1930 dust has 'dominated' surface temperature fluctuations [141] we have to reject any implications that this is due to anthropogenic dust. The latter, on a global scale, is puny compared with the dust from volcanic eruptions and the aerosol generated by sea spray [142], part of which reaches the stratosphere where its optical- and ultimate thermal-effects may be appreciable. No thermal effects of man-made dust beyond the 'noise' level are plausible at this time. Nor has there been a documented case made for global thermal effects of dust storms from deserts or soil blowing. Even alleged regional effects of such conditions have been discounted [143].

Actually there may be a more important climatic effect of dust, not in suspension, but settling on ice and snow. Such dust radically changes the albedo and can lead to melting [144, 145, 146, 147, 148, 149]. Davitaya [150] has shown that the glaciers of the high Caucasus have an increased dust content which parallels the development of industry in eastern Europe. Core analysis showed that up to 1920 the dust content of the glaciers was about 10 mg 1^{-1}. In the 1950s this increased to 235 mg 1^{-1}, an over 20-fold rise. While the dust is near the surface, it has an effect on the heat balance of the glacier. Other evidence shows that man-made dust has penetrated the polar regions and is found on the Greenland ice cap. Conceivably the dust could cause melting and over a very long time interval pave the way for a more radical climate change including rises in sea level. There has been some speculation along this line [151]. And although microclimatic effects are demonstrable there is, so far, no evidence that they have had any measurable influence on the Earth's climate. The possibility of deliberately causing changes in the albedo has figured prominently in the discussions of artificial modifications of climate. Some have advocated schemes to eliminate the arctic sea ice by spreading dust on it systematically. This seems technologically feasible [152, 153]. The consequences of an arctic ice-free ocean for the mosaic of climates in lower latitudes have not yet been assessed. Present computer models of world climate are still too crude to obtain sufficient details which would be needed to judge the ecological consequences of such a scheme.

All the preceding discussions refer to climate on a global scale. At that level natural influences definitely have the upper hand. Although monitoring and vigilance, in line with the SCEP report [84], are indicated, the evidence for any man-made effect on global climate is, at best, flimsy. This does not apply to the local scale, as will be related below.

6. Rural Effects

For nearly two centuries it has been said that man has affected the rural climates simply by changing vast areas from forest to agricultural lands. In fact, Thomas Jefferson suggested repetitive climatic surveys to measure the effects of this change in land use in the virgin areas of the United States [154]. Geiger has succinctly stated that man is the greatest destroyer of natural microclimates [155]. In discussing the change from mature forest to field below, I am drawing freely on points expressed by Tanner in an exchange of correspondence [156]. His contribution to this theme is gratefully acknowledged.

The changeover from forest to field locally alters the heat balance. The albedos generally increase in this process, and the aerodynamic roughness decreases. Exposure and cultivation of the soil greatly change the heat flux into and out of the soil, leading to greater temperature extremes at the soil surface. Low-level wind profile and turbulence change drastically. This leads occasionally to wind erosion. On the other hand, evapotranspiration is reduced. There are a number of reasons: drainage of fields prior to cultivation, rapid runoff, bare soil for considerable periods of the year, and shallower root systems of field crops compared with trees. Tanner communicates the following approximate annual values of evapotranspiration for Wisconsin: forest 700 mm. corn 510 mm, bluegrass 430 mm, and bare soil 380 mm. There are, as yet, no estimates as to what regional changes of climate may have been brought about by the conversion of vast areas of land to agricultural use.

Some of the obviously detrimental effects have been mitigated by planting hedges and shelter belts of trees. Special procedures have evolved to reduce evaporation, collect snow, and ameliorate temperature ranges by suitable soil covers and use of sheltering trees and shrubs [157, 158].

The classical case of a local man-made change is the conversion of a forest stand to pasture, followed by overgrazing and soil erosion, so that ultimately nothing will grow again. The extremes of temperature to which the exposed surface is subjected are often so detrimental to seedlings that they cannot become established. Geiger pointed this out years ago in connection with reforestation experiments. But not all grazing lands follow the cycle outlined above. Sometimes it is a change in the macroclimate that tilts the balance one way or another [159].

Beneficial microclimatic modification has been successfully practiced for years. The greatest effort has been devoted to fighting frost. Schemes to increase photosynthesis, reduce evapotranspiration, and promote germination have been employed. We can refer to these here only in passing [160].

The greatest efforts, since ancient times, have been devoted to various schemes of irrigation in order to compensate for the vagaries of natural rainfall. Irrigation not only offsets temporary deficiencies in precipitation but also affects the heat balance. It decreases the diurnal temperature ranges, raises relative humidities, and creates the so-called 'oasis effect'. Thornthwaite, only a decade and a half ago, categorically denied that man could deliberately cause any significant change in the climatic patterns of the Earth. Changes in microclimate seemed to him so local and trivial that special instrumentation was needed to detect them. However, "through changes in the water balance and sometimes inadvertently, he exercises his greatest influence on climate" [161].

What happens when vast areas come under irrigation? This has taken place over 62×10^3 km^2 of Oklahoma, Kansas, Colorado, and Nebraska since the 1930s. Some meteorologists have claimed that about 10% more rainfall occurs in the area during early summer, allegedly attributable to reevaporation of moisture from irrigated lands [162]. Synoptic meteorologists have generally made a good case for the importation of moisture for rainfall from marine sources, especially the Gulf of Mexico. Yet ^{3}H determinations have shown, at least for the Mississippi valley, that two thirds of the precipitated water derives from locally evaporated surface water. Anyone who has ever analyzed trends in rainfall records will be very cautious about accepting precipitation changes as real until many decades have passed. For monthly rainfall totals, 40 to 50 yr may be needed to establish increments, decrements, or trends because of the large natural variations. We will encounter this pesky theme again in other contexts subsequently. The main problem is the inadequacy of the ordinary rain gage as a sampling device. With about one gage per 75 km^2, we are actually sampling 5×10^{-10} of the area. Precipitation is usually unevenly distributed, especially when it occurs in form of showers. Then the sampling errors become very high. Even gages close to each other often show 10% differences in monthly totals. It takes, therefore, a long time to determine whether or not differences are significant or trends real. This problem is often conveniently overlooked by statisticians unfamiliar with meteorological measurements and by enthusiasts with favorite hypotheses [163, 164].

This century has seen, also, the construction of very large reservoirs. Soon after these fill they develop measurable influences on the immediate shore vicinity. These are typical lake or oasis effects. They generally do not reach much beyond the boundary layer [165, 166]. They include reduction in temperature extremes, an increase in vapor pressure and relative humidity, and – if the reservoir is large enough – local wind circulations of the land-and lake-breeze type. Rarely do we have long records as a basis for comparing conditions before and after establishment of the reservoir. One such study has been published by Zych and Dubianiewicz [167]. It describes the influence of the 30-yr-old reservoir of the Nysa Klodzka river in Poland with an area of about 30 km^2. At the town of Otmuchow, about 1 km below the newly created lake, a 50-yr temperature series was available (1881–1930). In the absence of a regional trend there has been an increase in the annual temperature of 0.7 °C at the town near the reservoir. It is now warmer below the dam than above it. This is in contrast to the earlier years when the higher stations were warmer because of the temperature inversions that used to form before the water exerted its moderating influence. It is estimated that precipitation has decreased, because of the stabilizing effect of the large body of cool water. Here, as elsewhere, the influence of a large reservoir does not extend more than 1 to 3 km from the shore. Satellite photographs indicate that Lake Nasser does not seem to affect local convective cloudiness. In contrast, Lake Volta remains relatively cloud-free, even in the rainy season, because the cooler water inhibits convection [168].

A deliberate form of man-controlled interference with microclimate, with potentially large local benefits, is suppression of evaporation by monomolecular films. Where wind speeds are low, this has been a highly effective technique for conserving water. The reduction of evaporations leads to higher water temperature. This may be beneficial for some crops, such as rice, if the technique is applied to the paddies [169, 170, 171].

The reduction of fog at airports by seeding of the water droplets also belongs into the category of man-controlled local changes. In the case of subcooled droplets, injection of suitable freezing nuclei into the fog will cause freezing of some drops. They grow at the expense of the remaining drops because of the vapor pressure difference over water and ice, fall out and this process gradually dissipates the fog. For warm fogs, substances promoting the growth or coalescence of droplets are used. In many cases dispersal of fog or an increase in visual range sufficient to permit flight operations can be achieved [172, 173]. Gratifying though this achievement is for air traffic, it barely qualifies as even a microclimatic change because of the brief time scale involved. Similarly the frost prevention measures in orchards and vineyards because of their very temporary nature hardly qualify as microclimatic changes.

In this context a brief note on general weather modification experiments is in order. Most of the past efforts have been devoted to attempts at augmenting rainfall and suppressing hail. The results have been equivocal and variously appraised [174, 175, 176, 177]. The common procedure has been cloud seeding by various agents. This produces undoubted physical changes in clouds, but usually the approach is too crude to permit prediction of the outcome. Thus, precipitation at the ground has been both increased and decreased [178]. Here we encounter again the problem of sampling. Huff has shown that in the past both experimental and statistical techniques have been very deficient [179]. He indicates that crossover tests, target–control, can get results in uncomplicated topographic settings in a reasonable period of time for the seeding of individual storms. An analysis of natural variability shows how readily an experimenter can be fooled. Computer simulation, using historical data, indicate that one could obtain target–control relations showing increases or decreases of precipitation at 0.001 level of significance simply by natural conditions. If a recording raingage is available every 25 km^2 the sampling error for a large number of storms can be reduced to below 3%. And if all storms are seeded, depending on the design of the experiment, a 20% rainfall increase can be judged at the 95% confidence limit at best in 2 and at worst in 16 yr. It is obvious that most past experiments do not meet these criteria.

Attempts are now being made to put cloud seeding on a more rational basis than heretofore. This involves the physical, and subsequent computer modeling, of small cloud systems. Most amenable to this treatment are convective cumulus clouds. These models permit a reasonable prediction which clouds may be expected to grow through the seeding process by liberating latent heat of fusion and subsequently cause rainfall. Although randomized experiments on cumulus seeding have not been encouraging in the past [180], the dynamic prediction may improve the chances materially. Subtropical areas may have the greatest potential. A series of experiments in Florida have shown promise. Radar observations of seeded clouds not only showed successful induction of the growth processes but substantial local rainfall increases [181].

Otherwise the most reliable results to induce precipitation have been achieved through seeding of clouds forming in up-slope motions of winds across mountains and cap clouds [182, 183]. The effects of cloud seeding downwind from target areas are not well known. No analysis has ever satisfactorily shown whether cloud seeding has caused a net increase in precipitation or only a redistribution. Here again the high local variability and terrain play a role in the assessment [184, 185].

Attempts to suppress hail by means of cloud seeding are also still in their infancy. Here the seeding is supposed to achieve the production of many small ice particles in the cloud. This is intended to prevent any of these particles growing to a size large enough to be damaging when they reach the ground. The seeding agent is introduced into the hail producing zones of cumulonimbus either by ground-fired or airborne pyrotechnical projectiles. Some successes have been claimed, but much remains to be learned before seeding can be considered a dependable technique for eliminating this climatic hazard [186, 187, 188].

Hurricane modification has also been attempted. The objective is reduction of damage caused by wind and storm surges. Seeding of the outer wall clouds around the eye of the storm is designed to accomplish this. The single controlled experiment that has been performed, albeit successfully in the predicted sense, provides too tenuous a base for appraising the potential of this technique [189]. A warning flag must be raised here before this is tried on storms approaching land because of the possibility of simultaneously changing the pattern of rainfall accompanying the storm. In many regions tropical storm rain is essential for water supply and agriculture. If storms are diverted or dissipated as a result of modification, the economic losses resulting from altered rainfall patterns may outweigh the advantages gained by wind reduction [190, 191]. A further comment of large-scale modification schemes will be made at the end of this review.

7. General Urban Effects

The most pronounced, and locally far-reaching, effects of man's activities on microclimate have occurred in the cities. Many of these have reached beyond the urban confines and may well be classified as mesoclimatic. Some of them were recognized during last century in the incipient metropolitan areas. Currently the sharply accelerated trend toward urbanization has accentuated these effects. From a problem that at first simply intrigued the meteorologists, it has in recent years shown aspects that are alarming. The literature in this field has grown rapidly and includes several reviews summmarizing the facts [192, 193, 194, 195, 196]. Chandler [197] has provided a comprehensive bibliography.

We are on the verge of having a satisfactory quantitative physical model of the effects of cities on the climate. It combines the two major features in which a settlement influences the atmosphere. One is the effect on heat and water balance; the other affects flow patterns and turbulence. To take the latter first, the major contributory factor is the increase in surface roughness. This influences the wind field and, in particular, causes a major adjustment in the vertical wind profile so that wind speeds near the surface are reduced. The varied skyline of the city also increases the number of eddies and produces changes in the turbulence spectrum.

The change in the heat balance is probably the most radical alteration produced by urbanization. We essentially convert the spongy, often moist, soil cover of rural areas of low heat conductivity and appreciable albedo into an impermeable surface layer with high capacity for conducting heat and, because of generally lower albedo, high absorptivity for radiation. These fundamental changes in surface conditions that accompany the change from rural to urban conditions lead to rapid runoff of precipitation and consequently to a reduction in local evaporation. This is, of course, equivalent to heat gain to which we can add the substantial heat increase due to the lowering of the albedo. This heat is effectively stored in stone, asphalt, concrete, and the deeper compacted soil layers of the city. In contrast, vegetated rural areas reflect more incoming radiation and store less in the soil where plant material may act as an effective insulator. Hence, the structural features alone favor a positive heat balance for the city. To this we can add the local heat production from combustion and metabolic processes. The end result is the so-called urban heat island. It induces convection over the city and leads to a converging wind field whenever the general atmospheric flow patterns are weak [198, 199, 200, 102].

Most of the features implied by this model have, over the years, been documented in the climatic observations near the surface by comparisons of measurements in cities and their rural surroundings, usually at airports. Such comparisons gave reasonably quantitative data on the magnitude of the urban effects. Some doubts, however, remained. These stemmed from the fact that many cities were located in special topographic settings which favored the founding of a city – such as a river valley, a lake front, a harbor, or an orographic trough. These would *a priori* have a micro-climate different from the environs. Similarly, airport sites were often chosen for microclimatic factors favorable for aviation. Some of the uncertainties can be removed by observing atmospheric changes as a town grows. An experiment along this line was started six years ago in the new town of Columbia, Maryland. It has already grown from a few hundred inhabitants to over twenty thousand. The results so far support the earlier results and have refined them [202, 203].

A very remarkable fact stands out, namely that a single block of buildings will already start the process of heat island formation. Both, air and infrared temperature measurements demonstrate this. Figure 1 illustrates this by observations made in a paved court surrounded by low buildings and surrounded by grass with a wood lot in proximity. On clear, relatively calm nights a small heat island develops in the court, fed by heat stored during a sunny day under the asphalted parking lot of the court and the brick building walls. This slows down the nocturnal radiative cooling process, relative to the rapid cooling from the grass surfaces. This keeps the air in contact with the paved area warmer than that over the grass [204]. Other experiments have shown that the daytime heat transport into the soil under asphalt pavement on a sunny day is over two and a half times as much into a clover-covered soil [203].

The heat island expands and intensifies as the city grows. In Columbia, Maryland, on calm, clear nights it increased in two years of rapid growth from 1 °C to 4 °C [203]. As it grows, stronger and stronger winds are needed to overcome it [205, 206]. But with the growth of the city the winds at street level decrease instead. In large metro-politan areas the heat island is present to some extent under many weather conditions but may nearly disappear on cloudy days with strong winds. It shows even in the long-term mean values. The Paris, France, region is a good example. There, as in other large cities, the temperature difference to the countryside may reach 6 to 8 °C in the early night hours during clear weather with low winds. But more remarkable is that a pronounced metropolitan heat island of 1.6 °C for the annual temperature value now exists in the multi-year mean. The Paris region is relatively flat so that topography plays no major role. The temperature rise has not only affected the mean air tempera-ture over the city but also the temperature of the soil. This has been documented by temperature measurements in a deep cellar under the city, which have been kept for two centuries [207]. Curiously enough the temperature in this cellar was once considered

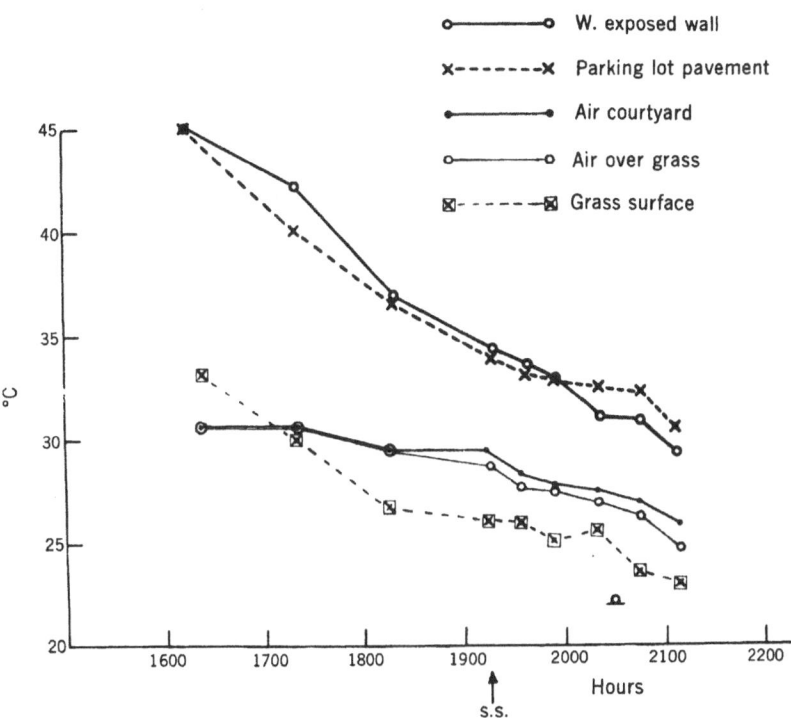

Fig. 1.   A typical example of microclimatic heat island formation in incipient urbanization. The top two curves show radiative temperatures of wall and parking lot pavement on a clear summer evening (6 August 1968). The two middle curves show air temperatures (at elevation of 2 m) in the paved courtyard and over an adjacent grass surface; from sunset (s.s.) onward, the courtyard is warmer than the air over the grass. The bottom (dashed) curve gives the radiative temperature of grass. The symbol at 2030 h indicates the start of dew formation.

constant enough that in the early days of thermometry it was proposed as one of the calibration points for thermometer scales. This man-made long-term trend of temperatures plays havoc with the old time series of observations from cities. They become suspect as guides for gaging the slow, natural fluctuations. For this reason the maintenance of climatic reference stations in undisturbed areas, with good prospects for maintaining this status for a long time to come, is imperative.

Part of the rise in temperature must be attributed to heat rejection from human and animal metabolism, combustion processes, and air-conditioning units. Energy production of various types certainly accounts for an appreciable fraction of it. In small towns, like Columbia, this element is as yet an order of magnitude smaller than the solar heat absorption and heat storage discussed above. In large cities this can become a considerably larger fraction. East reported for Montreal that it contributes, on an average, 22% of the heat island, about 8×10^{16} cal of a total heat island of 39×10^{16} cal [208]. This can reach to 300 m above the city level [209], indicating that a fairly sizeable parcel of air participates in the modification.

There have been projections of the energy rejection into the next decades. These lead to values we should ponder. One estimate indicates that in the year 2000 the Boston-to-Washington megalopolis will have 56 million people living within an area of 30000 km^2. The heat rejection is estimated at 65 cal cm^{-2} day^{-1}. In winter this is about 50%, and in summer 15%, of the heat received from the sun on a horizontal surface in the area [210]. An eminent French geophysicist has discussed the implications of doubling the energy consumption in France every 10 yr and concludes that this would lead to unbearable temperatures [211]. The advective forces of the general atmospheric circulation are likely to prevent catastrophic results; however, the world-wide increases of heat rejection induce uneasiness. Fedorov [212] guesses that, on a global scale, present heat rejection is 0.01% of solar radiation income but projects this figure to reach 1% in the foreseeable future. It is one of the large number of reasons for achieving, as rapidly as possible, a steady state in population and power needs.

An immediate consequence of the heat island of cities is increased convection over the built-up area, especially during daytime. This can be easily demonstrated by the drift of constant volume balloons over a city, that can be lifted several hundred meters [213]. The updraft leads, together with the large amount of water vapor released by combustion processes and wet cooling tower evaporation, to increased cloudiness over the city. It is also a potent factor in the increased rainfall reported from cities.

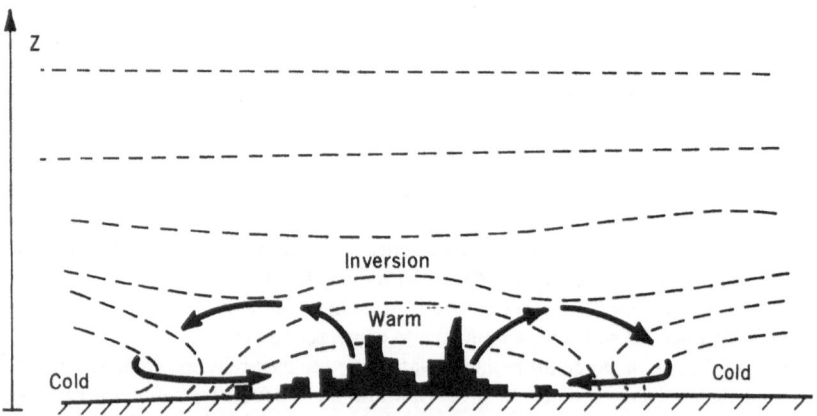

Fig. 2.   Idealized scheme of nocturnal atmospheric circulation above a city in clear, calm weather. The diagram shows the urban heat island and the radiative ground inversions in the rural areas, a situation that causes a 'country breeze' with an upper return current. (Dashed lines) Isotherms; (arrows) wind; Z, vertical coordinate.

This phenomenon will be discussed below in connection with air pollution problems. Even at night the heating from below will counteract radiative cooling and produce a positive temperature lapse rate in the lowest air layers over the city. At the same time inversions form over the undisturbed countryside. This, together with the surface temperature gradient creates a pressure field which will set a converging country breeze into motion [214, 215]. A schematic cross section of such a circulation system is shown in Figure 2.

Rapid runoff of rainfall caused by the imperviousness of the surfaces of roads, roofs, and parking lots is another major effect of cities. It is accentuated by the drainage system which channels the runoff quickly into the water courses. In minor rainfalls this has only the limited consequence of reducing the evaporation from the built-up area. This eliminates the cooling through vaporization which is common in rural areas. But let there be a major rainstorm and the swift rush of water into the draining streams and rivers will cause rapid rise. Problems of flash flooding have multiplied in recent years. Because of the unwise use of flood plains in urbanized areas this can lead to major damage. In the United States there are still 80 people killed annually and 75000 temporarily displaced. Money damages approximate $1 billion per year [216].

The flood height is linearly related to the amount of impervious area. For the 1- to 10-yr recurrence intervals, flood heights will be increased by 75% for an area that has become 50% impermeable, a value not at all uncommon in the usual urban setting. Observations in Hempstead, Long Island, have shown, for example, that, for a storm rainfall of 50 mm, direct runoff has increased from 3 mm in the interval from 1937 to 1943 to 7 mm in the interval from 1964 to 1966. This covers the time when the area changed from open fields to an urban community [217, 218, 219]. Stall *et al.* [220] estimate that the intense urbanization of 9 km² in a rural basin of Central Illinois will increase the flood peak for the 50-yr recurrence interval by a factor of 4 and for the mean annual flood by about 8 times.

It is very difficult to document the decrease of wind speed over cities. Long records obtained with unchanged anemometer exposures at representative heights are scarce. Reasonable interpretations of available records suggest a decrease of about 25% from the rural equivalents in the lowest 30 to 50 m. This is not unreasonable in the light of measurable increases in aerodynamic roughness. These are around 10 to 30 cm for meadows and cultivated fields and around 100 cm for woodland. There are several estimates for urban areas. It will give here a value calculated from the unique wind measurements on the Eiffel tower at the height of 316 m, and from other wind records in the Paris region [221]. These data yield values around 500 cm. They also suggest a decrease in wind speed at the top of the Eiffel tower from the interval 1890–1909 to the interval 1951–1960 of 0.4 m s^{-1}, or a 5% reduction in mean speed. In view of the height of this anemometer this is a quite notable adjustment of the wind profile to the increase in terrain roughness of the growing metropolis. Our work in Columbia showed in two years of urbanization an increase of the roughness parameter from about 5 to 100 cm and a mean decrease of wind speed in the urban sites, at 2 m, of 35% [203].

8. Air Pollution Effects

The most spectacular influence of cities upon their atmospheric environments is that
caused by air pollution. The inventory of pollutants put into the air by man is long
and has been commented about in so many contexts that we can refer here only to
collective works in the literature [222]. Also the much investigated special case of
Los Angeles where special topographic and climatic circumstances combine to create
complex interactions of pollutants and atmosphere can be mentioned only in passing
[223, 224, 225, 226]. Our focus here shall be on the rather universal effects of pollutants
on local climates.

Among these is the attenuation of solar radiation by suspended particles. Although
this affects the whole spectrum, it is most pronounced in the short wave-lengths. The
total direct radiation over major cities is weakened by about 15%, on an average,
usually more in winter and less in summer. The ultraviolet is, on an average, reduced by
30%, and in winter often no radiation below 390 nm is received. This extinction takes
place in a very shallow layer. This was shown by simultaneous measurements at the
surface and a tall steeple in Vienna, Austria [227]. We have already commented on
the observed general summer peak of turbidity as a possible effect of natural organic
interactions. However, in metropolitan areas other factors may enter into this. There
could be photochemical interactions creating particulates, larger diameters of hygroscopic nuclei because of the greater water vapor contents in summer, and lower wind
speeds for dispersal of the pollutant [228]. Much remains to be disentangled here.

Horizontally the particulate haze interferes with visual range in cities. When shallow
temperature inversions are present, the accumulation of aerosols can cause 80 or 90%
reduction of visual range compared with the uncontaminated environment. The haze
effect is accentuated by the formation of water droplets around hygroscopic nuclei,
even below the saturation point. This is noteworthy because relative humidities near
the surface are generally lower in cities than in the countryside. Our Columbia
experiments showed that this reduction of a few percent is half attributable to the
higher temperature and half to the reduced evaporation [203]. Nonetheless fog occurs
from two to five times as often in cities than in the surroundings [229, 230, 231, 232,
233]. Fortunately, the visibility and fog effects seem to be reversible. Recent clean-up
campaigns have shown that through the use of smokeless fuels considerable lessening
of particulate concentration can be achieved. Fog becomes again less in frequency and
the attenuation of light decreases. In London, for example, with change in heating
practices, winter sunshine has increased by 70% in the last decade and winter visibilities have improved by a factor of 3 by the switch in fuels [234].

The increase in cloudiness over cities has already been alluded to. It is likely that the
enormous number of condensation nuclei produced by human activities in and around
cities contributes to this phenomenon. Every set of measurements made has confirmed
early assessments that these constituents are more numerous by one or two orders of
magnitude in urbanized regions than in the country [235]. Every combustion process,
domestic, industrial, and principally motor vehicle exhaust, contributes to this part of

the particulates. As far as condensation processes are concerned there is little doubt that sulfur dioxide and its successor products play a major role. This effluent from human sources has doubled in the 25 yr 1940–1965 [236]. It now forms a substantial part of the background contamination [83, 237, 238]. There is good evidence that rainfall over cities has increased [239, 240] but the involvement of pollutants is not clearly substantiated. Without doubt the convection induced by the heat island can initiate or intensify showers. This has been demonstrated by Atkinson for London [241, 242], where thundershowers yield 30% more rain than in the surroundings. Orographic conditions would lead one to expect more showers in the hilly terrain but this is not the case. However, the buoyancy effect does not stand alone. In some towns there are observations of precipitation from subcooled winter stratus clouds over urban areas. Some well-documented, if isolated, cases of snow over highly industrialized towns suggest the possibility of a cloud seeding effect by some pollutants acting as freezing nuclei [243]. Also, the rather puzzling variation of urban precipitation in accordance with the human work week argues for at least a residual effect of nucleating agents produced in cities. The week is such an arbitrary division of time that artificial forces must be at work. Observations in various time intervals and various locations indicate increased precipitation for the days from Monday through Friday compared with weekend values. These increases usually parallel the increases in industrialization, and again there is evidence for a more pronounced effect during the cool season when seedable clouds are present [244, 245a].

Although most studies indicate that the overall increase in urban areas is around 10% – that is, very close to the limit of what could still be in the realm of sampling errors – there have been claims of substantially larger increases in a few cases. Most prominent is the alleged anomaly at LaPorte, Indiana [245b]. The increase there was not attributed to the effect of the town but rather to nucleating by the effluents of the nearby steelworks on the south shore of Lake Michigan. This case has not yet been lifted out of the umbra of scientific controversy, partially turning on the reliability of the observations [246, 247, 248, 249]. An analysis of precipitation downwind from steelworks in Australia does not show any effect, but the synoptic and climatological setting is quite different [250, 251, 252]. There is some weak, but independent, support of the LaPorte case by an increase in stream flow in the Kankakee River basin, which includes LaPorte [253, 254]. Tree ring studies throw little new light on the case. They show that 59% of the variance in growth rates can be attributed to general climatic fluctuations. But since 1940 a factor limiting tree growth became increasingly obvious. It could be air pollution. An increase in precipitation would have led to better growth. So the LaPorte anomaly remains in doubt [299].

On a larger scale, rainfall increases for the last quarter century have been claimed for the state of Washington. These have been attributed to the increase in pulp mills with 30% rainfall increases in the vicinity of the mills. This, too, is still contested [255, 256, 257, 258]. On the other hand, there are uncontrovertible observations of cloud banks forming for tens of kilometers in the plumes of power plants and industrial stacks. This is not necessarily associated with increased precipitation but raises the question how far downwind man's activities have caused atmospheric modifications.

In the absence of systematic three-dimensional observations we have to rely on surface data. An investigation by Band [259] has thrown some light on the conditions. He found that for a heat island 3 °C warmer than the surroundings, a small but measurable effect was still notable 3 km to leeward of the town. Similarly, a substantial increase in condensation nuclei was noted 3 km downwind from a small town. In case of a major traffic artery, an increased concentration of nuclei was measurable to 10 km downwind. For a major city, radiation measurements have suggested that the smoke pall may affect an area 50 times that of the built-up sectors. These values are probably conservative and there are increasing indications that urbanized complexes begin to modify the mesoclimate. Data collected aloft are very sparse and inconclusive as regards effects of anthropogenic nuclei on cloud and rain formation. The contaminating influence in Colorado, for example, is virtually undetectable a few tens of kilometers from small urban sources. Occasionally pollutants from distant sources seem to persist in aged aerosol at higher tropospheric altitudes but no quantitative assessment of frequency or untoward effects are available [260].

In all, it is very difficult to prove conclusively that any far-reaching climatic effects result from man's activities. If they exist or will come to pass, they will probably be a result of the vast man-made condensation and freezing nuclei. There is a large amount of recent literature concerned with this theme. A selection will be cited below but large-scale systematic investigations are called for to find unequivocal answers. One of these is now underway around St. Louis, Missouri [261]. One comforting aspect of this enigma is the fact that man-made aerosols often coagulate and become inactivated. Rain also has formidable efficiency in washing them out [262, 263]. Many of the low-level nuclei disappear this way. With an exponential decrease of nuclei with altitude, nature's cleansing process is, as yet, quite adequate. One side effect of the washout process is disquieting and that refers to increases in soil acidity reported from Sweden as an apparent result of rainfall derived from cloud droplets that picked up their nuclei from up-wind European sources of pollution.

Ice nuclei have been discovered in industrial effluents [264, 265, 266] but a particularly effective source has been claimed for automobile exhaust. Many present car fuels contain lead compounds. These have become ubiquitous, and if they combine with iodine or bromine can act as freezing nuclei. Schaefer and others have pointed out that they could have effects on precipitation far down-wind. In unstable air they have been found to 3 km altitude. In many places, especially near the seashores, iodine is present: nanograms per cubic meter. Calculations show that there is sufficient iodine to activate all available lead particles. However, in reality probably only few will achieve that state because the reactions at the small concentrations take place slowly [267, 268, 269, 270, 271]. This is one of the reversible man-made influences. As soon as lead is no longer used as a gasoline additive – hopefully soon – the supply of these nucleating agents will stop. Their effect, whatever its importance in the atmosphere, will promptly vanish because of the relatively short lifetime of these nuclei.

Perhaps more serious, and much more difficult to combat, is the oversupply of condensation nuclei. Gunn and Phillips pointed out years ago that, if too many hygroscopic particles compete for the available moisture, the cloud droplets will be small and the coalescence processes become inhibited [272]. This can lead to decreases in precipitation, a view that has recently been confirmed (273].

There remains one other area of concern: pollution caused by jet aircraft. Because of the fact that they operate in the upper troposphere and lower stratosphere their products are not as rapidly eliminated. The mean residence time of particulates in the lower troposphere is only about 10 days. In the stratosphere it is closer to 1 or 2 yr. The effective washout mechanism is absent. The most conspicuous effect of jet aircraft is condensation trails. These are occasionally quite persistent. According to one school of thought, these artificial clouds might increase the Earth's albedo and thus cause cooling [141]. Satellite pictures occasionally show cloud tracks that might have originated from these vapor trails. They seem to be so confined in time and space as to constitute a very minute fraction of the Earth's cloud cover. Some observations of the radiative effects of condensation trails have been made by Kuhn [274]. He calculates from his data that a trail 500 m thick with a 5 % time persistence would reduce surface temperatures about 0.159, a small effect indeed. Manabe [275] discusses the global equilibrium temperature as influenced by contrails as a double effect, namely an albedo increase and a decrease in outgoing terrestrial radiation. If these act like cirrus clouds and if one accepts Bryson's estimate of 0.8 % global sky cover by condensation trails – a guess at best – the effect on the Earth's temperature would be negligible.

The other view of the effect of these vapor trails, which change into cirriform clouds, is that ice crystals falling from them may nucleate lower cloud systems and cause precipitation [270, 276]. Dingle [277] points out that it is obviously difficult to document such events but shows some photographs where virga from condensation trails have induced precipitation in an altocumulus deck. He takes this as evidence that in the presence of thicker lower clouds a seeding effect would have affected the local rain pattern. If we accept this argument the question of how frequent such events are remains open. Lower clouds of appreciable thickness rarely require seeding to precipitate; they do it spontaneously.

Much alarm about stratospheric changes has been voiced as a possible consequence of a prospective fleet of supersonic transport aircraft. Military aircraft have operated at the altitudes proposed for the civilian fleet. No untoward effects have been reported. Nonetheless the question deserves serious study. Whether this can be done without actual observations of the injections of these craft and their subsequent history is doubtful. The public debate has so far been characterized by rather challengeable statements based on back-of-envelope calculations. The main arguments are again possible cloud formation of persistent type by water vapor injection, and chemical or photochemical effects on the stratospheric ozone layer, with various consequences. The cloud formation would only be a major problem if the mixing ratios were appreciably changed. We can dismiss any direct effect of the water vapor on the energy budget because a supersonic transport fleet would only add about 10^{-9} to the water vapor content of the atmosphere. There is presently appreciable transport of water vapor into the stratosphere through the general circulation and by thunderstorms breaking through the tropopause. It is also known that the water vapor mixing ratio near the tropopause and in the stratosphere shows marked horizontal variations. A fleet of supersonic transport aircraft flying on the usual commercial routes concentrated in narrow zones are not likely to affect the water vapor balance of the stratosphere seriously. A pessimistic estimate would place the stratospheric water vapor increase at 10% over that naturally injected. In narrow altitude ranges and along the routes this value may be temporarily higher but the dispersive properties of the atmosphere are likely to cause rapid horizontal and vertical diffusion [141, 278, 279, 280, 281, 282, 283, 284].

Considerably more difficult to answer than the possible cloud-producing effect of supersonic transport aircraft is the potentiality of an interaction of the water vapor produced with the atmospheric ozone. Laboratory experiments show that such dissociation can take place but there are so many other possible chemical and photochemical reactions in the middle stratosphere that deciding which one will actually take place on the basis of the known reaction rates is quite difficult. Some think that added water vapor is more likely to affect the atomic oxygen concentration than the ozone. Even if we assume that all the water vapor reacts with the ozone the global stratospheric ozone total would be diminished by less than 4%. With the considerable lateral transport of ozone the thermal effect would be minimal and the increase in ultraviolet radiation reaching the surface trivial. The last point should be stressed because of the dire warnings of the danger of increased skin cancer from this cause. With the genetically highly selective susceptibility to squamous cell cancer of the skin, the only variety apparently having ultraviolet solar radiation as etiology, the groups that might be afflicted are small and would be endangered by excessive exposure to solar radiation without the small possible increase if the ozone layer were really weakened. With the various restoring forces operating in the atmosphere the latter is highly unlikely [84, 285, 286, 287, 288, 289]. An effect claimed for the oxides of nitrogen in aircraft exhaust remains in doubt but some ozone reduction due to this is possible, although oxides of Nitrogen produced by nuclear air blasts have not shown a decisive effect on the ozone layer [300, 301, 302].

A short comment on the multitude of schemes that have been proposed to 'ameliorate' the Earth's climate seems to be in order. Most of them are either technologically or economically unfeasible. All of them would have side effects that the originators did not consider. Some of them are meteorologically sound. Bergeron [290] has given a good review of the meteorological principles that militate against weather and climatic modification, principally the dynamically stable processes of the atmospheric circulation. Interference on a larger scale would only be possible where meteorological fields are weak in low latitudes. For rainfall augmentation he thinks that irrigation, using water produced by nuclear power, would, through added evaporation, create favorable conditions for added shower activity. Diversion of some rivers might also be feasible. This is particularly talked about for some emptying into the Arctic Ocean.

For nearly all other schemes objective analysis offers little encouragement [291, 292]. Clearly, the demands for water of all types are rapidly increasing. Some predict that by the end of the century 40% of all surface flow will be needed by agricultural, industrial, and domestic uses [212]. This will increase the pressure to add to the precipitation income. How to do this without disturbing the ecology remains the question [293, 294]. All the questions of water supplies, air pollution, urbanization effects will only be resolved if man can cope with overpopulation and consequent overconsumption [295]. But whatever meteorological steps are taken to solve some of the problems by artificial means they should be analyzed for their consequences before hand. This has to include both the ecological implication and the assessment of meteorological effects elsewhere than the point of interference. Society will demand that [296, 297, 298]. All this points up the need for adequate numerical models of climate. Unfortunately, we are far from having them.

9. Conclusions

Natural climatic fluctuations, even those in recent years, cover a considerable range. They can be characterized as a 'noise' spectrum. This masks any possible effects of man-made additions to the atmosphere, such as carbon dioxide and dust. Volcanic eruptions seem to have played a greater role in these fluctuations than human activity. Local modifications, either deliberate or inadvertent, measurably affect the microclimate. Some small-scale alterations are definitely beneficial for agriculture. Among unplanned effects those produced by urbanization on local temperature and wind fields are quite pronounced. Runoff conditions are adversely affected by urbanization. The rainfall is increased by the heat-island effect but it is controversial if nucleating factors also operate and if these operate beyond the confines of metropolitan areas. The superabundance of various nuclei produced by human activity is certainly an alarming symptom.

Presently, we can state that man has produced local climatic changes and is on the verge of producing regional changes. His attempts at planned artificial modifications remain marginal. On the global scale, carbon dioxide increases should not create problems for the next few decades; dust and nuclei might. Their effect would not be appreciable on the thermal regime but their interference with natural precipitation processes is possible. Since these effects are reversible, control at the source is important.

To safeguard against any surprises and inadequacies of our analyses of climatic changes, monitoring of all atmospheric changes, including those of possible anthropogenic nature, is imperative. This should include safeguarding and expanding the system of climatic reference stations, and surveillance of atmospheric composition and radiation at remote sites. The present effort on this last point is woefully inadequate. Research on improvement of climatic models is also indicated.

Acknowledgment

The work discussed here has been supported in part by NSF (Atmospheric Science Section) grants GA-1104 and GA-13353.

References

1. Landsberg, H. E.: 'Man-Made Climatic Changes', *Science* **170**, 1265–1264 (1970).
2. Livingstone, D. A.: 'Speculations on the Climatic History of Mankind', *Am. Scientist* **59**, 332–337 (1971).
3. Robinson, G. D.: *Long-Term Effects of Air Pollution*, A Survey Center of Environment and Man, CEM, p. 4029 (1970).
4. National Center for Atmospheric Research: *Pollution, Radiation, and Climate*, NCAR Quarterly No. 27, pp. 1–10 (1970).
5. Council on Environmental Quality: *Man's Inadvertent Modification of Weather and Climate*, Chapter 5 in First Annual Report, pp. 93–104. (See also: *Bull. Am. Meteorol. Soc.* **51**, 1043–1047) (1970).
6. Berkner, L. V. and Marshall, L. C.: 'The Rise of Oxygen in the Earth's Atmosphere with Notes on the Martian Atmosphere', *Adv. Geophys.* **12**, 309–331 (1967).
7. Rasool, S. I.: 'Evolution of the Earth's Atmosphere', *Science* **157**, 1466–1467 (1967).
8. Valen, L. van: 'The History and Stability of Atmospheric Oxygen', *Science* **171**, 439–443 (1971).
9. Machta, L. and Hughes, E.: 'Atmospheric Oxygen in 1967–70', *Science* **168**, 1582–1584 (1970).
10. Broecker, W. S.: 'Man's Oxygen Reserves', *Science* **168**, 1537–1538 (1970).
11. Budyko, M. I.: 'The Effect of Solar Radiation Variations on the Climate of the Earth', *Tellus* **21**, 611–619 (1969).
12. Sellers, W. D.: 'A Global Climatic Model Based on the Energy of the Earth-Atmosphere System', *J. Appl. Meteorol* **8**, 392–400 (1969).
13. Budyko, M. I.: 'Comments on "A Global Climatic Model Based on the Energy Balance of the Earth-Atmosphere System"', *J. Appl. Meteorol.* **9**, 310–311 (1970).
14. Sellers, W. D.: 'Reply (to Budyko, 1970)', *J. Appl. Meteorol.* **9**, 311 (1970).
15. Budyko, M. I.: *Klimat i Zizn*, Gidrometeoizdat, Leningrad (1971).
16. Budyko, M. I.: *The Influence of Man's Activity on Climate*, W.M.O. lecture Congress VI (1971).
17. Brooks, C. E. P.: *Climate Through the Ages*, 2nd ed., Dover Reprint (1949).
18. Mannerfelt, C. M. ed., 'Glaciers and Climate' (Ahlmann Volume). *Geogr. Ann.* No. 1–2, 383 pp. (1949).
19. Shapley, H. ed., *Climatic Change*, Harvard Univ. Press (1953).
20. Mitchell, J. M., Jr., ed., 'Causes of Climatic Change', *Meteorol. Monogr.* **8** (30), 159 (1968).
21. Öpık, E. J.: 'Climatic Change and the Onset of the Ice Ages', *Irish Astron. J.* **8**, 153–157 (1968).
22. Kondratyev, K. Y. A.: 'Interaction Between Dynamic and Radiative Processes in Global Circulation of the Atmosphere', *W.M.O. Bull.* **20**, 78–84 (1971).
23. Milankovitch, M.: *Canon of Insolation and the Ice Age Problem*, Transl. 1969 of Kgl. Serb. Akad. Spec'l. Publ's. Vol. 132, Sect. Math. Nat. Sci., Belgrade (1941).
24. Emiliani, C.: 'Pleistocene Temperatures', *J. Geol.* **63**, 538–578 (1955).
25. Emiliani, C.: 'Paleotemperature Analysis of Core 280 and Pleistocene Correlations', *J. Geol.* **66**, 264–275 (1958).
26. Emiliani, C.: 'Isotopic Paleotemperatures', *Science* **154**, 851–857 (1966).
27. Emiliani, C.: 'Pleistocene Paleotemperatures', *Science* **168**, 822–825 (1970).
28. Emiliani, C. and Geiss, J.: 'On Glaciations and Their Causes', *Geol. Rundschau* **46**, 576–601 (1957).
29. Broecker, W. S., Thurber, D. L., Goddard, J., Ku, T. L., Matthews, R. K., and Mesolella, K. J.: 'Milankovich Hypothesis Supported by Precise Dating of Coral Reefs and Deep-Sea Sediments', *Science* **159**, 297–300 (1968).
30. Dansgaard, W. and Tauber, H.: 'Glacier Oxygen-18 Content and Pleistocene Ocean Temperatures', *Science* **166**, 499–502 (1969).

31. Shaw, D. M. and Donn, W. L.: 'Milankovitch Radiation Variations: A Quantitative Evaluation', *Science* 162, 1270–1272 (1968).
32. Sellers, W. D.: 'The Effect in the Earth's Obliquity on the Distribution of Mean Annual Sea-Level Temperature', *J. Appl. Meteorol.* 9, 960–961 (1970).
33. Bray, J. R.: 'Solar Activity Index: Validity Supported by Oxygen Isotope Index', *Science* 168, 571–572 (1970).
34. Bray, J. R.: 'Solar-Climate Relationships in the Post-Pleistocene', 171 1242–1243 (1971).
35. Purett, L.: 'Ice Cores: Clues to Past Climates', *Science News* 98, 369–370 (1970).
36. Baur, F.: 'Meteorologischer Nachweis von Strahlungsschwankungen der Sonne', *Meteorol. Abhdlg.* 50 (4), 88 pp. (1967).
37. Flohn, H.: 'Klimaschwankungen und grossräumige Klimabeeinflussung', *Bonner Meteorol. Abhdlg.* No. 2 (1963).
38. Fritts, A. C.: 'Tree-Ring Evidence for Climatic Changes in Western North America', *Monthly Weather Rev.* 93, 421–443 (1965).
39. Fritts, A. C.: 'Tree-Ring Analysis: A Tool for Water Resources Research', *IHD. Bull. EOS* 50, 22–29 (1969).
40. Lamb, H. H.: 'On the Nature of Certain Climatic Epochs which Differed from the Modern (1900–39) Normal in Changes of Climate', *UNESCO Arid Zone Research* 20, 125–150 (1963).
41. Yamamoto, T.: 'On the Climatic Change in XV and XVI Centuries in Japan', *Geophys. Mag.* 35, 187–206 (1971).
42. Rudloff, H. V.: *Die Schwankungen und Pendelungen des Klimas in Europa seit dem Beginn der regelmässigen Instrumenten-Beobachtungen (1670)*, Friedr. Vieweg & Sohn, Braunschweig, (1967).
43. Anderson, T.: *Temperature and Precipitation Records in Sweden*, Meteorol. Inst. Uppsala Univ. Repts, No. 13, (1969).
44. Anderson, T.: *Swedish Temperature and Precipitation Records Since the Middle of the 19th Century*, NH, Inst. Bldg. Res. Stockholm, Document D4, (1970).
45. Baur, F.: 'Langjährige Beobachtungsreihen', in F. Linke (ed.), *Meteorologisches Taschenbuch*, Neue Ausgabe, Vol. I, pp. 710–799 (1962).
46. Manley, G.: 'The Mean Temperature of Central England 1698–1952, *Quart. J. Roy. Meteorol. Soc.* 79, 558–567 (1953).
47. Konček, M. and Cehak, K.: 'Säkulare Temperaturschwankungen in Mitteleuropa während der letzten 190 Jahre', *Arch. Meteorol. Geophys. Bioklim.* Ser. B 16, 1–17 (1968).
48. Lamb, H. H.: *The Changing Climate*, London, Methuen & Co. Ltd., (1966).
49. Rubinstein, Ye. S. and Polozova, L. G.: *Sovremennoe Izmenenie Klimata*, Leningrad, (1966).
50. Arakawa, H.: 'On the Secular Variation of Annual Totals of Rainfall at Seoul From 1770–1944', *Ard. Meteorol. Geophys. Bioklim.*, Ser. B. 7, 205–211 (1956).
51. Arakawa, H.: 'Climatic Changes as Revealed by the Data from the Far East', *Weather* 12, 46–51 (1957).
52. Yamamoto, T.: 'On the Nature of the Japanese Climate in the So-Called "Little Ice Age" between 1750 and 1850', *Geophys. Mag.* 35, 165–185 (1971).
53. Landsberg, H. E.: 'Two Centuries of New England Climate', *Weatherwise* 20, 52–57 (1967).
54. Landsberg, H. E., Yu, C. S., and Huang, L.: *Preliminary Reconstruction of a Long Time Series of Climatic Data for the Eastern United States*, Univ. of Md. Inst. for Fluid Dynamics and Appl. Math., Tech. Note BN–57, (1968).
55. Baker, D. G.: 'Temperature Trends in Minnesota', *Bull. Am. Meteorol. Soc.* 41, 18–27 (1960).
56. Wahl, E.: 'A Comparison of Climate of the Eastern United States during the 1830s with the Current Normals', *Monthly Weather Rev.* 96, 73–82 (1968).
57. Eukui, E.: *The Recent Rise of Temperature in Japan*. Jap. Prog. in Climatol., pp. 46–65 (1970).
58. Langbein, W. B.: 'Comparison of Ground-Water Temperatures in the 1790s with Those of Recent Years', *U.S. Geol. Survey Water Res. Rev.* 8 (1968).
59. Hurley, C. W., Jr.: 'Recent Temperature and Snowfall Trends', *Mt. Washington Obs. New Bull.* 10, 13–14 (1969).
60. Carlson, W. S.: 'Ice Survey by the U.S. Coast Guard', *Science* 168, 396–397 (1970).
61. Bierknes, J.: 'Atlantic Sea-Air Interaction', *Adv. Geophys.* 10, 1–82 (1964).

62. Rasool, S. I. and Hogan, J. S.: 'Ocean Circulation and Climatic Change', *Bull. Am. Meteorol. Soc.* **50**, 130–134 (1969).
63. Yakovleva, N. I.: 'The Auto-Oscillatory Character of Climatic Change', *Izvest. Acad. Sci. U.S.S.R. – Atm. Ocean. Physics* (AGU Transl.) **5**, 699–700 (1969).
64. Namias, J.: 'Long-Range Forecasting of the Atmosphere and its Oceanic Boundary – an Inter-disciplinary Problem', *Calif. Mas. Res. Comm. Cal COFI Report.* **12**, 29–42 (1968).
65. Namias, J.: *Factors Associated with the Persistence and Termination of the Recent Northeast Drought*, Proceed. 4th Am. Water Resources Conf., pp. 582–594 (1968).
66. Namias, J.: 'Seasonal Interactions Between the North Pacific Ocean and the Atmosphere During the 1960's', *Monthly Weather Rev.* **97**, 173–192 (1969).
67. Namias, J.: 'On the Causes of the Small Number of Atlantic Hurricanes in 1968', *Monthly Weather Rev.* **97**, 346–348 (1969).
68. Namias, J.: 'Macroscale Variations in the Sea-Surface Temperature in the North Pacific', *J. Geophys. Res.* **75**, 565–582 (1970).
69. Namias, J.: 'Climatic Anomaly over the United States during the 1960s', *Science* **170**, 741–743 (1970).
70. Namias, J. and Born, R. M.: 'Temporal Coherence in North Pacific Sea-Surface Temperature Patterns', *J. Geophys. Res.* **75**, 5952–5955 (1970).
71. Arrhenius, S.: *Worlds in the Making*, Harper & Brothers, New York & London, pp. 51–54 (1908).
72. Callendar, G. S.: 'The Artificial Production of Carbon Dioxide and its Influence on Temper-ature', *Quart. J. Roy. Meteorol. Soc.* **64**, 223–237 (1938).
73. Plass, G. N.: 'The Carbon Dioxide Theory of Climatic Changes', *Am. J. Phys.* **24**, 376 (1956).
74. Brown, K. W. and Rosenberg, N. J.: 'Concentration of CO_2 in the Air Above a Sugar Beet Field', *Monthly Weather Rev.* **98**, 75–82 (1970).
75. Allen, L. H., Jr.: 'Variations in the Carbon Dioxide Concentration over an Agricultural Field', *Agr. Meteorol.* **8**, 5–24 (1971).
76. Bray, J. R.: 'An Analysis of the Possible Recent Change in Atmospheric Carbon Dioxide Concentration', *Tellus* **11**, 220–230 (1959).
77. Bischoff, W. and Bolin, B.: 'Space and Time Variations of CO_2 Content of the Troposphere and Lower Stratosphere', *Tellus* **18**, 155–159 (1966).
78. Callendar, G. S.: 'On the Amount of Carbon Dioxide in the Atmosphere', *Tellus* **10**, 243–248 (1958).
79. Harris, T. B.: 'Evidence of the Continuing Increase of Carbon Dioxide', *Bull. Am. Meteor. Soc.* **51**, 101 (1970).
80. Bolin, B. and Bischof, W.: 'Variations of the Carbon Dioxide Content of the Atmosphere in the Northern Hemisphere', *Tellus* **22**, 431–442 (1970).
81. Ariyama, T.: 'Carbon Dioxide in the Atmosphere and in Seawater in the Adjacent Seas of Japan', *Oceanog. Mag.* **21**, 53–59 (1969).
82. Junge, C. E. and Czeplak, G.: 'Some Aspects of the Seasonal Variation of Carbon Dioxide and Ozone', *Tellus* **20**, 422–434 (1968).
83. Georgii, H. W.: 'Neue luftchemische Untersuchungen in reiner Luft', *Bundesgesundheitsblatt*, No. 14., pp. 192–194 (1970).
84. *Man's Impact on the Global Environment*, Report of the Study of Critical Environmental Prob-lems, MIT Press, pp. 41–112 (1970).
85. Bolin, B.: *Large-Scale Atmospheric Pollution Problems and their Possible Repercussions on Global Climates*, WMO lecture Congress VI (1971).
86. Monteith, J. H.: 'Prospects for Photosynthesis from A.D. 1970 to A.D. 2000', *Weather* **25**, 456–462 (1970).
87. Leith, H.: 'The Role of Vegetation in the Carbon Dioxide Content of the Atmosphere', *J. Geophys. Res.* **68**, 3887–3898 (1963).
88. Peterson, E. K.: 'Carbon Dioxide Affects Global Ecology', *Env. Sci. Technol.* **3**, 1162–1169 (1969).
89. Revelle, R. and Suess, H. E.: 'Carbon Dioxide Exchange Between Atmosphere and Ocean and the Question of an Increase of Atmospheric CO_2 During the Past Decades', *Tellus* **9**, 18–27 (1957).

90. Suess, H. E.: 'Tritium Geophysics as an International Research Project', *Science* **163**, 1405–1410 (1969).

91. Berger, R. and Libby, W. F.: 'Equilibrium of Atmospheric Carbon Dioxide with Sea Water: Possible Enzymatic Control of the Rate', *Science* **164**, 1395–1397 (1969).
Berger, R. and Libby, W. F.: 'Exchange of CO_2 Between Atmosphere and Sea Water: Possible Enzymatic Control of the Rate', this volume, pp. 79–82.

92. Quinn, J. A. and Otto, N. C.: 'Carbon Dioxide Exchange at the Air-Sea Interface: Flux Augmentation by Chemical Reaction', *J. Geophys. Res.* **76**, 1539–1549 (1971).

93. Manabe, S. and Wetherald, R. T.: 'Thermal Equilibrium of the Atmosphere with a Given Distribution of Relative Humidity', *J. Atmos. Sci.* **24**, 241–259 (1967).

94. Manabe, S.: 'The Dependence of Atmospheric Temperature on the Concentration of Carbon Dioxide', this volume, pp. 73–78.

95. Rasool, S. I. and Schneider, S. H.: 'Effects of Large Increases in Atmospheric CO_2 and Aerosols on the Global Climate', *Science* **173**, 138–41 (1971).

96. Mitchell, J. M., Jr.: 'A Preliminary Evaluation of Atmospheric Pollution as a Cause of the Global Temperature Fluctuation in the Past Century', in S. F. Singer (ed.), *Global Effects of Environmental Pollution*, D. Reidel Publ. Co., pp. 139–155 (1970).

97. Molley, F.: 'The Influence of CO_2 Concentration in Air on the Radiation Balance of the Earth's Surface and on the Climate', *J. Geophys. Res.* **68**, 3877–3886 (1963).

98. London, J. and Sasamori, T.: 'Radiative Cooling as a Function of Temperature and Water Vapor in the Free Atmosphere', *J. Geophys' Res.* **75**, 6862–6863 (1970).

99. Newell, R. E. and Dopplick, T. G.: 'The Effect of Changing CO_2 Concentration on Radiative Heating Rates', *J. Appl. Meteorol.* **9**, 958–959 (1970).

100. Gates, D.: 'Weather Modification in the Service of Mankind: Promise or Peril?' in W. H. Helfrich, Jr. (ed.), *Environmental Crisis: Man's Struggle to Live with Himself*, Yale Univ. Press, New Haven, pp. 33–46 (1970).

101. Johnson, F. S.: 'The Oxygen and Carbon Dioxide Balance in The Earth's Atmosphere', this volume, pp. 49–56.

102. Symons, G. J. (ed.), *The Eruption of Krakatoa and Subsequent Phenomena*, Royal Society, London (1888).

103. Humphreys, W. J.: *Physics of the Air*, 3rd ed., McGraw Hill Book Co., pp. 587–618 (1940).

104. Volz, F. E.: 'Turbidity at Uppsala from 1909 to 1922 from Sjöström's Solar Radiation Measurements', *Meddel. Sver. Meteor. Hydrol. Inst. Ser.* B., **28**, 100–104 (1968).

105. Meinel, A. B. and Meinel, M. P.: 'Volcanic Sunset-Glow Stratum', *Science* **155**, 189 (1967).

106. Ebdon, R. A.: 'Possible Effects of Volcanic Dust on Stratospheric Temperatures and Winds', *Weather* **22**, 245–249 (1967).

107. Newell, R. E.: 'Modification of Stratospheric Properties by Trace Constituent Changes', *Nature* **227** 697–699 (1970).

108. Newell, R. E.: 'Stratospheric Temperature Change from the Mt. Agung Volcanic Eruption of 1963', *J. Atmos. Sci.* **27**, 977–978 (1970).

109. Cronin, J. F.: 'Recent Volcanism and the Stratosphere', *Science* **172**, 847–849 (1971).

110. Volz, F. E.: 'On Dust in the Tropical and Midlatitude Stratosphere from Recent Twilight Measurements', *J. Geophys. Res.* **75**, 1641–1646 (1970).

111. Volz, F. E.: 'Atmospheric Turbidity After the Agung Eruption of 1963 and Size Distribution of the Atmospheric Aerosol', *J. Geophys. Res.* **75**, 5185–5193 (1970).

112. Volz, F. E.: 'Stratospheric Aerosol Layers from Balloon-Borne Horizon Photographs', *J. Atmos. Sci.* reprint, (1971).

113. Unz, F.: 'Tropospheric and Stratospheric Concentration of Aerosol Obtained from Daytime Measurements of Sky Polarization in the Zenith', *Beitr. Phys. Atmos.* **42**, 1–35.

114. Lamb, H. H.: 'Volcanic Dust in the Atmosphere; With a Chronology and Assessment of its Meteorological Significance', *Phil. Transact. Roy. Soc.* **266**, 425–533 (1970).

115. Volz, F. E.: 'Spectral Skylight and Solar Radiance Measurements in the Caribbean: Maritime Aerosols and Sahara Dust', *J. Atmos. Sci.* **27**, 1041–1047 (1970).

116. Landsberg, H. E.: 'Observations of Condensation Nuclei in the Atmosphere', *Monthly Weather Rev.* **62**, 442–445 (1934).

117. Landsberg, H. E.: 'Atmospheric Condensation Nuclei', *Erg. Kosm. Phys.* **3**, 155–252 (1938)

118. Junge, C. E.: *Atmosphärische Spurenstoffe und ihre Bedeutung für den Menschen*, Sympos. 1966, Klimaphysiol. Stat. St. Moritz, Birkhäuser, Verlag, Basel, 9 (1967).
119. Abel, N., Jaenicke, R., Junge, C., Kanter, H., Garcia Prieto, R., and Seiler, W.: 'Luftchemische Studien am Observatorium Izana (Teneriffa)', *Meteorol. Rundschau* 22, 158–167 (1969).
120. Barrett, E. W., Pueschel, R. F., Weickmann, H. K. and Kuhn, P. M. *Inadvertent Modification of Weather and Climate by Atmospheric Pollutants*, ESSA. Tech. Report ERL 185-APCL 15, (1970).
121. Dinger, J. E., Howell, H. B. and Wojciechowski, T. A.: 'On the Source and Composition of Cloud Nuclei in a Subsident Air Mass over the North Atlantic', *J. Atmos. Sci.* 27, 791–797 (1970).
122. Cobb, W. E. and Wells, H. J.: 'The Electrical Conductivity of Oceanic Air and its Correlation to Global Atmospheric Pollution', *J. Atmos. Sci.* 27, 814–819 (1970).
123. McCormick, R. A. and Ludwig, J. H.: 'Climate Modification by Atmospheric Aerosols', *Science* 156, 1358–1359 (1967).
124. Volz, F. E.: 'Some Results of Turbidity Networks', *Tellus* 21, 625–630 (1969).
125. Bary, E. de and Rossler, F.: 'Compensation of the Particle Concentration from Measurements of the Scattered Radiation in the Stratospheric Dust', *Beitr. Phys. Atmos.* 43, 87–92 (1970).
126. Fischer, W. H.: 'Atmospheric Aerosol Background Level', *Science* 171, 828–829 (1971).
127. Lindenbein, B. and Wehry, W.: 'Bericht über die International Conference on Meteorology in Israel', *Beil. Berliner Wetterkarte* SO3/71, 7 pp. (1971).
128. Porch, W. M., Charlson, R. J., and Radke, L. F.: 'Atmospheric Aerosol: Does a Background Level Exist?', *Science* 170, 315 (1970).
129. Peterson, J. T. and Bryson, R. A.: 'Atmospheric Aerosols: Increased Concentration in the Last Decade', *Science* 162, 120–121 (1968).
130. Ellis, H. T. and Pueschel, R. F.: 'Solar Radiation: Absence of Air Pollution Trends at Mauna Loa', *Science* 172, 845–846 (1971).
131. Bryson, R. A.: '"All Other Factors Being Constant…" – A Reconciliation of Several Theories of Climatic Change', *Weatherwise* 21, 56–60, 94 (1968).
132. Barrett, E. W.: *Depletion of Total Solar Short-Wave Radiance at the Ground by Suspended Particulates*, paper presented at International Solar Energy Conf., Melbourne, Australia, preprint 6 pp. (1970).
133. Roach, W. T.: 'Some Aircraft Observations of Solar Radiation in the Atmosphere', *Quart. J. Roy. Meteorol. Soc.* 87, 346–363 (1961).
134. Bullrich, K.: 'Scattered Radiation in the Atmosphere and the Natural Aerosol', *Adv. Geophys.* 10, 99–260 (1964).
135. Quenzel, H.: 'Über den Einfluss der vertikalen Trübungsschicht auf den Energiegewinn der Atmosphäre durch Kontinuum-Absorption', *Pure Appl. Geophys.* 71, 149–163 (1968).
136. Charlson, R. J. and Pilat, M. J.: 'Climate: The Influence of Aerosols'. *J. Appl. Meteorol.* 8, 1001–1002 (1969).
137. Kattawav, G. W. and Plass, G. N.: 'Influence of Aerosols, Clouds, and Molecular Absorption on Atmospheric Emission', *J. Geophys. Res.* 76, 3437–3444 (1971).
138. Mitchell, J. M., Jr.: *The Effect of Atmospheric Aerosol on the Climate with Special Reference to Surface Temperature*, NOAA Tech. Memo. EDS 18, 28 pp. (1970).
139. Went, F.: 'Blue Hazes in the Atmosphere', *Nature* 187, 641–643 (1960).
140. Lovelock, J. E.: 'Air Pollution and Climatic Change', *Atmos. Environ.* 5 (1971).
141. Bryson, R. A. and Wendland, W. M.: 'Climatic Effects of Atmospheric Pollution', this volume pp. 139–148.
142. Blifford, I. H., Jr.: 'Tropospheric Aerosols', *J. Geophys. Res.* 75, 3099–3103 (1970).
143. Desai, B. N.: 'Is It Possible to Increase Rainfall in the Semi-Arid Regions of Rajasthan by Reducing Quantity of Dust Suspended in the Atmosphere?', *Indian J. Meteorol. Geophys.* 20, 377–380 (1969).
144. Landsberg, H. E.: 'The Use of Solar Energy for the Melting of Ice', *Bull. Am. Meteorol. Soc.* 21, 102–104 (1940).
145. Titlianov, A.: 'Accelerated Snow Melting in the Fields of Kamchatka Peninsula for Agricultural Purposes', *Dokl. Vses. Akad. Sel. skokhozy. Nauk* 6 (8), 8–11 (1941).
146. Kolchin, A. I.: 'Rapid Melting of Snow', *Les. Khoz.* 3 (3), 69–73 (1950).
147. Kolchin, A. I.: 'Artificial Snow Melting and its Utilization', *Les i Step.* 3 (1), 77–79 (1951).
148. Avsiuk, G. A.: 'Intensification of Glacier Melting', *Priroda* 43 (3), 82–86 (1954).

149. Atwater, M. A.: 'Planetary Albedo Changes Due to Aerosols', *Science* **170** 64–66 (1970).
150. Davitaya, F. F.: 'Possible Influence of the Atmospheric Dust on the Regression of Glaciers and the Warming of Climate', *Izv. Akad. Nauk S.S.S.R. Ser. Geogr.* **2**, 3–22. Transact. *Soviet Acad. Sci. Geogr.* (engl. transl.), **2**, 3–22 (1965).
151. Bloch, M. R.: 'Hypothesis for the Change of Ocean Levels Depending on the Albedo of the Polar Ice Caps', *Palaeogeogr. Paleocim. Palaeoecol.* **1**, 124–142 (1965).
152. Fletcher, J. O.: *The Polar Oceans and World Climate*, The Rand Corp. P–3801 (1968).
153. Fletcher, J. O.: *Managing Climatic Resources*, Rand Corp. P-4000-1, (1969).
154. Jefferson, T.: Letter to Dr. Lewis Beck of Albany, dated Monticello, July 16 (1824).
155. Geiger, R.: *Das Klima der bodennahen Luftschicht*, 4th ed. Friedr. Vieweg & Sohn, Braunschweig, p. 503 (1961).
156. Tanner, C. B.: personal communications, dated January 29 and March 30 (1971).
157. Eimern, J. van, *et al.*: *Windbreaks and Shelter Belts*, WMO Tech. Note No. 59 (1964).
158. Caborn, J. M.: *Shelter Belts and Windbreaks*, Faber and Faber, London (1965).
159. Campbell, I. A.: 'Climate and Overgrazing on the Shonto Plateau, Arizona', *Prof. Geogr.* **22**, 132–141 (1970).
160. Fournier D'Albe, E. M.: 'The Modification of Microclimates', in "Climatology", *UNESCO Arid Zone Res.* **10**, 126–146 (1958).
161. Thornwaite, C. W.: 'The Modification of Rural Microclimates', in W. L. Thomas, Jr. (ed.), *Man's Role in Changing the Face of the Earth*, Univ. of Chicago Press, pp. 567–583 (1956).
162. Joos, L. A.: *Recent Rainfall Patterns in the Great Plains*, paper presented at Am. Meteorol. Soc. Sympos., Madison, Wis. Oct. 21 (1969).
163. Landsberg, H. E.: *Physical Climatology*, 2nd ed., Gray Printing Co., Inc., Dubois, Pa., (1966).
164. Huff, F. A.: 'Sampling Errors in Measurement of Mean Precipitation', *J. Appl. Meteorol.* **9**, 35–44 (1970).
165. Korniyenko, V. I.: 'Influence of Evaporation from Lake Backal on Precipitation in the Surrounding Regions', *Soviet Hydrol., Selected Papers*, No. 2, 144–147 (1969).
166. Holmes, R. M.: 'Meso-Scale Effects of Agriculture and a Large Prairie Lake on the Atmospheric Boundary Layer', *Agron. J.* **62**, 546–549 (1970).
167. Zych, S. and Dubaniewicz, H.: 'Wplyw Zbiornika retencyjnego Otmuchow i hipoteza oddzialycvania zbiornika glebinow na klimat miejskowy', *Zeszyty Nauk. Univ. hodz.* Ser II, **32**, 3–20 (1969).
168. Ackermann, H. J. and Rabchevsky, G. A.: *Application of Nimbus Satellite Imagery to the Monitoring of Man-Made Lakes*, paper presented at Man-Made Lakes Symposium, Knoxville, Tenn. 3–7 May (1971).
169. Pushkarev, V. F. and Leochenke, G. P.: 'Use of Monomolecular Films to Reduce Evaporation from the Surface of Bodies of Water', *Soviet Hydrol. Selected Papers*, No. 3, 253–272 (1967).
170. Gangopadhyaya, M. and Venkataraman, S.: 'Evaporation Reduction', *Agr. Meteorol.* **6**, 339–345 (1969).
171. Kappessev, R., Greif, R., and Cornet, I.: 'Evaporation Retardation by Monolayers', *Science* **166**, 403 (1969).
172. Beckwith, W. B.: 'Impacts of Weather on the Airline Industry: The Value of Fog Dispersal Programs', in W.R.D. Sewell (ed.), *Human Dimensions of Weather Modification*, Univ. of Chicago, Dept. of Geog. Res. Paper No. 105, pp. 195–207 (1966).
173. Silverman, B. A.: 'An Appraisal of Warm Fog Modification', *Bull. Am. Meteorol. Soc.* **51**, 420 (1970).
174. National Academy of Sciences – National Research Council: *Weather and Climate Modification, Problems and Prospects*, Vol. I – Summary and recommendations, Vol. II – Research and development, NAS/NRC Publ. No. 1350 (1966).
175. Neiburger, M.: *Artificial Modification of Clouds and Precipitation*, WMO Technical Note No. 105 (1969).
176. Maybank, J. and Baier, W. (eds.): *Weather Modification: A Survey of the Present Status with Respect to Agriculture*, Res. Branch, Can. Dept. of Agric., Ottawa, (1970).
177. Tribus, M.: 'Physical View of Cloud Seeding', *Science* **168**, 201–211 (1970).
178. Lecam, L. and Neyman, J. (eds).: *Weather Modification Experiments*, Proceed. 5th Berkeley Symposium on Math. Stat. and Probabil., Univ. of Calif., Berkeley (1967).

179. Huff, F. A.: 'Evaluation of Precipitation Records in Weather Modification Experiments', *Adv. Geophys.* **15**, (in print) (1971).
180. Kornienko, E. E.: 'The Problem of the Effect of Convective Cloud Seeding with Solid Carbon Dioxide on Their Vertical Development (transl. title)', *Tr. Nauch Issled. Gidrometeorol. Inst. Ukraine*, No. **77**, 130–135 (1969).
181. Woodley, W. L.: 'Rainfall Enhancement by Dynamic Cloud Modification', *Science*, **170**, 127–132 (1970).
182. Stinson, J. R.: 'Project Skywater', in J. L. Gardner and L. E. Myers (eds.), *Water Supplies for Arid Regions*, Vol. 10 (1967).
183. U.S. Department of the Interior: Office of Atmospheric Water Resources. Project Skywater, Annual Reports, Denver, 1969, (1970, 1971).
184. Bergeron, T.: 'Mesometeorological Studies of Precipitation, IV, Oreigenic and Convective Rainfall Patterns', *Meteorol. Inst. Uppsala Rep.* **20**, (1970).
185. Sharon, D.: *A Real Pattern of Rainfall in a Small Watershed*, paper presented at Symposium on Results on Representative and Experimental Basins, IASH, Wellington, N.Z. (1970).
186. Schleusener, R. A.: 'Hailfall Damage Suppression by Cloud Seeding – a Review of Evidence', *J. Appl. Meteorol.* **7**, 1004–1011 (1968).
187. Sulakvelidze, G. K. (ed.): 'Metodi vozdeistviia na gradovye protsessy', *Vysokogornyi Geofiz. Tr.* **11**, (1968).
188. Battan, L. J.: 'Summary of Soviet Publications on Weather Modification', *Bull. Am. Meteorol. Soc.* **51**, 1030–1041 (1970).
189. Gentry, R. C.: 'Hurricane Debbie Modification Experiments, August 1969', *Science* **168**, 473–475 (1970).
190. Cry, G. W.: 'Effects of Tropical Cyclone Rainfall on the Distribution of Precipitation over the Eastern and Southern United States', ESSA Professional Papers, No. 1, (1967).
191. Sugg, A. L.: 'Beneficial Aspects of the Tropical Cyclone', *J. Appl. Meteorol.* **7**, 39–45 (1968).
192. Kratzer, A.: 'Das Stadtklima', *Die Wissenschaft* **90**, Friedr. Vieweg & Sohn, Braunschweig, (1956).
193. Landsberg, H. E.: 'The Climate of Towns', in W. L. Thomas, Jr. (ed.), *Man's Role in Changing the Face of the Earth*, Univ. of Chicago Press, pp. 584–603.
194. Landsberg, H. E.: City Air – Better or Worse? in *Symposium: Air over Cities*, H. S. Publ. Health Serv. R.A. Taft San. Engr. Ctr. Tech. Rep. A62–5, pp. 1–22 (1962).
195. Peterson, J. T.: *The Climate of Cities: A Survey of Recent Literature*, Nat. Air Pollut. Control Admin. Publ., AP-59, (1969).
196. Chandler, T. J.: 'Urban Climatology – Inventory and Prospect', in *Urban Climates*, WMO Tech. Note 108, pp. 1–14 (1970).
197. Chandler, T. J.: *Selected Bibliography on Urban Climate*, WMO-No. 276, TP. 155 (1970).
198. Estoque, M. A.: *A Numerical Model of the Atmospheric Boundary Layer*, GRD Scientif. Rep. No. 2 (Contract AF 19(604)–7484,) AFCR, 18 pp. (1962).
199. Myrup, L. O.: 'A Numerical Model of the Urban Heat Island', *J. Appl. Meteorol.* **8**, pp. 908–918, corrigendum, *J. Appl. Meteorol.* **9**, 541 (1969).
200. Tag, P. M.: 'Surface Temperatures in Urban Environment', in *Atmospheric Modification by Surface Influences*, Rep. No. 15, Dept. Meteorol., Pa. State Univ. pp. 1–71 (1969).
201. Delage, Y. and Taylor, P. A.: 'Numerical Studies of Heat Island Circulations', *Boundary-Layer Meteorol.* **1**, 201–226 (1970).
202. Landsberg, H. E.: 'Climatic Consequences of Urbanization', *J. Wash. Acad. Sci.* **60**, 82–87 (1970).
203. Maisel, T. N.: '*Early Micrometeorological Changes Caused by Urbanization*', Univ. of Maryland M.S. thesis, (unpublished) (1971).
204. Landsberg, H. E.: 'Micrometeorological Temperature Differentiation Through Urbanization', in *Urban Climates*, WMO Tech. Note No. 108, pp. 129–136 (1970).
205. Oke, T. R. and Hannell, F. G.: 'The Form of The Urban Heat Island in Hamilton, Canada', in *Urban Climates*, WMO Tech. Note No. 108, pp. 1130126 (1970).
206. Nicholas, F.: *Synoptic Climatology of Metro-Washington: A Meso-Scale Analysis of the Urban Heat Island under Selected Weather Conditions*, Univ. of Maryland M.S. thesis, (unpublished) (1970).

207. Dettwiller, J.: 'Deep Soil Temperature Trends and Urban Effects at Paris', *J. Appl. Meteorol.* 9, 178–180 (1970).

208. East, C.: *Chaleur Urbaine a Montreal.* paper presented at 2nd Can. Conference on Micrometeorology, (1971).

209. Oke, T. R. and East, C.: 'The Urban Boundary Layer in Montreal', *Boundary-Layer Meteorol.* 1, 411–437 (1971).

210. Jaske, R. T., Fletcher, J. F., and Wise, K. R.: *A National Estimate of Public and Industrial Heat Rejection Requirements by Decades Through the Year 2000 A.D.*, Batelle Pacific N.L.O. Lab., BNWL-SA-3052, 23 pp. (1970).

211. Coulomb, J.: *Good Use of Scientists.* Nat. Acad. Engr., Nat. Res. Council News Report, 20 (3), pp. 6–9 (1970).

212. Fedorov, E. K.: *The Interaction of Man and His Environment – the Present State of Affairs and Prospects for the Future*, WMO lecture Congress VI (1971).

213. Hess, W. A., Hoecker, W. H., Pack, D. H., and Angell, J. K.: 'Analysis of Low-Level Constant Volume Balloon (Tetroon) Flights over New York City', *Quart. J. Roy. Meteorol. Soc.* 93, 483–493 (1967).

214. Pooler, F., Jr.: 'Airflow over a City in Terrain of Moderate Relief', *J. Appl. Meteorol.* 2, 446–456 (1963).

215. Munn, R. E.: 'Airflow in Urban Areas', in *Urban Climates*, WMO Tech. Note 108, pp. 15–39 (1970).

216. Dougal, M. D. D. (ed.): *Flood Plain Management – Iowa's Experience*, Iowa State Univ. Press (1970).

217. Espey, W. H. K., Morgan, C. W., and Marsh, F. D.: *Study of Some Effects of Urbanization on Storm Run-Off from a Small Water Shed*, Texas Water Devel. Bd. Rep. 23, (1966).

218. Mateus, L. A.: *Flood Inundation and Effects of Urbanization in Metropolitan Charlotte, North Carolina*, U.S. Geol. Surv. Water-Supply Pap. 1591-C, (1968).

219. Teaburn, G. E.: *Effects of Urban Development on Direct Runoff to East Meadow Brook, Nassau County, Long Island, N.Y.*, U.S. Geol. Surv. Prof. Pap. 6217-B, (1969).

220. Stall, J. B., Terstriep, M. L., and Huff, F. A.: *Some Effects of Urbanization on Floods*, ASCE National Water Resources Engineering Meeting, Memphis, Tenn., Preprint 1130, (1970).

221. Dettwiller, J.: *Le vent au sommet de la tour Eiffel*, Monogr. Met. Nat. No. 64 (1969).

222. Stern, A. C.: *Air Pollution*, Vol. 3, Academic Press, N.Y. (1968).

223. Haagen-Smit, A. J., Bradley, E. E., and Fox, M. M.: 'Ozone Formation in Photochemical Oxidation of Organic Substances', *Ind. Eng. Chem.* 45, 2086 (1953).

224. Angell, J. K., Pack, D. H., Holzworth, G. C., and Dickson, C. R.: 'Tetroon Trajectories in Urban Atmospheres', *J. Appl. Meteorol.* 5, 565–572. (1966).

225. Neiburger, M.: 'The Role of Meteorology in the Study and Control of Air Pollution', *Bull. Am. Meteorol.* 50, 957–965 (1969).

226. Neiburger, M.: 'Diffusion Models of Urban Air Pollution', in *Urban Climates*, WMO Tech. Note No. 108, pp. 248–262 (1970).

227. Lauscher, F. and Steinhauser, F.: 'Strahlungsuntersuchungen in Wien and Umgebung', *Sitzber. Wien. Akad. Wiss. Math.-Natw. Kl. Abtza* 141, 15–32, 143, 175–196 (1932, 1934).

228. Flowers, E. C.: '*New Amplified Sun Photometers for Turbidity Measurements at* 3500 Å *and* 5000 Å', WMO Upper Air Instruments and Observations, pp. 533–542 (1970).

229. McNulty, R. P.: 'The Effect of Air Pollutants on Visibility in Fog and Haze at New York City', *Atm. Env.* 2, 625–627 (1968).

230. Charlson, R. J.: 'Atmospheric Visibility Related to Aerosol Mass Concentration', *Rev. Env. Sci. Technol.* 3, 913–918 (1969).

231. McCaldin, R. O., Johnson, L. W., and Stephens, N. T.: 'Atmospheric Aerosols', *Science* 166, 381–382 (1969).

232. Collier, C. G.: 'Fog at Manchester', *Weather* 25, 25–29 (1970).

233. Burt, E. W.: 'A Study of the Relation of Visibility to Air Pollution', *Amer. Ind. Hyg. Ass. J.* 22, 102–108 (1961).

234. Jenkins, L.: 'Increase in Averages of Sunshine in Greater London', *Weather* 24, 52–54 (1969).

235. Landsberg, H. E.: 'The Environmental Variation of Condensation Nuclei', *Bull. Am. Meteorol. Soc.* 18, 172 (1937).

236. Robinson, E. and Robbins, R. C.: 'Gaseous Atmospheric Pollutants from Urban and Natural Sources', this volume, pp. 111–124.
237. Georgii, H. W.: 'Das natürliche Aerosol in reiner und verunreinigter Luft', *Z. Angew. Bäder Klimaheilk.* **16**, 466–472 (1969).
238. Georgii, H. W.: 'Contribution to the Atmospheric Sulfur Budget', *J. Geophys. Res.* **75**, 2365–2371 (1970).
239. Chagnon, S. A., Jr.: 'Recent Studies of Urban Effects on Precipitation in the United States', in *Urban Climates*, WMO Tech. Note No. 108, pp. 325–241 (1970).
240. Reidat, R.: 'Über den Einfluss der Stadt auf die Niederschlagsverteilung bei starken Regenfällen in Hamburg', *Wetter Leben* **23**, 1–6 (1971).
241. Atkinson, B. W.: *A Further Examination of the Urban Maximum of Thunder Rainfall in London, 1951–1960*, Transact. Pap., The Inst. of Brit. Geogr., Publ. No. 48, p. 97.
242. Atkinson, B. W.: 'The Effect of an Urban Area on the Precipitation from a Moving Thunderstorm', *J. Appl. Meteorol.* **10**, 47–55 (1971).
243. Kienle, J. v.: 'Ein stadtgebundener Schneefall in Mannheim', *Meteorol. Rundschau* **5**, 132–133 (1952).
244. Detwiller, J.: *Incidence possible de l'activite' industrielle sur les precipitations a Paris*, WMO Tech. Note No. 108, pp. 361–362 (1970).
245a. Frederick, R. H.: 'Preliminary Results of a Study of Precipitation by Day-of-the-Week over Eastern United States', *Bull. Am. Meteorol. Soc.* **51**, p. 100; also *Am. Meteorol. Soc.* 2nd Nat. Conf. on Weather Modification, pp. 209–214 (1970).
245b. Changnon, S. A., Jr.: 'The LaPorte Anomaly – Fact or Fiction?', *Bull. Am. Meteorol. Soc.* **49**, 4–11 (1968).
246. Holzman, B. G. and Thom, H. C. S.: 'The LaPorte Precipitation Anomaly', *Bull. Am. Meteorol. Soc.* **51**, 335–337 (1970).
247. Changnon, S. A.: 'Reply (to Holzman & Thom)', *Bull. Am. Meteorol. Soc.* 337–343 (1970).
248. Holzman, B. G.: 'LaPorte Precipitation Fallacy', *Science* **171**, 847 (1971).
249. Changnon, S. A.: 'LaPorte Anomaly', *Science* **172**, 987–988 (1971).
250. Ogden, T. L.: 'The Effect on Rainfall of a Large Steelworks', *J. Appl. Meteorol.* **8**, 585–591 (1969).
251. Changnon, S. A., Jr.: 'Comments on "The Effect on Rainfall of a Large Steelworks"', *J. Appl. Meteorol.* **10**, 165–168 (1971).
252. Ogden, T. L.: 'Reply (to Changnon, 1971)', *J. Appl. Meteorol.* **10**, 168 (1971).
253. Hidore, J. J.: 'LaPorte Anomaly', *Science* **172**, 988 (1971).
254. Hidore, J. J.: 'The Effects of Accidental Weather Modification on the Flow of the Kankakee River', *Bull. Am. Meteorol. Soc.* **52**, 99–103 (1971).
255. Hobbs, P. V. and Radke, L. F.: 'Cloud Condensation Nuclei from Industrial Sources and their Influence on Clouds and Precipitation', *Bull. Am. Meteorol. Soc.* **51**, 81–89 (1970).
256. Hobbs, P. V., Radke, L. F., and Shumway, S. E.: 'Cloud Condensation Nuclei from Industrial Sources and their Apparent Influence on Precipitation in Washington State', *J. Atmos. Sci.* **27**, 81–89 (1970).
257. Elliott, W. P. and Ramsey, F. L.: 'Comments on "Cloud Condensation Nuclei from Industrial Sources and their Apparent Influence on Precipitation in Washington State"', *J. Atmos. Sci.* **27**, 1215–1216 (1970).
258. Hobbs, P. V., Radke, L. F., and Shumway, S. E.: 'Reply (to Elliot and Ramsey, 1970)', *J. Appl. Meteorol.* **27**, 1216–1217 (1970).
259. Band, G.: 'Der Einfluss der Siedlung auf das Freilandklima', *Mitt. Inst. Geophys. Meteorol. Univ. Köln*, Heft **9**, (1969).
260. Hidy, E. M., Bleck, R., Blifford, I. H., Jr., Brown, P. M., Langer, G., Ledge, J. P., Jr., Rosinski, J., and Shedlovski, J. P.: *'Observations of Aerosols over Northeastern Colorado'*, NCAR Tech. Note No. 49 (1970).
261. Changnon, S. A., Jr., Huff, E. A., and Semonin, R. G.: 'Metromex: An Investigation of Inadvertent Weather Modification', *Bull. Am. Meteorol. Soc.* **52**, 958–967 (1971).
262. Beilke, S.: *Untersuchungen über das Auswaschen atmosphärischen Spurenstoffe durch Niederschläge*, Ber. Inst. Met. Geophys., Univ. Frankfurt, No. 19, (1970).
263. Georgii, H. W. and Kaller, R. S.: *Über die Inaktivierung von Gefrierkernen durch Koagulations mit Aitken Kernen*, Ber. Inst. Met. Geophys. Univ., Frankfurt, No. 21, (1970).

264. Langer, G.: '*Ice Nuclei Generated by Steel Mill Activity*', Proceed. 1st Natl. Conf. Weather Modif., Am. Meteorol. Soc., pp. 220-227 (1968).

265. Schaefer, V. J.: 'The Inadvertent Modification of the Atmosphere by Air Pollution', *Bull. Am. Meteorol. Soc.* **50**, 199-206 (1969).

266. Lodge, J. P., Jr.: 'An Atmospheric Scientist Views Environmental Pollution', *Bull. Am. Meteorol. Soc.* **50**, 530-533 (1969).

267. Schaefer, V. J.: 'Ice Nuclei from Automobile Exhaust and Iodine Vapor', *Science* **154**, 1555-1557 (1966).

268. Hogan, A. W.: 'Ice Nuclei from Direct Reaction from Iodine Vapor with Leaded Gasoline', *Science* **158**, 800 (1967).

269. Parungo, F. P. and Rhea, J. O.: 'Lead Measurements in Urban Air as it Relates to Weather Modification', *J. Appl. Meteorol.* **9**, 468-475 (1970).

270. Schaefer, V. J.: 'The Inadvertent Modification of the Atmosphere by Air Pollution', this volume, pp. 177-196.

271. Moyers, J. L., Zoller, W. H., and Duce, R. A.: 'Gaseous Iodine Measurements and Their Relationship to Particulate Lead in a Polluted Atmosphere', *J. Atmos. Sci.* **28**, 95-98 (1971).

272. Gunn, R. and Phillips, B. B.: 'An Experimental Investigation of the Effect of Air Pollution on the Initiation of Rain', *J. Meteorol.* **14**, 272-283 (1957).

273. Allee, P.: 'Air Pollution and the Cloud Droplet Condensation Nuclei Concentration', *Bull. Am. Meteorol. Soc.* **51**, 102 (1970).

274. Kuhn, P. M.: 'Airborne Observation of Contrail Effects on the Thermal Radiation Budget', *J. Atmos. Sci.* **27**, 937-942 (1970).

275. Manabe, S.: 'Cloudiness and the Radiative Convective Equilibrium', this volume, pp. 175-176.

276. Murcray, W. B.: 'On the Possibility of Weather Modification by Aircraft Contrails', *Monthly Weather Rev.* **98**, 745-748 (1970).

277. Dingle, N.: 'Comment', *Science* (in print) (1971).

278. Chatham, G. N.: 'Will the SST Change the Weather?', *Mt. Washington Obs. News Bull.* **11** (1), 18-20 (1970).

279. Hall, F. F., Jr.: 'Pollution in the Upper Troposphere by Soot from Jet Aircraft and its Relation to Cirrus Clouds', *Bull. Am. Meteorol. Soc.* **51**, 101 (1970).

280. Kuhn, P. M.: 'Artificial Cloud Effect on the Thermal Radiation Budget', *Bull. Am. Meteorol. Soc.* **51**, 101 (1970).

281. Louis, J. F. and MacQueen, R. M.: 'Spectroscopic Observations of Water Vapor near the Tropopause', *J. Appl. Meteorol.* **9**, 722-724 (1970).

282. Scholtz, T. G., Ehhalt, D. H., Heidt, L. E., and Marfell, E. A.: 'Water Vapor, Molecular Hydrogen, Methane and Tritium Concentrations near the Stratopause', *J. Geophys. Res.* **75**, 3049-2054 (1970).

283. Muessle, V. and Holcomb, R. W.: 'Will the SST Pollute the Stratosphere?', *Science* **168**, 1562 (1970).

284. Singer, S. F.: personal communication (1971).

285. Urbach, F.: 'Geographic Pathology of Skin Cancer', in *the Biological Effects of Ultraviolet Radiation*, Pergamon Press, pp. 635-650 (1969).

286. Singer, S. F.: Personal communication to Russel Train, Chairman, CEQ.

287. Harrison, H.: 'Stratospheric Ozone with Added Water Vapor: Influence of High-Altitude Aircraft', *Science* **170**, 734-736 (1970).

288. Dütsch, H. U.: 'Photochemistry of Atmospheric Ozone', *Adv. Geophys.* **15**, 219-322 (1971).

289. Cadle, R. D. and Allen, E. R.: 'Atmospheric Photochemistry', *Science* **167**, 243-249 (1970).

290. Bergeron, T.: *Cloud Physics Research and the Future Fresh-Water Supply of the World*, Met. Inst., Uppsala Rep. No. 19, (1970).

291. Perrin de Brichambaut, C.: 'Modifications artificielles du temps et du climat: possibilités d'interventions humains', *Météorologie* **5**, (5), 15-57 (1968).

292. Lamb, H. H.: 'Climatic Variation and our Environment Today and in the Coming Years', *Weather* **25**, 447-455 (1970).

293. ESSA Research Laboratories, APCL contributions to the second conference on weather modification of the American Meteorological Society, April 1970, ESSA Tech. Memo ERLTM – APCL, 9, (1970).

294. Gates, D.: 'Weather Modification in the Service of Mankind: Promise or Peril?', in H. W. Helfrich, Jr. (ed.), *Environmental Crisis: Man's Struggle to Live with Himself*, Yale Univ. Press, pp. 33–46 (1970).
295. Ellsaesser, H. W.: 'Air Pollution: Our Ecological Alarm and Blessing in Disguise', *EOS* **52**, 92–100 (1971).
296. Danserau, P.: 'Ecological Impact and Human Ecology', in Y. F. Darling and J. P. Miller (eds.), *Future Environments of North America*, The Natural History Press, Garden City, N.Y., p. 425 (1966).
297. Dubos, R.: *A Theology of the Earth*, Smithsonian Institution Lecture, Oct. 2, 1969, (1969).
298. Bundy, McG.: *Managing Knowledge to Save the Environment*, Address 11th Ann. Mtg. Adv. Panel, House Committee on Science and Astronautics, Jan. 29, 1970, Ford Found. Reprint SR/41, (1970).
299. Ashby, W. C. and Fritts, H. C.: 'Tree Growth, Air Pollution, and Climate near LaPorte, Ind.', *Bull. Am. Meteorol. Soc.* **53**, 341–350 (1972).
300. Johnston, H.: Reduction of Stratospheric Ozone by Nitrogen Oxide Catalysts from Supersonic Transport Exhaust', *Science* **173**, 517–522 (1971).
301. Foley, H. M. and Ruderman, M. A.: 'Stratospheric NO Production from Past Nuclear Explosions', *J. Geophys. Res.* **78** 4441–4450 (1973).
302. Angell. J. K. and Korshover, J.: 'Quasi-Biennial and Long-Term Fluctuations in Total Ozone', *Monthly Weather Rev.* **101**, 426–443 (1973).
303. Deirmendjian, D.: 'On Volcanic and Other Particulate Turbidity Anomalies', *Adv. Geophys.* **16**, 267–296 (1973).

THE UPWARD TREND IN
AIRBORNE PARTICULATES THAT ISN'T

HUGH W. ELLSAESSER

Lawrence Livermore Laboratory, University of California, Livermore, Calif. 94550, U.S.A.

Abstract. Diverse data relating to natural and anthropogenic particulate emissions and trends in atmospheric turbidity are reviewed. It is estimated that man now contributes 13.6% of the 3.5×10^9 tons of primary and secondary particulates presently emitted to the atmosphere annually.

Values derived for the anthropogenic effect on the trend in total airborne particulates (if assumed monotonic) range from 0.06 to 0.4% increase per year. Since such an effect would be only marginally detectable in the available record, the much larger rates of increase which have been published as proof of an upward trend must be due to local or temporary effects and cannot be representative of a global trend. On the other hand almost all urban areas plagued by dirty skies prior to 1950 have shown in recent years measurable improvement in terms of visibility, dust fall, suspended particulates, or the ratio of sunshine in the central city to that at an outlying station. This leads to the conclusion that while an anthropogenic upward trend in airborne particulates existed in the past, it was halted and may even have been reversed over the past few decades.

1. Introduction

In recent years there have appeared a rash of papers claiming an upward trend in airborne particulates, which is presumed to have already reversed the alleged CO_2 induced heating of the atmosphere observed between the 1880's and 1940's and to pose the further threat of inducing another ice age. Allusions to the trend have become so common that many authors now cite it as an accepted reality requiring neither qualification nor attribution by reference [67]. Herein are collected and correlated much of the currently available information bearing on the reality and magnitude of any such trend.

2. Air Conductivity Measurements

Among the earlier claims for this increasing atmospheric turbidity was that of Wait [123] based on the apparent decline in air conductivity measured on cruises of the research ship *Carnegie* between 1913 and 1929. Subsequent studies and observations developed a correction factor for the later measurements used by Wait and indicated a significantly slower rate of decline in air conductivity over the North Atlantic [35] but still estimated to "be equivalent to perhaps a [60-yr] doubling of the fine particle aerosol pollution in the area" [17]. Gunn [35] and Cobb and Wells [17] also reported remarkably constant values of air conductivity (and therefore turbidity) in the South Pacific between 1907 and 1967. Table I tabulates a recent summary by Cobb [16] of data indicating trends in air conductivity in different geographical regions. These suggest downward trends in air conductivity in the paths of aerosol pollution extending

TABLE I

Trends in air conductivity (λ, $10^{-14} \Omega^{-1}m^{-1}$) by geographical areas as compiled by Cobb [16]

Geographical area	First period λ		Second period λ	
Central S. Pacific	1907–30	2.7–3.4	1967	3.31
Eastward from the U.S. in N. Atlantic	1909–29	3.5	1967	1.97
Eastward from Japan in N. Pacific	1915–29	2.7	1966	2.2
Southward from India in N. Indian Ocean	1911–20	2.9–4.2	1967	1.5–1.9
The Red Sea	1911	1.85	1967	1.82
Mauna Loa Observatory	1960	7.23	1969	7.25

eastward from the United States in the North Atlantic, from Asia and Japan in the North Pacific and southward from Asia in the northern Indian Ocean.

In view of the large geographical, seasonal and synoptic variations in turbidity the representativeness of measurements on a single ocean crossing becomes a very moot point. It is not clear that restriction to 'fair-weather' observations is sufficient to remove variations comparable to the 2- to 3-fold seasonal and geographical variations in turbidity reported over the United States by Flowers *et al.* [27]. Each of the quoted references noted many reservations concerning quantitative comparison of different air conductivity measurements. The indicated trend over the North Atlantic appears to be directly contradicted by eastern United States turbidity data which will be presented later. Also, as pointed out by Cobb [16] the regions indicating a downward trend in air conductivity comprise less than 10% of the Earth's ocean area. The overall significance of these air conductivity data must be weighed against other data presented herein.

3. The Arguments of Outstanding Proponents

The most vigorous proponents of increasing atmospheric turbidity due to anthropogenic air pollution have undoubtedly been Bryson [9] and Schaefer [98, 99]. Bryson has consistently supported his position with Davitaia's [23] data showing increased dust trapped in the glaciers of the Caucasus after 1930 and with Peterson and Bryson's [82] analysis of Mauna Loa Observatory data showing a 30% increase in turbidity between 1957 and 1967.

Mitchell [72] dismissed the Caucasus dustfall data as a local phenomena. The data themselves support Mitchell's argument. Using average annual glacial accumulation rates of 10 to 60 cm yr^{-1} (water equivalent) [131], Davitaia's value of approximately 230 mg l^{-1} in 1963 is equivalent to $1-7 \times 10^{10}$ tons yr^{-1} if uniform over the globe. Since this is at least an order of magnitude larger than the estimates of annual particulate production summarized below in Table V, 230 mg l^{-1} cannot be representative of a global fallout unless the rate of accumulation of the Caucasus glaciers is less than 1 cm yr^{-1}.

The analysis of the Mauna Loa turbidity data by Ellis and Pueschel [21] both fails to support a steady increase in turbidity between the start of measurements in 1957 and the Agung eruption of 1963 and indicates a return to pre-Agung values by 1970. They concluded that the 13 yr of Mauna Loa turbidity data gave no evidence that human activities affect atmospheric turbidity on a global scale. Volz [120], from an analysis of twilight data, also concluded that by 1970 the atmosphere at 42° N had returned to its pre-Agung state of transparency.

Undoubtedly, the most influencial argument that man is increasing the particulate burden of the atmosphere has been Schaefer's [97] disclosure that submicroscopic aerosol particles, presumed to be lead, are produced by auto exhaust. He recently compiled an extensive list of cases of inadvertent weather modification (due to this and other processes) substantiated largely by the 'intelligent eyes' [98, 99] of his own personal observations. In the natural sciences there is no substitute for observations by 'intelligent eyes' but one must distinguish between observation and interpretation.

In reading Schaefer's vivid descriptions of this personal observations from aircraft with which he supports his claim of increased air pollution, I cannot help but note the coincidence in time between the period of increase and the period of transition from propellor to jet aircraft. Under most circumstances the atmosphere indeed looks dirtier from 30000 than from 10000 ft.

Schaefer's [98, 99, 101] claim of a noticeable increase in air pollution during the decade 1958–1968 over and downwind of large metropolitan areas are directly contradicted by observations showing declines in settleable and filterable dust in urban areas [61] and by the counter claim that residual reduced tree-ring growth during the 1940's in trees along the southern end of Lake Michigan was due to a maximum in air pollution at that time [3].

Schaefer [98, 99] cites several observations purporting to show that man is inhibiting rainfall due to overseeding with cloud condensation nuclei in warm rain situations. These included Warner's [124] study showing a downward trend in precipitation downwind from areas of increased burning of sugar cane fields near Bundaberg, Australia. These must be weighed against subsequent investigations in Hawaii [132] and near Ayer, Australia [125] which failed to find associated downward trends in precipitation. On the other hand, Hobbs *et al.* [40] found an upward trend in precipitation in western Washington associated with increasing nuclei emissions from paper plants. The implied contradiction to Warner's [124] study appears inescapable since both cases involve addition of comparable quantities of cloud condensation nuclei to fresh maritime air masses under warm rain conditions.

If warm rain clouds are significantly stabilized by additional nuclei (Cadle and Junge [12] state that even estimates as to the direction and degree of such changes are not possible at the present time), such stabilization would occur naturally (even in the absence of man) in every region where maritime warm rain clouds invade a source of continental aerosol. The fact that only near Bundaberg, Australia has such a rain shadow been detected appears to argue against overseeding having a significant effect on the warm rain situation. Moreover, the effect even if significant is not progressive.

Additional particle concentrations can result in larger numbers of cloud droplets of smaller size (which do not coalesce as readily as bigger cloud droplets) only in relatively clean air. "If the initial aerosol concentration is higher than 2000 cm^{-3} or so, an increase will have little effect, because the cloud droplet concentration cannot be much higher than about 500 cm^{-3}, leaving most of the aerosol particles unused for condensation" [55]. From the statement by Cadle and Junge [12] "Under continental conditions, a large fraction of the particle population is left inactivated, since the total number of aerosol particles is much larger than the number of cloud droplets," it would appear that any anthropogenic effects in stabilizing warm clouds is limited by the degree to which man hastens the process of air mass transformation from marine to continental aerosol.

The dominant role in inadvertant weather modification was attributed by Schaefer [98, 99] to submicroscopic lead particles from auto exhaust activated (converted into ice nuclei through formation of PbI_2) by exposure to minute quantities of atmospheric iodine vapor. In air samples collected near urban areas and activated by exposure to iodine vapor he reported ice nuclei counts up to 10^5 l^{-1} at -20 °C. While not completely explicit he appears to claim no observations of increased concentration of ice nuclei in air samples not first activated by exposure to iodine vapor. Ice nuclei counts at Mt. Washington, New Hampshire begun by Schaefer [101] himself in 1948 are not cited to support an upward trend in ice nuclei and Junge [55] indicated that these data show no trend.

While observations in Cambridge, Massachusetts found gaseous iodine routinely present in quantities (10–18 ng m^{-3}) exceeding Schaefer's estimate of that needed to activate all lead containing particles to ice nuclei, no one has reconciled this with the fact that the corresponding quantity of ice nuclei are not observed [76]. Ice nuclei counts near urban areas have generally exceeded the background (1 l^{-1} active at -20 °C) by less than an order of magnitude except downwind of specific industrial activities such as steel mills [62, 109, 114, 39]. Hobbs et al. [40] made airborne measurements of ice nuclei from paper plants and industrial areas of Washington and reported: "No detectable increases over background levels were detected except in the dense plume from the ferroalloy plant where the ice nuclei concentration was about twice the normal level at a distance of about 1 km from the stack."

At 8 stations between Altadena and Palos Verdes, Alkezweeny and Green [2] found on November 4, 1969 the lowest ice nuclei count in downtown L. A. (0.31 l^{-1} at -20 °C and 1 % supersaturation) and the two highest counts downwind from a Long Beach refinery and steel plant (1.34 l^{-1}) and at Palos Verdes above the principal pollution layer where uncontaminated Pacific air was mixing with the polluted air of the Los Angeles basin (0.88 l^{-1}). Additional observations during a two-week period in October 1969 found ice nuclei counts in Altadena ranging from 0.5 to 1.5 l^{-1}. In describing these observations Smith et al. [108] reported a "tendency for lowest ice nuclei values to be associated with heavy pollution." They also quoted a private communication from Grant reporting no definable change in ice nuclei counts over the past 10 yr at Climax, Colorado.

Bigg and Stevenson [7] reported greater temporal than spatial variations in ice nuclei concentration in samples collected from 44 sites around the globe during January, February and March 1969. Mean counts were actually highest over Asia, South America and Africa and lowest over Europe, Australia and North America.

Perhaps the strongest argument against inadvertent weather modification by PbI_2 seeding through auto exhausted lead are recent observations by Ter Haar and Bayard [115]. They used an electron microprobe to physically identify lead compounds in particulates filtered from air samples and auto exhaust. PbI_2 did not appear among the 14 groups of lead compounds identified. Further, they found that the relative amount of lead halides identified declined significantly with ellapsed time following combustion. When auto exhaust collected directly in black plastic bags (to preclude photolytic decomposition) was sampled again after 18 h of aging, 75% of the lead bromides and 30 to 40% of the lead chlorides appeared to have been lost. The loss in lead halides was balanced by an increase in lead carbonates, oxycarbonates and oxides. X-ray fluorescence also revealed a rapid decrease in the ratio of bromine to lead with most of the loss occurring during the first hour.

These observations imply that while there appears to be enough I_2 to activate a significant fraction of auto exhaust lead aerosol, its concentration is insufficient to compete successfully with other possible anions present in the atmosphere. Not only is lead aerosol not generally converted to PbI_2 in the atmosphere; PbI_2 released directly to the atmosphere is presumably converted to other lead compounds at rates comparable to those observed for decay of the ice nucleation properties of AgI [107].

Additional doubt is cast on Schaefer's postulate of inadvertent weather modification by auto exhaust lead aerosol by:

(1) the ambiguity and lack of unanimity in results from efforts to date in artificial rainmaking.

(2) the serious gaps in our current basic knowledge of cloud physics revealed by our current inability to explain the well-documented observations (a) of clouds which produced ice crystals in concentrations around $10 \, 1^{-1}$ where the summit temperatures were no greater than -5 °C (pre-activation therefore ruled out) and (b) of concentrations of ice crystals in clouds as much as 10^4 times as great as the counted concentrations of available ice nuclei [75].

(3) the documented ice nucleating properties of many common and widely dispersed materials such as: kaolinite, mica, gypsum, calcite and 24 other natural mineral dusts active at temperatures above -18 °C [69, 89]; urea [58], phloroglucinol, methaldehyde (has highest temperature for ice nucleation so far known) and more than 75 other organic compounds active at temperatures above -10 °C [30]; amino-acid particles [70]. Anthropogenic emissions of lead aerosol are unlikely to be significant compared to the natural availability of aerosols of these materials.

Smith [106] recently reviewed precipitation changes which have been proposed as evidence of inadvertent weather modifications due to anthropogenic emission of particles. He concluded that only in the case of decreased rainfall near Bundaberg, Australia [124] did the changes appear to be consistent with our present understanding of the effect of the postulated increase in particulates.

If we restrict consideration to the effect of auto exhaust in producing airborne particulates (as opposed to ice nuclei), we find no lack of natural competing processes. Went [127] found large increases in condensation nuclei anytime a light absorbing material such as NO_2 or I_2 vapor was released in a sunlit atmosphere containing traces of organic matter. At temperatures above freezing the latter were always present. On cold days "it can be provided by α-pinene or turpentine vapors; [geyser] pools with algae growing in them provide an even faster acting nucleating matter." Vohra et al. [118] found that addition of trace organic vapors such as turpentine, eucalyptus, and ethyl and amyl alcohol to chambers containing moist air and nitric oxide caused an increase of two to five orders of magnitude in the number of nuclei formed, even in the dark. Schaefer [100] himself has reported that exposure of 'comparatively clean country air' to saturated iodine or turpentine vapor resulted in an increase of 2 to 3 orders of magnitude in the number of cloud and Aitken nuclei in the air sample. He found this effect to be extremely common; absent only in extremely pure and in heavily polluted air. In explanation of the latter effect he stated: "It is likely that in polluted air the clusters of pseudo-embryos that cause this reaction have become adsorbed on the Aitken nuclei, which in polluted air generally have concentrations of $50\,000$ cm^{-3} or more."

From these competing and even compensating processes I conclude that auto exhaust may well produce a local and temporary increase in the number of particulates including sublimation nuclei but is unlikely to lead to a significant overall increase in either. In support of this position numerous measurements of cloud condensation nuclei can be quoted which reflect only local and temporary anthropogenic effects. Twomey and Wojciechowski [117] measured concentrations of cloud nuclei over many areas of the globe and found them at least as numerous over Africa and Australia as over North America and Europe. Nor did they find an east-west gradient in concentration over North America as would be expected if there were a significant anthropogenic source. As reported above, Bigg and Stevenson [7] also failed to find evidence to support an anthropogenic source in the world wide distribution of ice nuclei. Kocmond and Mack [59] found that in spite of large increases in cloud and Aitken nuclei immediately downwind of urban pollution sources, the concentration of both quickly returned to lower values and approached the background level within 20 to 50 miles of the pollution sources. Selezneva [104] found that inhomogeneities in concentration of condensation nuclei from single powerful sources extended 20–25 km downwind and that those from industrial cities extended somewhat beyond 50 km. Hobbs et al. [40] described the increased concentrations of cloud condensation nuclei from the 'prolific' Kraft paper mills (up to 10^{19} s^{-1}) and other sources of Washington as 'extending up to ~100 km downwind'. Hogan et al. [44] on two traverses of the North Atlantic found Aitken nuclei concentration exceeding the mid-ocean median by at least 10-fold up to 500 miles offshore from the northeastern United States. Even this distance is small compared to both the dimensions of oceans and continents and to 3-day (usually quoted as lifetime of particles in the lower troposphere) mid-latitude trajectories. There are reports of individual sightings of clouds or plumes of pollutants far downwind of urban areas and continents. But, if these are representative they are difficult to reconcile with the essentially uniform worldwide distribution of particles reported by Twomey and Wojciechowski [117] and Bigg and Stevenson [7], with the unchanged electrical conductivity in the South Pacific and with the consistent classification of aerosols as background, oceanic and continental [54].

Landsberg [60b] cites earlier papers [60 and 60a] to support the statement "measurements made in the early 1930 [sic] gave values of around 100 of these [Aitken nuclei] per cm^3 as gloval background, a value which seems to have doubled in recent years." However, Landsberg [60] cites only two (out of 1051) observations of less than 200 cm^{-3}: a short series of 7 sets of observations on 4 days while crossing the North Atlantic showing an average of 950 cm^{-3} and a minimum of 150 cm^{-3}; and measurements through a subsidence inversion on Taunus Mountain showing 2200 cm^{-3} below and 110 cm^{-3} above the inversion. Landsberg [60a] tabulates average and average minimum Aitken counts per cm^3 for islands as 2900 and 460 and for oceans as 940 and 840. On the other hand Aitken nuclei counts of less than 200 cm^{-3} appear no less common today. Selezneva [104] reported that subsidence above inversions in winter anticyclones produces very clean and dry air often with nuclei counts of less than 100 cm^{-3} at heights as low as 500 m. In operating a continuous nuclei counter at Yaquina Head, Oregon, Hogan [43] found that during extended periods of rain, counts often fell below the threshold of sensitivity of the counter (150 cm^{-3}) for periods of several hours. Twomey *et al.* [116], discussing the conditions for anomalous cloud lines [18], stated: "Such concentrations [< 10 cm^{-3}] are well below the typical numbers found over the ocean, which are about 50–100 cm^{-3}, but nuclei concentrations well below 5 cm^{-3} have been found fairly regularly in polar maritime air around southeastern Australia and, more sporadically in the North Pacific and North Atlantic." Thus the recorded data do not appear to support Landsberg's [60b] claim of a possible doubling of the background concentration of Aitken nuclei.*

Because of the failure of observational data (such as visibility, dustfall and high volume filter measurements) to support the presumed upward trend in airborne particulates and because of the increasing use of control devices most effective on the larger particles, there has been a tendency for those claiming increased anthropogenic particulates to take the defensive position that it is the fine non-filterable submicron particles that are being increasingly released to the atmosphere. Such a claim is, of course, difficult to counter. However, if true and if man's emissions constitute a significant fraction of the total mass of such particles in the atmosphere then it becomes increasingly difficult to explain the numerous observations of particle size distributions approximating Friedlander's [29] self-preserving size distribution. Such a distribution is characterized by the relation $dN/dr \propto r^{-4}$ where N is the number of par-

* Landsberg (private communication, 1973) based his statement on a comparison of unpublished data collected above an inversion at Taunus in 1933 with contemporary data collected by Junge and collaborators on the Peak of Teneriffa.

ticles cm^{-3} of radius $> r$. Such a distribution has the same volume (or mass) of material at each radius examined, as would be expected for the equilibrium distribution if atmospheric aerosols were generated at the smallest sizes, grew by agglomeration and were removed by gravity when sufficiently large.

Junge [53, 54] rejects as the explanation for the recurrent observation of this distribution growth by agglomeration since Hidy's [36] computed growth times far exceed the observed residence times of a few days to more than three weeks for atmospheric particles [55]. He prefers to explain it as a statistical artifact of numerous source and sink mechanisms, the former tending to produce log normal distributions and the latter to remove comparable fractions of each size present.

In any event, a significant source restricted to a particular size range would be expected to stand out in the observed size distributions as do the sea salt particles generated by the bursting of bubbles of sea spray. These particles are observed only in the lowest two kilometers over and near the sea [54] despite the observed omnipresence of particles up to 100μ radius [47]. Additionally, all available observations including some recent ones in clean Atlantic air masses indicate continuous production of very small particles (increasing particle concentration with decreasing size) in the 0.001–0.01μ range [54, 15]. This is also the impression left by studies (some of which were cited earlier) of the processes by which particles are formed in the free atmosphere [74].

Thus, although we cannot claim to understand why, concentrations of atmospheric particles by size do appear to cluster about an r^{-4} distribution over most of the range 0.01–100μ in both clean and polluted air [54]. The only consistent departure from such a distribution so far recognized is the one due to sea salt particles which peaks around 0.3–0.4μ [79] and distorts the spectra up to 10μ [54]. Observations are beginning to reveal a peak in the mass (as opposed to number) spectra around 0.3μ [130] but interpretation remains controversial since relative maxima or lesser slopes between 0.1 and 1.0μ are also found in some of the few spectra from altitudes and areas not affected by continents or sea spray [54]. Even in the submicron mode of the bimodal mass spectra found in urban atmospheres most of the aerosol mass appears to result from the coagulation and condensation of very small chemically or photochemically formed aerosols [130].

If man's emissions of primary particles constitute a significant fraction of the total atmospheric loading and if his control efforts have merely truncated these to the submicron range then we should expect to find size spectra consistently peaking above r^{-4} in the 0.1 to 1.0μ range due to injection of primary particulates. In the absence of observational data showing consistent departures of this type it would appear that attempts to quantify man's emissions of 'fine' particles merely adds one more degree of uncertainty to estimates of the anthropogenic contribution. Such a position seems more acceptable than trying to reconcile estimates like that of Peterson and Junge [83] that man's contributions of 'fine ($< 5\mu$)' particulate emissions have reached 60% of natural continental ones in the Northern hemisphere (vs. 9% in the Southern hemisphere) with the observations of Twomey and Wojciechowski [117] and Bigg

TABLE II

Long-term trends in atmospheric turbidity (Schüepp coefficient) at various stations

Station	First period	B	Second period	B	References
Washington D.C.	1903–07	0.105	1962–66	0.165	McCormick and Ludwig [64]
Davos	1914–26	0.026	1957–59	0.046	McCormick and Ludwig [64]
McMurdo	1949–52		1966		
November		0.016		0.022	Fischer [26]
December		0.026		0.024	
Mexico City	1911–28	0.087	1957–62	0.181	Galindo and Muhlia [31]
Jerusalem	1930–34	0.043	1961–68	0.050	Joseph and Manes [49]
Table Mountain, California	1925–30	0.0185	1940–50	0.0081	Roosen et al. [94]
Mt. Montezuma, Chile	1920–30	0.0171	1940–48	0.0076	Roosen et al. [94]

and Stevenson [7] which indicate equal or even greater concentrations of condensation and ice nuclei over the continents of the southern hemisphere.

4. Long-Term Turbidity Comparisons

Since early turbidity observations were few and sporadic, not many opportunities to estimate long-term trends have been discovered. Table II summarizes those long-term comparisons encountered to date. All except the Antarctic and Roosen et al. [94] data appear to show upward trends and have been cited as indicative of a global increase in atmospheric turbidity.

Interpretation of these data must be tempered by at least two considerations: the extent to which they represent only local effects due to increased urbanization and the differences in volcanic dust loading of the stratosphere during the two observation periods. Flowers et al. [27] stated that the 1903–1907 Washington, D.C. value of 0.105 "is near the current [1962–1966] value for rural portions of the eastern United States." This would appear to indicate both that the 1962–1966 Washington, D.C. increased turbidity value reflected only local urban pollution and that the 60 yr trend in turbidity in the eastern United States was negligible. The Mexico City trend is also considered to be significantly influenced by local effects. Any recent visitor is made keenly aware that Mexico City has developed an air pollution problem that is both local and severe. Galindo and Muhlia [31] claim to have shown "how the industrial development has increased locally the atmospheric turbidity of the [Mexico] City." Their statement that "it can be stated that the minimum values of both series are without appreciable difference," further argues against an increase in background turbidity between 1911–1928 and 1957–1962. Junge [55] states that the turbidity increases at Washington, D.C. and Davos are primarily local. Increased local effects at Jerusalem are not easily documented. Joseph and Manes [49] cite the growth of human and auto populations for Jerusalem as 266562 to 296400 and 1000 to 15000 between 1930 and 1968.

Turning to volcanism as compiled by Mitchell [72], I find that both sets of Washington, D.C. observations were during periods of high volcanic dust loading of the stratosphere and should therefore be comparable in this respect. Wexler [129] reported turbidity values measured in 1932 at the American University about $5\frac{1}{2}$ miles northwest of the United States Capitol. The values (converted to B) for winter, spring and summer were 0.084, 0.106 and 0.071 (mean 0.087) compared to comparable values of 0.067, 0.145 and 0.060 (mean 0.091) for the 1903–1907 data. These also suggest that the Washington, D.C. values in Table II reflect either temporary or local effects. The first periods of observation at Davos, Mexico City and Jerusalem fell during the period of little if any volcanic penetration of the tropopause between Katmai in 1912 and Hekla in 1947. The first period Antarctica data were probably unaffected by the high northern latitude Hekla eruption. The second period of observation at Antarctica fell during the period of high volcanic dust loading of the stratosphere begun by Agung in 1963 and persisting until 1970 [120, 121]. The second period of observation at Davos followed closely the Bezymianny eruption of 1956. Cronin [19] described the latter as perhaps the most violent volcanic event of this century, expelling ash clouds to an altitude of 40 to 45 km. Stratospheric injection of dust at this altitude should assure hemispheric distribution and a long residence time. After seeing the dramatic effect of the Agung eruption on the Mauna Loa turbidity data as analyzed by Peterson and Bryson [82], Ellis and Pueschel [21] and Machta [66], I would expect stations of low natural turbidity such as Davos and Jerusalem to respond somewhat equivalently; Davos in 1957–59 to Bezymianny and Jerusalem in 1961–1968 to Agung. While Mexico City was probably influenced in 1957–1962 by the Bezymianny stratospheric dust it is unlikely that the latter would be noticeable above the high turbidity due to local tropospheric pollution. Galindo and Muhlia [31] claimed to identify in their data increased turbidity due to volcanic eruptions in 1912, 1913, 1927 and 1957, the last attributed to Volcan del Fuego or Colima rather than Bezymianny. Since only monthly departures from the respective annual means were presented, their claims are both unconvincing and impossible to evaluate without access to the original data. Manes [68] gives curves of mean seasonal and annual Linke turbidity factor for Jerusalem for 1961–1968. In a private communication he stated he was "not able to isolate any influence of the Agung eruption on the turbidity levels of the Jerusalem station during the years 1963, 1964 or 1965." His curves for the entire period show highest winter values in 1963 and 1964 but the lowest annual mean in 1964. This contrasts strongly with the response at Mauna Loa where the Linke turbidity factor increased some 20% or 0.6 Linke units following Agung [82]. Since the Linke factor during this period averaged about 1.5 at Jerusalem (well below the 2.98 average at Mauna Loa) this discrepancy suggests the existence of unresolved problems either in Manes' [68] Linke turbidity values or in our understanding of the atmosphere. This discrepancy cannot be attributed solely to the multiplicative effect of the high altitude of the Mauna Loa Observatory on the magnitude of the Linke turbidity factor. The decadic or

Schüepp turbidity coefficient (B) at Mauna Loa rose from around 0.04 to 0.06 following the Agung eruption [66], a change in the same range and more than double the 30-yr increase (0.043 to 0.050) in B at Jerusalem reported by Joseph and Manes [49].

It is also noteworthy that while Joseph and Manes [49] claimed a 10% per decade increase in turbidity at Jerusalem, Manes' [68] curve of mean annual Linke turbidity factor shows about a 2% (6% of the excess over unity) decline from 1961 to 1968.

The data from Roosen et al. [94] is discussed below in Section 6.

From this analysis I conclude that all the other data in Table II contain upward trend components due to either increased local urban pollution or differences in stratospheric dust loading or both. For the one station for which these local and temporary effects can be evaluated, Washington, D.C., the trend over the past 60 yr appears to have been negligible.

Flowers et al. [27] reported that the first six years (1961–1966) of the United States turbidity network data gave "no indication of any pronounced trend" even though the first arrival of the Agung cloud was clearly evident in the October 1963 values. The individual station curves presented suggest that a trend due to volcanic dust might have appeared had only stations of low turbidity such as Idaho Falls (mean annual B <0.05) been considered. These curves also revealed that even the turbidity values of the Agung cloud were dwarfed in the east and midwest by the anomalously high (0.30–0.65) values of July 1964. These occurred over a wide area during an extended anticyclonic period of extreme heat and dryness during the latter half of the month. Volz [119] also reported an area-wide synoptic increase in B from around 0.06 to over 0.3 in just a few days over central Europe in July 1963. He contrasted this rate of increase with the anthropogenic rate, $dB/dt \approx 0.01$ day^{-1}, which he had previously estimated under calm anticyclonic conditions and by comparing solar radiation measurements up- and down-wind of big cities. He hypothesized that this episode was caused by the advection of haze left by the evaporation of extended cloud masses which in their formative stage had trapped large amounts of trace gases. He also made a plea for additional case studies tracing the history of such events. I review these reports to counter the prevailing popular, and all to frequent professional, impression that high atmospheric turbidity is always of anthropogenic origin.*

5. Solar Radiation Data

Also quoted to support an upward trend in particulate loading of the atmosphere is Budyko's [11] curve of average direct solar radiation from stations "with the longest-period series of observations" in Europe and America. The curve shows a 4% decrease from the late 1930's to 1960. While generally not reported, the same data show values in 1887 and 1905–1910 comparable to that of 1960 and a sharp peak about 1897 comparable to the maximum around 1940. These data correlate fairly well with Mitchell's [72] chronology of stratospheric dust loading by volcanic activity. The mini-

* Went (private communication, 1973) cites as additional evidence on this point the haze in landscape paintings made in previous centuries.

mum in 1887 follows the series of eruptions began by Krakatoa in 1883; the decline after 1897 is premature but continues through the series began by Santa Maria and Mont Pelee in 1902; the rise after 1915 and essential leveling from 1922–1945 corresponds to the quiet period from Katmai in 1912 to Hekla in 1947; and the decline since 1945, while again premature, continues through the increasing volcanic dust loading from Hekla in 1947 to Agung in 1963. If Budyko's [11] curve were globally representative, the correspondence would raise questions concerning phasing. Since it contains stations from only Europe and America the correspondence appears to be as good as could be hoped for. In any event, stratospheric loading by volcanic dust appears to offer a possible explanation of the indicated decline in solar radiation since the 1940's.

6. Astronomical Data

Project ASTRA [14] is an attempt to use existing records of routine astronomical programs to study long-term anthropogenic pollution of the atmosphere. Hodge [41] compared his measurements of atmospheric extinction at Mt. Wilson Observatory on the rim of the Los Angeles basin with earlier Smithsonian measurements. They showed a 50-yr decrease in atmospheric transmission of 8 % at 0.55μ, 9 % at 0.44μ and 25 % at 0.35μ. Since these results agree in sign with expectation they will probably remain unquestioned. But their significance depends upon the representativeness and comparability of the two sets of selected observation times covering different portions of the day. The degree to which we are dependent on subjective human judgement (or caprice) in this respect is suggested by the following quotations. "In the years just before the advent of frequent occurrences of smog at these altitudes (1742 m above Los Angeles), which became important beginning in about *1962* [italics mine], I measured the atmospheric extinction in three wavelengths on 31 clear, smogless nights at the Mt. Wilson Observatory" [41]. "He [R. G. Roosen] related that Mt. Wilson was abandoned by the Smithsonian in *1920* [italics mine] because of atmospheric pollution." This last quote from Hodge *et al.* [42] continues with the following quotation from Abbot's [1] discussion of the removal of the Smithsonian solar-constant station which began on Mt. Wilson in 1905 to Mt. Harqua Hala in Arizona. "Mt. Wilson atmospheric conditions had proved unsuitable, and seemed steadily deteriorating for these exacting purposes as Los Angeles and neighboring towns expanded. Every day, towards midforenoon, the seabreeze rises there, and brings a haze over the mountain which is aggravated by the smoke and dust of the business of this immense population lying between the mountain and the Pacific."

One could use these statements to provide eyewitness evidence that the Los Angeles smog began before 1920 or after 1962 but not in the 1940's as commonly claimed. Alternatively, since the Smithsonian data reveal no increase in turbidity during the period of operation of the Mt. Wilson station, 1905–1920 [94], one could claim that removal of the station resulted from increasing awareness of a situation rather than an actual increase in urban pollution.

Jerzykiewicz [48] studied records of atmospheric extinction measured routinely at

Lowell Observatory (elevation 7250 ft) since 1955 and concluded: "(1) the long range extinction variations as observed from Flagstaff are due primarily to changes of the volcanic dust concentration in the stratosphere, and (2) there is no evidence of any increase of the extinction caused by human activity. It seems that for a long time to come such an increase, should it occur, will be completely concealed by the extinction variations due to volcanic dust loading into the stratosphere."

Roosen et al. [94], in an attempt to provide a historical baseline for measurements of long and short period variations in the nature and amount of atmospheric aerosols, studied Smithsonian measurements of atmospheric transmission made from 1902 to 1950 at various locations in North and South America and Africa over the spectral range from 0.35 to 1.6μ. They found the atmospheric extinction to be lowest in winter and highest in summer (except in Southwest Africa) and with regard to trend stated: "Including the results of more recent studies, our best estimate is that, except for sporadic perturbations due to volcanic activity, there has been no detectable change in the global atmospheric transmission measured from remote, high altitude sites in the last half century."

The Smithsonian observations at Table Mountain, California and Mt. Montezuma, Chile were not published from 1931 through 1939. When resumed, published observations were apparently restricted to selected clear days since there was a ten-fold reduction in frequency of published observations. Aerosol extinction coefficients calculated from these later observations were appreciably lower even than those from the lowest comparable fraction of observations from the earlier periods, see Table II. Roosen et al. [94] interpreted this as "presumably due to observational selection." For this interpretation to be correct would require either a change in criteria for selecting observation periods or that a significant number of the most favorable opportunities for solar observations had been missed during the earlier periods. In view of the Smithsonian's strenuous efforts to obtain optimum conditions for solar observations the latter possibility appears highly unlikely.

7. National Air Surveillance Network (NASN) Data

The most concrete data offered by those postulating increasing particulate pollution of the atmosphere is the NASN data for nonurban stations. The 9 yr of data for 1958–1966 averaged for 20 nonurban stations showed an average annual increase of 0.47 μg m^{-3} or about 2.5% yr^{-1} [110]. However, individually, only 6 stations revealed a significant upward trend, 1 a significant downward trend and 13 no significant change. This does not seem to imply a global trend. The nonurban stations uniformly displayed seasonal variations with summer maxima and winter minima in accordance with the turbidity data of Flowers et al. [27] and 180° out of phase with the seasonal variation in airborne particulates shown by the 58 urban stations. Further the amplitude of the annual variation increased during the period of observation, i.e., contrary to the effect proposed by Ellis and Pueschel [21] as indicative of increasing influence of anthropogenic contaminants.

8. Data Suggesting the Atmosphere is Getting Cleaner

While none of the above data offered to support the contention that the atmospheric particulate burden is increasing are free from ambiguity, a substantial body of data can be marshalled to suggest that the atmosphere has actually been cleansed by man over the past few decades. The most notoriously polluted cities of the past (in the western world at least) are claiming progress in cleaning their skies. These efforts have been aided by the progressive switch to cleaner and more convenient fuels, the proliferation of paving and horticultural and ornamental ground cover in urban and residential areas, and the spread of irrigation in arid regions. Of the 16×10^9 acres of potentially arable and grazable land on our planet, slightly over 50% is already so used [32]. Deliberate burning of stubble and cleared growth on part of this is unlikely to make up for the natural fires that would occur in the absence of man's husbandry. Wild fires in the remaining reduced areas of grassland and forest have been sufficiently reduced to raise cries for the necessity of controlled burning both to reduce the accumulation of combustible material and to restore the advantage to productive trees and meadow as opposed to nonproductive brush and shrubs [20]. It is also worth noting that the use of smudge pots to protect against frost damage has been drastically curtailed.

The apparent success of these measures in reducing the particulate loading of urban atmospheres is reflected in the available data. The NASN data of Spirtas and Levin [110] mentioned above show the mean particulate level at 58 U.S. urban stations declined between 1957 and 1966 at an average annual rate of $1.71 \mu g \, m^{-3}$ or about $1.6\% \, yr^{-1}$. Individually 23 stations showed significant declines, 2 significant increases and 33 no significant changes. Measurements of settleable dust in 6 major cities show dramatic multifold reductions from the beginning of observations until the early 1950's and less dramatic declines thereafter [61]. Chandler [13] reviews evidence which indicates that the atmosphere of central London, aside from a brief respite in the early 1950's, has become progressively cleaner since the 1880's. This conclusion is supported by Bonacina [8].

Visibility data offers one of the most voluminous sources of long-term homogeneous data for judging the trend in atmospheric transparency. Freeman [28] found for the period 1951–1964 at London/Heathrow Airport a progressive increase in the frequency of visibilities greater than 20 km. Holzworth [45] studied two periods of visibility data separated by 15 to 20 yr at 28 U.S. airport stations and Beebe [5] compared January 1945 and January 1965 visibilities at 8 U.S. airport stations. Both found a general downward trend in the frequency of hourly visibilities below 7 miles due to smoke and/or haze.

Not all visibility studies, of course, have indicated an upward trend in visibilities. I cite McNulty's [65] study of visibility observations for Battery Place, New York City because I believe it to be misleading and because it is not untypical of much of the literature documenting the 'air pollution crisis.' Using only daytime, non-precipitating observations from 1934 through 1960, he made two types of analyses. First he plotted the number of days of occurrence of haze alone, fog alone and haze and/or fog versus year (smoke and presumably smoke and haze occurrences were unconsidered). The analysis showed fewer than 10 days of haze a year prior to 1947 and a sharp rise to about 60 days yr^{-1} in 1954 jumping to over 150 days yr^{-1} in 1957 and subsequent years while the number of days of fog alone remained essentially unchanged. The second part of the analysis displayed percentage frequency of reports of haze by visibility ranges for successive 4 year periods beginning in 1946. These revealed that between 1946 and 1960 the percentage frequency of visibilities of less than 4 miles due to haze generally decline, those of 4–6 miles due to haze generally increased. McNulty's [65] main conclusion was "that [increasing] air pollutants in the atmosphere have affected atmospheric visibility by increasing the occurrences of haze with visibilities of 4 through 6 miles." From the abrupt drop in settleable dust in New York City subsequent to 1945 [61], I believe it is more likely that haze was replacing smoke as the predominantly reported obstruction to visibility due to the gradual decline of the latter and that visibilities were actually improving. (It is hoped that someone with access to New York City visibility data will be motivated to resolve this question.)

Another report of special interest is Keith's [57] study of downtown Los Angeles noon visibilities on weekdays for 1933–1969. Keith claimed upward trends in the frequencies of the visibility categories $1\frac{1}{2}$–$2\frac{1}{2}$ and 3–6 miles and downward trends in the frequencies of the categories 0–$1\frac{1}{4}$, 7–12, 13–35 and >35 miles; also a downward trend in average visibility from 12.6 miles for the period centered in 1937 to 7.9 miles in 1969. His data (normalized by 365 times the number of occurrences of each visibility category divided by the total observing days per year) appear in Figure 1. These reveal large secular fluctuations affecting almost all visibility categories with unusually good visibilities in 1938–1941 and 1951–1953 and unusually poor visibilities in 1943–1947 and 1962. These are believed to be largely of meteorological origin, as was implied by Keith [57]. The magnitude of these secular variations is such that removal of as few as three years of data would significantly alter the indicated trends.

In fact, only two features stand out prominently in the data; a decline in visibilities >35 miles during 1939–1943 and an abrupt increase in visibilities of 3–6 miles about 1964.

The U.S. Weather Bureau definition of visibility was changed in 1939 from greatest distance at which objects can be identified to the greatest distance prevailing over more than half of the azimuthal circle at which objects can be identified (National Weather Service historical file on Circular N). Since Catalina Island has always served as the primary check point for visibilities >35 miles, this change in definition alone can account for the abrupt decline at that time in the reporting of this visibility category. One might expect a definition change to result in a more abrupt decline than indicated by the upper line in Figure 1 but note that >35 miles visibilities declined in frequency 1933–1941 during the onset and through the period of best or greatest visibility of the entire period of record. It should also be noted that observing visibility is a highly subjective undertaking in which past practices might well prevail at times over definition changes. In July 1964 the observation site was moved from a height of 230 ft atop the Federal Building to a height of 80 ft atop the Air Pollution Control

HUGH W. ELLSAESSER

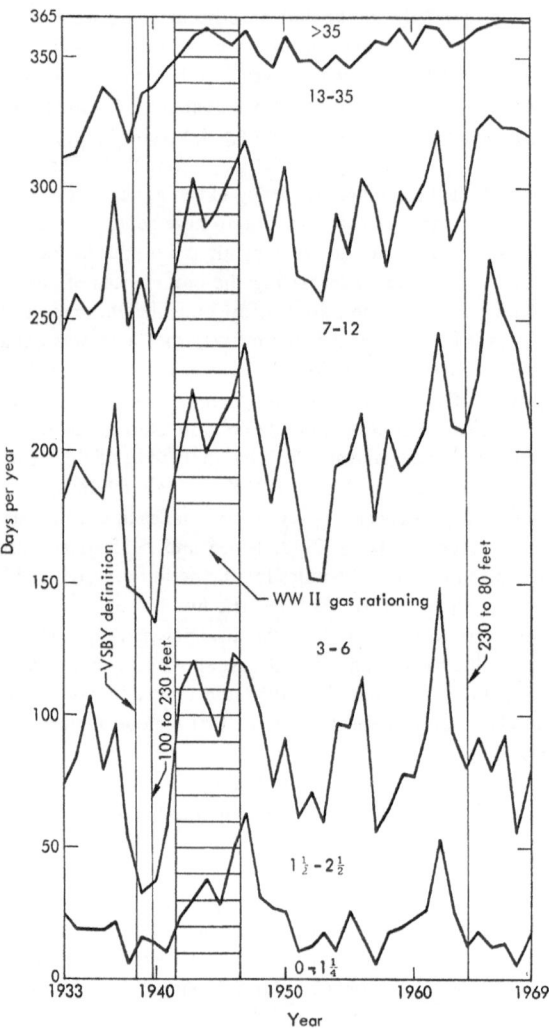

Fig. 1. Number of days per year (normalized to 365 days) that noontime observations of visibility in downtown Los Angeles from 1933 to 1969 were in the ranges 0–1¼, 1½–2½, 3–6, 7–12, 13–35, and greater than 35 miles. Vertical lines mark change in visibility definition in 1939; change in observation site from 100 ft AGL (above ground level) on Central Building to 230 ft AGL on Federal Building March 1, 1940; period of World War II gas rationing 1942–47, change in observation site from 230 ft AGL on Federal Building to 80 ft AGL on Air Pollution Control District Headquarters Building July 1, 1964. (Based on data compiled by Keith [57].)

District Headquarters Building. No allowance was made for this change in observing height even though Keith [57] admitted: "It would be expected that generally higher visibility values would have been reported during the period of March 1, 1940 to June 30, 1964, when observations were taken at 230 ft above ground level."

A second consideration is that visibility and particulate loading of the atmosphere are reciprocals. In averaging visibility categories the change from 40 miles (>35) to 24 miles (13–35) carries 16 times the weight of a change from 1 mile (0–$1\frac{1}{2}$) to 2 miles ($1\frac{1}{2}$–$2\frac{1}{2}$). In averaging dust loading the relative weights would be reversed and the ratio of the effects on the mean would be 1 to 3 rather than 16 to 1. Computation of average dust loading from the Class I Grouping of visibilities given in Keith's [57] Table II and reinversion for expression as visibility in miles yields 4.5, 3.4, 4.2, 3.6 and 3.9 miles for the four successive 9-yr periods and for 1969. Allowance for the two artifacts pointed out above would decrease the 4.5 value for 1933–1941 and increase the 3.6 and 3.9 values for 1960–1968 and 1969. This would leave the data essentially trendless aside from a period of poorest visibility in 1942–1950 and a suggestion of improvement in 1969. This is consistent with Neiburger's [78] earlier analysis which found slight deterioration to a minimum in 1944–1949 followed by slight improvement attributed to the imposition in 1947–1948 of controls on the emission of smoke and dusts by industry and sulfur dioxide by refineries. Had the above averaging of reciprocals been applied to the individual observations rather than to Keith's [57] Class I Groupings, the tendency for trend reversal would have been even greater.

TABLE III

Summary of estimates of anthropogenic emissions of primary particulates (10^6 metric tons yr^{-1})

Types of emissions	Area and reference								
	U.S. only				Worldwide				
	NAPCA [77]	SCEP [96]	Hidy and Brock [37]	MRI [71]	SCEP [96]	Hidy and Brock [37]	Peterson and Junge [83]	Robinson and Robins [92]	Adopted for this study
Stationary combustion	8.1	8.1		5.4		21	43.8	37.8	24
Transportation	1.1	1.1		1.1			2.2	0.8	3
Industrial (total)	6.8	40.6	16.1	11.2	158		56.4	33.6	50
Iron and steel	1.7	13.9	1.3	1.6	54.6			10.2	
Other metals	0.24	0.58		0.54				1.8	
Minerals	0.75	2.3		0.82				2.3	
Cement, sand, etc.	2.1	14.9		5.64	60.9	11.7		5.9	
Fuel processing	0.26	0.96		0.21				3.7	
Pulp and paper	0.65	4.3	3.1	0.53	9.5			0.4	
Grain, feed, etc.	1.0	3.3		1.65	33.0			8.4	
Chemicals	0.08	0.38		0.22				0.93	
Solid wastes	1.0	1.0		0.84			2.4	3.8	2
Agricultural burning	2.2	2.18		2.2	21.8				22
Slash and litter burning				15.4					
Miscellaneous	0.45			1.16			28.8	7.0	25
Totals	19.6	86.8		37	179.8	35–100	133.2	83	126

Measured concentrations of airborne particulates in downtown Los Angeles decreased from 265 μg m^3 in 1950 to 164.2 for the period 1957–1961 and to 124.5 for the period 1962–1966 [38].

These data suggest that the 'Los Angeles air pollution crisis' is due more to increasing public sensitivity to a situation than to any change in atmospheric conditions. In fact, the earlier quotation from Abbot [1] concerning removal of the Smithsonian solar-constant station from Mt. Wilson in 1920 and the Stanford Research Institute's [112] 1922 photograph from Mohlholland Drive depicting a smog-filled basin much like that seen today suggest that the Los Angeles smog existed for at least a quarter of a century before it's dramatic registration on public consciousness in the 1940's.

9. Man's Contribution to the Particulate Burden of the Atmosphere

Emission rates of anthropogenic particles are quite speculative and for the most part are based on production figures multiplied by emission factors. NAPCA [77], Hidy and Brock [37] and MRI [71] have made recent independent estimates primarily for the United States. Hidy and Brock [37], SCEP [96] and Peterson and Junge [83] have extrapolated global values from these using production, land area and energy use ratios as deemed appropriate. The resulting emission estimates are summarized in Table III. Several points should be noted. Original estimates were generally in short tons rounded to one or two digits; conversion to metric tons, rounding and summing can lead to systematic changes. In many reports this conversion of units has been ignored or confused. SCEP [96] refers to the NAPCA [77] figures as controlled emissions and presents substantially larger numbers as uncontrolled emissions attributed to unpublished NAPCA data. Since the different estimates do not include values for all, or the same, categories of emissions, comparison of totals is not straightforward. The last colum of Table III contains the estimates adopted for this study.

Estimates of natural emissions and anthropogenic emissions of secondary particulates are summarized in Table IV. Asterisks are used to denote values adopted for this

TABLE IV

Summary of estimates of worldwide emissions of particulates other than anthropogenic primary emissions (10^6 metric tons yr^{-1}). Asterisks identify estimates adopted for this study

Amount by type (10^6 tons yr^{-1})	Method of estimation and source
Natural primary	
Eroded dust	
1100	$20 \times$ MRI [71] estimate of 57×10^6 for U.S. based on fallout measurement at 15 stations.
500	Atmospheric dust content of 5 μg m^{-3} in first 5 km and 40 removals yr^{-1}. Ocean sedimentation rate of 0.1 g cm^{-2} 10^{-3} yr^{-1} [33].
500	. Extrapolated from Wadleigh's [122] estimate for natural and agricultural dust over U.S. (Arnason for SCEP, see [83]).
6–360	"Estimates from various sources", [50].

Table IV (continued)

Amount by type (10^6 tons yr^{-1})	Method of estimation and source
200	From steady state concentration of dust of 10 μg m^{-3} over land to 1 km, 0.1 μg m^{-3} elsewhere and residence time of 3 days below 1 km and 5–30 days above [92].
70–100	Glacier sediment data of Windom [131], an underestimate since the glaciers were 2–3 km above sea level [96, 33].
380*	Ocean sediment rate of 0.5 mm 10^{-3} yr^{-1} and density 1.5 (this study, see text).
Sea salt	
1000–2000	Range of values attributed to the analyses by Eriksson [22] and Woodcock [133].
960	One 1μ sea salt particle cm^{-2} s^{-1} from world's oceans [52].
300	3 × Eriksson's [22] estimate of land fallout of sea salt particles [105, 33].
1000*	Adopted as most reasonable value based on above estimates.
Forest Fires	
200*	6 × MRI [71] estimate of 34×10^6 for U.S. wildfires attributed to U.S. Forest service.
150	14 × 4.2×10^6 acres burned in U.S. in 1968 × 2×10^{22} particles acre^{-1} × 1.3×10^{-16} g particle^{-1} [37].
35	6 × NAPCA [77] estimate of 6.2×10^6 for U.S. [96].
3	4 × emission factor of 8.5 kg ton^{-1} × 18 tons acre^{-1} × 4.5×10^6 acres burned in U.S. in 1966 [92].
Volcanoes	
150*	Montmorillonite estimated to accumulate in ocean sediments at 0.15 g cm^{-2} 10^{-3} yr^{-1} [33].
42.5	Total of following stratospheric and tropospheric estimates.
6.5	2 × 12/14 × Mitchell's [72] estimate of northern hemisphere stratospheric loading of 3.8×10^6 tons based on assumption that 1 % of total ejected material reaches stratosphere and has 14 month residence time there.
36	Estimate of average tropospheric loading divided by residence time of 0.1 yr [96].
25	Estimate to represent interpeak average based on two above estimates [83].
8	0.25 × 1 eruption of 11.5×10^6 m^3 yr^{-1} of magma of density 2.75 [37].
Extraterrestrial	
10	Upper limit of Rosen's [95] estimate of $1–10\times10^6$ for the stratospheric flux of both terrestrial and extraterrestrial particles [83].
1.6	Average of estimates from iron, nickel and cobalt content of Greenland ice [63].
few × 10^{-6} – 3.6	Survey of available estimates [102].
0.018–2*	Probable upper limit and maximum possible upper limit, based on stratospheric balloon borne sampling [6].
Natural secondary	
Sulfate	
33–330	Conversion of 60 % of Robinson and Robin's [90, 91] estimate of natural emissions of SO_2 and H_2S [37].
244*	Conversion of $\frac{2}{3}$ of natural SO_2 and H_2S to $(NH_4)_2SO_4$ [83].
185	Conversion of biological H_2S [92].
170	244×10^6 of $(NH_4)_2SO_4$ less the NH_4 [105].
130	Assumed equivalent to anthropogenic rate [33].
Nitrate and Ammonium	
636	Conversion of 392×10^6 tons NO_3 from NO_x and 244×10^6 tons NH_4 from NH_3 [92].
570	Conversion of 42 % of NO_x to 330×10^6 tons NO_3 and 23 % of NH_3 estimate to 240×10^6 tons NH_4 [91].

Table IV (continued)

Amount by type (10⁶ tons yr⁻¹)	Method of estimation and source
75*	Secondary sulfates divided by 5.5, Stern's [113] ratio of sulfate to nitrate in high volume filter samples [83].

Hydrocarbon

1000*	Atmospheric content of volatile organics estimated as high as 10^6 tons with at least 5 turnovers per day on at least 200 days yr⁻¹. Estimated 20×10^6 tons condensation nuclei over vegetated parts of the world with a turn over rate of once per week [128].
438	Average concentration of volatile organics of 10 ppb exists up to 2 km with turn-over time of once per day during 270 days yr⁻¹ [86].
500	Mixing ratio of solid material in precipitation, $0.8 \times 10^{-5} \times 0.5$ (half organic) $\times 120$ cm yr⁻¹ (average precipitation) $\times 10^{18}$ cm² (land surface) [127].
175–200	Of 5×10^{10} tons yr⁻¹ of dry plant material produced on Earth 0.1% is carotenoids and phytol which are released on decomposition. Conservative estimate of volatiles such as terpenes released prior to decomposition set at 10^8 tons yr⁻¹ [126].
200	Total conversion of hydrocarbon value near Went's [126] lower estimate [92].
75	Assumes 50% conversion to particles of Went's [126] estimate of "154[sic] $\times 10^6$ tons yr⁻¹" of organic volatiles [83].

Anthropogenic secondary

Sulfate

220*	Conversion to $(NH_4)_2SO_4$ of $\frac{2}{3}$ of estimated 160×10^6 tons yr⁻¹ of pollutant SO_2 [83].
200	Attributed to Peterson and Junge [83] by SMIC [105].
133	Conversion of 70% of 132×10^6 tons of pollutant SO_2 [92].
130	Estimate of fossil fuel produced SO_4 from glacier content at present compared to periods unaffected by man [33].
100	Conversion of 60% of Robinson and Robins' [91] values for SO_2 (132×10^6) and H_2S (2.7×10^6) [37].

Nitrate

84	Conversion of 42% of man produced NO_x to NO_3 [91].
35*	Anthropogenic secondary sulfates divided by 5.5, Stern's [113] ratio of sulfate to nitrate in high volume filter samples [83].
30	Conversion of 42% of 48×10^6 tons pollutant NO_2 [92].

Hydrocarbon

90	Attributed to SCEP [96] by SMIC [105].
25	Reactive fraction of Robinson and Robins [90] estimate of total pollutant hydrocarbon [37, 92].
15*	Conversion of 25% of $2 \times$ NAPCA [77] estimate of U.S. emission of 29×10^6 tons of pollutant hydrocarbons in 1968 [83].

study and further summarized in Table V. While selection was admittedly subjective, an attempt was made to choose estimates best supported by data and resulting in a total particulate burden matching available data on fractional composition of collected aerosols.

Despite the wide range of estimates for volcanic and forest fire emissions (almost two orders of magnitude) maximum figures were adopted in both cases. Accumulation

TABLE V

Comparative summary of worldwide natural and anthropogenic particulate loading of the atmosphere. Values in 10^6 metric tons yr^{-1} are summarized from Tables III and IV

Natural	10^6 tons yr^{-1}	Anthropogenic	10^6 tons yr^{-1}
Primary emissions			
Eroded dust	380	Stationary combustion	24
Sea salt	1000	Transportation	3
Forest fires	200	Industrial	50
Volcanoes	150	Solid wastes	2
		Agricultural burning	22
		Miscellaneous	25
Primary subtotals	1730		126
Secondary emissions			
Sulfate	170		160
Nitrate	75		35
Ammonium	74		60
Hydrocarbon	1000		15
Secondary subtotals	1319		270
Total particulates	3049		396

rates of montmorillonite in ocean sediments is by far the least arbitrary source of the volcanic estimates. For forest fires the MRI [71] estimate from the U.S. Forest Service and the SCEP [96] multiple of 6 to obtain the worldwide estimate appeared most authoritative even though the details of the estimate were unavailable. Another reason for adopting this figure was to allow for emissions from slash and burn fires which Flohn (1971, unpublished, see [105]) apparently estimated as high as 80×10^6 tons yr^{-1}. The argument for tabulating slash and burn emissions as natural rather than anthropogenic is given above in the section on Data Suggesting the Atmosphere is Getting Cleaner.

The Goldberg [33] and SMIC [105] argument that sea salt particles have a short

TABLE VI

Percentage composition of natural and total particulates as summarized in Table V

Component	% of natural particulates	% of total particulates
Water soluble	45.5	47.6
Sulfate	8.1	11.8
Nitrate	2.5	3.2
Ammonium	2.4	3.9
Combustible	35.0	31.4
Insoluble inorganic	19.6	21.0

lifetime appears neither unique to these particles nor valid when considering the total global burden of particulates. Accordingly Eriksson's [22] estimate of 10^9 tons was retained and 7.7% of it counted as sulfate in Table VI. Went's estimate of natural organic emissions essentially stands alone, increasing steadily with further study and subsequent papers. While the basis for his estimates is lacking in breadth, there is substantial evidence that the blue hazes of summer must constitute a significant fraction of the total atmospheric burden of particulates. No other component appears capable of explaining the marked seasonal and geographical variations in atmospheric turbidity reported by Flowers *et al.* [27] and Ellis and Pueschel [21]. Went [127, 128] ascribes the blue hazes to natural organic material consisting of terpenes; phenolic, carotenoid and phytyl compounds; black carbon; high boiling point oils without resolution in a gas chromatograph and black semiliquid tar-like droplets. Earlier, Went [126] described this material as nondegradable except by combustion and photochemistry and suggested that the more than 10^{16} tons estimated to have been produced during the geological ages could easily be the sources of our some 10^{12} tons of planetary petroleum deposits.

Robinson and Robins' [91, 92] high estimates for nitrate particulates cannot be reconciled with available data on aerosol composition. Thus Peterson and Junge's [83] estimates were accepted. The inconsistancy resulting from not separating the NH_4 from the SO_4 before dividing by 5.5, Stern's [113] ratio of sulfate to nitrate in high volume filter samples, was left unchanged in Table IV but was considered in computing fractional composition in Tables V and VI.

Table V gives a comparative summary of natural and anthropogenic particulate emission estimates adopted for this study. To the extent to which they are valid they indicate that man now contributes 13.6% or approximately one-seventh of the 3.5×10^9 tons yr^{-1} of particulates injected into the Earth's atmosphere. The validity of such estimates is hard to determine but the following comparisons tend to support this conclusion. Separation by composition of the Table V estimates for natural and

TABLE VII

Summary of estimates of the anthropogenic contribution to total airbone particulates

Anthropogenic fraction (percent)	Source
5	Selezneva [104]
5	Squires [111]
20	SCEP [96]
6	Hidy and Brock [37]
20–30	Mitchell [73]
5–45	SMIC [105]
18	Peterson and Junge [83]
11	Robinson and Robins [92]
14.4	Average of above
13.6	Table V, this study

total emissions, as shown in Table VI, yields values consistent with those which have been reported for atmospheric aerosols and the nonaqueous material found in precipitation. (For Table VI forest fire emissions were split equally among the three categories.) Previous estimates for man's share of total particulate emissions, summarized in Table VII, range from 5 to 45% and average 14.4%. Note that the ratio of anthropogenic to total particulate sulfate indicated by Table VI (31%) is three times the estimate arrived at by Selezneva [103]. This suggests that natural sources of particulate sulfate have been underestimated, as does the failure of Junge [51] and Federer [24] to find any recent upward trend in sulfate in polar snows.

Moreover, in a recent study Junge [56] estimated that sulfur is removed from the atmosphere (half by rainout of sulfate and half by surface absorption of SO_2) at five times the rate it is injected by man.

By projecting both growth and emission controls through the year 2000, Hidy and Brock [37] arrived at a future growth rate for anthropogenic particulates of 0.4% yr^{-1}. On the other hand, for the largest (SO_2) component of man's contribution, Robinson and Robbins [91] determined a past doubling time of roughly 26 yr or a growth rate of about 2.75% yr^{-1}. Since the anthropogenic contribution constitutes about one-seventh of the total, these figures indicate a growth in total atmospheric burden of particulates of 0.06 to 0.4% yr^{-1}. If there is a global upward trend in atmospheric particulates due to man this is the magnitude of effect we should be looking for.

Continuation of the above growth rate over 60 yr would amount to a 4–27% effect. While there are two sets of data spanning 60 yr (the South Pacific air conductivity and the rural eastern United States turbidity) which do not appear to support a growth rate of even this small magnitude, neither is sufficiently direct or clear-cut to preclude such a trend. However, the higher growth estimate based on man's SO_2 emissions does appear inconsistent with data on the trend in sulfate content of polar snows: Junge [51] could detect no increase for the period 1915 to 1957 and Federer [24] none for the period 1880 to 1968.

Quantizing of any upward trend in airborne particulates at the above low level allows its proponents to take refuge in the lack of a sufficient period of suitable data for corroboration. To do so, however, they must first repudiate the substantiating data so far introduced as reflecting only local or short period variations rather than a global trend.

10. Longer Period Data

A possible way of extending the range of observational data is to examine rates of accumulation of atmospheric sediments. According to Griffin et al. [34] the under 2μ size fraction of pelagic clay deposits constitutes about 60% by weight of the non-biogenous phases and is composed predominantly of the four mineral groups chlorite (high latitude glacial origin), montmorillonite (typically of volcanic origin), illite (mica group) and kaolinite (low lattitude origin). The maximum eolian fraction of these clay deposits can be estimated by measuring accumulation rates in areas protected from river runoff such as midocean ridge valleys and sea mounts. From his

TABLE VIII

Summary of accumulation rates of atmospheric sediments.

(a) Eolian accumulation rates as reported by Griffin *et al.* [34], minimum accumulation rates as estimated by Windom [131], and most probably values (see text).

Ocean	Area	Transport Path	Accumulation rates (mm $10^{-3}yr^{-1}$)		
			Griffin *et al.*	Windom	Most probable
Atlantic	15° N–50° N	Jet stream	0.3–0.5	0.09–0.15	0.21–0.35
	15° N–15° S	Trade winds	0.7–0.8	0.21–0.25	0.49–0.56
	15° S–50° S	Jet stream	0.2–0.4	0.06–0.12	0.14–0.28
Pacific	15° N–15° N	Jet stream	0.4–2.0	0.12–0.6	0.28–1.4
	15° N–15° S	Trade winds	0.3–0.6	0.09–1.8	0.21–4.2
	15° S–50° S	Jet stream	0.3–0.8	0.09–0.24	0.21–0.56
Indian	15° S–20° N	Trade winds	1	0.3	0.7
	20° S–50° S	Jet stream	0.5–0.8	0.15–0.24	0.35–0.56

(b) Global component of particulate matter in snowfields from Windom [131].

Location	Transport path	Accumulation rate (mm 10^{-3} yr^{-1})
Greenland	Polar easterlies	0.14
St. Elias Range, Yukon Territory	Westerlies	0.11
Mt. Olympus, Washington	Westerlies	0.21
Mt. Orizaba, Mexico	Trade winds	0.09
Mt. Popocatepetl, Mexico	Trade winds	<0.01
Tasman Glacier, New Zealand	Westerlies	0.80
New Byrd Station, Antarctica	Polar easterlies	<0.01

study of dust in permanent snow fields Windom [131] estimated that only 25 to 75% of the pelagic clay sediments were actually of eolian origin. Thus, 50% of the 60% of these pelagic clays in the under 2μ size fraction represents a minimum eolian accumulation rate.

Table VIII summarizes these data giving both the maximum and minimum accumulation rates as described above for the eolian fraction of ocean sediments. These figures indicate a minimum global atmospheric fallout rate of something over 0.1 mm 10^{-3} yr^{-1}. The ocean data are representative of the past several thousand years while the snowfall data are representative of the past 50 to 200 yr. For a marine sediment density of 1.5 g cc^{-1} as used by Windom [131], 0.1 mm 10^{-3} yr^{-1} is equivalent to 76×10^6 tons yr^{-1} over the globe. This represents only the insoluble inorganic fraction of atmospheric particulates. Of the possible sources; meteors, volcanos and surface erosion, the latter is considered to be by far the most predominant. Judson [50] refers to the annual rate of wind erosion of the continents as about 10^8 tons yr^{-1} but in a table gives $0.6–3.6 \times 10^8$ tons yr^{-1} which he attributes to "estimates from various sources". His quote of eolian ocean sediment accumulation rates of 0.25–1 mm 10^{-3} yr^{-1} and density of 0.7 g cc^{-1} correspond to $0.9–3.6 \times 10^8$ tons yr^{-1} over the globe and thus probably indicates the source of the upper limit given. (Since Judson was estimating soil erosion rather than atmospheric fallout one would have expected these figures to be reduced by 25% to correct for that fraction falling back on the continents, if this is indeed the source of his upper estimate.)

There are valid reasons for believing that accumulation rates as derived by Windom [131] are underestimates. For each permanent snow field sampled except the Tasman Glacier in New Zealand, most fallout occurs from an air mass which has traversed an entire ocean since last exposed to a significant surface source and should accordingly be significantly depleted of surface eroded dust. Secondly, sampled snow fields are located 2–3 km above sea level and thus receive fallout only from the upper and generally cleaner fraction of the atmosphere. To obtain an estimate free of these two biases I used Windom's [131] finding for the North Pacific (from comparative analyses of pelagic and glacial sediments) that 50% of the $<2\mu$ size fraction and essentially all of the $>2\mu$ material was eolian. Since the $<2\mu$ fraction represents 60% of the total mass, this leads to 70% as the eolian fraction. This is the basis for the most probable values given in Table VIII.

Considering now that the trade winds cover about half the globe and that the continents (assumed to have a fallout rate at least as large) cover roughly another quarter, 0.5 mm 10^{-3} yr^{-1} does not seem an unreasonable global average. It could be argued that this 0.5 mm 10^{-3} yr^{-1} includes the 0.15 g 10^{-3} yr^{-1} of montimorillonite attributed by Goldberg [33] to volcanos. This is not done and the apparent inconsistency is justified by the following arguments. Counting the montmorillonite twice compensates in total mass (while not in fractional composition) for the neglect of the soluble and organic fraction of volcanic and eroded dust. Secondly there are studies suggesting that the data compiled by Windom [131] underestimates sediment accumulation rates. Opdyke and Knowles [80] reported South Atlantic accumulation rates of 3.5–27 mm 10^{-3} yr^{-1}, an order of magnitude higher than those previously reported. Recent observations by Prospero and Carlson [85] indicate that between 10N and 25N the trade winds transport through the longitude of Barbados over $6\frac{1}{2}$ times as much Saharan dust as would need to be deposited enroute to maintain an eolian sedimentation rate of 0.5 mm 10^{-3} yr^{-1} over this latitude band of the Atlantic.

Using Windom's [131] density value of 1.5 g cc^{-1}, 0.5 mm 10^{-3} yr^{-1} is equivalent to a global fallout of 380×10^6 tons yr^{-1} which is the estimate used in Table IV.

TABLE IX

Uniform airborne concentrations of insoluble inorganic fraction of airborne particulates implied by a global sedimentation rate of 0.5 mm 10^{-3} yr^{-1}, a density of 1.5 [131] and the deposition velocity indicated

Airborne concentration μg m^{-3}	Deposition velocity cm s^{-1}	Source of deposition velocity
2.9	0.83	Complete rain scavenging of lowest 5 km of atmosphere in 1 week [25]
4.1	0.58	Complete rain scavenging of lowest 5 km of atmosphere in 10 days [81]
80	0.03	Settling velocity for 2μ particles [84]

One can work backwards from accumulation rates to airborne concentrations using the formula: airborne concentration = (total particulates/sedimentary component) × (density × sedimentation rate/deposition velocity). An accumulation rate of 0.5 mm 10^{-3} yr^{-1} and a density of 1.5 are equivalent to 2.4 (μg m^{-3}) (cm s^{-1}). Table IX gives the corresponding airborne concentrations of the sedimentary particulates resulting from the cited deposition velocities. These should be multiplied by 9, the ratio of total to sedimentary particulates, to obtain airborne concentrations of total particulates. This yields airborne concentrations of particulates of 26 to 720 μg m^{-3} for the past several thousand years. The lower limit is approximately equal to the average measured at 20 nonurban U.S. stations 1958–1966 [110]. This presents a comforting argument against any recent increase in airborne particulate concentrations.

It is also comforting to note that the above estimates for past airborne concentrations are almost independent of the eolian sedimentation rates on which they are based. This occurs because the sedimentation rate also determines the sedimentary component of particulates. The appearance of the latter in total particulates is all that prevents sedimentation rate from dropping out of the equation.

It is more meaningful to use this equation to obtain an estimate of deposition velocity since this is both the most critical and most arbitrary parameter in the equation.

Robinson and Robins [92] constructed mean atmospheric profiles of particulate concentration beginning at 20 μg m^{-3} for the first 1.5 km over land and 10 μg m^{-3} for the first 0.5 km over the oceans and both declining to 0.6 μg m^{-3} above 9 km. This corresponds to 6.6×10^3 tons of particulates in the lowest meter of the atmosphere. Uniform deposition of 3.5×10^9 tons yr^{-1} from this layer would require a deposition velocity of 1.7 cm s^{-1}. Strict application of the model allowing oceanic fallout only from air with a marine aerosol content of 10 μg m^{-3} requires a deposition velocity of 1.2 to 2.1 cm s^{-1} to deposit 0.5 mm 10^{-3} yr^{-1} (of density 1.5) of insoluble inorganic sediment, depending on whether the later component constitutes 20% or 11% of the total aerosol.

Robinson and Robins [92] also used aerosol atmospheric residence times increasing from 3 days at 0–1 km to 33 days at 9–14 km to arrive at an annual aerosol production rate of 1.34×10^9 metric tons yr^{-1}. If the total annual production is rather 3.5×10^9 tons yr^{-1} as indicated by this study, mean residence times for airborne particulates must be 2.5-fold smaller or about 1.2 days in the lowest kilometer of the atmosphere. While this is significantly shorter than the residence times which have been computed from radioactive tracers used as the basis for Robinson and Robin's [92] model, and while it may represent a mass weighted average, it is believed to be more representative of the turnover rate of the total atmospheric burden of particulate matter. The consistent distinction between continental and marine aerosols, the association of characteristic skyscapes with deserts, mountains and seashore, and the small scale patterns of atmospheric turbidity revealed by Flowers et al. [27] are all difficult to explain if the mean residence time of atmospheric particulates is 3 days or more.

11. Discussion

The above estimate of 13.6% for the current anthropogenic component of total atmospheric particulates suggests man has increased this burden by 16% since the beginning of the industrial revolution in 1760. This yields a growth rate of 0.07% yr^{-1} or only slightly above the lower limit quoted above from Hidy and Brock's [37] projection. To claim that this represents a real current upward trend ignores the actions taken in the most industrialized areas of the world over the past 20 to 50 yr to produce cleaner skies. Since the results of these cleansing actions are noticeable in atmospheric measurements (while a global trend of the above magnitude is not) the current trend in anthropogenic airborne particulates may well be downward, even on a global basis.

The proponents of a continuing upward trend could retreat to the more logical position that man, through local and regional increases, is increasing the *mean* (rather than *background*) turbidity of the atmosphere and that such an increase could not be expected to be detected in observations from sites such as Antarctica, the South Pacific, Mauna Loa and remote astronomical observatories. However, unless such a claim is to be based on faith or on the post-1940 cooling of the Earth (which appears to be its ultimate source), supporting it will require far more than the citation of a single set of observations used to support most such claims to date. Such an argument would also be hard pressed to counter the mounting evidence that industrialized population centers notorious for smog in the past have been making progress in cleaning their skies despite increases in both population and per capita energy consumption.

Even if we assume a continuing upward trend at the maximum estimated rate, a doubling time of 26 yr for the anthropogenic component will require 75 yr to double and 120 yr to quadruple the present atmospheric burden of total particulates. The most recent analyses of the effects of particulates on climate [4, 87] suggest that a quadrupling of present levels, estimated to cause a cooling of the surface temperature of approximately 3 °C, is about the condition which would have to be "sustained over a period of a few years" to trigger an ice age.

This might be cause for concern except for at least four opposing factors. Man's increase in fossil fuel consumption of 4% yr^{-1} is producing an increase in the CO_2 content of the atmosphere of about 0.2% yr^{-1} and an increase in heat released at the surface of the Earth of about 4% yr^{-1}. If continued these would lead in about 130 yr to a doubling of the CO_2 content of the atmosphere [96] and to a surface heat release equivalent to 1% of that deposited by the Sun [29a]. While the estimates vary as to the resulting effects on the mean surface temperature of the Earth, most suggest that the combined effects would, as a minimum, be sufficient to significantly reduce the hypothesized 3 °C cooling due to particulates. The third opposing factor is man's current efforts to reduce his particulate and particulate-generating emissions to the atmosphere. I see no reason to attempt to improve on Hidy and Brock's [37] estimate that these lead to a growth rate in anthropogenic particulates almost an order of magnitude smaller than that considered here. The fourth opposing factor is the exhaustion of the fossil fuels themselves which will bring all of the above effects to a halt. (Surface release of heat from man's energy conversions will remain a problem if man is successful in developing one or more of the alternate energy sources offered by breeder reactors, controlled thermonuclear fusion and solar energy.) Various estimates are available as to how long our fossil fuels will last. One of the more objective, authoritative [88] and ominous concludes: "*When widespread industrialization is assumed*, an analysis of fuel use versus time shows that the middle 80% of the world's supply of crude oil and natural gas would be consumed during a period of 15–20 yr and that of coal in less than a century" [17a]. Without the assumption of widespread industrialization, Hubbert [46] arrived at 58–65 and 300 yr respectively.

On the basis of his present knowledge and understanding man can conclude that, while many difficult problems must be solved if he is to maintain both his growing population and his industrialized standard of living, avoidance of an ice age initiated by anthropogenic airborne particulates is *not* one of them. I might add that his survival as a species does not appear threatened. Man has amply demonstrated that he is capable of coping with almost any circumstance except his own success. He has developed what he calls advanced civilizations many times in many places. But in each case as he freed himself from the ecological constraints of his environment and developed new patterns of response to the artificial environment of his own creation he reached a stage at which he could or would no longer work to maintain his created environment. Obversely, when reexposed to the ecological constraints of a natural environment he not only survived but was again able to build a new advanced civilization.

12. Summary

Its proponents have in general been content to find *one* set of data indicating an upward trend in particulates and to offer it as sufficient evidence of a global trend. They do not appear to realize that a global trend requires that *every* set of data not compensated by decreases in local effects must respond to the trend. It is rather the counterargument which requires only one set of representative data denying the trend to establish its point. I have found the following sets of data denying an upward trend due to anthropogenic sources.

(1) Turbidity values in the nonurban eastern United States in 1962–1966 were comparable to that measured in Washington, D.C. in 1903–1907 [27].

(2) Conductivity of the air over the South Pacific has remained remarkably constant since 1908 [17].

(3) Minimum turbidity values at the Mauna Loa Observatory showed no trend between 1958–1962 and by 1970 had returned to pre-Agung values [21].

(4) Early summer turbidity values in Antarctica indicated no trend between 1949–1952 and 1966 [26].

(5) Routine measurements of atmospheric extinction made at Lowell Observatory since 1955 reveal "no evidence of any increase of the extinction caused by human activity." [48].

(6) Atmospheric aerosol extinction values calculated from Smithsonian solar observations made from 1902 to 1950 reveal that "there has been no detectable change in the global atmospheric transmission measured from remote, high altitude sites in the last half century" [94].

7. Almost all major western urban areas plagued by dirty skies by 1950 have shown improvements in recent years in terms of visibilities [5, 28, 45], dust fall [61], suspended particulate matter [110], and ratio of sunshine in the central city to that at an outlying station [13]. Even at the 20 United States nonurban sites for which suspended particulates were measured 1958–1966, 13 showed no significant trend and 1 a down trend [110].

(8) Observers of cloud condensation nuclei generally report their concentrations and spectra to show little if any anthropogenic influence. Twomey and Wojciechoski [117] found higher concentrations over Australia and Africa than over North America and Europe. No significant east to west gradient in concentration was found over North America itself. Kocmond and Mack [59] found that urban increases in cloud and Aitken nuclei concentrations were temporary and were undetectable above background 20 to 50 miles downwind of the pollution sources. Selezneva [104] found that condensation nuclei from single powerful sources extended to distances of 20–25 km downwind and those from industrial cities somewhat beyond 50 km.

(9) Working backward from accumulation rates for the eolian fraction of pelagic sediments and estimates of settling and deposition velocities yields airborne particulate estimates of 26 to 720 μg m^{-3} for the past several thousand years. This range exceeds most nonurban measurements of today.

Evaluation of the airborne particulate trend rate expected to result from anthropogenic sources yields a value of 0.06–0.4% yr^{-1}. This weakens the above arguments against a trend since this rate of change is below that which can be resolved with the record length and data quality now available. However, it also destroys the arguments of the trend's proponents since cited trends have exceeded this range severalfold and therefore must be due to local or short period variations rather than to a global trend.

Regardless of how one feels about the reality of an anthropogenic upward trend in airborne particulates, projections of currently estimated growth rates do not indicate any significant effect on global climate before exhaustion of our fossil fuels would bring such a trend to a halt.

13. Epilogue

This study contains ominous portent for our air pollution control program. Most atmospheric scientists as well as the lay public have been led by years of preconditioning to believe that the ratio of anthropogenic to natural emissions of airborne particulates are quite different than indicated by this study. A 13.6% reduction in airborne particulates would be barely detectable in our observational net and imperceptible to the general public. It would appear reasonable to assume that the distribution of anthropogenic aerosols is sufficiently non-uniform that their removal would result in noticeable improvement in areas of greatest concentration. But Keith's [57] compilation of Los Angeles visibility data back to 1933 does not provide much support for such an assumption. Neither does Table V which indicates that anthropogenic emissions of primary aerosols account for less than 4% of the atmospheric total. Almost 70% of man's contribution is in the form of gaseous precursors which require time to condense into or onto particulates and may not become detectable as aerosol until far removed from its source. As a consequence even 100% success in the control of anthropogenic particulate emissions will be regarded as abject failure by the public. And even if this situation can be made pallatable by a reeducation program it will be quite a different task to convince them that such miniscule and non-self-evident results are worth the tremendous costs which will become all too self-evident by 1975–76.

Also, those who propose gas rationing as a palliative for Los Angeles smog should first explain the coincidence of World War II gas rationing with the mid-1940's worsening of visibility usually identified as the beginning of L.A. smog (see Figure 1).

Acknowledgements

This work was preformed under the auspices of the United States Atomic Energy Commission and has been supported (in part) by the Climatic Impact Assessment Program, Office of the Secretary, Department of Transporation. Uncited comments from R. A. Bryson, W. E. Cobb, H. E. Landsberg, J. M. Mitchell, Jr., G. D. Robinson, R. G. Roosen, and V. J. Schaefer are gratefully acknowledged.

References

1. Abbot, C. G., *The Sun and the Welfare of Man*, Vol. 2, *Smithsonian Scientific Series*, New York p. 68, 1929.
2. Alkezweeny, A. J. and Green, W. D.: 'The Effect of Air Pollution on the Concentration of Ice and Condensation Nuclei', *J. Rech. Atmos.* 4, 31–33 (1970).
3. Ashby, W. C. and Fritts, H. C.: 'Tree Growth, Air Pollution and Climate near LaPorte, Indiana', *Bull. Amer. Meteorol. Soc.* 53, 246–251 (1972).
4. Barrett, E. W.: 'Climate Change', *Science* 173, 983 (1971).
5. Beebe, R. G.: 'Changes in Visibility Restrictions over a 20 Year Period', *Bull. Amer. Meteorol. Soc.* 48, 348 (1967).
6. Bhandari, N., Arnold, J. R., and Parkin, D. W.: 'Cosmic Dust in the Stratosphere', *J. Geophys. Res.* 73, 1837–1845 (1968).
7. Bigg, E. K. and Stevenson, C. M.: 'Comparison of Concentrations of Ice Nuclei in Different Parts of the World', *J. Rech. Atmos.* 4, 41–58 (1970).
8. Bonacina, L. C. W.: 'London Fogs – Then and Now', *Weather* 5, 91–93 (1950).
9. Bryson, R. A.: 'All Other Factors Being Constant...', *Weatherwise* 21 56 (1968).
10. Bryson, R. A. and Baerris, D. A.: 'Possibilities of Major Climatic Modifications: Northwest India, a Case for Study', *Bull. Amer. Meteorol. Soc.* 48, 136–142 (1967).
11. Budyko, M. I.: 'The Effect of Solar Radiation Variations on the Climate of the Earth', *Tellus* 21, 611 (1969).
12. Cadle, R. D. and Junge, C. E.: 'Atmospheric Nuclei', in W. H. Matthews, W. W. Kellogg and G. D. Robinson (eds.), *Mans Impact on the Climate*, pp. 396–399, MIT Press, Cambridge, Mass. (1971).
13. Chandler, T. J.: 'The Climate of London', Hutchinson, London (1965).
14. Charlson, R. J., Hodge, P. W., Lucke, P. B., Mannery, E. J., Porch, W. M., and Snow, T. P.:

University of Washington Project ASTRA, *Project ASTRA Publication No. 2*, University of Washington, Seattle (1970).

15. Clark, W. E. and Whitby, K. T.: 'Concentration and Size Distribution Measurements of Atmospheric Aerosols and a Test of the Theory of Self-Preserving Size Distributions', *J. Atmos. Sci.* **24** (6), 677–687 (1967).

16. Cobb, W. E.: 'Oceanic Aerosol Levels Deduced from Measurements on the Electrical Conductivity of the Atmosphere (Abstract)', *EOS* **53**, 387 (1972).

17. Cobb, W. E. and Wells, H. J.: 'The Electrical Conductivity of Oceanic Air and its Correlation to Global Atmospheric Pollution', *J. Atmos. Sci.* **27**, 814–819 (1970).

17a. Committee on Natural Resources: 'Energy Resources', *Nat. Acad. Sci – Nat. Res. Counc.* Publ. 1000-D, Washington D.C. (1962).

18. Conover, J. H.: 'Anamolous Cloud Lines', *J. Atmos. Sci.* **23**, 778–785 (1966).

19. Cronin, J. F.: 'Recent Volcanism and the Stratosphere', *Science* **172**, 847–849 (1971).

20. Dodge, M.: 'Forest Fuel Accumulation – A Growing Problem', *Science* **177**, 139–142 (1972).

21. Ellis, H. T. and Pueschel, R. F.: 'Solar Radiation: Absence of Air Pollution Trends at Mauna Loa', *Science* **172**, 845–846 (1971).

22. Eriksson, E.: 'The Early Circulation of Chloride and Sulfur in Nature; Meteorological, Geochemical and Pedological Implications', Part I, *Tellus* **11**, 375–443 (1959), Part II, *Tellus* **12**, 63–109 (1960).

23. Davitaia, F. F.: 'On the Possible Influence of Atmospheric Dust Cover on the Diminution of Glaciers and the Warm-Up of Climate', *Trans. Sov. Acad. Sci. Geogr. Ser.* **2** (1965).

24. Federer, B.: *Investigation of the Aerosol Content of Greenland Snow Between 1880 and 1968*, Suppl. Vol. Proc. 7th Int. Conf. on Condensation and Ice Nuclei September 18–24, 1969, Prague and Vienna, Academic, Prague (1971).

25. Ferguson, W. S., Griffin, J. J., and Goldberg, E. D.: 'Atmospheric Dusts from the North Pacific – A Short Note on a Long-Range Eolian Transport', *J. Geophys. Res.* **75**, 1137–1139 (1970).

26. Fischer, W. H.: 'Some Atmospheric Turbidity Measurements in Antarctica', *J. Appl. Meteor.* **6**, 958–959 (1967).

27. Flowers, E. C., McCormick, R. A., and Kurfis, K. R.: 'Atmospheric Turbidity over the United States, 1961–1966', *J. Appl. Meteorol.* **8**, 955–962 (1969).

28. Freeman, M. H.: 'Visibility Statistics for London/Heathrow Airport', *Meteor. Magazine* **97**, 214–218 (1968).

29. Friedlander, S. K.: 'Theory of Self-Preserving Size Distributions in a Coagulation Dispersion', *Radioactive Fallout from Nuclear Weapons Tests*, pp. 253–259, U.S. Atomic Energy Commission (1965).

29a. Frisken, W. R.: 'Extended Industrial Revolution and Climate Change', *EOS* **52**, 5000–508 (1971).

30. Fukuta, N.: 'Experimental Studies of Organic Ice Nuclei', *J. Atmos. Sci.* **23**, 191–196 (1966).

31. Galindo, I. G. and Muhlia, A.: 'Contribution to the Turbidity Problem in Mexico City', *Arch. Meteorol. Geophys. Bioklimatol.* **B18**, 169–186 (1970).

32. Gates, D. M.: 'The Flow of Energy in the Biosphere', *Sci. Amer.* **224**, 88–100 (1971).

33. Goldberg, E. D.: 'Atmospheric Dust, the Sedimentary Cycle and Man', *Earth Sci. Geophys.* **1**, 117–132 (1971).

34. Griffin, J. J., Windom, H., and Goldberg, E. D.: 'The Distribution of Clay Minerals in the World Ocean', *Deep-Sea Res.* **15**, 433–459 (1968).

35. Gunn, R.: 'The Secular Increase of the World-Wide Fine Particle Pollution', *J. Atmos. Sci.* **21** 168–181 (1964).

36. Hidy, C. M.: 'On the Theory of the Coagulation of Noninteracting Particles in Brownian Motion', *J. Colloid. Sci.* **20**, 123–144 (1965).

37. Hidy, G. M. and Brock, J. R.: 'An Assessment of the Global Sources of Tropospheric Aerosols', in H. M. Englund and W. T. Beery (eds.), *Proceedings Second International Clean Air Congress*, p. 1088, Academic Press, New York and London (1971).

38. Hidy, G. M. and Friedlander, S. K.: 'The Nature of the Los Angeles Aerosol, in H. M. Englund and W. T. Beery (eds.), *Proceeding Second International Clean Air Congress*, p. 391, Academic Press, New York and London (1971).

39. Hobbs, P. V. and Locatelli, J. D.: 'Ice Nucleus Measurements at Three Sites in Western Washington', *J. Atmos. Sci.* **27**, 90–100 (1970).

40. Hobbs, P. V., Radke, L. F., and Shumway, S. E.: 'Cloud Condensation Nuclei from Industrial Sources and Their Apparent Influence on Precipitation in Washington State', *J. Atmos. Sci.* **27**, 81–89 (1970).

41. Hodge, P. W.: 'A Large Decrease in the Clean Air Transmission of the Atmosphere 1.7 km above Los Angeles', *Nature* **229**, 549 (1971).

42. Hodge, P. W., Lanlainen, N., and Charlson, R. J.: 'Astronomy and Air Pollution Summary of Informal Symposium', *Project ASTRA Publication No. 16*, University of Washington, Seattle (1972).

43. Hogan, A. W.: 'Experiments with Aitken Counters in Maritime Atmosphere', *J. Rech. Atmos.* **3**, 53–57 (1968).

44. Hogan, A. W., Bishop, J. M., Aymer, A. L., Harlow, B. W., Klepper, J. C., and Lupo, G.: 'Aitken Nuclei Observations over the North Atlantic Ocean', *J. Appl. Meteorol.* **6**, 726–727 (1967).

45. Holzworth, G. C.: 'Some Effects of Air Pollution on Visibility in and near Cities', Symposium: Air Over Cities, p. 69, *SEC Tech. Rept. A62-5*, Public Health Service, Washington, D.C. (1962).

46. Hubbert, M. K.: 'The Energy Resources of the Earth', *Sci. Amer.* **224**, 60–70 (1971).

47. Jaenicke, R. and Junge, C. E.: 'Studien zur oberen Grenzgrösse des natürlichem Aerosols', *Beitr. Phys. Atmo* **40**, 129–143 (1967).

48. Jerzykiewicz, M.: 'A Study of Atmospheric Extinction Observations Made at Flagstaff, Arizona, 1955–1972', in N. Laulainen and P. W. Hodge (eds.), *Project ASTRA Publication No. 15*, University of Washington, Seattle (1972).

49. Joseph, J. H. and Manes, A.: 'Secular and Seasonal Variations of Atmospheric Turbidity at Jerusalem', *J. Appl. Meteorol.* **10**, 453–462 (1971).

50. Judson, S.: 'Erosion of the Land – Or What's Happening to Our Continents?' *Amer. Sci.* **56**, 356 (1968).

51. Junge, C. E.: 'Sulfur in the Atmosphere', *J. Geophys. Res.* **65**, 227–237 (1960).

52. Junge, C. E.: *Air Chemistry and Radioactivity*, Chapter 2, Academic Press, New York (1963).

53. Junge, C. E.: 'Comments on "Concentration and Size Distribution Measurements of Atmospheric Aerosols and a Test of the Theory of Self-Preserving Size Distributions" ', *J. Atmos. Sci.* **26**, 603–608 (1969).

54. Junge, C. E.: *The Physical and Chemical Properties of Atmospheric Aerosols and Their Relation to Condensation Processes*, Suppl. Proc. 7th Int. Conf. on Condensation and Ice Nuclei, September 18–24, 1969, Prague and Vienna, pp. 31–49, Academic, Prague (1971a).

55. Junge, C. E.: 'The Nature and Residence Times of Tropospheric Aerosols', in W. H. Matthews, W. W. Kellogg and G. D. Robinson (eds.), *Man's Impact on the Climate*, pp. 302–309, MIT Press, Cambridge, Mass. (1971b).

56. Junge, C. E.: 'The Cycle of Atmospheric Gases – Natural and Man Made', *Quart. J. Roy. Meteorol. Soc.* **98**, 711–729 (1972).

57. Keith, R. W.: *Downtown Los Angeles Noon visibility Trends*, Air Quality Report No. 65, Los Angeles Air Pollution Central District, Los Angeles (1970).

58. Knollenberg, R. G.: 'Urea as an Ice Nucleant for Supercooled Clouds', *J. Atmos. Sci.* **23**, 197–201 (1966).

59. Kocmond, W. C. and Mack, E. J.: 'The Vertical Distribution of Cloud and Aitken Nuclei Downwind of Urban Pollution Sources', *J. Appl. Meteor.* **11**, 141–148 (1972).

60. Landsberg, H. E.: 'Observations of Condensation Nuclei in the Atmosphere', *Mon. Weather Rev.* **62**, 442–445 (1934).

60a. Landsberg, H. E.: 'Atmospheric Condensation Nuclei', *Erg. Kosm. Phys.* **3**, 155–252 (1938).

60b. Landsberg, H. E.: *Man-Made Climatic Changes*, this volume, pp. 197–234.

61. Ludwig, J. H., Morgan, G. B., and McMullen, T. B.: 'Trends in Urban Air Quality', *EOS* **51** 468–475 (1970).

62. MacCready, P. B., Smith, T. B., Lockhart, T., and Diamond, R. J.: *Investigations of Silver Iodide Decay and Transport and Natural Nuclei*, Report for Advisory Committee on Weather Control, Meteorological Research Inc., Altadena (1955).

63. McCorkell, R., Fireman, E. L., and Langway, C. C., Jr.: 'Dissolved Iron, Nickel and Cobalt Greenland Ice (Abstract)', *Trans. Amer. Geophys. Union* **48**, 158 (1967).

64. McCormick, R. A. and Ludwig, J. H.: 'Climate Modification by Atmospheric Aerosols', *Science* **156**, 1358–1359 (1967).

65. McNulty, R. P.: 'The Effects of Air Pollutants on Visibility in Fog and Haze at New York City', *Atmos. Environ.* **2**, 625–629 (1968).
66. Machta, L.: 'Mauna Loa and Global Trends in Air Quality', *Bull. Amer. Meteorol. Soc.* **53**, 402–420 (1972).
67. Malone, T. F.: 'The Environmental Context', *EOS* **52**, 508–512 (1971).
68. Manes, A.: *Atmospheric Turbidity over Jerusalem*, Meteorol. Notes Series A, No. 26, Israel Meteorol. Svc., Bet-Dagan (1972).
69. Mason, B. J. and Maybank, J.: 'Ice-Nucleating Properties of some Natural and Mineral Dusts', *Quart. J. Roy. Meteorol. Soc.* **84**, 235–241 (1958).
70. Maybank, J. and Barthakur, N.: 'The Ice Nucleation Behavior of Aminoacid Particles', *Can. J. Phys.* **44**, 2431–2445 (1966).
71. Midwest Research Institute (MRI): *Particulate Pollutant System Study, Vol. I – Mass Emissions*, Report to Air Pollution Control Office, EPA under contract No. CPA 22-69-104, Project No. 3326-C, MRI, Durham, North Carolina (1971).
72. Mitchell, J. M., Jr.: 'A Preliminary Evaluation of Atmospheric Pollution as a Cause of the Global Temperature Fluctuation of the Past Century', in S. F. Singer (ed.), *Global Effects of Environment Pollution*, p. 139, D. Reidel Publ. Co. Dordrecht-Holland (1970).
73. Mitchell, J. M., Jr.: *Air Pollution and Climate Change*, presented at 64th annual meeting, American Institute of Chemical Engineers, San Francisco, Dec. 1, 1971.
74. Mohnen, V. A. and Lodge, J. P., Jr.: *General Review and Survey of Gas-to-Particle Conversions*, pp. 69–91, Proc. 7th Int. Conf. on Condensation and Ice Nuclei, September 18–24, 1969, Prague and Vienna, Academia, Prague (1969).
75. Mossop, S. C.: 'Concentrations of Ice Crystals in Clouds', *Bull. Amer. Meteorol. Soc.* **51**, 474–479 (1970).
76. Moyers, J. L., Zoller, W. H., and Duce, R. A.: 'Gaseous Iodine Measurements and their Relationship to Particulate Lead in a Polluted Atmosphere', *J. Atmos. Sci.* **28**, 95–98 (1971).
77. National Air Pollution Control Administration (NAPCA): *Nationwide Inventory of Air Pollutant Emissions, 1968*, Publication No. AP-73, NAPCA, Raleigh, North Carolina (1970).
78. Neiburger, M.: *Visibility Trends in Los Angeles*, Report No. 11, Air Pollution Foundation, Los Angeles (1955).
79. Noll, K. E.: 'A Procedure for Measuring the Size Distribution of Atmospheric Aerosols', *Trend Engin.* **19**, 21–27 (1967).
80. Opdyke, N. D. and Knowles, R.: 'Paleomagnetism of Cores from the South Atlantic', *EOS* **53**, 170–171 (1972).
81. Parkin, D. W., Phillips, D. R., Sullivan, R. A. L., and Johnson, L.: 'Airborne Dust Collections over the North Atlantic', *J. Geophys. Res.* **75**, 1782–1793 (1970).
82. Peterson, J. T. and Bryson, R. A.: 'Atmospheric Aerosols: Increased Concentrations during the Last Decade', *Science* **162**, 120–121 (1968).
83. Peterson, J. T. and Junge, C. E.: 'Sources of Particulate Matter in the Atmosphere', in W. H. Matthews, W. W. Kellogg and G. D. Robinson (eds.), *Man's Impact on the Climate*, pp. 310–320. MIT Press, Cambridge, Massachusetts and London, England (1971).
84. Prospero, J. M.: 'Atmospheric Dust Studies on Barbadoes', *Bull. Amer. Meteorol. Soc.* **49**, 645–652 (1968).
85. Prospero, J. M. and Carlson, T. N.: 'Saharan Dust in the Atmosphere of the Northern Equatorial Atlantic Ocean: a Major Constituent of the Marine Aerosol (Abstract)', *Bull. Am. Meteorol. Soc.* **52**, 1138 (1971).
86. Rasmussen, R. A. and Went, F. W.: 'Volatile Organic Material of Plant Origin in the Atmosphere', *Proc. Nat. Acad. Sci.* **53**, 215–220 (1965).
87. Rasool, S. I. and Schneider, S. H.: 'Atmospheric Carbon Dioxide and Aerosols: Effects of Large Increases on Global Climate', *Science* **173**, 138 (1971).
88. Reardon, W. A., Merrill, J. A., Jacobson, L. D., and Bathke, W. L.: *A Review and Comparison of Selected United States Energy Forecasts*, prepared by Pacific Northwest Laboratories for OST, Government Printing Office, Washington, December (1969).
89. Roberts, P. and Hallett, J.: 'A Laboratory Study of the Ice Nucleating Properties of Some Mineral Particulates', *Quart. J. Roy. Meteorol. Soc.* **94**, 25–34 (1968).
90. Robinson, E. and Robbins, R. C.: *Sources, Abundance, and Fate of Gaseous Atmospheric*

Pollutants, Report to American Petroleum Institute, Project PR-6755, Stanford Research Institute, Menlo Park, California (1968).

91. Robinson, E. and Robbins, R. C.: *Sources, Abundance, and Fate of Gaseous Atmospheric Pollutants – Supplement*, Report to American Petroleum Institute, Project PR-6755, Stanford Research Institute, Menlo Park, California (1969).

92. Robinson, E. and Robbins, R. C.: *Emissions, Concentration, and Fate of Particulate Atmospheric Pollutants*, Report to American Petroleum Institute, Project SCC-8507, Stanford Research Institute, Menlo Park, California (1971).

93. Roosen, R. G. and Angione, R. J.: 'Worldwide Variations in Atmospheric Transparency during the Twentieth Century', in N. Laulainen and P. W. Hodge (eds.), *Project ASTRA Publication No. 15*, University of Washington, Seattle (1972).

94. Roosen, R. G., Angione, R. J., and Klemcke, C. H.: 'Worldwide Variations in Atmospheric Transmission 1 – Baseline Results From Smithsonian Observations', *Bull. Amer. Meteorol. Soc.* **50**, 307–316 (1973).

95. Rosen, J. M.: 'Stratospheric Dust and its Relationship to the Meteoric Influx', *Space Sci. Rev.* **9**, 58–89 (1969).

96. SCEP: 'Report on the Study of Critical Environmental Problems', in *Man's Impact on the Global Environment*, MIT Press, Cambridge, Mass. (1970).

97. Schaefer, V. J.: 'Ice Nuclei from Automobile Exhaust and Iodine Vapor', *Science* **154**, 1555 (1966).

98. Schaefer, V. J.: 'The Inadvertent Modification of the Atmosphere by Air Pollution', *Bull. Amer. Meteorol. Soc.* **50**, 199–206 (1969).

99. Schaefer, V. J.: 'The Inadvertent Modification of the Atmosphere by Air Pollution', this volume pp. 177–196.

100. Schaefer, V. J.: 'Condensation Nuclei: Production of very Large Numbers in Country Air', *Science* **170**, 851 (1970b).

101. Schaefer, V. J.: 'The Measurement of Aerosol Particles on a Global Scale', in W. H. Matthews, W. W. Kellogg and G. D. Robinson (eds.), *Man's Impact on the Climate*, pp. 364–368, MIT Press, Cambridge, Mass. (1971).

102. Schmidt, R. A.: *A Survey of Data on Microscopic Extraterrestrial Particles*, Research Report Series 63-2, Dept. of Geology, University of Wis. Madison (1963).

103. Selezneva, E. S.: 'Estimation of the Background Contamination of the Atmosphere from the Chemical Composition of Precipitation, *Tellus* **24**, 122–127 (1972).

104. Selezneva, E. S.: *Atmosfernye Aerozoli, (Atmospheric Aerosols)*, Gidrometeorologischeskoi Izdatel'stvo, Leningrad, 1966, translated and published in English by Indian National Scientific Documentation Center, New Dehli (1970).

105. (SMIC): Report of the Study of Man's Impact on Climate, in *Climate Modification*, MIT Press, Cambridge, Mass. and London, England (1971).

106. Smith, T. B.: 'Pollutants and Cloud Condensation Processes', in W. H. Matthews, W. W. Kellog and G. D. Robinson (eds.), *Man's Impact on the Climate*, pp. 400–409, MIT Press, Cambridge, Mass. and London, England (1971).

107. Smith, E. J., Heffernan, K. J., and Seely, B. K.: 'The Decay of Ice-Nucleating Properties of Silver Iodide in the Atmosphere', *J. Meteorol.* **12**, 379–385 (1955).

108. Smith, T. B., Weinstein, A. I., and Alkezweeny, A. J.: *Nucleation Aspects of Inadvertent Weather Modification*, Report to Travelers Research Corp., Meteorology Research, Inc., Altadena (1970).

109. Soulage, G.: 'Contributions des fumées industrielles á l'enrichissement de l'atmosphère en noyaux glacogènes', *Bull. Observ. Puy Dome* **4**, 121–124 (1958).

110. Spirtas, R. and Levin H. J.: 'Patterns and Trends in Levels of Suspended Particulate Matter', *J. Air Pollut. Cont. Asc.* **21**, 329–333 (1971).

111. Squires, P.: 'An Estimate of the Anthropogenic Production of Cloud Nuclei', *J. Rech. Atmos.* **2**, 297–308 (1966).

112. Stanford Research Institute: *The Smog Problem in Los Angeles County*, Western Oil and Gas Association, Los Angeles (1954).

113. Stern, A. C. (ed.): *Air Pollution*, Vol. I, Academic Press, New York (1968).

114. Telford, J. W.: 'Freezing Nuclei From Industrial Processes', *J. Meteorol.* **17**, 676–679 (1960).

* See also this volume pp. 111–124.

115. Ter Haar, G. L. and Bayard, M. A.: 'Composition of Airborne Lead Particles', *Nature* **232**, 553–554 (1971).
116. Twomey, S., Howell, H. B., and Wojciechowski, T. A.: 'Comments on "Anomalous Cloud Lines" ', *J. Atmos. Sci.* **25**, 333–334 (1968).
117. Twomey, S. and Wojciechowski, T. A.: 'Observations of the Geographical Variation of Cloud Nuclei', *J. Atmos. Sci.* **26**, 684–688 (1969).
118. Vohra, K. G., Vasudevan, K. N., and Nair, P. V. N.: 'Mechanisms of Nucleus-Forming Reactions in the Atmosphere', *J. Geophys. Res.* **75**, 2951–2960 (1970).
119. Volz, F. E.: 'Some Results of Turbidity Networks', *Tellus* **21**, 625–630 (1969).
120. Volz, F. E.: 'Atmospheric Turbidity After the Agung Eruption of 1963 and Size Distribution of the Volcanic Aerosol', *J. Geophys. Res.* **75**, 5185–5193 (1970a).
121. Volz, F. E.: 'On Dust in the Tropical and Midlatitude Stratosphere from Recent Twilight Measurements', *J. Geophys. Res.* **75**, 1641–1646 (1970b).
122. Wadleight, C. H.: *Wastes in Relation to Agriculture and Forestry*, Miscellaneous Publication No. 1065, U.S. Department of Agriculture, Washington, D.C. (1968).
123. Walt, G. R.: 'Some Experiments Relating to the Electrical Conductivity of the Lower Atmosphere', *J. Wash. Acad.* **36**, 321 (1946).
124. Warner, J.: 'A Reduction in Rainfall Associated with Smoke from Sugarcane Fires – an Inadvertent Weather Modification?' *J. Appl. Meteorol.* **7**, 247–251 (1968).
125. Warner, J.: *Smoke from Sugar-Cane Fires and Rainfall*, Proc. International Conference on Weather Modification, Canberra, Australia, September 6–11, 1971, pp. 191–192, Amer. Meteorol. Soc., Boston (1971).
126. Went, F. W.: 'Organic Matter in the Atmosphere and its Possible Relation to Petroleum Formation', *Proc. Nat. Acad. Sci.* **46**, 212–221 (1960).
127. Went, F. W.: 'The Nature of Aitken Condensation Nuclei in the Atmosphere', *Proc. Nat. Acad. Sci.* **51**, 1259–1267 (1964).
128. Went, F. W.: 'On the Nature of Aitken Condensation Nuclei', *Tellus* **18**, 549 (1966).
129. Wexler, H.: 'Turbidities of American Air Masses and Conclusions Regarding the Seasonal Variation in Atmospheric Dust Content', *Mon. Weather Rev.* **62**, 397–402, 1934.
130. Whitby, K. J.: 'Aerosol Measurements: New Data on Urban Aerosols', in I. H. Blifford, Jr. (ed.), *Particulate Models: Validity and Application*, pp. 3–12, TN/PROC-68, NCAR, Boulder (1970).
131. Windom, H. L.: 'Atmospheric Dust Records in Permanent Snowfields: Implications to Marine Sedimentation', *Geol. Soc. Amer. Bull.* **80**, 761–782, 1969.
132. Woodcok, A. H. and Jones, R. H.: 'Rainfall Trends in Hawaii', *J. Appl. Meteorol.* **9**, 690–696 (1970).
133. Woodcok, A. H.: 'Solubles', in M. N. Hill (ed.), *The Sea*, Vol. I., Interscience (Wiley), New York, (1962).

PART IV

WORLDWIDE OCEAN POLLUTION
BY TOXIC WASTES

INTRODUCTION

The oceans have been termed the 'ultimate sink' for the natural wastes of the world. Most of these are carried there by rivers. But, with the increase of technology, new types of wastes are being generated, some of which, like lead, add a considerable load to the naturally-occurring content; others, like DDT, are a completely foreign substance to the ocean environment. Incidentally, both lead and DDT are partially transported to the oceans by the atmosphere.

Goldberg traces the routes of introduction, and catalogs the amounts of exotic materials which are introduced through pollution. The United States' contribution to the world total is remarkably high, and both rates are increasing. New kinds of chemicals are added almost daily, and radioactive discharges will undoubtedly rise. The fate of the pollutants is no longer under control. "The solution to pollution is dilution!" becomes the slogan. The actual fate is quite complicated and may involve slow chemical reactions or uptake by organisms.

Biologically active materials released into the environment, whether on land, in lakes, or oceans, lead to general ecological effects conforming to well-known patterns. These are described by Woodwell, Craig and Johnson, with particular reference to DDT.

Marine pollution also has quite direct implications on the harvest of food from the sea. As related by Ketchum, this problem has become especially pronounced in estuaries which are more subject to pollution than the open ocean. It will become even more important when we try to enhance food yields through aquaculture in the coastal waters.

Finally, the ocean is a pervasive influence on the environment of the globe and plays a central role in the interaction of all ecosystems. Lundholm describes some of these interactions – the transfer of pollutants between atmosphere and hydrosphere.

MAN'S ROLE IN THE MAJOR SEDIMENTARY CYCLE*

EDWARD D. GOLDBERG

Scripps Institution of Oceanography, La Jolla, Calif., U.S.A.

Abstract. Natural inputs to the sedimentary cycle result from river runoff into the oceans, atmospheric transport, and glacial movement. Although direct injection into the cycle by man is at least an order of magnitude less than that of nature, he has enhanced some of the natural processes. Landscape alterations, for instance, have increased by a factor of 2 or 3 the rate of natural erosion, which represents the single most important input to the cycle.

Perhaps more important than the total cycle are studies of injections of specific pollutants. Man is the major or sole agent for the input of many constituents, such as organic and certain exotic chemicals, and a rival to nature in others, such as heavy metals.

1. Introduction

The activities of human society are measurably changing the chemical makeup of the Earth's waters and airs. The rates of such alterations will increase with the expanding world population and with the growing per capita consumption of materials and energy. During the acquisition and in the utilization of matter and fuels, there are losses and disposals in places other than those from where the materials were originally obtained. When a domain undergoes compositional changes that result in losses or restricted uses of its resources, a state of pollution has come into being. Such contaminated zones will persist until the supply of the pollutant is exhausted or curtailed or until its leakage to the environment is regulated.

Disasters can and have occurred before the pollutant supply becomes limited; the concerns of this presentation will be with the problems of leakages. The ability to predict future levels of pollutants in the atmosphere and in natural waters, coupled with an ability to define tolerance levels for the environment, can form the basis of effective environmental management. Clearly no part of our surroundings will be free of man's wastes. The maximization of waste accommodations with minimum disturbance to natural resources and to living systems should be one of the guiding concepts for environmental managers.

A rewarding entry to understanding the fates of man-introduced materials to the environment is a consideration of the behavior of substances during weathering cycles where there are movements of materials between and within the lithosphere, hydrosphere and atmosphere, transfers often mediated by organisms other than man. For synthetic substances that have little resemblances to naturally occurring species predictions of their flows through geospheres often can be made on the basis of their chemical and physical properties.

* Reprinted with changes from 'The Changing Chemistry of the Ocean', *Nobel Symp.* **20** (ed. by D. Dyrssen and D. Jagner), Almqvist and Wiksell, Upsala, 1972.

A comparison of the fluxes of materials in the sedimentary cycle with those of substances introduced by man to his surrounding will yield a quantitative appreciation of society's role as a geological agent. Mobilization rates of materials as a result of agricultural, industrial or social usages now compete with those that occur during weathering. These impingements have taken place primarily over the last several hundred years.

2. The Major Sedimentary Cycle

Rock and soil particles are transported from the continents to the marine environment by three agencies: rivers, winds and glaciers. Man has added two others: sewer outfalls and ships.

The river movements of rock and soil materials exceed by at least an order of magnitude their transport through the atmosphere (Table I). While river discharges predominate in the Atlantic Ocean and its adjacent seas, the solids carried by the principle wind systems, the trades, mid-latitude westerlies (the jets) and the polar easterlies, are transported within latitudinal belts.

Man's injections into the environment also display a latitudinal zonation. The principle utilization of energy and materials takes place in the mid-latitudes of the northern hemisphere. The flow of this energy and matter through a society, and consequently the leakage to the surroundings, can be measured by the Gross National Prod-

TABLE I

Some fluxes in the major sedimentary cycle (adapted in part from Goldberg [12]

Material	Geosphere receiving material	Flux in 10^{14} g yr^{-1}
Suspended river solids	Oceans	180
Dissolved river solids	Oceans	39
Continental rock and soil particles	Atmosphere	1–5
Sea salt	Atmosphere	3
Volcanic debris	Stratosphere	0.036
Volcanic debris	Atmosphere	<1.5
Wastes of society (excluding fossil fuels)	Hydrosphere, lithosphere, atmosphere	30
Wastes dumped from ships	Oceans	1.4
Carbon and fly ash from fossil fuel combustion	Atmosphere	0.25
Industrial particulates (Total)	Atmosphere	0.54
<2 μ		0.12

uct (GNP) of a country [13]. The nine countries with the highest GNPs, U.S.A., U.S.S.R., Japan, Federal Republic of Germany, France, Great Britain, Canada, China and Italy, have a total GNP of 1900 billion dollars with individual values all greater than 70 billion dollars. These nations are at mid-latitudes in the northern hemisphere. The next twelve countries in rank, with GNPs above 15 billion dollars each, have a combined value of the GNP of 200 billion dollars and include India, Poland, Australia, Sweden, Netherlands, Spain, Democratic Republic of Germany, Brazil, Mexico, Belgium, Switzerland and Czechoslovakia. Of these, India and Brazil are in the tropics, while only Australia is in the southern hemisphere. Thus, as a first approximation, man's atmospheric injections take place under a single transporting agency, the prevailing westerly flow in the northern hemisphere.

The other important parameter governing the dispersion of solids about the atmosphere is the height of injection. Sea salt particles enter at the water-air interface while dusts from volcanoes are estimated to achieve heights of 50 km or more.

The salts of the sea are injected into the atmosphere with the spray and Eriksson [11] estimates the amount to exceed 1000 megatons yr^{-1}. Most of these particles rapidly return to the sea surface. Eriksson computes that about 100 megatons fall upon land and are returned to the oceans in runoff. Assuming both that this latter value represents those salts transported over long distances and that a similar fallout on an areal basis occurs in oceanic domains, then the total amount of sea salt involved in the major sedimentary cycle is about 300 megatons yr^{-1}.

There have been many estimates of the fluxes through the rivers based upon their flow and their measured dissolved and suspended burdens. On the other hand, only recently has data become available to ascertain the rates of mobilization of rock and soil debris through the atmosphere. Present approaches involve the rates of accumulation of solids in permanent snow fields, rates of accumulation of eolian materials in deep sea sediments, and the standing crops of dusts in the atmosphere.

Windom [33] determined the accumulation rates of detrital solids in the Greenland and Antarctic ice sheets and in five temperate glaciers, all of which had their upper levels dated by Pb-210 geochronologies. On the bases of size* and mineral distributions, locally derived solids could be distinguished from those that had been transported from distant places. An average value for the accumulation rate of the global component appeared to be around 0.1 mm 1000 yr^{-1}. Taking this rate as representative of the Earth's surface and giving such materials a density of 2 g cm^{-3}, the annual rate of mobilization of continental soil and rock particles to the atmosphere is computed to be 10^{14} g.

* The distinction between globally dispersed particles and those introduced locally is most difficult to make. Many investigators base their definitions upon an upper limit in diameter with the concept that particles below some value, say 10 μ or 1–2 μ, have a high probability of being involved in long-range transport. Windom [33] introduced the refinement of considering both the size distributions of minerals and the composition of exposed surface solids near the sampling area. Here there will be no attempt to obtain a uniformity in the published values of globally transported dusts in the atmosphere. Any such effort would be difficult, if not impossible, to make. Also, the uncertainties in the reported values usually exceed any differences that might result from altering the size criterion.

Another estimate involves the average atmosphere dust burden, estimated by Goldberg [12] to be of the order of 5 μg m^{-3}. If this is representative of dust concentrations to heights of 5000 m into the atmosphere, and if from this zone there are 40 removals by rain each year, then a flux of 5×10^{14} g yr^{-1} is found.

The rates of sedimentation in the open ocean provide means to evaluate the fluxes both of continental and of volcanic debris into and out of the atmosphere. The contributions of wind-borne terrestrial solids to the non-biogenous phases of deep-sea deposits are estimated to be about 50% [33, 14]. Their accumulation rates vary between 0.01 and 0.1 cm 1000 yr^{-1}, or between 0.01 and 0.1 g cm^{-2}/10^3 yr, since their average density is near 2 g cm^{-3} and the water content of the sediments averages around 50%. If 0.1 g cm^{-2}/10^3 yr is a representative value for both land and sea accumulation of wind-borne solids, then a world flux of 5×10^{14} g yr^{-1} is found for the mobilization of such continental debris to the atmosphere.

Recent volcanic activity appears to have latitudinal zonations [8]. Over the past several decades, volcanic explosions which have ejected ash and gases into the stratosphere have taken place in two latitudinal belts. One is located in a narrow zone just below the Arctic Circle (56°–65° N) and includes the volcanoes of Kamchatka, the Aleutian Islands, Alaska and Iceland. The other, a broader region, spans the equator from 8° S to 15° N and includes the volcanoes of Indonesia, the Phillipines, Central America, Ecuador, the Galapagos and the Caribbean.

There is a periodicity to these events. During the last two decades there have been two periods of intense volcanic activity, 1950–1956 and 1963–1970, with a quiet period intervening. In the former period, seven major events occurred, including perhaps the most violent eruption of the century, the Bezymianny explosion in central Kamchatka. In 1963, the major eruption of Agung on the Island of Bali placed great quantities of material into the stratosphere. Such periods of activity appear to influence the atmosphere turbidity.

Volcanic ashes rapidly degrade in the marine environment to montmorillonite as well as to some other less abundant phases. The montmorillonite contents in the sediments allow estimates to be made of the amounts of volcanic material annually injected into the atmosphere. The zones of highest montmorillonite concentration, the eastern equatorial and the south Pacific, have their deposits composed almost entirely of volcanic debris and their degradation products. Their measured rates of sedimentation, several tenths to a half of a mm 1000 yr^{-1}, represent an upper limit to the accumulation of atmospherically introduced volcanic debris. In other areas, wind, river and glacially transported continental materials dilute the volcanic solids. If the fallout of volcanic materials in pelagic sediments (density of solids taken as 2 g cm^{-3} in a sediment composed of 50% water) is characteristic of that over the surface of the Earth, then up to 1.5×10^{14} g is calculated to be the amount of volcanic dust annually introduced to the atmosphere.

Mitchell* found a stratospheric flux of 3.6×10^{12} g y^{-1} for volcanic debris with the

---

* These numbers have recently been revised; see pp. 149–173.

following model: (a) the dust that reaches the stratosphere is 1% of the material erupted and it has a residence time there of 14 months; (b) the remaining 99% of the mass, after introduction to the atmosphere as larger particles, falls back to the Earth's surface in much shorter time periods; and (c) the contribution of each eruption is taken as one half of its stratospheric injection mass in the first year. He derived a loading of volcanic debris in the stratosphere of 4.2×10^{12} g.

The major fluxes in the sedimentary cycle range between tenths and tens of billions of tons per year (Table I). Although the individually calculated numbers rest upon tenuous assumptions or data, nevertheless, their containment within these two orders of magnitude provides a basis to understand man's influence upon the cycle.

Detailed compilations of material and energy utilization data have been published in the United States. Such considerations can be extended globally in the following approximate way. The U.S. is responsible for about 35% of the world's energy consumption. A similar figure applies to resource utilization. As a first approximation, then, the world's utilization of energy and matter can be taken as thrice that of the U.S. For example, the U.S. society utilizes 5 tons of mineral, food, and forest products per person per year. This can be extrapolated to a world usage of 3×10^{15} g y^{-1}. Since the world society disposes of, rather than accumulates, the materials necessary for its ways of life, this utilization is somewhat equivalent to the annual global waste problem – a cube about a kilometer on a side. But more important, its magnitude is comparable to fluxes in the major sedimentary cycle.

In addition, landscape alteration may have altered the intensity of the sedimentary cycle. Judson [17] offers several lines of reasoning to indicate that the rate of continental erosion (and consequentially the delivery of suspended and dissolved solids to the oceans) has increased by a factor of 2 or 3 due to man-induced changes in land use. The transformation of forests to croplands and grazing areas seems to increase the erosion rate tenfold. With one fourth of the U.S. area now so involved, the erosion rate may be larger by a factor of slightly greater than 3.

Judson approaches the global problem in another way by noting that for areas unaffected by the activities of man the erosion rate appears to be about 3.6 cm 1000 yr^{-1}. Continental areas of 151 million km^2 include 100 million km^2 whose weathered products are carried to the sea by rivers (areas like the interior of Australia and the western U.S. have no rivers to carry erosion products to the oceans). With a density of eroded materials of 2.6 g cm^{-3}, this weathered material brought to the marine environment is 93×10^{14} g. The total flux of materials to the oceans is about 220×10^{14} g (Table I; river bed load is not included here but it is probably about 5% or less of the total carried by rivers). Thus, man appears to have more than doubled the rate of erosion.

About 48 million tons of wastes were dumped from ships at sea in the U.S. during 1968 (Table II). Dredge spoils account for the bulk of this disposal and include besides sediments, industrial, domestic and agricultural wastes.

Only a small portion of the domestic solid wastes, 26000 out of 190000 tons, is subject to oceanic disposal. Trebling these figures, a global value of 1.4×10^{14} g

TABLE II

Ocean dumping in the U.S. for 1968 [5]

Waste type	Amount in tons	Percent
Dredge spoils	38 428 000	80
Industrial wastes	4 690 500	10
Sewage sludge	4 477 000	9
Construction and demolition debris	574 000	~1
Solid waste	26 000	<1
Explosives	15 200	<1
Total	48 210 700	100

of oceanic dumping per year is estimated. This may be compared with a suspended river input of 180×10^{14} g yr^{-1}. Whereas the river discharge areas are fixed, the ship disposal sites can vary. The increasing volumes of oceanic dumping [5] may cause extensive damage to the inshore marine environment.

Man's contribution to atmospheric dust loads has been investigated by Parkin et al. [25] in airs above the temperate north Atlantic. The eolian solids were grey to dark grey in color resulting from contents of fly ash and carbon from the combustion of fossil fuels. These materials constituted 60% of the dusts near land and 5% over the open ocean. Goldberg [12] used the data to attain a value of 0.25 μg of fly ash m^{-3} of air over the north Atlantic. A sedimentation rate to the Earth's surface may be based on a removal of the particles every ten days with rain. For a 5 km column, 1 m^2 in cross section, there would be a standing crop of 1250 μg of man-introduced solids. This yields a sedimentation rate of 0.05 mm/10^3 yr, where the solid

TABLE III

Sources of U.S. particulate emissions from stationary operations. In percent [29]

Fuel combustion	33
Crushed stone, sand and gravel	25.6
Agriculture	9.8
Iron and steel	7.5
Cement	5.2
Forest products	3.7
Lime	3.2
Clay	2.6
Primary non-ferrous metals	2.6
Fertilizer and phosphate rock	1.9
Asphalt	1.2
Ferroalloys	0.9
Iron foundries	0.8
Secondary non-ferrous metals	0.7
Carbon black	0.5
Coal cleaning	0.5
Petroleum	0.25
Mineral acids	0.1

phases have a density of 2 g cm^{-3} and the sediments contain fifty percent water. This is much lower than the rates of accumulation of continental and volcanic solids which amass at levels of millimeters or fraction of millimeters per thousand years.

About 18×10^{12} g of particulates/year are broadcast about the atmosphere from stationary sources* in the U.S. [29]. The fine particle component (sizes less than 2μ) is estimated to be about 22% of this figure, 4×10^6 tons yr^{-1}. The major sources are fuel combustion, the crushed stone industry, agriculture and related operations, the iron and steel industry and the cement industry (Table III). If the U.S. represents one third of the world's inputs of such solids, then on a global basis the total and fine particle emissions are 54×10^{12} and 12×10^{12} g yr^{-1} respectively. The smaller particles are preferentially emitted from operations involving combustion, condensation and vaporization as compared to those from mechanical processes. The most abundant particles are oxides of silicon, iron, aluminum, calcium and magnesium, and calcium carbonate.

3. Fossil Fuels

The direct introduction of crude oil and petroleum products to the marine environment through human activity far exceeds natural seepages (Table IV). The annual influxes due to man appear to be of the order of 2.1×10^{12} g. On the other hand, an

TABLE IV

The involvement of petroleum with the marine environment (Data from SCEP, [28])

World oil production (1969)	1.82×10^{15} g yr^{-1}
Oil transport by tanker (1969)	1.18×10^{15} g yr^{-1}
Direct injections into marine environment through man's activities	
Tanker operations	0.5×10^{12} g yr^{-1}
Other ship operations	0.5×10^{12} g yr^{-1}
Offshore oil production	0.1×10^{12} g yr^{-1}
Accidental spills	0.2×10^{12} g yr^{-1}
Refinery operations	0.3×10^{12} g yr^{-1}
Industrial and automotive wastes	0.45×10^{12} g yr^{-1}
Total	2.1×10^{12} g yr^{-1}
Torrey Canyon Discharge	0.118×10^{12} g
Santa Barbara Blowout	$0.003–0.011 \times 10^{12}$ g
Atmospheric input from continents through vaporization of petroleum products	90×10^{12} g yr^{-1}
Natural seepage into ocean	$< 0.1 \times 10^{12}$ g yr^{-1}

* The particles from moving sources, primarily transportation, would increase these figures almost 5% [28].

upper limit of natural seepages, 10^{11} g yr^{-1}, was obtained by Revelle *et al.* [27] with the argument that, if such an amount did enter the marine environment, the total estimated oil reserves would be depleted within a few million years.

The direct hydrocarbon entries to the world's waters may be dwarfed by those from the atmosphere following the vaporization of petroleum products on the continents, estimated to occur at about 90×10^{12} g yr^{-1}. Although a substantial portion of such gases may be oxidized in the atmosphere, the resistant part can fall to the land and sea surface subsequent to sorption on dust particles or with rain.

The heaviest influxes occur near or between man's habitats. Hydrocarbon fallout from the atmosphere takes place primarily in the mid-latitudes of the northern hemisphere, the band of heavily industrialized populations and the zone of the transporting system, the jet streams. The leakages of oils from ship and nearshore drilling operations have stained the waters of our coastal areas and shipping lanes quite conspicuously. The ofttimes spectacular ship disasters or blowouts can introduce amounts of oil which are substantial fractions of the chronic injection rate. For example, the breakup of the tanker *Torrey Canyon* dispersed oil about the sea at a level equal to 5% of the annual injection rate (Table IV). It is revealing to note that the annual world oil production of 1.82×10^{15} g, if spread over the entire oceanic area of 3.6×10^{18} cm^2, would produce a film of but 0.5 mg cm^{-2}.

The veneer of petroleum accumulating at the sea surface is extending the natural film of fatty acids and alcohols. Such a coating, perhaps a hundred to a thousand ångströms thick, may act as a concentration site for oil soluble substances, such as DDT, freon and the polychlorinated biphenyls, being dispersed by human activities to our surroundings.

The combustion of fossil fuels, coal, oil, lignite and natural gas, can potentially mobilize many elements into the atmosphere at rates in general less than, but comparable to, their rates of flow through natural waters during weathering cycles [4]. The world production of fossil fuels, and presumably their utilization, in 1967 was [28]:

coal	1.75×10^{15} g yr^{-1}
lignite	1.04×10^{15} g yr^{-1}
fuel oils	1.63×10^{15} g yr^{-1}
natural gas	0.66×10^{15} g yr^{-1}

The computed fluxes assume (1) about 10% of the total ash produced escapes to the atmosphere (fly ash) with the remainder (bottom ash) collected in the combustion apparatus; and (2) about half of the coal mined is burned with the other half utilized in the manufacture of coke (Table V).

The amount of an element yearly entering the world's oceans from rivers can be obtained from water discharge and composition data or from rates of sedimentation in marine areas. Both numbers are usually given as in Table V. The river data consider only the transfer of dissolved phases, while the sedimentation numbers take into account both the dissolved and the particulate loads of rivers. Both methods depend upon obtaining global averages of geological parameters, numbers which have some uncertainty in them.

Selective volatilization can introduce the readily distillable materials at levels as much as twenty times greater than those indicated in Table V. Observations of emissions from the d.c. arc spectrograph both for elemental states and for the oxides, sulfates, carbonates, silicates, phosphates and sulfides indicate a preferential transfer of As, Hg, Cd, Sn, Sb, Pb, Zn, Tl, Ag and Bi to the atmosphere during fuel burning.

The mid-latitudes of the northern hemisphere are the principle sites of fossil fuel combustion and the consequential alterations in the compositions of rivers, lakes and coastal sea waters should evidence themselves there.

About 30×10^{12} g of sulfur dioxide are discharged annually into the U.S. atmosphere as a result of fossil fuel burning and a world figure of around 100×10^{12} g yr^{-1}

TABLE V

The mobilization of elements during the combustion of fossil fuels and during weathering processes (10^9 g yr^{-1}) [4].

Element	Fossil fuel mobilization	Weathering mobilization	Element	Fossil fuel mobilization	Weathering mobilization
Be	0.41	5.6	Ga	1	3 —
B	10.5	360			30
Na	280	57 000 —	Ge	0.7	12
		230 000	As	0.7	72
Mg	280	42 000 —	Se	0.45	7.2
		148 000	Sr	70	600 —
Al	1400	14 000 —			1800
		140 000	Y	1.4	25 —
S	3400	140 000			60
Ca	1400	70 000 —	Mo	2.3	28 —
		540 000			36
Sc	0.7	0.14 —	Ag	0.07	0.03 —
		10			11
Ti	70	108 —	Sn	0.28	11
		9000	Ba	70	360 —
V	12	32 —			500 —
		280	La	1.4	7.2 —
Cr	1.5	36 —			40
		200	Ce	1.6	2.2 —
Mn	7	250 —			90
		2000	Er	0.085	1.8 —
Fe	1400	24 000 —			5.0
		100 000	Hg	1.6	1.0 —
Co	0.7	7.2 —			2.5
		8	Pb	3.6	21 —
Ni	3.7	11 —			110
		160	U	0.14	8 —
Cu	2.1	80 —			11
		250			
Zn	7	80 —			
		720			

is a reasonable estimate. Such an input appears to have been identified in atmospheric fallout over the natural SO_2 background by Koide and Goldberg [18]. The sulfate contents in Greenland glacial ices (after correction for contributions from sea salt) doubled within the past decade, a phenomenon attributed to fuel combustion. Thus, society seems to compete now with nature on an equal basis with respect to sulfur emissions. The natural injections come from volcanic activity and from the bacterial decomposition of organic matter in soils and coastal sediments. Sea spray, on the basis of the chloride contents of these glacial ices, does not contribute significant quantities of sulfate compared to these other two sources. The sulfur dioxide has a short residence time in the atmosphere; estimates range from hours to several days.

On the other hand, selenium, the vertical periodic table neighbor of sulfur, is mobilized to a much lesser extent through the atmosphere, a difference most probably related to the chemical behavior of their respective tetravalant oxides [32]. Sulfur dioxide is a moderately stable gas; selenium dioxide, a white solid, reacts readily in air to form red elemental selenium. There has been no evident increase of selenium in the Greenland ice sheets during the last decade, indicating it has not been globally dispersed in the manner of sulfur following fossil fuel burning.

The input of nitrogen oxides from fuel burning is 20.6×10^{12} g into the U.S. atmosphere or perhaps around 60×10^{12} g on a worldwide basis [28]. NO is very rapidly oxidized by ozone to NO_2 which is one of the building blocks of photochemical smog through its reactions with hydrocarbons in the presence of sunlight. NO_2 can be decomposed by the adsorption of visible radiation. The influence on the nitrogen cycle by man-made oxides is as yet undetermined.

The fate of the 13.4×10^{15} g of carbon dioxide produced by world fossil fuel combustion has been considered in the SCEP report [28] upon which the following summary is based. The CO_2 content in the atmosphere is increasing at about 0.5% annually, i.e. the present concentration of about 320 ppm on a volume basis rises 0.7 ppm each year. There is about 1 ppm more CO_2 in the most northerly samples (Point Barrow, Alaska) than in the Antarctic samples. Distinct season oscillations occur, higher at higher latitudes (about 9 ppm at Point Barrow, Alaska) and at higher altitudes.

About 50% of the CO_2 that has been added to the atmosphere in the past decade has remained there; the reservoirs for the other 50% are probably identified properly as the biosphere, the mass of living and non-living organic matter in the oceans and on land, and the oceans. The biosphere could take up additional CO_2 by growing faster but the quantitative aspects of this have not yet been developed. The relative importance of these two sinks remains to be resolved. It is interesting to note that the input of carbon dioxide by fossil fuel burning is about equal to the amount fixed annually by photosynthetic marine organisms, 50×50^{15} g yr^{-1}.

With respect to carbon monoxide our ideas have undergone substantial changes in the past few years. We no longer believe that the injection is due mainly to human activities, but have discovered that natural sources are more important by an order of magnitude.* The total injection into the atmosphere is estimated to be over 4×10^{15} g yr^{-1} and is believed to arise largely from the oxidation of methane. The lifetime of CO in the troposphere is only about 1–2 months. Various removal mechanisms, chemical and biological, must be active to maintain the observed CO concentration at a constant level.**

About 9×10^{13} g of uncombusted hydrocarbons annually enter the atmosphere and can be involved in the photochemical formation of aerosols. The absorption of solar ultraviolet light by gaseous molecules including nitrogen dioxide produce ozone and nitric oxides which react with the hydrocarbons to produce 'photochemical smog'.

4. Halogenated Hydrocarbons

Of the thousands of synthetic organic compounds made each year, only data on the chlorinated hydrocarbon pesticides are comprehensive enough to allow flux calculations. Such substances as DDT and its degradation products (hereafter referred to as DDT residues), dieldrin, endrin, hetachlor, hepoxide and benzene hexachloride, are not produced in measurable amounts by marine organisms, but have been detected in all levels of the food web. The DDT residues have been extensively analyzed and this treatment will emphasize them. The polychlorinated biphenyls (PCBs) have been used much longer (since the 1930s whereas the pesticides came into use after 1946) and have been detected in many members of the marine biosphere. However, not enough measurements on them have been carried out to formulate mass balances.

The total amount of DDT so far produced in the world is about 2×10^{12} g with a present yearly production rate of 10^{11} g [28].

The atmosphere is most probably the major route of transfer of the DDT into the oceans from the continents [24]. DDT residues can be enveloped within wind systems during their aerial or ground application to plants or can volatalize from surfaces which have accommodated them. Once in the atmosphere they primarily return to the Earth's surface in rains.

Analyses of western U.S. rivers give 100 parts per trillion as a maximum concentration of DDT residues. If the total world river discharge of 3.7×10^{19} cc yr^{-1} contained this amount, then an upper limit of 3.7×10^9 g of DDT residues would be annually transported by the rivers to the sea, about 3% of the world production.

The DDT residue movement through the atmosphere can be estimated by considering its content in rains. The most extensive measurements were carried out in Great Britain where the average of 80 parts per trillion was found in samples measured at 7 stations between August 1966 and July 1967. DDT residues in south Florida precipitation averaged 1000 parts per trillion in 18 samples taken at 4 sites between June 1968 and May 1969. If the annual precipitation over the ocean, 3.0×10^{20} cc, contained the English content of DDT residues, a total of 2.4×10^{10} g yr would be

* See review by Jaffe, pp. 83–110.
** Idem.

transported to the oceans, about one quarter of the estimated DDT annual production.

The PCBs are ubiquitously distributed in the members of the marine and terrestrial biosphere and on the basis of present analyses may be the most abundant group of synthetic organic chemicals in the environment. They are used industrially as insulating fluids in electrical equipment, as heat exchangers, plasticizers and in a variety of other ways. They exist in a large number of homologues and isomers with 210 different chlorinated biphenyls possible. The following discussion is based upon the results of A Study on Marine Environmental Quality, held under the Auspices of the Ocean Affairs Board of the National Academy of Sciences-National Research Council at Durham, New Hampshire, August 9–13, 1971.

Production figures are in general not released by manufacturers. Recent estimates of global production are based upon Japanese production figures. A yearly output of these chemicals in the range of 0.05 to 0.1 megaton now appears reasonable with an integrated production of a megaton. These figures are very similar to those for DDT.

Leakage to the environment probably takes place through discharges to water systems such as rivers and sewage outfalls, through the atmosphere following volatilization or incineration, or through dumping. The relative importance of these methods of dispersion is not as yet known.

Analyses of plankton and fish indicate that the PCBs now exceed the DDT residues in concentration. Such was the case with freshwater fish monitored by the U.S. Bureau of Sport Fisheries and Wildlife [15] and with marine zooplankton and fish from the Atlantic analyzed at the Woods Hole Oceanographic Institution (Vaughan Bowen, personal communication) and from the Pacific analyzed at the Fishery Oceanography Center, La Jolla, California (Izadore Barrett, personal communication).

5. Metals

The metals that are disseminated to the atmosphere in discharges from manufacturing, mining operations, power production and agriculture can have altered environmental concentrations on both global and regional scales. There are increased lead concentrations in surface ocean waters and elevated levels of mercury in recent glacial strata (see following sections). Most probably the atmospheric and hydrospheric burdens of other metals have been similarly affected, although there is little evidence to strengthen this argument. An entry to the problem may be found in an examination of atmospheric aerosol compositions, which can provide indications, if not estimates, of man's environmental impact.

A useful model assumes that atmospheric particulates are removed by rain every nine days (or forty times a year) from a column 5000 m in height. Although the midlatitudes of the northern hemisphere are the sites of broadcast for most of man's wastes, nonetheless, a fallout over the surface of the Earth is assumed. This does overestimate global fallout somewhat, but it does take into account the effects of atmospheric mixing. Examples of such calculations are given in Table VI.

TABLE VI

Atmospheric mobilizations of some metals

Metal	Concentration in air as atmospheric particulates[a] ($ng\ m^{-3}$)	Concentration in seawater ($mg\ l^{-1}$)	Atmospheric rainout (Megatons yr^{-1})
Na	3500 ± 2000	10 500	350
Mg	410 ± 240	1 350	40
Ca	160 ± 90	400	16
K	120 ± 60	380	12
Sr	2.4 ± 1.5	8	0.24
Fe	12 ± 9	0.01	1.2
Pb	3.0 ± 2.8	0.00003	0.3
Cu	1.9 ± 2.1	0.003	0.2
Mn	0.19 ± 0.14	0.002	0.02
Al	6.7 ± 7.2	0.01	0.7
V	0.16 ± 0.11	0.002	0.02

[a] Collected from a tower on the windward coast of Oahu, Hawaii when the tradewinds had been blowing at least 24 h before the start of sampling and during the entire sampling period. From unpublished Ph.D. thesis of G. L. Hoffman, 'Trace Metals in the Hawaiian Marine Atmosphere', University of Hawaii, 1971. An average of at least 56 samples for all elements.

The Na, Mg, Ca, K and Sr concentrations in the atmospheric particulates bear a strong resemblance to their seawater values, strongly indicating a marine source for these metals. On the other hand, the other metals are markedly enriched in the atmospheric dusts and clearly must have different sources. Compared to the aluminum and iron concentrations, the Pb, Cu, Mn and V are too high for an origin in soils or crustal rocks. It is reasonable that these latter metals had sources in human activities – lead in lead alkyl combustion, vanadium from the burning of diesel oils, etc.

It is comforting to see the agreement between these computations with independent estimates of atmospheric mobilizations. For example, the computed sodium washout (350 megatons of NaCl yr^{-1}) is near the value previously obtained from sea salt fluxes (300 megatons yr^{-1}). The lead value of 0.3 megatons yr^{-1} is in conformity with the lead alkyl production and presumably lead discharge to the atmosphere of 0.31 megatons yr^{-1} (Table VIII).

6. Heavy Metals: Mercury

There is one indication that substantial quantities of mercury are being mobilized over large expanses of the Earth by the activities of man: the mercury contents in snows deposited on the Greenland glacier have increased in recent years [31]. This mercury, presumably removed from the atmosphere in precipitation, has levels ranging from 30 to 75 ng kg^{-1} of water (average 60) during the period 800 BC to 1952 and from 87 to 230 ng kg^{-1} of water (average 152) between 1952 and 1965. If this

TABLE VII

Environmental mercury fluxes [31,32]

	Flux in g yr^{-1}
Natural flows	
Continents to atmosphere	
Basis of precipitation with rain	8.4×10^{10}
Basis of atmospheric content	1.5×10^{11}
Basis of content in Greenland glacier	2.5×10^{10}
River transport to oceans	$<3.8 \times 10^9$
Flows involving man	
World production (1968)	8.8×10^9
Entry to atmosphere from fossil fuel combustion	1.6×10^9
Entry to atmosphere during cement manufacture	1×10^8
Losses in industrial and agricultural usage	4×10^9

increase is man-caused, how did he bring it about? An examination of environmental mercury fluxes (Table VII) suggests that the mercury burden of the atmosphere arises from the degassing of the Earth's crust and that, if there is an impact by man, it must be through an enhancement of this degassing process.

The flow of mercury from the continents to the atmosphere is calculated to be in the region of 2.5 to 15×10^{10} g yr^{-1} (Table VII). The mercury content in unpolluted airs ranges from less than 1 to 10 ng m^{-3}. A conservative average of 1 ng m^{-3} yields a total atmospheric burden of 0.4×10^{10} g. Rain effectively washes the mercury out of the air. Taking an average time between rains of 10 days, the annual flux of mercury is 1.5×10^{11} g. A complementary calculation involves the mercury content of rain. The mercury concentration in rain water averages 0.2 parts per billion. An annual world precipitation of 4.2×10^{17} l gives a flux of 8.4×10^{10} g yr^{-1}. Using the pre-1952 glacier waters as representative of typical fallout, a flux of 2.5×10^{10} g yr^{-1} is calculated.

The principal transfer of mercury from the continents to the oceans most probably takes place through the atmosphere. The mercury carried by the rivers to the oceans is at least one order of magnitude less than the amount volatilized from the Earth to the atmosphere (Table VII).

A survey of industrial activities has not revealed any mercury releases to the atmosphere that rival that of the natural degassing rate (Table VII). Thus, increased fluxes through the actions of man probably come about as a result of the enhancement of this degassing process. Landscape alterations that result in disturbances to surface solids, agriculture, mining, construction, etc., can allow more mercury vapor and more gaseous mercury compounds to enter the atmosphere.

7. Heavy Metals – Lead

Lead, like mercury, is being introduced to the marine environment through man's activities in amounts that begin to rival those brought in by the rivers (Table VIII).

TABLE VIII

Annual lead budget [23]

	10^{12} g yr^{-1}
World lead production (1966)	3.5
Northern hemisphere production	3.1
Lead burned as alkyls	0.31
River input of soluble lead to marine environment	0.24
River input of particulate lead to marine environment	0.50

The primary source of lead from society is its emission from internal combustion engines where it is used as an anti-knock additive in the fuel. Fourteen percent of the lead consumed in the U.S. is released in this way (CEQ, 1971).

The dispersion of lead to the surroundings by man has been taking place for at least several millenia in measureable amounts. The concentrations of lead that precipitate from the atmosphere are recorded in the annual layers of permanent snowfields. Murozumi et al. [23] found lead increasing in a north Greenland glacier 25-fold from 800 BC to 1750 as a result of smelting operations. By 1940 the lead levels were 175 times greater than the pre-historic values, whereas in 1966 they were 500 times greater. Up to 1940, the increases are attributed to smelting, and subsequently to the combustion of lead tetra-ethyl. The lead emitted from internal combustion engines, in the micron and sub-micron ranges, is removed primarily by precipitation from the atmosphere.

Surface ocean concentrations of lead have dramatically increased as a consequence of the rainout of man-introduced lead from the atmosphere. Contents in surface waters of the northern hemisphere today are about 0.07 μg of Pb kg^{-1} of seawater compared with estimated pre-historic values of 0.01 to 0.02 μg kg^{-1} [6].

Most probably the greater portion of the lead burned in internal combustion engines eventually enters the marine environment. That which rains down upon the continents can be transferred to the ocean by the rivers. The effect of this increased lead upon surface marine organisms in unknown as yet. If the use of lead in gasolines is curtailed, the concentration in surface waters will probably revert to much lower values within a decade, as the residence time of this element in surface waters appears to be of the order of several years, with removal primarily through biological activity.

8. Artificial Radioactivities

The natural radioactive background of the atmosphere and of the oceans has been increased by man's introduction of artificially produced radionuclides through fission and fusion explosive devices and through the production of energy by nuclear fission. The bomb detonations have not occurred uniformly with time, although there has been an increasing number of nuclear power plants established. The nuclear explosions have introduced far more artificial radioactive species to the oceans than have the reactors or nuclear fuel reprocessing plants up to the present. The oceanic burdens in 1970 are [26]:

Source of radioactivity	Amount in oceans in Curies
Nuclear explosions:	
Fission products exclusive of H-3	$2-6 \times 10^8$
H-3	10^9
Reactors and fuel reprocessing	
Fission and activation products	
exclusive of H-3	3×10^5
H-3	3×10^5
Total	10^9
Natural radioactivities K-40	5×10^{11}

The artificial radioactivities attain but 1/1000 of the total natural radioactivity in the oceans. This level will probably not increase very much in the foreseeable future [26]. Weapon tests appear to be decreasing in number. Although nuclear power programs and the concommittant fuel reprocessing operations will increase, improvements in the management of radioactive wastes and their containment on land will lessen the chances of any global contamination. These remarks do not apply to some individual isotopes like tritium, H-3, whose environmental levels are expected to rise primarily as a result of leakage from nuclear reprocessing plants, and plutonium.

9. Synthetic Organics Other than Halogenated Hydrocarbons

Volatile synthetic organics can escape to the atmosphere at rates of the order of millions of tons per year during their production or utilization. For example, about 2.5% of the total U.S. production of gasoline is lost by volatilization during transfer processes, from production site to vehicles and to storage tanks, and from carburetors. This amounts to 1.75×10^{12} g yr^{-1} [9]. The evaporation of dry-cleaning solvents reaches levels of 3.3×10^{11} g yr [9] about equal to the production rate of the most widely used one, perchloroethyelene, 2.1×10^{11} yr^{-1}. Global values might reach two to three times the above values. The fates of such vapors in the atmosphere are not known.

The production of synthetic organics involves about 5% of the annual petroleum production or about 10^{14} g yr^{-1} on a worldwide basis. Many of these substances have low boiling points and high stabilities and can be expected to enter and to persist in the atmosphere and oceans. Table IX lists the production of such substances in the U.S. in order of rank and with boiling points and water solubilities.

There are already evidences that such substances may be entering our surroundings in substantial amounts. With production rates of 10^{10} to 10^{12} g yr^{-1}, losses in the percent range could contribute measureable amounts to the environment. Some recent

TABLE IX

U.S. Production of synthetic organic chemicals and their properties [2]

Substance	Annual production $\times 10^{12}$ g	Boiling point point °C	Water solubility[a]
	Boiling points between < 0 to 30 °C		
Formaldehyde	2.0	−21	s
Vinyl chloride	1.4	−14	ss
Ethylene oxide	1.2	13–14	s
Acetaldehyde	0.72	21	vs
Ethyl chloride	0.26	13	ss
Dichlorodifluoromethane	0.15	−29	s
Methyl chloride	0.14	−24	s
Glycerol tri-ether	0.10	0.95	
Trichlorofluoromethane	0.09		
Methyl amine	0.050	− 6	vs
	Boiling points between 30 and 60 °C		
Acetone	0.62	56	vs
Propylene oxide	0.44	35	vs
Carbon disulphide	0.36	45	s
Methylene chloride	0.14	40–41	ss
Diethyl ether	0.050	35	s
	Boiling points between 60 and 100 °C		
1, 2-dichloroethane	2.2	84	ss
Methanol	1.7	65	vs
Ethanol	0.96	78.5	vs
Cyclohexane	0.93	81	i
Isopropyl alcohol	0.93	82	vs
Acrylonitrile	0.46	90–92	s
Carbon tetrachloride	0.35	77	i
Vinyl acetate	0.33	71–72	i
Trichloroethylene	0.24	87	ss
2-Butanone	0.21	80	vs
Methyl chloroform	0.14	74	i
Chloroform	0.082	61	s
Ethyl acetate	0.081	77	s
Ethyl acrylate	0.075	100	s

[a] vs, very soluble; s, soluble; ss, slightly soluble; ı, insoluble.

work indicates that such losses are detectable. Trichlorofluoromethane, one of the freons used as a propellant in dispensers and fire extinguishers and in refrigerant fluids, has been measured in winds over southwest Ireland in July and August 1970 with contents of 2×10^{-10} and 1×10^{-11} by volume per volume of air coming from the directions 225°–315° and 45°–135° respectively [20]. Corwin [7] found acetone, butyraldehyde and 2-butanone (methyl-ethyl ketone) in surface waters from the Florida Straits, eastern Mediterranean (no butyraldehyde found) and the Amazon estuary. These low molecular weight species, with relatively low boiling points and high water solubilities, may leak to the surroundings during production or utilization by man.

10. Discussion

The amounts of materials that man is introducing into the sedimentary cycle are approaching, and in a few cases exceeding, those of geological processes. Sometimes the impacts of man are readily detectable such as lead in surface sea waters, increasing contents of carbon dioxide in the atmosphere, or petroleum in marine organisms. For substances alien to the surroundings, such as DDT or artificial radioactivities, man's invasion can easily be identified.

On the other hand, global scale impingements of society may be difficult to observe, especially where there are large background fluctuations in the material under consideration. The site of the measurements also can be most significant as both man's and nature's injections can have zonal or regional distributions.

Such may be the case with atmospheric turbidity. Ludwig *et al.* [21] indicates the intrusion of man-injected detritus on nonurban air quality in the U.S. with a statistically significant increase from 25.4 μg m^{-3} for the period 1962–1966. On the other hand, for U.S. cities a 7% decrease was observed over about the same period. This latter effect was attributed to effective control measures applied by local agencies.

At Mauna Loa, Hawaii, measurements of solar radiation over a 13 yr period (1957–1970) indicated no evidence of man's alteration of atmospheric turbidity [10]. This observation is not surprising. The site of observation (19° 31′ N and 155° 35′ W) is at the lower limit of the jet stream which would transport matter from the human activities in Europe and Asia. A decrease in insolation occurred in 1963, an event which coincided with the eruption of Mount Agung in Bali. The return of atmospheric transmission to the 1963 level took 7 yr. This long recovery period was attributed to a continued fallout of particles from the stratosphere, a continuous production of atmospheric aerosols from volcanic gases, such as sulfuric acid from SO_2, an input of particles from subsequent volcanic explosions, or a combination of any of the three.

The changes in composition of three U.S. rivers, the Illinois, the Mississippi, and the Ohio, and of Lake Michigan during the past century have been chartered by Ackerman *et al.* [1]. There has been a significant increase in their dissolved loads since the beginning of the century as follows:

Species	Average % change per year – time interval given in parenthesis
Chloride	2.0 (52 yr) – 4.3 (70 yr)
Sulfate	2.0 (43 yr) – 2.5 (62 yr)
Nitrate	1.1 (50 yr) – 2.3 (70 yr)
Dissolved solids	0.17 (108 yr) – 0.7 (71 yr)

Lake Michigan has shown a decline in nitrate concentrations in recent years, a phenomenon that may be related to the change in sampling sites. The long-term changes in alkalinity were not of a substantial nature in any of the four areas.

TABLE X

Sulfate fluxes in rivers. Chloride, sulfate and discharge data from Livingston [19]. Table adapted from Berner [3]

	Total Cl⁻ (mg/l)	Total SO₄⁻⁻ (mg/l)	Pollutant SO₄⁻⁻	Flux to Oceans in 10^{12} g yr⁻¹		
				H_2O	SO_4 (natural)	SO_4 (pollution)
Europe	6.9	24	17	2.5×10^6	17	45
N. America	8.0	20	12	4.6×10^6	37	55
S. America	4.9	4.8	0	8.2×10^6	39	0
Africa	12.1	13.5	0	6.0×10^6	81	0
Asia	8.7	8.4	0	11.2×10^6	94	0
			Total	32.5×10^6	268	100

A similar sense was reached by Berner [3] through an examination of the dissolved sulfate concentrations in rivers flowing through areas of high and low levels of industrialization. The argument is made that most of the world's sulfur pollution arises in the northern hemisphere, primarily Europe and North America. On the other hand, the rivers flowing through Asia, Africa and South America are assumed to receive trivial amounts of sulfur as a result of man's activities. Thus, by accepting a SO_4^{--}/Cl ratio for unpolluted waters equal to the average of the rivers of Asia, Africa and South America, a correction factor can be applied to European and North American rivers to take into account pollution caused by man (Table X). Utilizing the discharge data of Livingston [19], Berner calculated the flux of natural (non-pollution) and pollution discharge for each continent (Table X). On a global basis, it appears that 1.0×10^{14} g of the 3.68×10^{14} grams of sulfate annually brought by the rivers to the oceans can be attributed to man. Since many of the analyses for average river values were obtained before 1900, the sulphate pollution value of 27% $[(1.00/3.68) \times 100)]$ is probably a lower limit.

It is interesting to note that the 100 megatons of sulphate pollution presumably carried by the rivers compares favorably with the 150 megatons of sulphate (100 megatons of SO_2) released annually to the atmosphere through the combustion of fossil fuels.

References

1. Ackermann, W. C., Marmeson, R. H., and Sinclair, R. A.: 'Some Long-Term Trends in Water Quality of Rivers and Lakes', *EOS* 51, 516 (1970).
2. Anonymous: *Synthetic Organic Chemicals*, U.S. Production and Sales, 1968, U.S. Tariff Commission Publication N. 327 (1970).
3. Berner, R. A.: 'Worldwide Sulfur Pollution of Rivers', *J. Geophys. Res.* 76, 6589 (1971).
4. Bertine, K. K. and Goldberg, E. D.: 'Fossil Fuel Combustion and the Major Sedimentary Cycle', *Science* 173, 233 (1971).
5. CEQ: *Ocean Dumping. A National Policy*, U.S. Council on Environmental Quality, (1970).
6. Chow, T. J. and Patterson, C. C.: 'Concentration Profiles of Barium and Lead in Atlantic Waters off Bermuda', *Earth Planet. Sci. Letters* 1, 397 (1966).

7. Corwin, J. F.: 'Volatile Organic Materials in Sea Water', in Donald W. Hood (ed.), *Organic Matter in Natural Waters*, Institute of Marine Science Occasional Publication No. 1, University of Alaska, pp. 169–180 (1970).
8. Cronin, F. J.: 'Recent Volcanism and the Stratosphere', *Science* 172, 847 (1971).
9. Duprey, R. L.: *Compilation of Air Pollutant Emission Factors*, U.S. Public Health Service Publication No. 999-Ap-42 (1968).
10. Ellis, H. T. and Pueschel, R. F.: 'Solar Radiation: Absence of Air Pollution Trends at Mauna Loa', *Science* 172, 845 (1971).
11. Eriksson, E.: 'The Yearly Circulation of Chloride and Sulfur in Nature; Meteorological, Geochemical and Pedological Implications I', *Tellus* 11, 375 (1959).
12. Goldberg, E. D.: 'Atmospheric Dust, the Sedimentary Cycle and Man', *Comm. Earth Sci. Geophys.* 1, 117 (1971).
13. Goldberg, E. D. and Bertine, K. K.: 'GNP/Area Ratio as a Measure of National Pollution', *Mar. Pollut. Bull.* 2, 94 (1971).
14. Griffin, J. J., Windom, H., and Goldberg, E. D.: 'The Distribution of Clay Minerals in the World Ocean', *Deep-Sea Res.* 15, 433 (1968).
15. Henderson, C., Inglis, A., and Johnson, W. L.: 'Organochlorine Insecticide Residues in Fish – Fall, 1969', *Pestic. Monit. J.* 5, 11 (1971).
16. Inman, R. E., Ingersol, R. B., and Levy, E. A.: 'Soil: a Natural Sink for Carbon Monoxide', *Science* 172, 1229 (1971).
17. Judson, S.: 'Erosion of the Land', *Am. Scientist* 56, 356 (1968).
18. Koide, M. and Goldberg, E. D.: 'Atmospheric Sulfur and Fossil Fuel Combustion', *J. Geophys. Res.* 76, 6589 (1971).
19. Livingston, D. A.: *Chemical Composition of Rivers and Lakes*, U.S. Geol. Surv. Prof. Paper 440-G (1963).
20. Lovelock, J. E.: 'Atmospheric Fluorine Compounds as Indicators of Air Movements', *Nature* 230, 379 (1971).
21. Ludwig, J. H., Morgan, G. B., and McMullen, T. B.: 'Trends in Urban Air Quality', *Trans. Amer. Geophys. Union* 51, 468 (1970).
22. Mitchell, J. M., Jr.: this volume, pp. 149–173.
23. Murozumi, M., Chow, T. J., and Patterson, C.: 'Chemical Concentrations of Pollutant Lead Aerosols, Terrestrial Dusts and Sea Salts in Greenland and Antarctic Snow Strata', *Geochim. Cosmochim. Acta* 33, 1247 (1969).
24. NASCO: *Chlorinated Hydrocarbons in the Marine Environment*, National Academy of Sciences Report (1971).
25. Parkin, D. W., Phillips, D. R., Sullivan, R. A. L., and Johnson, L.: 'Airborne Dust Collections over the North Atlantic', *J. Geophys. Res.* 75, 1782 (1970).
26. Preston, A., Fukai, R., Volchok, H., Yamagata, N., and Dutton, J.: *Radioactivity*, in Seminar on Methods of Detection, Measurement and Monitoring of Pollutants in the Marine Environment, FAO, Rome, Dec. 4–10, 1970.
27. Revelle, R., Wenk, E., Corino, R., and Ketchum, B. H.: *Ocean Pollution by Petroleum Hydrocarbons*, manuscript presented at SCEP (1970).
28. SCEP: *Man's Impact on the Global Environment*, MIT (1970).
29. Shannon, L. J., Vandegrift, A. E., and Gorman, P. G.: *Assessment of Small Particle Emissions*, Midwest Research Institute Report on Contract CPA-22-69-104 with the Air Pollution Control Office, Environmental Protection Agency (1970).
30. Swinnerton, J. W., Linnenbom, V. J., and Lamontagne, R. A.: 'The Oceans: a Natural Source of Carbon Monoxide', *Science* 167, 984 (1970).
31. Weiss, H. V., Koide, M., and Goldberg, E. D.: 'Mercury in a Greenland Ice Sheet: Evidence of Recent Input by Man', *Science* 174, 292–4 (1971a).
32. Weiss, H. V., Koide, M., and Goldberg, E. D.: 'Selenium and Sulfur in a Greenland Ice Sheet: Relation to Fossil Fuel Combustion', *Science* 172, 261 (1971b).
33. Windom, H. L.: 'Atmospheric Dust Records in Permanent Snowfields – Implications to Marine Sedimentation', *Bull. Geol. Soc. Amer.* 80, 761 (1969).

DDT IN THE BIOSPHERE: WHERE DOES IT GO?

GEORGE M. WOODWELL, PAUL P. CRAIG and HORTON A. JOHNSON

*Brookhaven National Laboratory, Upton, N.Y., U.S.A.**

Abstract. A global model traces the movement of DDT and its residues through the biosphere. Assuming curtailed production, and no use after 1974, the model suggest maximum concentrations in the atmosphere in 1966 and in the mixed layer of the ocean in 1971. The biota appears to have absorbed only a small fraction of the DDT available in the biosphere.

DDT has been used in large quantities as an insecticide since 1942. Its residues [1] are sufficiently persistent and mobile to have a worldwide distribution, appearing in the lipids of most organisms [2–7], in air [7–9], and occasionally in meltwaters of Antarctic snows [10]. Concentrations in certain of the Earth's biota have reached toxic levels, causing spectacular declines in populations of certain carnivorous and scavenging birds and fish, aggravating the problems of pollution, and threatening significant contamination of human food chains [2, 4, 11–14]. Recognition of the seriousness of these problems has led to recent restrictions in the use of DDT in the United States and abroad. There is at least a possibility that most of the DDT that has been or will ever be produced has already been used and that little, if any, will be applied after the mid-1970's [15]. The persistence of DDT residues is great enough, however, that the residues will continue to be redistributed for many years after use of the pesticide has stopped, presumably presenting a continuing hazard to all the biota. The extensive data available on the distribution and effects of DDT make it, together with radioactive substances, the best known of the biospheric pollutants and a valuable subject for a case history study [12, 16, 17]. But basic questions remain, among them the following: What becomes of DDT released into the biosphere? How serious are the hazards? and, How long will the hazards persist?

In an effort to answer these questions we have attempted to develop a model of the circulation of DDT in the biosphere. We have done this on the basis of two limiting assumptions: (1) that use of DDT will decline to zero by 1974 and, alternatively, (2) that, between now and then, use will increase throughout the world.

Certain physical properties of DDT are important in determining its behavior in the biosphere. First, because of the high solubility of DDT in fats together with its low solubility in water [18, 19], DDT residues tend to accumulate in lipids and therefore in plants and animals. Second, the residues are very persistent in nature: estimates of their half-life range upward to 20 yr, perhaps longer under certain circumstances [20–24]. Third, DDT has a vapor pressure high enough to assure direct losses from plants and soil into the atmosphere, which can carry residues worldwide [12, 13]. Thus

---

* Reprinted with changes from *Science* **174** (10 Dec. 1971), 1101–1107.

S. Fred Singer (ed.), The Changing Global Environment, 295–309. All rights reserved.
Copyright © 1975 by D. Reidel Publishing Company, Dordrecht–Holland.

soils, air, the water of the oceans, and the biota are all potential reservoirs for DDT residues, and the hazard to the biota, including man, hinges on the distribution of DDT residues among these reservoirs. How large are the reservoirs, and what are the rates of exchange between them?

The answers are not available in any simple or absolute sense. Most are available, however, at least by inference. First, we must know how much DDT has been produced and something about its distribution.

1. Input: DDT Production

The amount of DDT produced in the United States each year is reported by the U.S. Tariff Commission [25]. In the crop year 1963 the amount of DDT produced reached a maximum 8.13×10^{10} g (179×10^6 lb) (Figure 1). Production has dropped in the United States since 1963, but more than 6.0×10^{10} g of DDT were produced in 1969. Preliminary figures for 1970 reveal that DDT production declined by more than 50%. About 70% of the amount of DDT produced appears to have been used outside the United States. The total amount of DDT produced in the United States, integrated over the entire period through 1974, when we have assumed that DDT will no longer be used, is estimated to be 1.4×10^{12} g. No data on world production are available. We have assumed that the amount of DDT produced in the world, including the U.S. fraction, is twice the amount produced in the United States, or about 2.8×10^{12} g in total through 1974.

This DDT has been distributed widely around the world, most heavily in humid temperate and tropical zones. It is commonly applied by spraying a liquid suspension or solution from mobile ground equipment or from aircraft. The fraction of the spray that lands on the target varies, but some fraction of both aerial and ground applications remains airborne. Aerial applications of DDT to forests in the northeastern United States show that 50% or less of the amount emitted from the planes is depo-

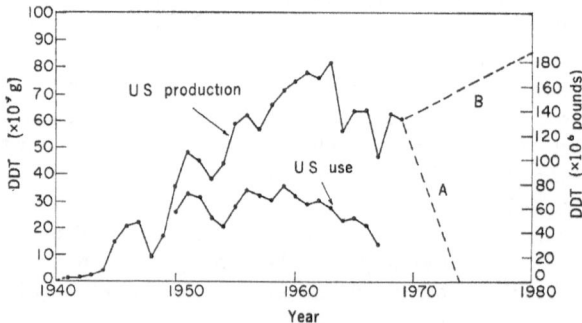

Fig. 1.   DDT production and use in the United States: curve A, based on the assumption of declining use through 1974; curve B, based on the assumption of increasing use. Dotted lines indicate projections.

TABLE I

DDT residues in soils of agricultural and nonagricultural land in the United States. Data were selected because of large sample size or because they are the only data available. For more detailed tabulations see Edwards [21, 38]

| Soil sites | Sites sampled (No.) | DDT residues (g m^{-2}) | | Reference |
		Range	Mean	
Agricultural				
Orchards	14	0.34 –22.1	6.0	[29]
Crops	24	0 –0.87	0.24	[29]
Root crops	48	0.045– 5.73	1.25	[30]
Vineyards	2	2.13 – 3.18	2.69	[31]
Orchards	2	8.18 –14.60	11.4	[32]
Vegetable crops	10	0.07 – 9.52	2.62	[33]
Randomly selected	41	0.002– 1.30	0.148	[33]
Alfalfa crops	12	0.06 – 0.98	0.336	[34]
Soybean crops	43	0.004– 4.03	0.986	[35]
Nonagricultural				
Boreal forests (sprayed)	3	0.179– 0.258	0.213	[22]
Boreal forests (unsprayed)			0.0045	[23]
Forest in Pennsylvania (unsprayed)		0.0003–0.0006	0.0004	[28]

sited on the forest. The rest is dispersed into the air. Much of the airborne fraction returns to the ground nearby, but small droplets or particles are likely to remain aloft, to become associated with other particles, and may be carried great distances [7–9, 14, 26].

2. DDT in Soils

The persistence of DDT led to early recognition that residues might accumulate in soils. On the basis of a review of published data (Table I), we have estimated that agricultural soils in the United States contain an average content of DDT approaching 0.168 g m^{-2} (1.50 lb per acre) [22, 23, 27–35]. Nonagricultural soils were estimated to contain an average of 4.5×10^{-4} g m^{-2} [36].

These estimates can be used to calculate the rates of loss of DDT residues from soils in the United States. Within the United States, the total contiguous land area is 7.7×10^{12} m^2 (1.9×10^9 acres), about 11 % of which (0.85×10^{12} m^2) is kept in crops on which insecticides are used [37]. The agricultural land apparently retains about 1.42×10^{11} g of DDT. These data were obtained in the early 1960's when DDT production was high. The rate of use of DDT in the United States during this period was 2.7×10^{10} g y^{-1} (6×10^7 lb yr^{-1}) (Figure 1) or 0.0318 g yr^{-1} for each square meter of agricultural cropland (0.28 lb acre^{-1} yr^{-1}).

The mean lifetime of DDT in soils must be about 10^{11} (2.7×10^{10}) = 5.3 yr. This estimate approximates Edwards' earlier estimate of 4.3 yr [21, 38]; it is substantially less than the estimated lifetime of 10 yr for DDT in certain soils [22, 24]. We have assumed a mean residence time of 4.5 yr for DDT on land.

Four mechanisms probably account for most losses of DDT residues from soils: (i) volatilization (including losses by wind erosion of small particles from the soil surface), (ii) removal by harvest of organic matter, (iii) water runoff, and (iv) chemical (including biotic) degradation.

The occurrence of DDT residues in rainwater suggests that large amounts of DDT may move through the atmosphere either adsorbed to particles or as vapor. The small number of data available on the volatilization of DDT suggest a time constant for volatilization of several years, but the evidence supports the conclusion that vaporization is more important than such a long time constant would indicate [14, 21, 38, 39]. The rates of disappearance of residues of dieldrin [1] by volatilization have been shown to be a function of the rate of air movement through soils [40]. Residence times for DDT in organic soils where air movement is slow are greater than residence times in mineral soils, a relationship that could only be the case if biotic degradation of DDT residues proceeds slowly in these soils in comparison with volatilization. Apparently evaporation is a major mechanism for the removal of residues of the persistent pesticides from soils [21, 38], and, despite its slowness, proceeds faster than chemical breakdown.

DDT residues are also removed from soils by the harvest of organic matter, but the evidence suggests that this is not a major route of transport. The net amount of crop harvested, even for highly productive agricultural crops, seldom exceeds 5000 g m^{-2} and more often approaches 1000 g m^{-2} [41]. If we assume an annual harvest of 3000 g m^{-1} containing 1 part per million (ppm) of DDT, currently the maximum concentration allowed in many foods, the harvest would remove 0.003 g m^{-2} of DDT, about 1 % of an annual application of 0.33 g m^{-2}, or about 10 % of the estimated annual average amount of DDT used for all crop land (see above). Thus, even a harvest of 100 % of the primary production would remove only a small fraction of an annual application of DDT. Actual harvests that remove the organic matter from the site and might transport DDT residues to urban areas or to watercourses are much less efficient. We have assumed a removal of 1 % of the total DDT used on the crop.

DDT is not normally applied to forage crops or in places where it can contaminate tissues of animals used for food, but farm animals do, nonetheless, become contaminated, as do most animals. It is difficult to visualize under present levels and patterns of use a circumstance in which harvest of farm or range animals would account for more of the DDT residues than would the annual harvest of the plant crops, despite the possibility of concentration of the residues through food chain effects.

The transport of DDT residues from agricultural soils in surface waters has often been assumed to be of major importance in the accumulation of residues in the oceans. Heavy rains do remove DDT either adsorbed to soil particles or in solution but the amounts are small in comparison with the total amounts of DDT produced. Surface runoff over the entire United States is about one-third of the annual precipitation, or 23 cm (9 in.) yr^{-1} out of an average precipitation estimated as 76 cm [42].

The total volume of this water is 2×10^{12} m^3 y^{-1}. If it were saturated with DDT residues, it would contain about 1 part per billion (ppb) [18, 43] and would remove 2×10^9 g of DDT per year in solution. DDT is applied only at certain times of the year in certain areas and seldom directly to bodies of water. Observed river water concentrations of pesticides (including particles) range from concentrations below the limit of detection (less than 10 parts per trillion (ppt)) to almost 100 ppt [18, 43]. An average concentration of 50 ppt implies an annual runoff of about 10^8 g yr^{-1}, which accounts for about 0.1 % of the amount of DDT produced per year. Similar conclusions have been reached by Risebrough *et al.* [9], who also decided that movement of residues in the atmosphere is the most important transport route, and by the ocean pollution group of the M.I.T. Study of Critical Environmental Problems [14].

3. DDT in the Atmosphere

The vapor pressure of DDT at 20 °C is 1.5×10^{-7} mm of mercury [44], producing an equilibrium concentration of DDT in the atmosphere of about 3×10^{-6} g m^{-3} or about 2 ppb by weight. The vapor pressure drops with decreasing temperature. If we assume that DDT in the atmosphere remains as vapor, the saturation capacity of the atmosphere to the tropopause would be about 10^{12} grams of DDT, or about as much as has been produced to date. But DDT residues also exist in association with atmospheric particles, and the Earth's atmosphere can probably contain very much larger quantities than the saturation capacity alone would indicate. This means that the atmosphere is potentially a large reservoir in addition to being a major means of transport for the residues.

Residues are removed from the atmosphere by rainfall, diffusion across the air-sea interface, and chemical degradation. The dominant mechanism for the removal of DDT from the atmosphere is probably rainfall. In England, DDT concentrations in the range from 73 to 210 ppt have been reported in rain in areas close to regions where DDT has been used, and similar concentrations have been reported in the United States [45]. A DDT concentration in meltwaters from Antarctic ice of 40 ppt has been reported recently [10]. Earlier measurements were less sensitive [46].

DDT concentrations in rainfall vary appreciably throughout the year. The variation is related to the seasonal application of DDT. The fact that there is a seasonal variation suggests that the time constant for removal of DDT from the atmosphere probably does not exceed a few years. If the average DDT concentration in rainfall were 60 ppt and precipitation averaged 1 m yr^{-1}, rainfall would remove a total of 3×10^{10} g of DDT residues from the atmosphere annually, most of that into the oceans. The annual amount of DDT produced throughout the world in the mid-1960's was about 10^{11} g, approximately $3\frac{1}{2}$ times the amount that would be removed worldwide by rain containing 60 ppt of DDT. Thus an average rainfall concentration of DDT of 60 ppt gives an upper limit for the mean time for removal of DDT residues from the atmosphere by rainfall of about 3.3 yr. Residues deposited on the ground are, of course, available for reevaporation.

Measurements of the transfer of carbon dioxide into the oceans suggest a time constant of 7 yr [47, 48]. This period is extremely long as compared to estimates for diffusive transport. Although one might expect atmospheric DDT to be transferred more slowly than carbon dioxide because of DDT's low solubility in water, direct measurements seem to be lacking. Accumulation of lipids at the ocean surface [49] would probably increase the rate of transfer as would association of DDT residues with particles in air. We have assumed a time constant of 4 yr for the transfer of DDT residues from the atmosphere to the Earth's surface. The mean residence time is probably not longer than this.

Chemical degradation of DDT in the atmosphere may also be important. Efficient photodegradation of DDT vapor occurs primarily at wavelengths shorter than 2700 Å [50]. These wavelengths are heavily absorbed by the atmospheric ozone layer. Residues adsorbed on particles are probably considerably more resistant, however. We have assumed atmospheric degradation to be unimportant as compared to transport, but this topic is obviously in need of further study.

4. DDT in the Oceans

DDT residues circulate initially in the 'mixed' layer, which frequently extends to a depth of 75 to 100 m. They are transferred slowly below the thermocline into the much larger volume of the abyss [51]. Sedimentation of organic matter removes DDT residues from the upper layers, but direct measurements of sedimentation of DDT residues seem to be lacking. We assume that the virtual insolubility of DDT in water combined with its solubility in fat assures the association of DDT with organic matter and that the rate of transfer of carbon to the abyss would approximate the rate of transfer of DDT residues. Biological mixing may also be important in DDT transport within the ocean, but direct evidence is lacking. As an estimate of the time for the transport of DDT from the mixed layer to the abyss, we have used a result from studies of carbon dioxide that indicate a mean transfer time of about 4 yr [47, 48].

Within the abyss, transfer rates for carbon dioxide and presumably for other substances such as DDT are very slow indeed, ranging up to hundreds and even thousands of years for certain segments. Because of the fact that the volume of the abyss is immense and because of the possibility that DDT residues may be lost to sedimentation, the abyss is a very large reservoir, virtually infinite for the purposes of this discussion.

5. DDT in the Biota

The total amount of DDT retained within the biota is small by comparison with the totals that can be retained in other pools within the biosphere. It is also small by comparison with the annual amount of DDT produced. Liberal assumptions with respect to the concentrations of residues in various segments of the biota, including man, lead to an estimate of 5.4×10^9 g of DDT held within the biota worldwide (Table II). Estimates of the world biomass are notoriously variable but they are probably correct

TABLE II

DDT residues in the biota in the late 1960's. Concentrations are expressed to the nearest order of magnitude only; parts per million

Location	Dry biomass ($\times 10^9$ metric tons)	DDT content (ppm)[a]	Total DDT ($\times 10^8$ g)
Plant biomass[b]			
On land			
Lakes and streams	0.04	0.010	0.004
Swamps and marshes	24	0.001	0.240
Terrestrial vegetation (forests, desert,			
savanna, grassland, tundra)	1814	0.0001	1.814
Agriculture	14	0.1	14.000
Total land	1852		15.058
In oceans			
Open ocean algae	1.0	0.1	1.0
Continental shelf algae	0.3	1.0	3.0
Attached algae	2.0	1.0	20.0
Total ocean	3.3		24.0
Total plants	1855		39.06
Animal biomass[c]			
On land			
Feral mammals	0.009	1.0	0.09
Domestic mammals	0.17	1.0	1.7
Man	0.30[d]	1.0	3.0
Birds	0.00024	1.0	0.002
In oceans			
Fish	0.65	1.0	6.5
Mammals	0.055	1.0	0.55
Others (protozoa, coelenterates, annelids,			
nematodes, mollusks, echinoderms,			
arthropods)[e]	3.02	0.1	3.02
Total animals	4.20		14.86
Total DDT in the biota:		5.4×10^9 g	

[a] Estimates based on values in the literature and the experience of G.M.W. Sources include [2–7, 11–14, 20–24, 36, 55, 56] and others. All estimates are to the nearest order of magnitude only; questionable data have been resolved toward the higher number.
[b] Adapted from Whittaker [60].
[c] Adapted from Bowen [61].
[d] Bowen used 0.03×10^9 tons, which seems to be about 10 times too low (G.M.W.).
[e] Listed by Bowen [61]; obviously incomplete, but indicative.

to within a factor of 2 to 3, almost certainly to within less than a factor of 10. DDT analyses that are specifically appropriate for compilation of such an inventory are few, and the data of Table II are, at best, crude estimates. The data on residues have been expressed to orders of magnitude only to avoid a false indication of precision. Questionable estimates have been resolved in the direction of the higher order of magnitude, thus giving a bias toward a higher estimate. The analysis indicates that there may now be between 10^9 and 10^{10} g of DDT circulating in the biota, about 1/30 of the amount produced in 1 yr during the mid-1960's. We consider this an estimate of the maximum amount of DDT that could be in the biota; the only other estimate available is 6×10^8 g for the biota of the oceans alone published by the M.I.T. study group [14]. This means that, despite the importance of the biota and the effects of DDT on it, the capacity of the biota for holding DDT residues is small enough that we can ignore it for the moment in our attempt to appraise the worldwide movements of DDT.

6. A Model of DDT Circulation in the Biosphere

The number of pathways that are clearly important in the worldwide movement of residues appears to be small. The primary reservoirs are the land surface, the troposphere, the mixed layer of the ocean, and the abyss. In the previous sections, we have estimated time constants for the dominant physical processes. The constants have been used in a set of first-order rate equations to yield estimates of DDT loads in the various reservoirs as functions of time. The rate equations have the form:

$$\frac{dN_i}{dt} = R_i(t) - \sum_{j=1}^{m} \frac{N_j}{\tau_{ij}^{(l)}} + \sum_{j=1}^{m} \frac{N_i}{\tau_{ij}^{(g)}} \quad i = 1 \dots m$$

Here $R_i(t)$ is the rate at which newly produced DDT is introduced into the ith reservoir and N_i is the amount of DDT in the ith reservoir. The sums represent losses from and gains to the ith reservoir; $\tau_{ij}^{(l)}$ is the time constant for DDT loss from the ith to the jth reservoir, and $\tau_{ij}^{(g)}$ represents inputs from the jth to the ith reservoir. There are m reservoirs, and there are thus m simultaneous, first-order, differential equations to be solved.

We have used the time constants estimated above to solve the rate equations over the period from 1940 to 2000 on a digital computer. The DDT input for each year has been taken to be twice the amount produced in the United States per year and has been projected in two ways (Figure 1, curves A and B). Except for the distinction between land and ocean, no attempt has been made to include geographical variation. Local fluctuations may be expected to be large.

The calculated average DDT concentrations in the atmosphere and in the mixed layer of the oceans for the period from 1940 to 2000 are shown in Figure 2. If the world DDT production becomes zero in 1974, the concentration in the lower atmosphere would have reached a peak in 1966 at about 72 ppt (84×10^{-9} g m^{-3}). The mixed layer of the ocean would contain its maximum of 15 ppt in 1971. The concentrations in both reservoirs can be expected to decline gradually, with the concentration in air reaching 10% of its peak value in 1984. The concentration in the mixed layer will not decline to 10% of its peak value until 1993. The total load of DDT on the land surface reached a maximum concentration of 6.34×10^{11} g in 1964 to 1966, and will decline if DDT production slows.

There is reason to assume that the worldwide production of DDT will not drop, but may increase, despite U.S. restrictions. The increase will be in response to an in-

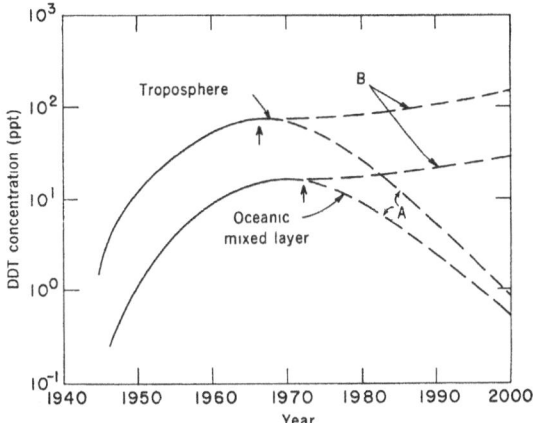

Fig. 2. DDT concentrations in the mixed layer of the oceans, in the troposphere, and in the biota projected through the year 2000 on the basis of assumptions A and B of Figure 1.

creasing demand for an inexpensive means of pest control in agriculture and for control of vectors of disease, especially malaria. If we assume that foreign production replaces U.S. production and that the total use of DDT in the world increases after 1969 (curve B, Figure 1), the concentrations of DDT in air and water will follow the curves marked B in Figure 2, continuing to increase until after the year 2000.

7. Implications for Life

The physical processes we have discussed dominate the transfer of DDT residues throughout the biosphere. Living systems retain quantities of DDT that are small by proportion, and living things appear, at least superficially, to play a minor role in the world budget of DDT. Yet it is the residues that are available to living systems that are the hazard, and we must examine their behavior with special care. The total quantities of DDT residues in the biota are but 1/30 or less of the annual amount of DDT produced in recent years; they are also a small fraction of the annual transfers estimated from soils to air and the oceans. The quantities are small enough that the transfer from land to water by surface runoff, small as it is, must be assumed to contribute to contamination of the coastal biota. The residues presently held in the biota, and the maximum quantity that the biota could hold (not very greatly different), are so small in proportion to the total amount of DDT produced that we wonder why the biota has not been affected much more drastically than it has been – and what the future holds.

The answers are far from clear. DDT residues are accumulated in living systems and recycled in much the same ways that certain elements essential for life are recycled. Just as phosphorus is recycled from sediments by various means, so DDT residues may reenter complex food webs from organic sediments. One such route is by direct consumption of detritus [52]. Others must include oxidation of the sediments. Concentrations of DDT in the biota may ultimately reach as much as 10^6 times the concentrations in the general environment [36]. The effects of high concentrations are clear enough: food webs are reduced, carnivores eliminated, and hardy, small-bodied organisms favored [53]. The changes are similar to those that occur in eutrophication; the sedimentation of organic matter is probably increased, often speeded by a shift toward an increasingly anaerobic benthos. We assume that under such extreme conditions DDT residues tend to accumulate in anaerobic sediments and are removed from circulation. Thus one of the effects of DDT is to reduce the biota and to increase the rate of removal of DDT residues into sediments. The process tends to restrict the movement of residues in the larger circulations of the biosphere, accentuating the importance of local contamination. There is, however, not much question that the oceans, as well as lakes and estuaries, are vulnerable to such effects. How much more DDT would it take to degrade the biota significantly?

The answer hinges on both the rate of movement of residues through the major reservoirs of the biosphere and on the coupling between the biota and the environment. How rapidly does the biota absorb DDT? Is there a possibility of the biota's achieving an equilibrium in which inputs of DDT residues are exactly balanced by losses?

DDT residues enter the biota both through food webs and by direct absorption. The relative importance of these routes varies between land and water and among species, and the time for the biota to come to equilibrium with residues in the environment must also vary. One attempt at appraising the time for a food web to reach equilibrium led to an estimate of between "four times the average life span of the longest-lived species and the sum of the life spans for all trophic levels" [54, p. 506]. Such an analysis suggests that equilibrium for the entire biota would be reached only after many decades. Movement of DDT residues into the abyss appears much more rapid than this.

On the other hand, plankton in water would be expected to reach equilibrium with residues in solution in the water very rapidly, and small-bodied, warm-blooded carnivores that have high rates of metabolism and feed from water-based food webs might be expected to accumulate high concentrations of residues rapidly and to be affected by them. This circumstance, of course, is what we see: aquatic carnivorous birds accumulate high concentrations of DDT and then their numbers rapidly diminish. So, although the biota as a whole may not have achieved equilibrium with the residues circulating now, certain segments of the biota are being reduced, a fact that indicates that the biota may be appreciably changed before an 'equilibrium' with present rates of DDT production and present world inventories is reached. Under these circumstances the concept of an equilibrium becomes elusive; there is no true equilibrium, only a constant state of flux through a pool that probably grows smaller as the biota is reduced. The coupling between DDT residues in the environment and in aquatic food webs would seem to be reasonably close, especially for lower trophic levels, and a decline in the amount of DDT used should be reflected almost immediately in lower concentrations in the biota.

A variety of evidence favors this conclusion. Fauna of salmon streams in the Miramichi River in New Brunswick, Canada, and fish in Sebago Lake, Maine, responded year by year to a reduction in the use of DDT [55]. More recent experience on Long Island with osprey populations seems to be indicating a similar response, although the observations are far from conclusive. Populations dropped abruptly during the mid-1960's. The decline seems to have been arrested, perhaps reversed, after cessation of use of DDT for mosquito control in 1967 [56]. Terrestrial ecosystems probably respond more slowly, but the patterns of circulation of residues and their effects are similar.

These examples and others, such as the observation that the effects of pesticides on reproduction of bird populations in England are related to the intensity of use of chlorinated pesticides [57], emphasize that most conspicuous effects on the biota in the past have not been the result of worldwide movements of DDT residues but rather have been attributable to local concentrations identifiable with some local or regional use. Restriction in the use of pesticides has usually reduced the effects within a few years.

Residues in other segments of the biota, however, are very much more closely related to larger patterns of circulation. Oceanic birds such as the sooty and slender-billed shearwaters [4, 5] and the Bermuda petrel [6] that feed in the open ocean, the California mackerel, the penguin, and the crabeater seal of the Antarctic [58] must obtain their residues from patterns of circulation that are close to being 'worldwide'. All of these organisms are contaminated with DDT, some with concentrations occasionally exceeding 10 ppm. These are the organisms that are most closely coupled to the major nonliving pools of DDT circulating in the biosphere. Our analyses suggest that such organisms should reflect world use of DDT within a few years, with residues in the biota increasing or decreasing as world use rises or falls.* The fact that none of these organisms has yet become extinct from DDT effects is mere good fortune: the total amounts of DDT estimated to be circulating in the biosphere are many times greater than the amounts required to eliminate most such animals. We know that the residues can be concentrated by many factors of 10 into the biota – and affect it catastrophically. Yet, although there is no question about the devastation wrought by DDT locally and even regionally, the worldwide component seems not yet to have reached the point of widespread extinctions. (The difficulties of measuring effects are so great as to make most biologists who examine this question in any depth suspicious that effects may be occurring unobserved or masked by other causes). If we assume that the lower trophic levels of aquatic food webs are more closely coupled to their environment by dint of the two pathways for entry of residues than the analyses of Harrison *et al.* [54] suggest, then reduction in the use of DDT should be reflected within a few

---

* Recent correspondence with Mr David Wingate of Bermuda seems to lend support to this conclusion. Mr Wingate has observed a sharp increase in the reproductive success of the Bermuda petrel during recent years since the curtailment of production of DDT in the U.S.

years in a reduction in the DDT residues in the biota identified with the worldwide distribution.

8. Where Has the DDT Gone?

The physical and chemical characteristics of DDT might lead one to assume that the biosphere should behave as a giant separatory funnel, gradually partitioning the lipid-soluble residues into the lipid-rich biota. Although there is no question that this process does occur, there is also no escape from the conclusion that it does not work well on the biospheric level. Most of the DDT produced has either been degraded to innocuousness or sequestered in places where it is not freely available to the biota. Recent work seems to support the latter assumption and the assumptions of our model. A preliminary report of detailed analyses of DDT residues in the air of nine U.S. cities in 1967–1968 [59] shows concentrations in winter, when DDT is not used locally and residues might be expected to be mixed throughout the troposphere, commonly falling between 10^{-9} and 100×10^{-9} g m^{-3}. The range approximates our prediction of 84×10^{-9} g m^{-3} in air based on the assumption of declining use. The observation supports our assumptions on the routes of movement and sizes of pools. The fact remains, however, that, despite the abundance, persistence, and worldwide distribution of DDT residues, they are not as freely available to the biota as might be assumed. How and precisely where they are held is not yet clear, but the biosphere appears to have a large capacity for holding them apart from the biota. What is clear is that large quantities of DDT were introduced into use before any appraisal was made of the capacity of the biosphere for receiving them. In this instance man seems to have been blessed with extraordinary good fortune.

9. Summary

The worldwide pattern of movement of DDT residues appears to be from the land through the atmosphere into the oceans and into the oceanic abyss. Calculations based on the fragmentary data available on rates of movement and sizes of various pools of DDT residues lead to the conclusion that concentrations in the atmosphere and in the mixed layer of the oceans lag by only a few years behind the amounts of DDT used annually throughout the world. A model suggests that maximum concentrations of DDT residues occurred in air in 1966 and will occur in the mixed layer of the oceans in 1971. The biota probably contains in total less than 1/30 of 1 year's production of DDT during the mid-1960's, a very small amount in proportion to the total potentially available. The reason for the biota's failure to absorb larger quantities and to be affected much more severely is unclear. The analysis suggests that mere good fortune has protected man and the rest of the biota from much higher concentrations, thus emphasizing the need to determine the details of the movement of DDT residues and other toxins through the biosphere and to move swiftly to bring world use of such toxins under rational control based on firm knowledge of local and worldwide cycles and hazards.

References and Notes

1. DDT residues include DDT and its decay products, DDD and DDE: DDT, 1,1,1,-tri-chloro-2,2-bis(p-chlorophenyl)ethane; DDE, 1,1-dichloro-2,2-bis(p-chlorophenyl)ethylene; DDD, 1,1-dichloro-2,2-bis(p-chlorophenyl)ethane; di-eldrin, 1,2,3,4,10,10-hexachloro-6,7-epoxy-1,4,-4a,5,6,7,8,8a-octahydro-endo-exo-1,4:5,8-dimethanonaphthalene.

2. For a discussion of residues in man, see U.S. Department of Health, Education and Welfare, *Report of the Secretary's Commission on Pesticides and Their Relation to Environmental Health*, Government Printing Office, Washington, D.C. (1969).

3. Some of the most recent studies establishing the extent of contamination are: Risebrough, R. W. [4]; Risebrough, R. W., Menzel, D. B., Martin, D. J., and Alcott, H. S. [5]; Wurster, C. F., Jr. and Wingate, D. B. [6]; Tabor, E. C. [7].

4. Risebrough, R. W.: in M. W. Miller and G. G. Berg (eds.), *Chemical Fallout*, p. 5, Thomas, Springfield, Ill. (1969).

5. Risebrough, R. W., Menzel, D. B., Martin, D. J., and Olcott, H. S.: *Nature* **216**, 589 (1967).

6. Wurster, C. F., Jr. and Wingate, D. B.: *Science* **159**, 979 (1968).

7. Tabor, E. C.: *Trans. N.Y. Acad. Sci.* **28**, 569 (1966).

8. Antommaria, P., Corn, M., and DeMaio, L.: *Science* **150**, 1476 (1965); Abbott, D. C., Harrison, R. B., Tatton, J. O., and Thompson, J.: *Nature* **211**, 259 (1966).

9. Risebrough, R. W., Huggett, R. J., Griffin, J. J., and Goldberg, E. D.: *Science* **159**, 1233 (1968).

10. Peterle, T. J.: *Nature* **224**, 620 (1969).

11. Moore, N. C. W. (Ed.): *J. Appl. Ecol.* **3** (Suppl.), (1966).

12. Woodwell, G. M.: *Sci. Amer.* **216**, 24 (March 1967).

13. Frost, J.: *Environment* **11**, 14 (1969).

14. Massachusetts Institute of Technology, *Man's Impact on the Global Environment: Report of the Study of Critical Environmental Problems*, pp. 126–131, MIT Press, Cambridge, Mass. (1970).

15. In the United States the Secretary of Health, Education, and Welfare, the Secretary of Interior, the Secretary of Agriculture, and the administrator of the Environmental Protection Agency have moved recently toward removing DDT from most uses by 1972. These moves are being challenged, and the question remains unresolved. Certain other countries, including Canada and England, have restricted use of persistent pesticides. There is no basis for expecting an international ban and little basis for expecting restraint in the use of DDT outside of these countries.

16. Woodwell, G. M.: *BioScience* **19**, 884 (1969).

17. Commoner, B.: *Science and Survival* (Viking, New York, 1963).

18. Bowman, M. C., Acree, F., Jr., and Corbett, M. K.: *Agr. Food Chem.* **8**, 406 (1960).

19. West, T. F. and Campbell, G. A.: *DDT and Newer Persistent Insecticides*, Chemical Publishing Company, New York (1952).

20. There is no completely satisfactory way of appraising the persistence of DDT in nature; residues have different degrees of persistence in different places. The concept of 'half-life' implies a systematic degradation that may not occur universally. Herman, S. G., Garrett, R. L., and Rudd, R. L. [in Miller, M. W. and Berg, G. G. (eds.) *Chemical Fallout*, Thomas, Springfield, Ill., 1969, p. 24] show that the population of western grebes of Clear Lake, California, included individuals having substantially the same range of concentrations of DDT residues over a 10-yr period despite the cessation of spraying; mean values dropped by about one-half in that period. Residues are known to remain for many years in soil, especially in organic soils. See the review by Edwards, C.A. [21]; see also Woodwell, G. M. and Martin, F. T. [22], Dimond, J. [23] and Nash, R. G. and Woolson, E. A. [24]. These observations all suggest that residues persist with mean lives of many years. The assumption of a 10-yr half-life for residues within the biosphere as a whole has long appeared reasonable to one of us (G.M.W.) [16]; the assumption may be in error in that residues tend to be stored or cycled in places where they are not degraded chemically and may persist longer.

21. Edwards, C. A.: *Residue Rev.* **13**, 83 (1966).

22. Woodwell, G. M. and Martin, F. T.: *Science* **145**, 481 (1964).

23. Dimond, J.: *Maine Agr. Exp. Sta. Misc. Rep. 125* (1969).

24. Nash, R. G. and Woolson, E. A.: *Science* **157**, 924 (1967).

25. U.S. Tariff Commission, *Synthetic Organic Chemicals, U.S. Production and Sales, 1969*, Government Printing Office, Washington, D.C. (1970); *U.S. Dep. Agr. Econ. Rep. No. 158* (April 1969).

26. Carson, R.: recognized in 1963 the significance of aerial transport of DDT residues in contaminating the oceans. She cited observations in Maine and elsewhere indicating that only half the spray emitted from aircraft lands on the ground, the rest being dispersed in the atmosphere. Recent measurements of DDT attached to particulate matter in the atmosphere confirm the earlier conclusions (Carson, R., statement to the Subcommittee on Reorganization and International Organizations of the Committee on Government Operations, U.S. Senate, May–June 1963, part 1, p. 207); Woodwell, G. M.: *Forest Sci.* **7**, 194 (1961); Risebrough, R. W., Huggett, R. J., Griffin, J. J., and Goldberg, E. D., [9]; Cope, O. B.: *Trans. Amer. Fish. Soc.* **90**, 239 (1961).

27. Most measurements have been in agricultural soils where there was a real question about hazards after long use. Published data are, therefore, heavily skewed toward high values and any tabulation such as that of Table I may be misleading. The data of Table I are representative of published reports: Cole, H. *et al.* [28]; Lichtenstein, E. P. [29]; Seal, W. L. *et al.* [30]; Taschenberg, E. F. *et al.* [31]; Terriere, L. C. *et al.* [32]; Trautman, W. L. *et al.* [33]; Ware, G. W. *et al.* [34]; U.S. Department of Agriculture [35]. The maximum in this tabulation was in an orchard in Indiana that contained DDT residues totaling 22.0 grams per square meter. Almost all agricultural soils contain detectable DDT residues, most in excess of 0.056 g m^{-2}. Averages of several samples from a region are rarely less than 0.224 g m^{-2}, according to this tabulation. Nonetheless, Table I contains few data from crops that are rarely sprayed, such as many grains and fodder crops. Any average calculated for agricultural soils must account for the extensive acreages devoted to these crops as well. The average contamination of agricultural land is probably greater than the mean reported for 31 randomly sampled agricultural and forest soils in Wisconsin and eight states west of the Mississippi River (Table I). In that tabulation, 22 of the soils contained less than 0.0022 g of DDT m^{-2}, but others had high enough concentrations to produce an average concentration of 0.148 g m^{-2}. It seems very unlikely that the average amount of DDT in agricultural soils, including all agricultural soils, would approach the value of 0.99 g m^{-2} reported for soybeans in 1968 (Table I). A reasonable estimate of the average value for agricultural soils where DDT is used would seem to be in the range of 0.140 to 0.56 g m^{-2}. The average would be skewed toward the lower end of the range by the inclusion of crops that are sprayed irregularly. We have assumed for our calculations an average contamination of agricultural soils of 0.168 g m^{-2}.

28. Cole, H., Barry, D., Frear, D. E. H., and Bradford, A.: *Environ. Contam. Toxicol.* **2**, 127 (1967).

29. Lichtenstein, E. P.: *J. Econ. Entomol.* **50**, 545 (1957).

30. Seal, W. L., Dawsey, L. H., and Cavin, G. E.: *Pesticide Monit. J.* **1**, 22 (1967).

31. Taschenberg, E. F., Mark, G. L., and Gambrell, F. L.: *Agr. Food Chem.* **9**, 207 (1961).

32. Terriere, L. C., Kiigemagi, U., Zwick, R. W., and Westigard, P. H.: in R. F. Gould (ed.), *Organic Pesticides in the Environment*, p. 263, American Chemical Society, Washington, D.C. (1966).

33. Trautman, W. L., Chesters, G., and Pionke, H. B.: *Pesticide Monit. J.* **2**, 93 (1968).

34. Ware, G. W., Estensen, B. J., and Cahill, W. P.: *ibid.*, p. 129.

35. U.S. Department of Agriculture, Plant Pest Control Division, Agricultural Research Service, *ibid.*, p. 58.

36. Most of the nonagricultural soils for which data on DDT residues are available are also soils from areas that have received repeated applications of DDT. The highest values range from several tenths of a gram per square meter for a marsh along the eastern coast [Woodwell, G. M., Wurster, C. F., Jr. and Isaacson, P. A.: *Science* **156**, 821 (1967)] and for other organic soils that had been heavily treated, to zero, or at least to values below the limits of detection. Unsprayed forested areas in Pennsylvania (Table I) contained in 1967 4.5×10^{-4} g m^{-2}. Soils in Maine forests that had never been sprayed contained ten times that quantity (Table I). Organic soils and sediments contain more residues, but usually not more than a few tenths of a pound per acre. Any world average estimated from such data is most tenuous. We estimate that nonagricultural soils contain a minimum of 1.1×10^{-4} g m^{-2}. A reasonable estimate of the mean concentration of DDT in the temperate zone of the Northern Hemisphere appears to be 4.5×10^{-4} g m^{-2} as reported for the soils of Pennsylvania forests (Table I).

37. *Hammond Citation World Atlas*, p. 192. Hammond, Maplewood, N.J. (1966).

38. Edwards, C. A.: *Persistent Pesticides in the Environment*, Chemical Publishing Company, New York (1970).

39. Nazarov, G. S.: *Tr. Saratov. Zootekh. Vet. Inst.* **7**, 319 (1958).

40. Spencer, W. F. and Cliath, M. M.: *Environ. Sci. Technol.* **3**, 670 (1969).

41. 'Net primary production' usually refers to the net amount of dry matter produced per unit of land area: Woodwell, G. M. and Whittaker, R. H.: *Amer. Zool.* **8**, 19 (1968); Odum, E. P.: *Fundamentals of Ecology*, 3rd ed., Saunders, Philadelphia, (1971); Woodwell, G. M.: *Sci. Amer.* **223**, 64 (September 1970).

42. Long, L. H. (Ed.): *World Almanac* p. 273, Newspaper Enterprise Association, Inc., New York (1967); Piper, A. M.: *U.S. Geol. Surv. Water Supply Pap. No. 1797* (1965).

43. Weaver, L., Gunnerson, C. G., Breidenbach, A. W., and Lichtenberg, J. J.: *Public Health Rep.* **80**, 481 (1965).

44. Standen, A. (Ed.): *Encyclopedia of Chemical Technology*, 2nd ed., vol. 11, p. 691, Interscience, New York (1966).

45. Tarrant, K. R. and Tatton, J. O. G.: *Nature* **219**, 725 (1968); Wheatley, G. A. and Hardman, J. A.: *ibid.* **207**, 486 (1965).

46. George, J. L. and Frear, D. F. H.: *J. Appl. Ecol.* **3** (Suppl.), 155 (1966).

47. Craig, H.: *Tellus* **9**, 1 (1957).

48. Revelle, R. and Suess, H. E.: *ibid.*, p. 18.

49. Garrett, W. D.: *Deep-Sea Res.* **14**, 221 (1967); Seba, D. B. and Corcoran, E. F.: *Pesticide Monit. J.* **3**, 190 (1969).

50. Lipne, H. and Kearns, C. W.: *J. Biol. Chem.* **234**, 2129 (1959); Goldberg, L.: in Kuiper, G. P. (ed.), *The Earth as a Planet*, p. 434; Univ. of Chicago Press, Chicago (1954), Koller, L. R.: *Ultraviolet Radiation*, 2nd ed., Wiley, New York (1965); Mosier, A. R., Guenzi, W. D., and Miller, L. L.: *Science* **164**, 1083 (1969); Bhandari, N. and Lal, Rama, D.: *Tellus* **18**, 391 (1966).

51. Wooster, W.: quoted by Craig, H. [47].

52. Odum, W. E., Woodwell, G. M., and Wurster, C. F.: *Science* **164**, 576 (1969).

53. Woodwell, G. M.: *ibid.* **168**, 429 (1970).

54. Harrison, H. L., Loucks, O. L., Mitchell, J. W., Parkhurst, D. F., Tracy, C. R., Watts, D. G., and Yannacone, Jr., V. J.: *ibid.* **170**, 503 (1970).

55. Keenleyside, M. H. A.: *Can. Fish. Cult.* **24**, 17 (1959); *J. Fish. Res. Board Can.* **24**, 807 (1967); Ide, F. P.: *ibid.*, p. 769; Anderson, R. B. and Euerhart, W. H.: *Trans. Amer. Fish. Soc.* **95**, 160 (1966); Anderson, R. B. and Fenderson, O.: *J. Fish. Res. Board Can.* **27**, 1 (1970); Anderson, R. B.: *Sebago's Bright Future* (Maine Department of Inland Fisheries and Game, Augusta, 1966); DeRoche, S.: unpublished data on fish size in Sebago Lake, 1957–1967, taken by the Maine Department of Inland Fisheries and Game; Rudd, R. L.: *Pesticides and the Living Landscape*, Univ. of Wisconsin Press, Madison (1964).

56. Counts of nesting pairs and young by Puleston, D. (Brookhaven Lecture Series No. 104, *Brookhaven Nat. Lab. Publ.* 50309 (15 September 1971)) and his colleagues over more than 20 yr have documented the catastrophic decline of the osprey in the early 1960's, and the slight recovery of reproductive success in recent years. Whether the small residual population will recover remains doubtful, however.

57. Ratcliffe, D. A.: *J. Appl. Ecol.* **7**, 67 (1970).

58. Sladen, W. J. L., Menzie, C. M., and Reichel, W. L.: *Nature* **210**, 670 (1966).

59. Stanley, C. W., Barney, J. E., II, Helton, M. R., and Yobs, A. R.: *Environ. Sci. Technol.* **5**, 430 (1971).

60. Whittaker, R. H.: *Communities and Ecosystems*, Collier-Macmillan, New York (1970).

61. Bowen, H. J. M.: *Trace Elements in Biochemistry*, Academic Press, New York (1966).

62. Research was carried out under the auspices of the U.S. Atomic Energy Commission.

BIOLOGICAL IMPLICATIONS OF
GLOBAL MARINE POLLUTION*

BOSTWICK H. KETCHUM

Woods Hole Oceanographic Institution, Woods Hole, Mass. 02543, U.S.A.

Abstract. The living resources of the sea provide a substantial part of the world population with an essential source of animal protein, and the marine environment is a valuable resource for recreational facilities. In some of our estuaries, these resources are endangered at present by marine pollution, which has already limited our harvest of sea food in polluted estuaries. The remaining areas of the sea suitable for marine life must be protected from additional pollution if we are to maintain and increase our harvest of protein from the sea and to retain the amenities of the marine waters for recreational purposes. Critical marine pollutants are identified, and various methods of assessing the impact of these pollutants on marine life are discussed. Although additional research is needed before we can make absolute recommendations for acceptable levels of pollutants in the marine environment, it is also emphasized that we already know enough to recommend limits which appear, on the basis of present available knowledge, to provide assurance of minimal risk of damage to the marine environment. Some examples of these acceptable limiting concentrations are presented.

1. Introduction

The marine environment is subjected to many conflicting demands and uses, some of which are not compatible with others. Historically, the estuaries and harbors were selected for early settlement because of the ease of marine transportation of goods and materials. Even today, our densest centers of population are located either on the sea or in locations where ready access to the sea is available. Today, more than 50% of the population of the United States live in counties bordering the seacoast or the Great Lakes, and the concentration of population in these areas will probably increase in the future. Similar concentrations of coastal populations are found throughout the world. Concomitant with the growth of these coastal populations, and for similar reasons, has been the concentration of industries along our seacoast. Thus, the marine environment has been subjected for generations to intensive population pressures and the demands of our ever-increasing technology.

These combined pressures have led to activities which have greatly modified the marine environment in many places – although some parts of our seashore are still relatively unaffected by man's activities. Harbor development has required extensive dredging and the construction of breakwaters and other structures to protect the shipping. The demand for coastal land and docking space has led to the filling of large areas of salt marsh and wetlands. The waste products of dense populations and of industries have been discharged into these waters with little understanding or appreciation of the possible effects. All of these activities have had impacts on the environment and have produced stresses on the marine ecosystem. So long as the receiving

* Contribution No. 2974 from the Woods Hole Oceanographic Institution.

S. Fred Singer (ed), The Changing Global Environment, 311–328. All rights reserved.
Copyright © 1975 by D. Reidel Publishing Company, Dordrecht–Holland.

TABLE I

Photosynthetic productivity (carbon fixation) and fish production for various oceanic provinces [32, 49]

Province	Area of ocean (%)[a]	Productivity 10^9 tons C yr^{-1}	Trophic levels	Ecological efficiency (%)	Fish production tons (fresh weight)
Open ocean	90	16.3	5	10	16×10^5
Coastal zone[b]	9.9	3.6	3	15	12×10^7
Upwelling areas	0.1	0.1	1.5	20	12×10^7
Total	100	20.0	—	—	24×10^7

[a] Total area $= 365.6 \times 10^6$ km^2.
[b] Includes offshore areas of high productivity.

capacity of the body of water is not exceeded, the ecosystem is able to recover from these additional stresses without permanent damage. Today, however, in the most densely populated parts of this country and the entire world, the stresses resulting from man's activity have increased to the point where the marine ecosystem is not able to absorb them without damage, and we now realize that man's mode of action must be drastically changed if we are to retain a marine ecosystem of value to man.

The necessity of maintaining the marine ecosystem in a desirable condition is emphasized by two facts: (1) the animal protein derived from marine products is of great importance in maintaining adequate nutrition for many of the densely populated, developing countries of the world and, (2) the increasing demand for adequate recreational opportunities in aquatic environments supports one of the fastest growing industries in the United States' economy and that of other developed nations. Water of high quality is essential both for recreation and for the survival of the majority of the marine organisms of economic importance.

The annual world harvest of sea food is now nearly 60 million metric tons [11] and this supplies more than half of the animal protein for about half of the population of the world [7, 19]. How much can this marine harvest grow? Numerous estimates of the photosynthetic production of organic matter in the oceans have been made. Table II compiled by Sisler* shows wide variations in different localities and at various times of year. This organic matter is produced by microscopic plants, the phytoplankton, which are present in such low concentrations that direct harvest is impractical. Man uses the larger forms, and the amount available depends upon the number of steps (trophic levels) in the food chains and the efficiency of utilization of the food. With a three-step food chain and a 10% efficiency at each step a thousand pounds of phytoplankton are required to produce one pound of fish for the table.

On the basis of productivity, trophic levels and efficiencies, Schaefer [49] and Ryther [32] have estimated the potential food production of the sea, with the results shown in Table I. We harvest only the top carnivores, such as tuna and swordfish, from the open ocean, and it is estimated that five trophic levels are required in their

* This volume, p. 62.

food chain. The efficiency is low since the population density is sparse and much energy must be expended in seeking and capturing prey. As a result, fish production in the open sea is less than 1 % of the total, even though this province constitutes 90 % of the area of the world oceans. The coastal waters, including offshore areas of relatively high productivity, are estimated to produce approximately half of the total harvest assuming three trophic levels and a somewhat higher efficiency. The upwelling areas, such as those off the coast of Peru, are unusually rich since the essential plant nutrients from the deep water are brought up into the surface layers where photosynthetic production is possible. Ryther [32] estimates the rate of photosynthetic production in upwelling areas to be six times greater than that found in the open sea. Furthermore, the small, herbivorous anchovy are large enough to be directly utilizable by man and these feed on trophic levels very near the primary producers. Consequently, the food chain is short and, because of the high density of the prey species in the area, the efficiency of conversion is comparatively high. These fertile areas, although they only constitute 0.1 % of the surface areas of the sea produce nearly half of the world harvest of sea food.

The estimates of fish production in Table I are those of Ryther [32]; Schaefer [49] estimated somewhat higher values than these using a similar approach but different assumptions concerning food chain lengths and efficiencies. Schaefer and Alverson [36] estimate that the rate of harvest could increase four fold as a result of more effective utilization of all of the fish caught and the discovery of new fish stocks not now being exploited. Ryther [32] is less optimistic and estimates that about half of the 240 million tons of fish potential in the sea could be harvested by man on a sustainable yield basis. Other predators, such as the guano birds off the coast of Peru, and other carnivorous marine animals would account for the balance of the annual production and, of course, enough of the crop must be maintained if the estimated annual production is to continue indefinitely. Neither of these estimates takes account of potential damage to the fisheries by increasing pollution impacts, though the deleterious effects of pollution are already obvious in most of the world estuaries.

The value of the marine environment for recreation is also high. It has been estimated by Winslow and Bigler [41] that, in the United States alone, about 112 million people spent about $14 billion dollars seeking recreation in the coastal zone. Using the multiplier suggested by Goldberg [13, p. 271] the world value for recreation may be about triple this amount. Clearly, maintaining the quality of marine waters is economically essential, both as a source for food and for recreation.

2. Critical Marine Pollutants*

To give perspective to the problems of global marine pollution, a conceptual framework for water quality evaluation is presented in Figure 1. The imports to the marine ecosystem and the exports from it are important to an understanding of the entire

---

* Some pollutants, domestic sewage and waste heat for example, are not discussed. This is not because they are unimportant but because their source and impact is local, rather than global.

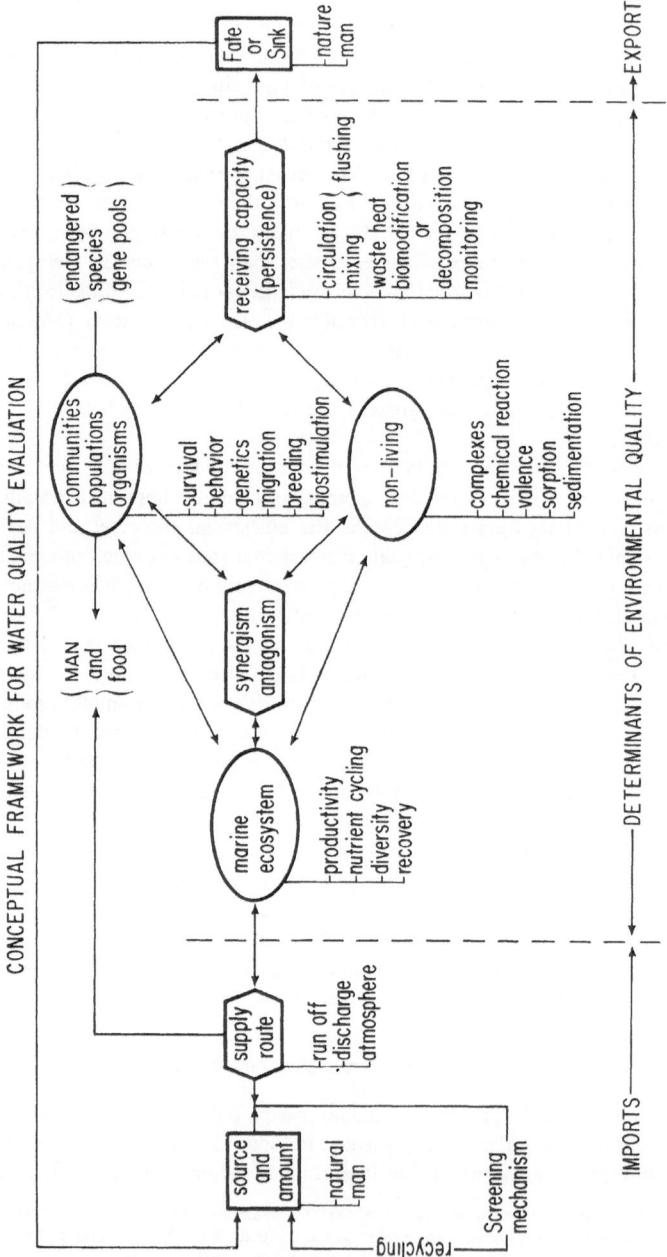

CONCEPTUAL FRAMEWORK FOR WATER QUALITY EVALUATION

Fig. 1.

problem [13]. A toxic chemical, which is produced in large amounts, reaches the sea and persists for a long period of time in the marine environment, will have a vastly different impact than a chemical of equivalent toxicity which is produced in relatively small amounts and/or decays rapidly in the marine system. For the former type of pollutant, a screening mechanism should be available which could confine it to the source and not permit it to reach the environment, after which man loses most or all of his ability to control the impact of the pollutant.

Examples of the persistent pollutants which have been known to reach the sea in substantial amounts are the toxic elements mercury, cadmium, nickel, zinc, lead, manganese, copper, chromium, selenium, silver, and arsenic. These elements are listed in the approximate order of their deleterious effects as a function of their toxicity and the rate of mobilization by man and by natural weathering as shown in Table II. Data from Bertine and Goldberg [4] on the rate of mobilization by combustion of fossil fuels and by the present-day transport of the element to the sea in river flow are combined with data from Waldichuk [26, 40] on the comparative toxicities. Combined, these data produce an index of how critical the element may be as a marine pollutant. The supply from the combustion of fossil fuels provides a significant part of these materials carried by the atmosphere to the sea. The river transport is based upon analyses of dissolved material in river waters and known rates of flow, and reflects natural weathering, agricultural runoff and pollution sources. There are unquestionably other sources of supply for these elements, but the data used, although uncertain on a global scale, are adequate to rate the relative hazards.

Actually, this index gives the volume of water which would receive an annual increment of the element equal to the concentration shown in column D (toxicity) at the

TABLE II

Toxic elements of critical importance in marine pollution based on potential supply and toxicity

Element	Rate of mobilization (10^9 g yr^{-1})[a]			Toxicity[b] D mg l^{-1}	Relative critical index (10^{12} l yr^{-1})	
	A(man) Fossil fuels	B(natural) River flow	C total		A/D	C/D
Mercury	1.6	2.5	4.1	10^{-4}	16000	41000
Cadmium	0.35[c]	0.5	0.85	2×10^{-4}	1750	4250
Nickel	3.7	160	164	2×10^{-3}	1350	82000
Zinc	7	720	727	2×10^{-2}	350	36350
Lead	3.6	110	113.6	1×10^{-2}	360	11360
Manganese	7.0	250	257	2×10^{-2}	350	12850
Copper	2.1	250	252.1	1×10^{-2}	210	25210
Chromium	1.5	200	201.5	1×10^{-2}	150	20150
Selenium	0.45	7.2	7.7	5×10^{-3}	90	1540
Silver	0.07	11	11.1	1×10^{-3}	70	11100
Arsenic	0.7	72	72.7	1×10^{-2}	70	7270

[a] After Bertine and Goldberg [4].
[b] Water Quality Criteria: Concentration considered to pose minimal risk of deleterious effect. After Waldichuk [40], NAS [26].
[c] Data from CEQ [50].

given rates of mobilization. Since the total water volume of the oceans is about 1.4×10^{21} l, the time to raise the concentration of the whole ocean by this amount could also be calculated, but this is really meaningless. The oceans are not uniformly mixed, and the concentration at the locality where the pollutant is introduced will invariably increase more rapidly than the average for the whole ocean.

A wide variety of synthetic organic chemicals are also reaching the environment, particularly the chlorinated hydrocarbons such as DDT and its decomposition products, and polychlorinated biphenyls (PCB's). These are not readily biodegradable, and the ocean is the ultimate sink for such compounds. Woodwell et al. [42] have modeled the circulation of DDT in the biosphere and they conclude that the largest reservoir for DDT is in the atmosphere, but also that the amount not decomposed by ultraviolet rays in the troposphere will ultimately be added to the surface of the sea. If production of DDT stops in 1974 the model predicts maximum concentrations in the mixed layer of the sea (upper 100 m) in 1971, after which it would decrease to 10 % of the maximum by 1993. If production were to increase, however, the concentrations in both the sea and the atmosphere would also increase.

Harvey et al. [14] found substantial concentrations of DDT and its breakdown product, DDE (up to 100 μg kg^{-1} wet weight in a shark) and even higher levels of PCB's (up to 1056 μg kg^{-1} wet weight in a dolphin) in a variety of organisms collected from the open sea many miles from land, confirming the probability of atmospheric transport. As expected, these compounds are concentrated in the lipid pool of the organisms with a maximum concentration of 3300 for DDT and DDE and of 21 100 μg kg^{-1} lipid for PCB's. None of the concentrations observed by Harvey et al. [14] were as great as those assumed by Woodwell et al. [42] for oceanic fish or plankton to estimate the accumulation in the biota (assuming dry weight to be 25 % of wet weight). The lower accumulations in marine organisms could be caused by a shorter atmospheric half life of DDT, by faster degradation in the marine environment or by greater accumulation in sediments than the estimates used by Woodwell et al. [42] in their computations.

A variety of synthetic organic chemicals, including other pesticides, detergents and pharmaceuticals are also undoubtedly reaching the marine environment with impacts which are virtually unknown. The detrimental effect of DDT on bird breeding potential is well documented and some experiments have been done on a few forms of marine life, but the information is still inadequate for a complete evaluation of the impact on the marine biota of DDT and even less adequate for the other synthetic organic compounds [24, 26, 35].

Petroleum, including crude oil, refined products and petrochemicals are now polluting the sea in large amounts. Revelle et al. [29] estimated the total direct oil pollution of the oceans to be 2.2 million tons annually. The sources were accidental spills, tanker operations, other ship operations, offshore production, refinery operations and industrial and automotive wastes. Oil slicks and tar balls have been observed on the high seas, and the abundance of tar balls is now greater than the normal sargassum weed in the open Atlantic [45, 47]. Although accidental oil spills, such as the grounding of the *Torrey Canyon* or the Santa Barbara oil well blowout, are spectacular events and attract the most public attention, they actually contribute less than 15% of the total amount of oil entering the marine environment annually. Revelle *et al.* [29] estimated that most of these sources of oil pollution can be expected to increase in proportion to the total world production of oil unless they are more adequately controlled.

Numerous studies of toxicity and effects of oil pollution have been made, but more careful studies of selected fractions of this complex mixture of hydrocarbons are needed. It is apparent that these hydrocarbons are degraded in sea water, but virtually nothing is known about the rate of turnover of this material in the marine environment. Our technology is based upon an expanding energy use which will require additional petroleum supplies including those from submarine reservoirs and large amounts transported in tankers from distant oil fields. If the rate of loss in our utilization and transportation of oil cannot be radically improved by application of adequate controls wherever possible, the amount of petroleum hydrocarbons entering the sea will increase.

Because of the increase in oil tanker traffic and of ships burning fuel oil and the resultant pollution of the high seas by oil, this has become an international problem. The Intergovernmental Maritime Consultative Organization (IMCO), a specialized agency of the United Nations, convened an international conference in London in 1954 which drew up the International Convention for the Prevention of Pollution of the Sea by Oil. This came into force in July, 1958, and was subsequently amended by an IMCO-convened conference in 1962. Further resolutions provide for the prohibition of deliberate oil discharge from ships at sea and for the establishment of an international compensation fund for oil pollution damage [46].

Solid waste disposal has become one of the most urgent and difficult problems in crowded urban centers. The types and amounts of waste materials dumped at sea in the coastal waters of the United States in 1968 is presented in Table III [44]. Nearly 50 million tons of waste material was dumped in United States' coastal waters, most of which was dredging spoils resulting from channel and harbor development. The Council estimated that 34% of these dredging spoils could be considered polluted.

TABLE III

Ocean dumping: types and amounts, 1968 (in tons)

Waste type	Atlantic	Gulf	Pacific	Total	Percent of total
Dredge spoils	15808000	15300000	7320000	38428000	80
Industrial wastes	3013200	696000	981300	4690500	10
Sewage sludge	4477000	0	0	4477000	9
Construction and demolition debris	574000	0	0	574000	~1
Solid waste	0	0	26000	26000	<1
Explosives	15200	0	0	15200	<1
Total	28887400	15966000	8327300	48210700	100

Pearce [48] has presented data to show that both the polluted dredging spoils and the sewage sludge from waste treatment plants which have been dumped in the New York Bight have caused damage to the bottom dwelling populations in the area. In the United States, on October 23, 1972, the Marine Protection, Research and Sanctuaries' Act came into effect (PL 92–532). This law regulates the transportation and dumping of materials into the oceans, coastal zones and other waters. A permit system is established to be administered by the Army Corps of Engineers for dredging and filling and by the Environmental Protection Agency for all other purposes.

Ocean dumping is also a subject of international concern. An intergovernmental conference convened by the United Nations was held in London, 30 October to 10 November, 1972. A Convention on the Dumping of Wastes at Sea was adopted and will be open for signature from 29 December, 1972, until 1 December, 1973. It will come into force when it has been ratified by 15 nations. The Convention prohibits the dumping of some materials; requires a special permit for the dumping of other identified substances; and a general permit for all other substances. In brief, the prohibited substances include organohalogen compounds, mercury and cadmium and compounds of these metals, persistent plastics and other persistent synthetic materials, crude oil and other petrochemicals, high level radioactive wastes and materials produced for biological and chemical warfare. The prohibition does not include substances which are rapidly rendered harmless by physical, chemical or biological processes in the sea provided they do not make edible marine organisms unpalatable or endanger human health. The list of materials which require special permits include arsenic, lead, copper, zinc and their compounds, organosilicon compounds, cyanides, fluorides and those pesticides not covered in the previous list. In addition to these, the disposal of industrial wastes containing large quantities of acids and alkalis should also be screened for berylium, chromium, nickel, vanadium and their compounds.

These recent international developments on the disposal of petroleum and the dumping of materials at sea offers promise of ultimate control and preservation of the marine environment. A number of scientific studies are needed for each of the pollutants listed above in order to evaluate adequately the determinants of environmental quality shown in Figure 1.

3. Biological Effects of Pollution

A marine pollutant may have subtle indirect effects on the ecosystem, which may be the result of a more direct effect on a certain species or group of species essential to the ecosystem (Figure 1). The indirect effects may not be predictable from studies of individual parts of the ecosystem, and a change in the characteristics of the ecosystem may be an early warning indicator of deleterious effects. Laboratory micro-ecosystems, and experiments in ponds or tanks which include several components of the ecosystem can be useful in extending bioassays to longer time scales. Knowledge of ecosystem processes is necessary as a foundation for such models, which are valuable in defining critical studies or observations [28].

3.1. ECOSYSTEM ANALYSIS

The marine ecosystem includes all of the biological, chemical, geological and physical components of the environment and their highly complex interactions. The impact of water quality on the total ecosystem may be greater than the sum of the impacts on individual parts of the system. For example, eliminating or decreasing the abundance of an organism or life stage which is essential as food for another organism might disturb the entire pattern of energy flow throughout the system. It is essential to understand the interrelationships among organisms, and between organisms and their environment, in order to evaluate these subtle and secondary effects.

Studies of ecosystems must also include the imports to and exports from it. In the marine environment, imports and exports continually occur from coastal runoff, tidal action, oceanic currents, atmospheric fallout and from exchanges with adjacent water bodies, with the benthos or the atmosphere. Each local environment is somewhat different from all others, but the species inhabiting any given environment have evolved over long periods of time and each individual species in a community plays its own role. Any additional stress, whether natural or man-made, will tend to eliminate some species, leaving only the more resistant and tolerant forms to survive. Stresses which are transient may permit replenishment of the species by recruitment from adjacent unaffected areas. In such a case the impact may be reversible and the ecosystem can reestablish itself. A change which is chronic and permanent, such as the excessive pollution of some of our harbors, will never permit the recovery of the original ecosystem until the source of unusual stress is removed. Although the effects of water quality on the marine ecosystem are difficult to establish and evaluate, methods are now available which make the interpretation of ecosystem studies much more meaningful than it was a few years ago.

In order to assess the impact of any new pollutant on a body of water, it is necessary to acquire information on the conditions existing before, during and after the addition. Information on the physical and chemical characteristics of the system, on the distribution and abundance of species, and on the normal variations of these characteristics over at least an annual cycle will be necessary. Evaluations of productivity, nutrient cycling, diversity and recovery potential should be emphasized.

3.1.1. *Productivity and Energy Flow*

The productivity of marine ecosystems varies from place to place and from time to time depending upon a variety of environmental factors including the availability of essential plant nutrients, the clarity of the water, and consequently the depth of penetration of sunlight of adequate intensity to permit photosynthesis, the stability of the water column, and the import and export of various materials into the system [31]. Methods for the study of primary production or photosynthesis, particularly of the planktonic populations, depend either upon the use of Carbon 14 and its assimilation by the plant in the production of new organic material [39], or upon the evolution of oxygen in the process of photosynthesis [30]. Similar methods can be used in evaluating the productivity of the sedentary plant populations in shallow waters. It should be emphasized, however, that a high rate of productivity is not necessarily an indication that the ecosystem is healthy. The statement that Lake Erie is 'dead' is biologically absurd since it is the most productive of the Great Lakes. It is not, however, necessarily a healthy ecosystem, since the species produced in Lake Erie today are not the ones most desirable for man's use [3, 16].

3.1.2. *Nutrient Cycling*
The marine ecosystem provides a natural mechanism for the recycling of those elements which are essential for photosynthesis and growth of both plants and animals. Unlike energy, which follows a one-way path from its fixation in photosynthesis to its ultimate dissipation as heat regardless of the number of transfers through the food-web, nutrients are returned to the system as a result of decomposition and can be used over and over again [18]. The productivity of the marine ecosystem is frequently limited by the lack of nutrients which are commonly reduced to very low concentrations in the illuminated part of the water column. Various types of pollution, such as domestic or food processing wastes, enrich the waters with nutrients, and when present in excessive amounts can result in the replacement of the normal phytoplankton population with noxious species and change the entire biotic structure of the marine ecosystem in the process known as 'cultural eutrophication' [23].

We need to develop more precise methods to evaluate the mechanisms and efficiency of nutrient recycling under a variety of marine conditions. In the shallow, inshore waters the exchanges of nutrients with the bottom sediments is an important part of this process, though in the waters with depths greater than 100 or 200 m, most of the organic material formed in the illuminated surface layers decomposes before reaching the bottom [18, 22]).

The efficiency of the nutrient cycling system and the concentrations of nutrients which produce objectionable growths in the water require further and intensive study under a wide variety of marine environmental conditions.

3.1.3. *Diversity*
A wide variety of indices have been developed to evaluate the diversity of ecosystems [33, 34, 37]. These all relate the number of different species with the total abundance of organisms. In general the ecosystems with high diversity are also the most stable, in part because there are large numbers of pathways for the transfer of food and energy from one part of the system to another. Even natural and undisturbed ecosystems have, however, different degrees of diversity depending upon the natural stresses on the system. High diversity is generally associated with minimum stress such as is found in the tropics, where the annual cycle of change is small, or in the deep sea where it is continuously cold and dark. Low diversity is generally found in strongly stressed ecosystems such as the Arctic, where seasonal changes are severe, or in estuaries where the ebb and flow of the tide and fluctuating river flow produce severe stresses on time scales ranging from hourly to seasonally. Consequently,

before diversity can be used as an indicator of change, the natural diversity of the environment being studied must be known.

Pollution, or any other man-made modification of the environment, can be considered as an additional stress imposed on the ecosystem and will tend to eliminate some species, leaving only the more resistant and tolerant forms to survive. The diversity will tend to be decreased by this additional stress, even under situations where the total productivity is increased. When the natural diversity of the system is known, the disappearance of sensitive species and the consequent decrease in diversity may be the first indication of impending deterioration of the environment. We need to study the diversity of a wide variety of marine ecosystems and learn to interpret changes in diversity as indicators of overall water quality.

3.1.4. *Recovery of Ecosystems*

Very little is known about the ability of ecosystems to recover if the stresses imposed by man are modified or removed. Some man-made modifications are irreversible. A salt marsh which is filled, paved over and built upon, or which is dredged for a navigation channel, cannot be expected to return to its original condition. There is clear evidence, however, that the removal of sources of pollution, either by improved treatment processes or by transferring the pollutant to a water mass of greater receiving capacity can permit recovery [9, 10]. This is especially true if the original species can be recruited from nearby unmodified areas. Much remains to be learned about the potential for deliberate restocking of a cleaned-up water body with imported species. For anadromous species, such as salmon, which return to their birthplace to spawn, intentional restocking may be the only way to re-establish the population in the ecosystem.

When pollution has continued for many years, the sediments may form a large reservoir and may take years to recover. However, for pollutants which remain in solution, and are not changed by biological, chemical or geological processes, the recovery time will be a function of the flushing time of the system (see Figure 1). The rates of exchange, which determine the flushing time (Ketchum, [17]), need both fundamental studies to improve our understanding of the mechanisms, and application to many local situations which have not yet been investigated.

3.2. EFFECTS ON COMMUNITIES, POPULATIONS AND ORGANISMS

Knowledge of and an ability to work with organisms, populations, and communities which make up the living part of the marine ecosystem is fundamental to the understanding of the system and the effects of potential pollutants on the system. The biological effects of pollution may be reflected by acute toxicity to individual organisms, or by chronic, sub-lethal effects on the survival of populations through modification of behavior, genetics, migration or breeding potential. Some pollutants may stimulate or accelerate biological processes, for example, fertilizers can increase the rate of photosynthesis, and increased temperatures can accelerate the rate of respiration. These effects can modify the functioning of the ecosystem, but cannot be evaluated directly without full evaluation of the impact on the system.

3.2.1. *Acute Toxicity*

The acute toxicity of a pollutant is generally evaluated by means of a bioassay test on a selected organism. These tests determine the concentration of a substance which will kill one half of a population of test organisms within a given period of time, usually 48 or 96 hr. The results are reported as the lethal concentration for 50% of the test organism (LC50), or median tolerance limit (TLm or TL50). The bioassay may be a static test, in which the organism is exposed to the test solution in a tank, or a flow-through system, in which the test solution is continuously renewed. Both may determine sub-lethal effects as well as mortality. Methods for routine bioassays are described in standard methods for the examination of water and waste water [1], and their application is evaluated by Sprague [38]. Clearly, the LC50 is not a 'safe' concentration and water quality criteria are set at a fraction (application factors of 0.1, 0.01 or less) of the LC50.

Because organisms, and life stages of organisms, may vary considerably in their resistance to a given toxic substance, the selection of the least tolerant life stage or organism is critical to the sensitivity and validity of the test. Research is needed to develop more effective bioassay methods for potential toxicants. The most sensitive marine organisms should be identified and cultivated on an egg-to-egg basis under laboratory conditions to provide genetic variability information on the test species and to form a basis for intercomparison of experiments and results.

Equipment and methods are needed for the collection, isolation, and production of all life stages of important aquatic organisms so that they may be available at all times in sufficient numbers for bioassay studies. The effects of organic and inorganic pollutants on marine organisms should be investigated with controlled variation of environmental conditions, including salinity, temperature, dissolved oxygen, and pressure.

An advantage of bioassays which duplicate existing conditions of the area being studied is that they measure the effect of the substance including the valence state and natural complexing with materials in sea water, and synergistic or antagonistic effects (cf. Figure 1). Thus, the total effect can be estimated without a clear understanding of all mechanisms involved. This does not, however, obviate the need for such an understanding because only with firm knowledge of the mechanisms involved can transfer of information to other systems and prediction of effects be achieved. Application factors relating acute toxicity data (bioassays) to long-term effects must be developed in order to define more precisely tolerable environmental levels.

3.2.2. *Chronic, Sub-Lethal Effects*

Many biological effects of pollution may not show up in the bioassay test for acute toxicity. This would be true if the effect were slow to develop or if the effect were to produce a general debility which might interfere with some of the normal life functions of the organism rather than killing it directly. Long-term exposures to sub-lethal concentrations may be necessary to produce the effect, and evaluation of this type of action is difficult in a laboratory analysis. Long-term exposures to sub-lethal concentrations may produce effects on migration, behavior, incidence of disease, life cycle, physiological processes, nutrition or food chains, or genetics which are not revealed in tests shorter than complete life cycles or numbers of generations. Unfortunately, little is known about these chronic, sub-lethal effects. Behavioral scientists should be encouraged to seek methods for evaluating the effect of pollutants on activities of marine organisms and to identify low levels of pollutants which may have undetectable short-term effects. Long-term studies on sensitive life stages of organisms should be undertaken to provide a basis for application factors.

3.2.2.1. *Migrations.* Sub-lethal concentrations may interfere with the normal migration patterns of organisms. The mechanisms used for orientation and navigation by migrating organisms, though extensively studied, are still not well understood [2, 15]. In some cases chemotaxis plays an important role. Salmon and many other anadromous fishes have been excluded from their home streams by pollution though it is not always known whether the reason is that a chemical cue has been masked or because the general polluted environment is offensive to the fish.

3.2.2.2. *Behavior.* Much of the day-to-day behavior of species may also be mediated by means of chemotactic responses [2]. The finding and capture of food, or the finding of a mate during the breeding season would be included in this category of activity. Again, any pollutant which interfered with the chemoreceptors of the organism would interfere with behavioral patterns which are essential to the survival of the population.

3.2.2.3. *Incidence of disease.* Long-term exposure to sub-lethal concentrations of pollutants may make an organism more susceptible to a disease. It is also possible that some pollutants which are organic in nature may provide an environment suitable for the development of disease-producing bacteria or viruses. In such cases, even though the pollutant is not directly toxic to the adult organisms it could still have a profound effect on the population of the species over a longer period of time.

3.2.2.4. *Life cycle.* The larval forms of many species of organisms are much more sensitive to pollution than are the adults which are commonly used in the bioassay. In many aquatic species millions of eggs are produced and fertilized, but only two of the larvae produced need to grow to maturity and breed in order to maintain the standing stock of the species. For these species the pre-adult mortality is enormous even under the best of natural conditions. Because of an additional stress on the developing organisms enough individuals might fail to survive to maintain the population. Interrupting any stage of the life cycle can be as disastrous for the population as would death of the adults because of acute toxicity.

3.2.2.5. *Physiological processes.* Interference with various physiological processes, without necessarily causing death in a bioassay test, may also interfere with the survival of a species. If photosynthesis of the phytoplankton is inhibited, algal growth will be decreased and the population may be grazed to extinction without being directly killed by the toxic. DDT has been shown by Wurster [43] to depress photosynthesis in planktonic algae, but only at concentrations greater than its solubility in water (about 1 ppb). Menzel *et al.* [21] demonstrated that different phytoplankton species showed various responses to DDT, dieldren and endrin ranging from no effect to toxicity at concentrations of 0.1 to 1 ppb.

Respiration might also be adversely affected as could various other enzymatic processes by sub-lethal concentrations of pollutants. The effect of DDT and its decomposition products on the shells of bird eggs is probably the result of interference with enzyme systems. Mercury is a general protoplasmic poison but has its most damaging effect on the nervous system. In humans this is known to be serious; how it affects fish and other aquatic organisms in sub-lethal concentrations is still unknown.

3.2.2.6. *Nutrition and food chains.* Pollutants may interfere with the nutrition of organisms by affecting the ability of an organism to find its prey, by interfering with digestion or assimilation of food, or by contaminating the prey species so that it is not accepted by the predator. On the other hand, if predator species are eliminated by pollution the prey species may have an improved chance of survival. An example of the latter effect was shown for the kelp resurgence after the oil spill in Tampico Bay, California (North, [27]). The oil killed the sea urchins which used young, newly developing kelp as food. When the urchins were killed, the kelp beds developed luxurious growth within a few months.

3.2.2.7. *Genetic effects.* Many pollutants produce genetic effects which can have long-range significance for the survival of species. Radioactive contamination can cause mutations directly by the action of radiation on the genetic material. Also oil and other organic pollutants may include both mutagenic and carcinogenic compounds [6]. From genetic studies in general it is known that a large majority of mutations are detrimental to the survival of the young, and many are lethal. Little is known about the intensity or frequency of genetic or developmental effects of pollutants, except for radioactive materials [8]. The doses required for serious damage greatly exceed the radioactivity present as pollution in the marine environment [25].

3.2.2.8. *Food value for human use.* Sub-lethal concentrations of pollutants can so taint sea food that it becomes useless as a source of food. Oil can be ingested by marine organisms, pass through the wall of the gut and accumulate in the lipid pool. Blumer [5, 6] states that oil in the tissues of shellfish has been shown to persist for several months after an oil spill in West Falmouth, Mass. The polluted area was closed for shellfishing for a period of 18 months. Sea food may be rendered unfit for human consumption because of the accumulation of pollutants. California mackerel and coho salmon from Lake Michigan were condemned because they contained more DDT than the permissible amount in human food (5 ppm). Likewise tunafish and swordfish were removed from the market because the mercury content of the flesh exceeded the allowable concentration (0.5 ppm). There was no evidence that these concentrations had any adverse effect on the fish, but their removal from the market has adversely affected the economics of the fisheries.

Sub-lethal effects on chemotactic responses, reproductive efficiency, survival, growth, vigor, behavior, and general activity should be studied. A debilitation index developed from this type of research could provide indications of incipient toxicity. Studies of metabolic rates under multifactorial combinations of various environmental components in the presence of pollutants are needed.

4. Discussion and Conclusions

This paper has stressed the biological impact of pollution and consequently several of the processes identified in Figure 1 have not been discussed or have been mentioned only briefly. Sea water is a complex solution, containing numerous elements at various concentrations and substantial amounts of naturally occurring organic materials. Any pollutant added to the sea can, thus, be expected to undergo a variety of chemical reactions including the formation of complexes or changes in the valence state of the element. Pollutants can be sorbed on living or dead particulate material and removed from the water column by sedimentation. Whether this is an undesirable or beneficial process will depend upon the characteristics of the pollutant and upon its effect on the organisms inhabiting the bottom.

The sea has a finite receiving capacity for any given type of pollutant. The receiving capacity will depend upon the rate of circulation and mixing of the water masses involved (flushing) and can be predicted with a reasonable degree of certainty for pollutants which are dissolved in the water column and are not modified by biological, chemical or geological processes [17, 26]. If the rates of change as a result of these other processes are known, the fate of the pollutant can also be predicted, but the model becomes progressively more complex as additional variables are considered. Locally, in many places, the receiving capacity of the marine system has already been exceeded. Ketchum [20] estimated, for example, that the Hudson River estuary is now receiving 5 to 10 times more domestic pollution than can be adequately handled within a forty-mile length of the estuary. Beyond these limits, dilution reduces the concentration to acceptable limits. The disposal of wastes to the marine environment has long been an accepted practice but it has been done without understanding or consideration of the possible consequences. Some pollutants, such as mercury, are naturally present in sea water at concentrations which have already produced deleterious effects. In other cases, particularly the chlorinated hydrocarbons, the material is not naturally produced and organisms capable of metabolizing it have not evolved, so that these compounds persist for long periods of time in the environment. In cases such as these, further additions as a result of man's activities should be prohibited. The concept of zero discharge rate for pollutants which can be assimilated by the marine ecosystem, however, is not reasonable, but it is essential to know the receiving capacity and, by further treatment or diversion, maintain the ambient concentrations within acceptable limits of the ecosystem.

Throughout this paper, the need for additional scientific investigations has been repeatedly emphasized. We should recognize, however, that we already know a great deal about the marine ecosystem and that reasonable models already exist and can be applied in many cases to evaluate discharge rates which will maintain acceptable limits of pollutant concentrations. We can evaluate the impact of such pollutants on the living marine resources, and can specify the treatment or approaches necessary to stay within acceptable limits of water quality [12, 26]. It is abundantly clear, however, that we can no longer continue to be indiscriminate in our actions and that drastic changes will have to be made in man's activities in order to maintain the marine environment in a suitable condition. It is not a question whether we can do it or can afford to do it; we must do it if we are to leave a marine environment of desirable quality for posterity.

References

1. APHA: *Standard Methods for Examination of Water and Wastewater*, American Public Health Association, American Water Works Association, and Water Pollution Control Federation: 13th Ed. (APHA Washington, D.C.), (1971).
2. Bardach, John E. and Todd, John H.: 'Chemical Communication in Fish', in J. W. Johnson, Jr., D. G. Moulton and Amos Turk (eds.), *Advances in Chemoreception, Volume 1, Communication by Chemical Signals*, Appleton-Century-Crofts, New York, Educational Division, Meredith Corporation, pp. 205–240 (1970).
3. Beeton, A. M.: 'Changes in the Environment and Biota of the Great Lakes', in *Eutrophication: Causes, Consequences, Correctives*, National Academy of Sciences, Washington, D.C., pp 150–187 (1969)
4. Bertine, K. K. and Goldberg, E. D.: 'Fossil Fuel Combustion and the Major Sedimentary Cycle', *Science* 173, 233–235 (1971).
5. Blumer, M.: 'Oil Pollution of the Ocean', in D. P. Hoult (ed.), *Oil on the Sea*, Plenum Press, New York, 5–13 (1969).
6. Blumer, M.: *Oil Contamination and the Living Resources of the Sea*, FAO Technical Conference on Pollution and Its Effects on Living Resources and Fishing, Rome, Italy, December 9–18, 1970 (1970).
7. Borgstrom, Georg: 'New Methods of Appraising the Role of Fisheries in World Nutrition, *Fishing News Int.* 1, 33–37 (1961).
8. Donaldson, L. R. and Foster, R. F.: 'Effects of Radiation on Aquatric Organisms', in *The Effects of Atomic Radiation on Oceanography and Fisheries*, Publ. 551, National Academy of Sciences National Research Council, Washington, D.C., pp. 96–102 (1957).
9. Edmondson, W. T.; 'Eutrophication in North America', in *Eutrophication: Causes, Consequences, Correctives*, National Academy of Sciences, Washington, D.C., pp. 124–149 (1969).
10. Edmondson, W. T.: 'Phosphorus, Nitrogen and Algae in Lake Washington after Diversion of Sewage', *Science* 169, 690–691 (1970).
11. FAO: *Yearbook of Fishery Statistics*, Vol. 26, Food and Agricultural Organization of the United Nations, Rome (1968).
12. FWPCA: *Water Quality Criteria*, Federal Water Pollution Control Administration, U.S. Department of the Interior (1968).
13. Goldberg, E. D.: in D. Dyrssen and D. Jagner (eds.), 'The Changing Chemistry of the Oceans', *Nobel Symposium* 20, Almqvist & Wiksell, Stockholm, and John Wiley & Sons, Inc., New York, pp. 267–288 (1972); this volume, pp. 275–294.
14. Harvey, G. R., Bowen, V. T., Backus, R. H., and Grice, G. D.: in D. Dyrssen and D. Jagner (eds.),

'The Changing Chemistry of the Oceans', *Nobel Symposium* 20, Almqvist & Wiksell, Stockholm and John Wiley & Sons, Inc., New York, pp. 177–186 (1972).

15. Hasler, A. D.: *Underwater Guideposts, Homing of Salmon*, University of Wisconsin Press, Madison, Milwaukee and London, (1966).

16. Hubschman, J. H.: 'Lake Erie: Pollution Abatement, Then What?', *Science* 171, 536–540 (1971).

17. Ketchum, B. H.: 'Distribution of Coliform Bacteria and Other Pollutants in Tidal Estuaries', *Sewage Ind. Wastes* 27, 1288–1296 (1955).

18. Ketchum, B. H. and Corwin, Nathaniel: 'The Cycle of Phosphorus in a Plankton Bloom in the Gulf of Maine', *Limnol. Oceanogr.* 10, Supplement, pp. R148–R161 (1965).

19. Ketchum, B. H.: 'Man's Resources in the Marine Environment', in T. A. Olson and F. J. Burgess (eds.), *Pollution and Marine Ecology*, Interscience Publishers, pp. 1–11 (1967).

20. Ketchum, B. H.: 'Eutrophication of Estuaries', *Eutrophication: Causes, Consequences, Correctives*, National Academy of Sciences, Washington, D. C., pp. 197–209 (1969).

21. Menzel, D. W., Anderson, J., and Randke, A.: 'Marine Phytoplankton Vary in Their Response to Chlorinated Hydrocarbons', *Science* 167, 1724–1726 (1968).

22. Menzel, D. W. and Ryther, J. H.: 'Distribution and Cycling of Organic Matter in the Oceans', in D. W. Hood (ed.), *Symposium on Organic Matter in Natural Waters*, Institute of Marine Science, Occ. Pap. #1, Univ. of Alaska, College, Alaska, pp. 31–54 (1970).

23. NAS: *Eutrophication: Causes, Consequences, Correctives*, National Academy of Sciences, Washington, D.C., (1969).

24. NAS: *Chlorinated Hydrocarbons in the Marine Environment*, National Academy of Sciences, Washington, D.C., (1971).

25. NAS: *Radioactivity in the Marine Environment*, National Academy of Sciences-National Research Council, Washington, D.C., (1971).

26. NAS: *Water Quality Criteria*, National Academy of Sciences, Washington, D.C. (in press).

27. North, W. J.: 'Tampico: A Study of Destruction and Restoration', *Sea Frontiers* 13 212–217 (1967).

28. Odum, H. T.: *Environment, Power and Society*, Wiley-Interscience, New York, London, Sydney, Toronto, (1971).

29. Revelle, R., Wenk, E., Ketchum, B. H., and Corino, E. R.: 'Ocean Pollution by Petroleum Hydrocarbons', in W. H. Mathews, F. E. Smith and E. D. Goldberg (eds.), *Man's Impact on Terrestrial and Oceanic Ecosystems*, MIT Press, Cambridge, Mass, pp. 297–318 (1971).

30. Riley, G. A., Stommel, H., and Bumpus, D. F.: 'Quantitative Ecology of the Plankton of the Western North Atlantic', *Bull. Bingham Oceanogr. Coll.* 12, 1–169 (1949).

31. Ryther, J. H.: 'IV: Geographic Variations in Productivity', in M. N. Hill (ed.), *The Sea*, Vol. 2, Interscience Publishers, New York, London, pp. 347–380 (1963).

32. Ryther, J. H.: 'Photosynthesis and Fish Production in the Sea', *Science* 166, pp. 72–76 (1969).

33. Sanders, H. L.: 'Marine Benthic Diversity: A Comparative Study', *Amer. Natur.* 102, 243–282 (1968).

34. Sanders, H. L.: 'Benthic Marine Diversity and the Stability – Time Hypothesis', in G. M. Woodwell and H. H. Smith (eds.), *Diversity and Stability in Ecological Systems*, Brookhaven Symposia in Biology, No. 22, pp. 71–81 (1969).

35. SCEP Task Force: 'Chlorinated Hydrocarbons in the Marine Environment', in W. H. Mathews, F. E. Smith and E. D. Goldberg (eds.), *Man's Impact on Terrestrial and Oceanic Ecosystems*, MIT Press, Cambridge, Mass., pp. 275–296 (1971).

36. Schaefer, M. B. and Alverson, D. L.: 'World Fish Potentials', *The Future of Fishing Industry of the United States*, Univ. of Washington, Publications in Fisheries, New Series, Vol. 4, 81–85 (1968).

37. Slobodkin, L. B. and Sanders, H. L.: 'On the Contribution of Environmental Predictability to Species Diversity', in G. M. Woodwell and H. H. Smith (eds.), *Diversity and Stability in Ecological Systems*, Brookhaven Symposia on Biology, No. 22, pp. 82–95 (1969).

38. Sprague, J. B.: 'Measurement of Pollutant Toxicity to Fish. III: Sub-Lethal Effects and "Safe" Concentrations', *Water Res.* 5, 245–266 (1971).

39. Steemann Nielsen, E.: 'The Use of Radioactive Carbon (^{14}C) for Measuring Organic Production in the Sea', *J. Cons. Explor. Mer.* 18, 117–140 (1952).

40. Waldichuk, M.: (personal communication), to be published in *Water Quality Criteria*, National Academy of Sciences (in press).

41. Winslow, E. and Bigler, H. B.: 'A new Perspective on Recreational Use of the Ocean', *Undersea Technol.* **10**, 51–55 (1969).
42. Woodwell, G. M., Craig, P. C., and Johnson, H. A.: 'DDT in the Biosphere: Where Does it Go?', this volume, pp. 295–310.
43. Wurster, C. F., Jr.: 'DDT Reduces Photosynthesis by Marine Phytoplankton', *Science* **159**, 1474–1475 (1968).
44. CEQ: *Ocean Dumping – A National Policy*, A Report to the President Prepared by the Council on Environmental Quality, October, 1970, x+45 pp., U.S. Govt. Printing Off., Washington, D.C. 20402 (1970).
45. Horn, M. H., Teal, J. M., and Backus, R. H.: 'Petroleum Lumps on the Surface of the Sea', *Science* **168**, 245–246 (1970).
46. IMCO: *International Convention for the Prevention of Pollution of the Sea by Oil*, Sales No. 1967.4 (English/French) 1967.3 (Spanish) 101–104, Piccadilly, London WIVOAE (1967).
47. Morris, B. F.: 'Petroleum: Tar Quantities Floating in the Northwestern Atlantic Taken with a New Quantitative Neuston Net', *Science* **173**, 430–432 (1971).
48. Pearce, J. B.: *The Effects of Waste Disposal in the New York Bight*, Interim Report, Sandy Hook Marine Laboratory, U.S. Bur. Sport Fisheries and Wildlife (1970).
49. Schaefer, M. B.: 'The Potential Harvest of the Sea', *Trans. Amer. Fish. Soc.* **94**, 123–128 (1965).
50. CEQ: *Environmental Quality – 1972*, Third Annual Report of the Council on Environmental Quality, U.S. Govt. Printing Off., Washington D.C. 20402 (1972).

INTERACTIONS BETWEEN OCEANS AND TERRESTRIAL ECOSYSTEMS

BENGT LUNDHOLM

Swedish Ecological Research Committee,
Natural Science Research Council, Stockholm, Sweden

Abstract. The interactions between the different ecosystems are of importance, when pollution problems are penetrated. Even if oceans often are the sink for many pollutants, it is, however, possible that a transport takes place from the sea to the air. This material is later washed out over the continents. DDT may assemble in the thin oilslicks on the oceans and may be airborn in the breaker zones along the coasts. Also sulphur may leave the marine environment and by forming sulphuric acid contribute to the acidity in the rainwater.

In Sweden the interest is focused on two main pollution problems: the mercury situation and the acid rains. The lesson learnt is, that the pollutants move freely between air, water, soil and living organisms. There are also indications that these two pollution complexes are interlocked. The total environment must always be considered and the global pollution dynamics is of utmost importance.

1. Introduction

The oceans are often regarded as the final sink for many pollutants, but the toxic wastes may return to man in his food products. It is, however, possible that we also have other important influences from the sea on the terrestrial ecosystems. I want to give two examples, which may illustrate how the pollutants of the sea affect the terrestrial conditions.

DDT and its metabolites are assembled in the oceans, transported by rivers and winds. It has – as Dr Goldberg mentioned – (pp. 275–294) been found that African pesticides appear in the Caribbean winds. When this DDT reaches the aquatic environment, things happen very quickly. It has been shown that DDT added to an algal culture will be adsorbed by the cells in less than 15 s. The lipid soluble DDT passes quickly through the cell wall into cells [2]. (The solubility of DDT in water is extremely low – 1.2 μg l^{-1} – while its solubility in lipids is in the order of 100 g l^{-1}. The DDT is thus trapped in the organic material and is concentrated in the lipids, either in plants or animals. What happens when the cells are broken down? It has been noticed that very often areas with thin films of lipid materials appear on the sea surface. These slicks, which originate from broken-down cells may also contain the fat soluble pesticides. These films are broken up and the material is airborne especially in the breaker zones along the coasts. Dr Erik Eriksson at the International Meteorological Institute in Stockholm has told me about this mechanism. He estimates that of the particles found in the air, 50% is organic material and half of this is fat or lipids, mostly transported from the sea. Even if it is not proved, the possibility of a transport of DDT in the same way seems likely.

In Europe, chlorinated hydrocarbons transported by rain have only been analyzed in Great Britain. It is rather remarkable that, during winter (November–January), the

S. Fred Singer (ed.), The Changing Global Environment, 329–336. All rights reserved.
Copyright © 1975 by D. Reidel Publishing Company, Dordrecht-Holland.

two coastal stations have an increased amount of DDT metabolites and higher amounts than the inland stations [3].

Another interesting detail is mentioned in the paper about 'Pesticides: Transatlantic Movements in the North East Trades', [1] among others. The landward winds at La Jolla have a high amount of pesticides, but, unfortunately, we here have "an unknown admixture of air from neighboring agricultural areas". The theory about marine origin of some DDT may be very easy to test at La Jolla by analyzing the contents from different wind directions separately.

2. Marine Eutrophication

The other example of interactions between a marine and a terrestrial environment is from the Baltic. I have some graphs illustrating the situation. We have a very marked trend in the Baltic towards decreased oxygen and increased phosphorus in the deeper parts (Figure 1). We see on the graph that H_2S has appeared this year. These changes have enormous importance to the ecosystem in the water as anaerobic conditions .prevent the life of higher organisms. The reasons for these changes are not known, but is it possible that pollution and especially eutrophication is one of the reasons. As a result of these changes, some of the hydrogen sulphide may leave the water; in the atmosphere, it can be oxidized to sulphuric acid. When this acid is washed out by the rains, the precipitation will be more acid with a lowered pH. In Sweden we have recorded a very marked increase of the acidity in the rains the last 10 yr (see Figure 2). The main reason for this is, of course, the increased industrial air pollution by SO_2, but we are now also discussing other sources such as H_2S from polluted marine areas, both the Baltic and the North Sea. This may be a problem in all polluted coastal areas along the continents.

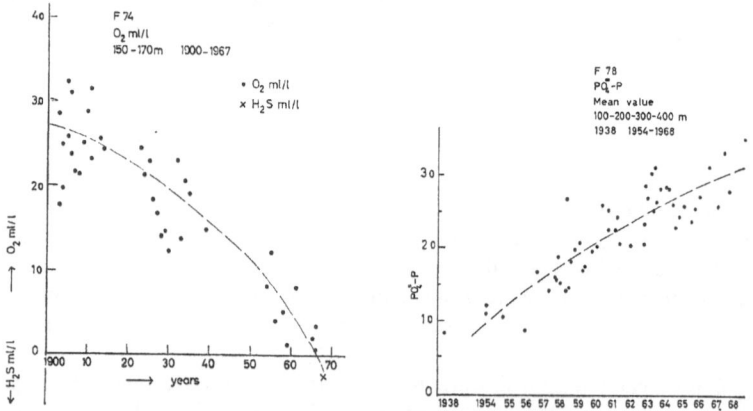

Fig. 1.   Oxygen and phosphorus in different deep parts of the Baltic. (After Fonselius, *Modern kemi* **11** (1968).)

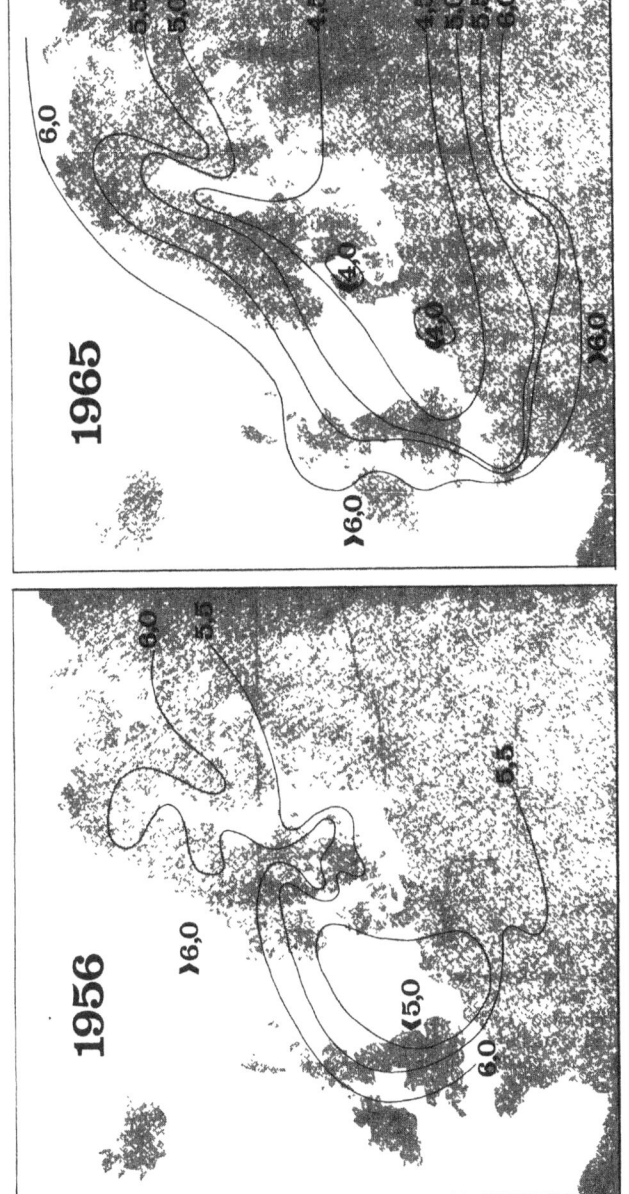

Fig. 2. pH in precipitation over Europe. The annual values for pH were obtained from monthly observations from the network of atmospheric chemistry stations. All data are available from the International Meteorological Institute, Tulegatan 41, Stockholm. (From S. Odén, *Forskning och Framsteg* **1** (1969).)

Even if the acid rains have nothing directly to do with the oceans, I think I would like to follow the chain of causality with a few words. The acid rains affect the surface waters, especially in areas where the buffering capacity is low in the water. In oligo-trophic waters, we get a marked influence resulting in a decreased pH, which has been recorded in large areas in western Sweden.*

3. Heavy Metals Pollution

These events might be connected with another Swedish pollution problem; the occur-rence of mercury in the natural environment. Mercury has been used as a pesticide and in connection with different industrial activities. We are especially concerned with the aquatic environment as both lakes and coastal areas have been blacklisted and commercial fishing is forbidden, as the mercury content in the fish is too high. We have now found that mercury or any mercury compound, poured out in the water is transferred to the very toxic methyl mercury compounds. This methylation is per-formed by microorganisms in the sediments. A monomethyl compound (CH_3Hg)

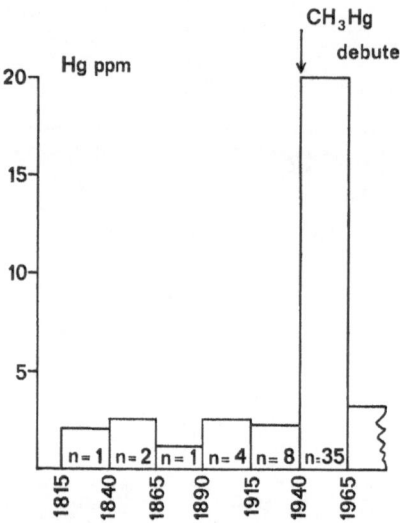

Fig. 3. Mercury levels in feather of goshawk (*Accipiter gentilis*) ♀♀ shot in May–June. (After A. Johnels and T. Westermark, Viltforskningsrådets Nordiska konferens 3–5/3 1966 and with results from the contribution by T. Westermark *et al.* at the symposium on mercury arranged by Nordforsk at Lidingö 10–11.10 1968) *n* refers to size of sample and 'CH_3Hg debute' marks the intro-duction of these compounds for seed dressing.

* In the Swedish case study to the UN Conference on the Human Environment it was claimed that more than 50 % of the excess acid deposited comes from foreign sources. New investigations are, however, indicating that the foreign contribution is far less and that the deposit of sulphur is rather local [4].

is formed in the water and in an alkaline environment this is transferred to a dimethyl compound $(CH_3)_2Hg$, which leaves the water as it is very volatile. By these mechanisms, the mercury is removed from the sediments to the atmosphere. In an acid environment, the dimethyl compound is not formed and the mercury is trapped in the water as a monomethyl compound and is available to the ecosystems. An increased acidity in the surface waters will thus give a higher level of mercury. As a result of this, it is possible to find high mercury contents in fishes in unpolluted lakes very remote from both agricultural and industrial activities, especially as the biomass in these lakes is small and the amount of available mercury is fairly high. The mercury may have been transported to these lakes by air. As a result of these high mercury levels in fresh water and fresh water ecosystems and by the runoff to the sea, the coastal marine areas are also polluted by mercury. And the pollution circle is closed.

The problems in Sweden began with methylmercury compounds used in seed dressing. This resulted in very high mercury levels in terrestrial food chains. In 1966, this compound was forbidden in Sweden and now we can with satisfaction record that the mercury levels have dropped (see Figure 3). In aquatic environments, however, we still have very high mercury levels (see Figure 4). Here the sources are different and

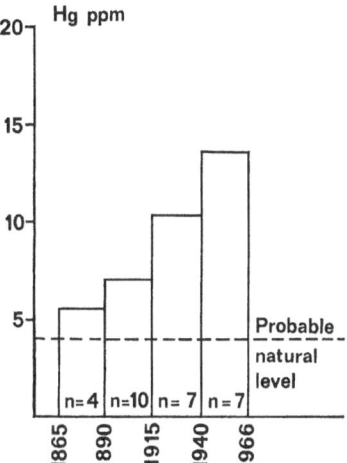

Fig. 4. Mercury levels in feather of great crested grebe (*Podiceps cristatus*). (Redrawn from A. Johnels and T. Westermark, Viltforskningsrådets Nordiska konferens 3–5/3 1966.)

not the same as in the terrestrial ecosystems. The increase of the levels started much earlier than in the terrestrial food chains. These high mercury levels especially in fish from fresh waters and some polluted coastal waters have caused trouble, as the Swedish government has been forced to blacklist several areas where fish is not regarded as suitable for human consumption.

BENGT LUNDHOLM

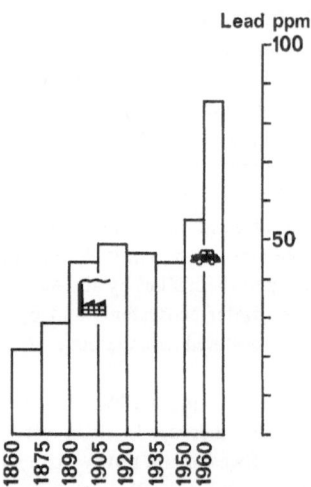

Fig. 5. Lead content in three different species of mosses (*Hylocomium, Pleurozium and Hypnum*). (After Å. Rühling and G. Tyler, *Botaniska Notiser* 3 (1968).)

In these two cases, birds of prey have been used as "indicator species" to the accumulation of mercury. In Sweden, we have also tried to evaluate the importance of the increased outflow of lead into the environment. We have tried to find lead accumulation in birds of prey and also predatory fish, but we have so far not succeeded in getting a clear picture of the situation. Recently, however, certain species of mosses were shown to be good indicators for lead accumulation (see Figure 5). I think it is very important to realize that the different species react very differently to the same pollutant and that we have to find specific indicators to every pollutant.

4. Pollution from Pesticides

Toxic wastes – such as pesticides – transported by the atmosphere are falling out in rains over both the oceans and the continents. On the continents, the wastes are accumulated in soils. Depending on the climatic conditions, some of the substance is removed from the soil. This transportation may be measured by the removal time for half of the amount. I prefer the term "half removal time" to "half life" as some of the compounds are still very much "alive" after their removal from the soil.

These differences in removal time result in an increased accumulation in the northern soils. This has been shown for Swedish soils by Svante Oden at the University of Agriculture in Upsala (see Figure 6). It would be of great interest to find out if we have similar latitude differences in the seas. Is this the explanation to the occurrence of DDT in the penguins? We now have to focus our scientific investigations on the global transport systems.

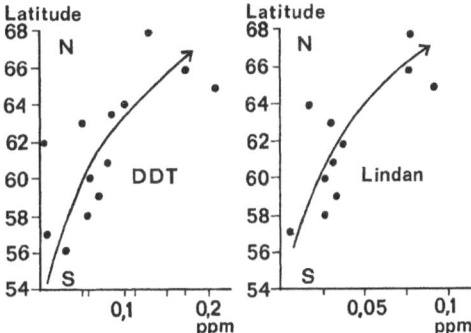

Fig. 6. Residue contents from north to south in Swedish soils, which have not been treated with pesticides. The material is based on 500 soil samples. (After S. Odén, *Handelstidningen* 21.10.1968.)

The marine ecosystems are extremely useful if we want to study the impact of pollution on the biosphere. In the oceans we find an integration of the large-scale processes. There are no local disturbances which mostly complicate the studies of terrestrial ecosystems. By using the oceans as study objects, many basic problems concerning global pollution may be solved. To take one example, DDT and other persistent pesticides are man-made and new to nature. These substances can now be used as labelled compounds. Their geographic distribution in the seas, their position in the marine ecosystems, and their transport ways to and from the oceans will give us a picture of the global pollution.

The international status of the seas makes them easily available to scientific investigators. By using existing resources and facilities for scientific research, it would not be too difficult to plan and carry out a model study of marine pollution. Even with rather limited funds it would be possible to reach outstanding result in a short time.

If we start immediately with the investigations, it would be possible to get fairly good ideas about the general situation concerning global pollution in time for the United Nation's conference about the human environment in 1972. At this conference, the scientists will be able to state their case to the politicians.

The urgency of the situation, or rather the urgency of getting facts which allow us to evaluate the situation, is a challenge to the scientific community. I also think this is an area for cooperation between scientists from different nations, but also an area when even just a few cooperating nations could make valuable progress*.

---

* The Conference on the Human Environment 1972 approved recommendations for global monitoring, and a special Earth Watch will be established. The objectives of this Earth Watch are to identify and determine the extent of environmental problems of international importance and to warn against impending crises. The components will be: research, monitoring, review and information, and information exchange. The principles and the scientific background to global monitoring were worked out by a special commission on monitoring which reported to the U.N.-conference.

References

1. Goldberg, E. D. *et al.*: 'Pesticides, Transatlantic Movements in the North East Trades', *Science*
   159, 1233–1235 (1968).
2. Søldergren, A.: *Oikos* **19**, 126–138 (1968).
3. Tarrant, K. R. and Tatton, J. O'G.: *Nature* **219**, 725–727 (1968).
4. Air Pollution Across National Boundaries. The Impact on the Environment of Sulphur in Air and
   precipitation, Sweden's Case Study for the United Nations Conference on the Human Environment, Stockholm (1971).
5. *Global Environmental Monitoring*. Report Submitted to the U.N. Stockholm Conference on the
   Human Environment by the Commission on Monitoring of the Scientific Committee on Problems
   of the Environment (SCOPE) appointed by the International Council of Scientific Unions (ICSU),
   Stockholm, Editorial Service, Swedish Natural Science Research Council (1971).

PART V

NITROGEN COMPOUNDS IN SOIL, WATER,
ATMOSPHERE AND PRECIPITATION

INTRODUCTION

One of the vexing problems is the conflict between the demands for more food and the demands for a clean environment. Modern, scientific and highly-productive agriculture requires fertilizers, principally nitrogen and phosphorus. It is very susceptible to damage by various pests. It usually requires irrigation, which often involves a transfer of water, and a return flow which carries off nutrients, chlorides, pesticides – all undesirable substances – into streams and rivers.

The addition of fertilizers and pesticides also affects the character of the soil. Over-application, especially of nitrates, may lead to harmful effects on cattle and produce risks to human health. These and other problems are discussed in the papers by Commoner and by Byerly. Keeney and Gardner deal particularly with nitrogen transformations in the soil.

The effects of pollution, whether produced by agriculture, by industry, or by municipal sewage, are most noticeable in the case of lakes, the subject of Hasler's paper. Examples exist throughout the world; it may be appropriate here to detail the case of Lake Erie.

Lake Erie, one of our largest lakes, the world's twelfth largest, provides a resource to 11.5 million people in the U.S.A. and Canada in terms of water supply, recreation, commercial fishing and shipping. In addition, the annual value added by manufacturing in the Erie Basin stands at more than $17 billion. By the year 2000, the population will have doubled and so will the industry of the Basin. These people and industries depend on Lake Erie, which forms a priceless national heritage whose quality must be maintained and enhanced, so that it can be passed on to future generations in a condition of unlimited usefulness.

Lake Erie constitutes a warning to mankind, as Professor Barry Commoner points out. Other lakes, both large and small, are experiencing or will soon experience the same fate, as will many of the Nation's important estuaries. Referred to technically as eutrophication, this aging of lakes is accelerated by an oversupply of nutrients which are generated by human activities. It can bring about the premature death of the lake, making it less and less useful, until it finally becomes a stinking nuisance.

Our strategy is twofold: prevention and restoration. Prevention means greatly slowing down eutrophication by removing key nutrients from the wastewater before it enters the lake. Restoration means removing or inactivating the nutrients after they have reached the lake. Restoration techniques have to be carefully researched in order to find an economically acceptable method likely to succeed. Examples are: mechanical harvesting of algae, or harvesting with organisms which eat algae (preferably edible fish), tying up the nutrients by chemical methods (the Swedes are treating lakes experimentally with pelletized aluminum sulphate), flushing with water of low nutrient content, or bringing oxygen to the deeper layers by mechanical or thermal mixing. All of these techniques, and many others, are being studied intensively by the National Eutrophication Research Program of the Federal Water Pollution Control Administration of the Department of the Interior.* The work is done in government laboratories, university laboratories, and by private industry and, in many cases, uses small lakes in different parts of the country as field laboratories.

Restoration methods presently proposed are not only expensive but may not be able to do the whole job. Prevention, on the other hand, is less costly, technically simpler and, on the basis of our present knowledge, will improve the condition of the lake – without restoration. A specific approach is the removal of phosphorus – a key nutrient for the growth of algae – from the wastewater released by municipal treatment plants. Nitrogen is also a key nutrient for algal growth. However, its removal is more difficult and, in any event, it can enter the lake in rainfall or be captured from the atmosphere by certain blue-green algae that can fix nitrogen.

Based on past research, we know that chemicals can precipitate phosphates out of waste water, and do the job quite economically. For example, FWPCA figures show that phosphate precipitation by lime would add a cost of less than 3 cents per thousand gallons in a 10 million gallon per day plant.

Fortunately, most major sources of phosphorus are "point sources"; i.e., the wastewater comes out of a pipe and can be treated. About $\frac{3}{4}$ of the load comes from municipal wastes, which contribute about 3 lb per person per year; $\frac{2}{3}$ of this comes from detergents. (Interestingly enough, a domestic duck produces nearly 1 lb yr^{-1}). Certain industries, such as potato processing, add 1.7 lb per ton. It is more difficult to control non-point sources; for example, agricultural drainage, unless appropriate land management measures are practised, especially proper methods of applying fertilizers.

Some recommend more extreme measures and would make the restoration of Lake Erie a national project, similar in scope and size to the space program. It could involve tremendous engineering projects including even the physical removal of the polluted sediments. But before we undertake such tasks and commit ourselves to vast expenditures involving many billions of dollars, we must take immediate steps which we think have a good chance of success, and which cost less.

It is rather unlikely that Lake Erie can feasibly be returned to the condition which existed prior to man's appearance, or even to the condition which existed at the turn of the century. It can return to some intermediate stage of aging, but the exact stage cannot be predicted. However, we can expect a major improvement, as well as protection of water quality.

* Now in the U.S. Environmental Protection Agency.

THREATS TO THE INTEGRITY OF THE NITROGEN CYCLE:
NITROGEN COMPOUNDS IN SOIL, WATER, ATMOSPHERE
AND PRECIPITATION

BARRY COMMONER

Center for the Biology of Natural Systems, Washington University, St. Louis, Mo., U.S.A.

Abstract. Intensified application of nitrogenous fertilizers in the United States has produced severe stress of the natural nitrogen cycle by introducing unprecedented amounts of inorganic nitrogen. Thus, whereas the annual nutritional turnover of nitrogen amounts to 7 or 8 million tons, technology and agriculture introduce into the cycle a nearly equivalent amount of 10 million tons of nitrogen compounds. Soil bacteria rapidly convert nitrogenous fertilizer to nitrates, a large proportion of which drain off into streams and rivers. The result is increased eutrophication and potential hazard from nitrite poisoning, or methemoglobinemia. Surveys show that: (1) nitrate acquired by the Missouri River from Nebraska farmlands has increased in parallel with the increasing annual use of nitrogenous fertilizer in Nebraska since 1955 and that (2) high nitrate content of rivers in Illinois farmland is traceable almost entirely to fertilizer that drains into groundwater. Increased use of inorganic N fertilizer also appears to be associated with certain instances of dangerously high nitrate levels in vegetable foods. In the U.S. nitrogen oxides from automobile exhausts also represent an important stress on the N cycle.

1. Introduction

Four chemical elements make up the bulk of living matter – carbon, hydrogen, oxygen, and nitrogen – and they move in great, interwoven cycles in the surface layers of the Earth: now a component of the air or water, now a constituent of a living organism, now part of some waste product, after a time perhaps built into mineral deposits or fossil remains. Nitrogen occupies a special place among these four elements of life because it is so sensitive an indicator of the quality of life. Nitrogen deprivation is a first sign of human poverty; a certain outcome is poor health, for so much of the body's vital machinery is made of nitrogen-bearing molecules: proteins, nucleic acids, enzymes, vitamins, and hormones. Nitrogen is, therefore, closely coupled to human needs. Indeed, as the world population grows, nitrogen will become, increasingly, the crucial element in our efforts to avert catastrophic famine.

All living things, by their very life processes, can alter the precise balance of the natural cycles. In nature these processes affect the chemistry of the environment slowly, on the time scale of geological events, and the systems of living organisms have time to adjust to them. Only one living thing, man, has the power – through technology – to make changes that are so rapid, and so large compared with the natural cycles as to stress them to the point of collapse.

The nitrogen cycle is very vulnerable to human intervention. In the U.S., the overall annual nutritional turnover of nitrogen amounts to about 7 or 8 million tons. At the present time technology and agriculture introduces into this cycle about 7 million tons

S. Fred Singer (ed.), The Changing Global Environment, 341–366. All rights reserved.

of nitrogen compounds from chemical fertilizers, and about 2–3 million tons of nitrogen compounds generated by automobile exhausts and power plants.

The purpose of this paper is to discuss what we know and need to know about the consequences of this intrusion into the natural nitrogen cycle. The evidence will, I believe, support the conclusion that modern technology has intruded into the cycle at its most vulnerable point. This intrusion has produced stresses which have already contributed to a serious deterioration of the quality of the environment, If continued they threaten to collapse important segments of the cycle itself. Correction of these trends will require us to solve very grave economic, social, and political problems.

In the ecosphere – the Earth's thin skin of soil, water, air, and living things – nitrogen is found in relatively few basic chemical forms. A striking feature of nitrogen chemistry is that combinations of nitrogen and oxygen are relatively rare. The great bulk of the Earth's nitrogen is represented by the nitrogen gas in the air. In living things nitrogen almost invariably occurs in combination with hydrogen rather than oxygen (i.e., the nitrogen compounds tend to be in relatively reduced rather than oxidized states); only a few natural biochemical substances, among the many thousands known, contain nitrogen in combination with oxygen. In nature there are only relatively small amounts of the molecular forms – nitrate, nitrite, and several gaseous nitrogen oxides – in which nitrogen and oxygen are combined. Thus, we live in an environment in which nitrogen, as it passes through the successive stages of its cycle rarely finds itself, for long, (except in inorganic mineral deposits) combined with oxygen.

It is precisely this slender span of the nitrogen cycle which is most affected by human intrusion, for the nitrogen introduced into the environment by technology is almost entirely in oxidized forms. Regardless of its original chemical state, nitrogen fertilizer is rapidly converted to nitrate by soil bacteria; automobile exhausts and power plants generate various oxides of nitrogen. These intrusions upon the natural nitrogen cycle have developed very rapidly. In the United States, during the last 25 yr the annual consumption of inorganic nitrogen fertilizer has increased about 14-fold and automotive activity has increased about three-fold. In this period, *biological* intrusion on the nitrogen cycle due to human activity has risen more slowly; nitrogen in sewage has increased about 70 %.

The maintenance of the naturally low concentrations of oxidized forms of nitrogen is essential to the integrity of the Earth's life system. Important hazards to this system are generated when the concentrations of these nitrogen compounds are artificially increased. One hazard is pollution of surface waters by excessive amounts of nitrate. When the normally low level of nitrate in natural waters is increased, the growth of algae may be sharply enhanced. The resulting 'algal blooms', which soon die, overburden the water with organic matter, which on being oxidized by micro-organisms depletes the oxygen content of the water, causing the natural cycles of self-purification to collapse. The Spilhaus report estimates that by 1980 the burden of organic matter imposed on surface waters will be sufficient to consume the total oxygen content of the summertime flow of every river system in the U.S. Excessive amounts of nitrate may contribute to this potential collapse of the self-purifying processes of the nation's water systems.

A second hazard is due to the potential toxicity of nitrate to human beings and domestic animals. While nitrate is itself relatively innocuous when ingested, it may be converted by intestinal bacteria into nitrite – which, on combining with hemoglobin in the blood (methemoglobinemia) destroys the oxygen-carrying power of the blood, resulting in asphyxia, and possibly death. Another hazard is that nitrite, on reaction with certain organic constituents of food, or even of the body itself, may form nitroso compounds, some of which are powerful carcinogens. A third hazard of imbalance in the nitrogen cycle is that certain of the gaseous nitrogen oxides are powerful poisons and pharmacological agents. Finally, such oxides, when activated by sunlight, can react with waste hydrocarbons in the air to produce the noxious constituents of smog.

In general, then, we must regard any appreciable elevation in the normally low concentrations of oxidized forms of nitrogen as a serious hazard to the environment. Since it is precisely this most vulnerable segment of the nitrogen cycle which is stressed by recent increases in the nitrogen produced by human activities, it becomes urgent to determine to what extent such stresses are present or potential threats to the quality of the environment.

2. The Natural Nitrogen Cycle

The soil is a useful place to begin. Nitrogen is crucial in the soil economy. In nature, let us say a plant growing in a wood or meadow, most of the soil's nitrogen – usually well over 80% – is in the form of complex and still poorly-understood organic substances, especially humus. Slowly, bacteria release nitrate from humus and decaying organic wastes – manure and the residues of animals and plants. The resultant concentration of nitrate in the soil water is very low and the roots need to work to pull it into the plant. For this work the plant must expend energy which is released by biological oxidation processes in the roots. These processes require oxygen, which can reach the roots only if the soil is sufficiently porous. Soil porosity is governed by its physical structure; in particular a high level of organic nitrogen, in the form of humus, is required to maintain a porous soil structure. Thus, soil porosity, therefore its oxygen content, and hence the efficiency of nutrient absorption, are closely related to the organic nitrogen content of the soil.

When the United States was settled, the soil system was in this natural condition; the soil cycle was in balance, maintaining its nitrogen reserve in the stable organic form. Some nitrogen from the air is fixed by legumes and helps build up the organic reserve. In virgin land only a very small amount of nitrate escapes the plant's root systems and leaches out into surface waters, or escapes to the air in volatile compounds: nitrogen gas, ammonia, and nitrogen oxides; and the last two are soon returned to the Earth in rain and snow.

In natural waters a similar nitrogen cycle prevails, except that the large reserve of organic nitrogen represented by the soil humus is lacking. Living things, for example fish, contribute organic nitrogen to the water: waste, and, on death, their bodies. This organic nitrogen is quickly converted to an inorganic form; the bacteria of decay free nitrogen from its organic combination with carbon and hydrogen, and unite it with oxygen to form, ultimately, nitrate. The needed oxygen reaches the water from the air and from the photosynthetic activity of plants. Nitrate in turn nourishes the aquatic plants, chiefly algae; these in turn furnish food for fish and other animals, and the cycle is complete. In a balanced natural system the amounts of organic nitrogen and nitrate dissolved in the water remain low, the population of algae and animals is correspondingly small, and the water is clear and pure. And because the natural nitrogen cycle in the soil is tightly contained, relatively little nitrogen is added to the water in rainfall, or in drainage from the land.

The nitrogen gas that makes up about 80% of the air is very stable, chemically. It will react with oxygen to form nitrogen oxides, and eventually, nitrate, only under intense heat, which in natural circumstances is obtainable only from lightning; however recent evidence suggests that the amounts of nitrogen oxides produced by this natural agency are very slight. In natural conditions animals that breathe air carry to their lungs little else, besides the air's nitrogen, oxygen, water vapor and carbon dioxide. In particular nitrogen oxides, which are incompatible with animals' respiratory physiology, are normally absent.

3. Agricultural Stresses on the Soil

We must introduce into this balanced system, man – not only a rapidly proliferating animal, but also the one living thing that has generated the vast environmental intrusions of technology, agriculture, and urbanization. How has the natural nitrogen cycle been faring under these stresses?

When the United States was settled, the soil system was in the natural condition that I have already described. As this natural system was taken over for agricultural purposes, plants were grown on the soil in amounts much greater than it would sustain in its virgin state. The organic store of nutrients was gradually depleted and crop yields declined year by year. (In the Midwest, the organic nitrogen of the soil is now about one-half of what it was in the virgin soil.) With new lands always available, farmers moved westward repeating the process of skimming from the soil the most available nutrient and leaving it when its productivity fell below a point which made westward migration more attractive. This process came to an end at about 1900 and from then on as agriculture became intensified to meet the demands of a growing population, more and more of the original store of organic nutrients was withdrawn from the soil in the form of crops.

At the turn of the century, then, with the nation's need for food rapidly rising, and the first easy fruits of the soil's wealth already taken, we were confronted with the vast problem of intensifying the declining agricultural productivity of the soil. Much agricultural research was devoted to sustaining productivity by adding chemical fertilizers to the soil. At several agricultural research stations, such as the Missouri

Agricultural Experiment Station, long-term experiments were undertaken to study the effects of different agricultural practices on crop yield, and on the nature of the soil. In 1942, this Station published a remarkably revealing account of 50 yr of patient study of their experimental plots, Sanborn Field.

The report showed that nitrogen, provided as nitrate, was an effective means of maintaining good crop yields. But the report also showed that the soil suffered important changes:

The organic matter content and the physical conditions of the soil on the chemically-treated plots have declined rapidly. These altered conditions have prevented sufficient water from percolating into the soil and being stored for drought periods. Apparently a condition has developed in the soil whereby the nutrients applied are not delivered to the plant when needed for optimum growth ... Evidently most of the nitrogen not used by the immediate crop is removed from the soil by leaching or denitrification .... From these figures it is evident that heavy application of chemical fertilizers have given a very low efficiency of recovery. [1]

The story is clear and well supported by numerous laboratory and field investigations: While nitrate fertilizer sustains crop growth, it fails to rebuild the humus nitrogen lost from the virgin soil. As a result soil porosity – which depends on humus content – deteriorates, aeration becomes more difficult and the plant roots, which must have oxygen if they are to absorb nutrients, are unable to take up all of the nitrate made available to them by the added fertilizer. The rest of the nitrate is lost from the soil. Some nitrate leaches out in the ground water, ultimately into rivers and lakes. Some nitrate is converted to ammonia, nitrogen gas, and nitrogen oxides by bacterial denitrification processes – which are stimulated when a humus-depleted soil becomes water-logged.

In effect the Sanborn Field studies, and similar investigations elsewhere, were a warning that in humus-depleted soil, fertilizer nitrate tends to break out of the natural self-containment of the soil system and to penetrate the air and the water. But the warning was ignored. In 1942, when the 50-yr results of the Sanborn Field studies were published, the annual U.S. consumption of chemical fertilizer amounted to somewhat less than 0.5 million tons of nitrogen. It is now about 7 million tons – a 14-fold increase in about 25 yr. The vast use of nitrogen and other chemical fertilizers, together with other means of intensifying crop production, has, of course, given us a remarkable increase in yields. But as shown by the Sanborn tests, we must look for some of the 7 million tons of nitrogen added to U.S. soil during the last year not in the soil, or in the harvested crops, but in the nation's lakes and rivers, and in the rain and snow that fall on the land.

4. The Fate of Fertilizer Nitrogen

What is the present nitrogen balance in the U.S.? It should be stated at the outset that this question cannot now be answered, for we simply lack the data needed to draw up a total balance sheet. Overall statistics are misleading. The total amount of nitrogen removed from U.S. soils in crops annually is now about 9–10 million tons [2]. Since this figure is about 2–3 million tons greater than the total amount of nitrogen added to the soil in fertilizer, it is evident that some nitrogen is being withdrawn from the organic store in the soil. But it would be a mistake to conclude that *all* of the fertilizer nitrogen appears in the crops and that only 2–3 million tons of nitrogen is withdrawn from the soil's organic store. Numerous laboratory and field experiments show that only part of fertilizer nitrogen is taken up by crops. Efficiency of nitrogen uptake varies with the crop, growing conditions and nature of the soil; in no case is efficiency greater than about 80%, and values of the order of 50% are common [3]. It must be assumed, therefore, that an appreciable portion of fertilizer nitrogen may fail to be absorbed by the crop and appears in surface waters and the air. In the absence of overall data capable of evaluating the total distribution of nitrogen, the best we can do is to examine the entry of nitrogen into surface waters and the air and to determine the extent to which such nitrogen originates in fertilizer and in other possible sources.

A particularly useful set of data is available on the nitrate content of the rivers of Illinois where for the last 25 yr the Illinois State Water Survey has been making increasingly detailed studies of the quality of surface waters [4]. Illinois is intensively farmed (about 60% of the state's area is in crop land) and the system of farming involves very heavy use of inorganic nitrogen fertilizer. As in most parts of the U.S., the use of inorganic nitrogen fertilizer has increased by an order of magnitude in the last 20–25 yr.

Figure 1 summarizes the results of analyses of most of the rivers of Illinois for nitrate content*. The general trend is quite evident; since 1949 there has been an appreciable rise in the average nitrate concentrations, which by 1969 had approached, as a maximum, the limit established by the U.S. Public Health Service for potable water, 10 ppm of nitrate-nitrogen. Figure 2 shows a typical monthly pattern of variation in nitrate concentration and water discharge. The rivers which exhibit the highest nitrate levels show a striking monthly variation in nitrate concentration; nitrate levels reach a maximum in May and June and a minimum in September and October. This effect is closely associated with the amount of water discharged into the rivers, suggesting that that movement of water precipitated on the land into the river is responsible for the appearance of nitrate in the stream. There is a two-month delay between the period of peak water discharge (March–April) and peak nitrate concentration (May–June). This suggests that the nitrate reaches the river by way of percolation into the soil and as a result of drainage, rather than in the immediate runoff from precipitation. This is in keeping with expectations based on the known effects of precipitation on the movement of nitrate in soil; as rain begins to penetrate the soil it carries free nitrate down into the soil

---

* We wish to take this opportunity to correct an error made in the original text. Figure 1 shows the seasonal variation of nitrate nitrogen in the Kaskaskia River during two periods of time (1945 to 1950 and 1956 to 1966). This comparison was inappropriate because the sampling during these two periods were in different locations. The values of nitrate nitrogen concentration in the period 1945–1950 are for the Kaskaskia River at New Athens, while the values for 1956–1966 are for the Kaskaskia River at Shelbyville. The latter drainage area is heavily cropped and fertilized, while the New Athens area which is over 100 miles downstream/from Shelbyville contains a considerable proportion of grass and timberland. We would like to thank Dr S. R. Aldrich who called this error to my attention.

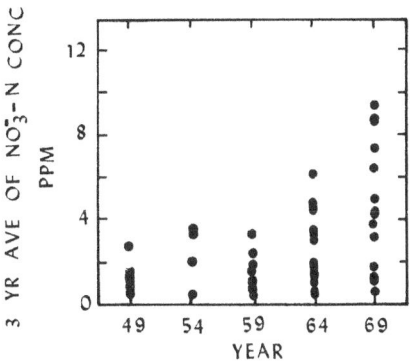

Fig. 1.

toward ground water, so that, when the soil becomes water-logged and rains runs off at its surface, little nitrate is carried off to the rivers.

The possible sources of the increased burden in Illinois rivers are now under intensive investigation. The use of inorganic nitrogen fertilizer in Illinois has risen in the same period during which river nitrate concentrations have increased. Moreover, no increase in nitrate concentration has been observed in a stream, the Skillet Fork River, which does not traverse heavily fertilized agricultural land. Detailed studies, now under way, regarding the relationship between river nitrate concentrations and the factors which might effect them (rate of fertilizer application, rainfall and crop patterns, etc.) are expected to provide a definitive answer to the question of the sources of river nitrate.

Other possible sources of stream nitrogen are animal wastes from feedlots, urban sewage, and nitrogen compounds present in rain and snow. Feedlots are relatively scarce in Illinois and may be disregarded as a significant source of stream nitrogen. Precipitation can also be ruled out as an important source of river nitrate, since the

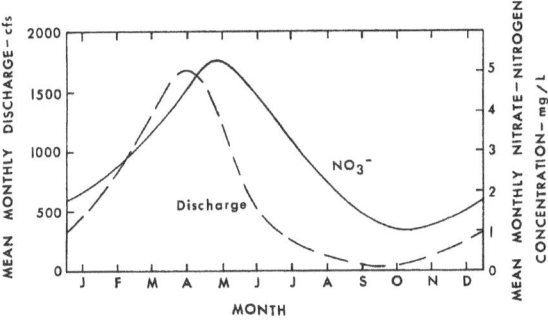

Fig. 2.

average nitrate-nitrogen content of precipitation in that area is about 0.1–0.2 ppm [5], whereas the Kaskaskia and the Sangamon exhibit nitrate-nitrogen concentrations in excess of 5 ppm.

The recent interim report from the Illinois State Water Survey [4] includes data which enables one to estimate the contribution of sewage nitrogen to the nitrate content of the Sangamon River. Since the population involved in sewage effluent delivered to the Sangamon River is known (7000–8000 persons), as is the average nitrogen contribution to sewage per capita per day (about 0.11 lb of nitrate), it is possible to calculate the expected daily contribution of sewage nitrogen to the total nitrate load of the Sangamon River: 175–200 lb of nitrate-nitrogen per day. Figure 3 reports the calculated nitrate content carried by the Sangamon River at Monticello in pounds of nitrate-nitrogen/day. It is evident that sewage can account for the Sanga-mon River nitrate load only in October. The total annual nitrate-nitrogen content of the Sangamon River water is about 50 million lb. Hence, the total annual nitrate-nitrogen contributed by sewage (ca. 65000 lb) represents an insignificant proportion of the total nitrate-nitrogen entering the river. Similar calculations for the Kaskaskia River show that in 1968 nitrogen due to sewage effluent entering the river can account for less than 1 % of the total nitrate-nitrogen carried by the river.

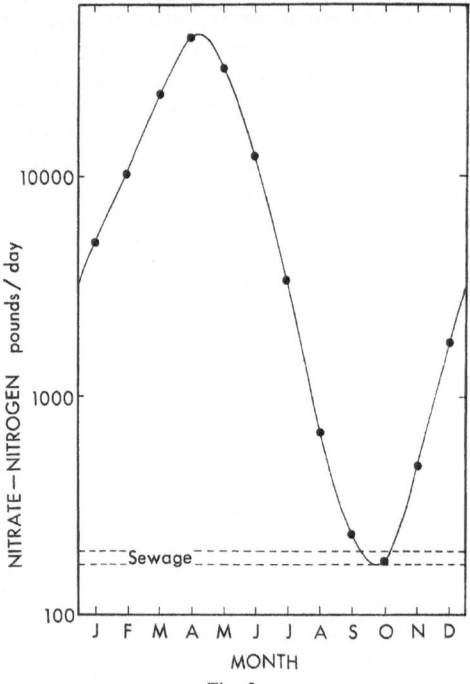

Fig. 3.

What are the consequences of the nitrate levels which have been induced in Illinois surface waters by intensive nitrogen fertilization? The Illinois Water Survey data show that all of the State's rivers which traverse heavily farmed land – and this comprises all but a few of the State's streams – have nitrate levels which are far in excess of those which lead to heavy algal growth (this level is usually estimated at about 0.3 ppm of nitrogen) [6]. Indeed, the survey has recently announced that the State's streams, as a whole, have now become eutrophic – so burdened with nutrients as to support algal blooms, which impose a sufficient organic load on the water to deplete its oxygen supply. This situation is dramatically illustrated by estimates of the nitrate content expected in a new reservoir which is about to be filled at Shelbyville on the Kaskaskia River. The interim report points out that had the dam been closed early in 1968, "it could have been filled very rapidly during April, May, and June of 1968 – with water in which the nitrate content ranged between 19.5 and 56 mg^{-1} (i.e., 4.5–13.0 ppm of nitrate-nitrogen)." When this new reservoir is filled, there will be an immediate problem from nitrate-induced eutrophication. From the evidence cited above, it is apparent that eutrophic pollution of surface waters in Illinois, which is now critical, involves nitrogen which is largely due to fertilizer.

Illinois also faces a health hazard due to the high nitrate levels of most of its rivers. The acceptable limit for the nitrate-nitrogen concentrations of potable water established by public health agencies is 10 ppm. As noted in the statement quoted above, this level has already been exceeded by the Kaskaskia River. According to information provided to me by the Department of Health of Decatur, Illinois, the city water supply, which is taken from the Sangamon River, has approached this level during the spring months of the last two years. It has also been reported that about 25% of Illinois groundwaters from shallow wells of 25 ft depth or less, contain more than 10 ppm of nitrate-nitrogen and that groundwater derived from fertilized fields frequently contains over 14 ppm of nitrate-nitrogen in the spring months [7]. It would appear, therefore, that in rural regions of Illinois the intensive use of nitrogen fertilizer has created a public health problem due to high levels of nitrate in potable water (the medical aspects of this problem are to be discussed below).

Fig. 4.

The U.S. Geological Survey reports water quality data, including nitrate concentrations, from a large number of river sampling stations throughout the U.S. These data represent a vast resource of information applicable to our problem, but unfortunately little has been done, thus far, to analyze them in terms which are relevant to the nitrogen cycle. As an initial step in this direction I have made a preliminary analysis of the USGS data on part of the Missouri River drainage basin [8].

At its headwaters, the Missouri River drains forested maintain regions which merge into pasture land in the Dakotas. Further downstream, the river drainage basin includes intensively farmed and heavily fertilized crop lands in Nebraska. The contributions of the two types of land areas to the river drainage can be separated by comparing data taken at Nebraska City, Nebraska, with data from upstream points at the boundary between the two areas, Williston, N.D. and Overton, Nebraska (see map, Figure 4). By this means, one can calculate both the water discharge and the nitrate specifically arising from the Nebraska agricultural drainage basin.

Calculations have been made of the mean annual nitrate concentration (weighted according to water discharge rate) and the mean annual water discharge for the Nebraska region for the period 1950–64. In general, nitrate concentration is correlated with total water discharge, a relationship which is also evident in the Illinois data. Thus, in the period 1950–55 there was a sharp decline in water discharge and concurrently a sharp decline in mean nitrate concentration. In the period 1955–64 both discharge rate and nitrate concentration increased.

From such data it is possible to calculate the total nitrate input into the river from the Nebraska drainage area. This is shown in Figure 5, both for water leaving the river at Williston, and for the Nebraska drainage area. One striking result is that the nitrogen drainage from the *upstream* parts of the Missouri River has been declining

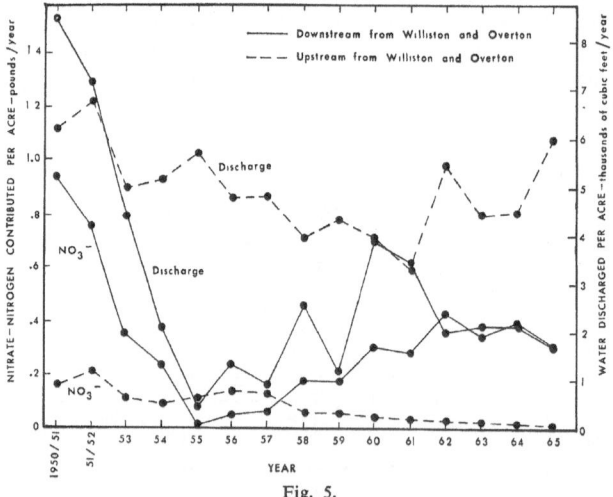

Fig. 5.

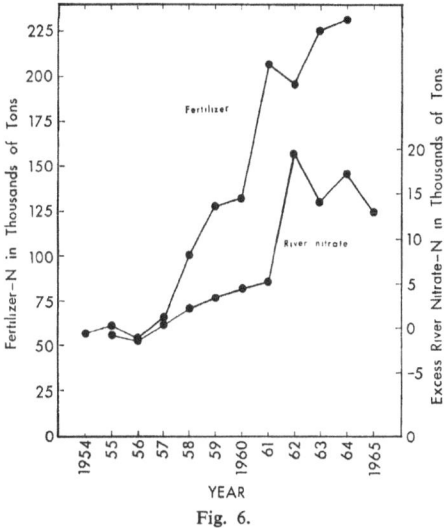

Fig. 6.

steadily since 1950. I am informed that pasture land in this region has been considerably improved by conservation practices in this period; the decline in nitrogen drainage from this area very likely reflects this effect. In any case, this curve shows that in an area which does not receive an appreciable man-made nitrogen input (it is sparsely populated and uses relatively little fertilizer), nitrogen losses from the soil are minimal, and are even capable of being reduced. In contrast the nitrate entering the Missouri River from the Nebraska area has increased sharply since 1955, following the decline due to falling water discharge. Thus, somewhere in this area there must be significant sources of inorganic nitrogen which have increased since 1955.

Figure 6 compares the annual fertilizer nitrogen tonnage used in the area with the nitrate load in the river (corrected for variations in water discharge). It is evident that in general the nitrate load in excess of that which can be accounted for by the concurrent rate of water discharge increases during the period of rising fertilizer usage. Examination of this figure also suggests that there is a one-year delay between the effect of a change in fertilizer use and the change in nitrate load of the river. This effect is examined explicitly in Figure 7 which is a plot of the annual *change* in nitrogen fertilizer use against the annual *change* in river nitrate load *during the following year*. There appears to be a positive correlation between the two values, with the regression line passing through the origin. The regression coefficient for these data indicates, at the 99 % level of confidence, that the change in nitrate load acquired by the Missouri River in the Nebraska region is proportional to the change in the amount of nitrogen-fertilizer used in the preceding year. This is the relationship expected if fertilizer nitrogen percolates into groundwater and reaches the river by that route, for this process involves a delay of some months.

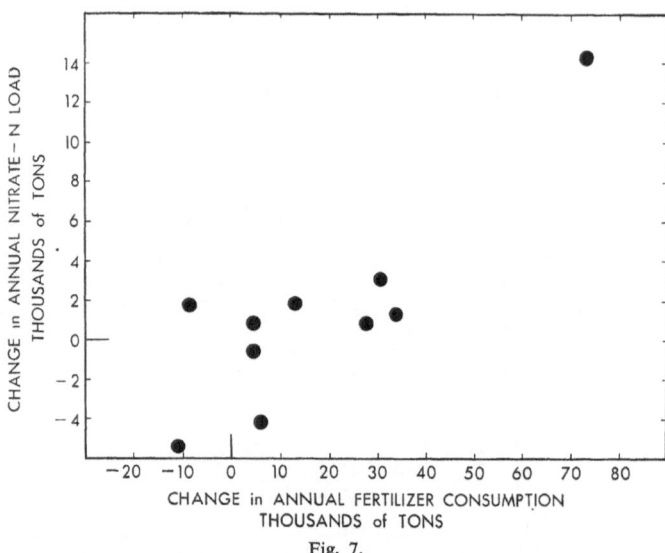

Fig. 7.

Apart from fertilizer, other possible inputs of nitrogen to river water are precipitation, sewage, and livestock manure. Nitrogen concentrations in precipitation are about an order of magnitude too low to account for river nitrate levels. The relative importance of the contribution of nitrogen from sewage and manure to the river can be evaluated by examining the ammonia content of the river, for it is well-known that this substance reflects the input of organic waste. Unfortunately, ammonia analyses are usually not carried out by the USGS. Such values are obtained by public health agencies (which, for their part, usually fail to analyse river nitrate levels). However, the U.S. Public Health Service has reported [9] ammonia values at river stations quite close to the USGS stations at Williston, Overton, and Nebraska City, so that the concentrations of ammonia and nitrate in the water added to the river in the Nebraska drainage area can be compared.

Seasonal changes in these values, averaged for the period 1957–63 are shown in Figure 8. The ammonia concentration is *inversely* related to the seasonal rate of water flow. This is the relationship expected if the ammonia were derived from material (such as sewage and runoff manured lands) added directly to the river at a rate *independent* of water flow, and therefore diluted by the latter. In contrast, nitrate concentration is highest during the period of peak water flow – a relationship consistent with the conclusion that nitrate is carried into the river by the groundwater contribution to water flow. This is evidence that an appreciable part of the ammonia and nitrate entering the river in the Nebraska drainage area originates from different sources. Since ammonia is largely due to sewage effluent, livestock manure, and other organic sources, it would appear from these considerations, that a significant part of

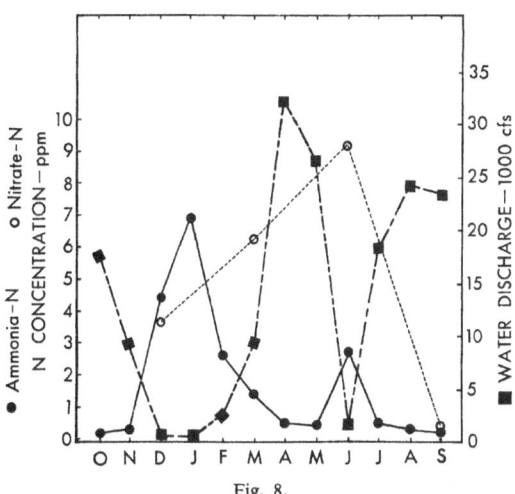

Fig. 8.

the river nitrate must originate from some other source. The only known source which remains to account for the observed nitrate content added to the Missouri River in the Nebraska region is fertilizer.

Additional evidence in support of this conclusion is shown in Figure 9, which compares the ammonia and nitrate concentrations of the water added to the river in the Nebraska region, annually, for the period of 1957 to 1963. Ammonia concentrations do not change significantly in that period; in contrast, nitrate concentrations increase appreciably. Thus, while changes in annual nitrate concentration tend to parallel

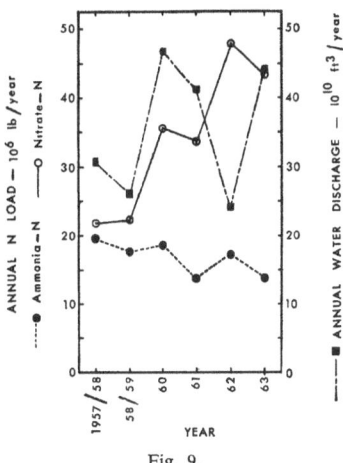

Fig. 9.

concurrent changes in fertilizer consumption, changes in ammonia concentration do not. Again, this indicates that river nitrate levels are related to input from fertilizer rather than from sewage and manure.

Thus, as in the case of the rivers of Illinois, this analysis leads to the conclusion that inorganic nitrogen fertilizer contributes significantly to the nitrogen content acquired by the Missouri River from drainage arising in the Nebraska region.

An important consequence of these analyses is that they provide specific evidence that nitrate originating in fertilizer reaches surface waters by a process which carries in from the soil in proportion to the rate of flow of surface waters. In contrast, nitrogen originating in sewage effluents is delivered to the stream at a rate independent of water flow.

Against this background, it is useful to take note of the results of a detailed survey of the nitrogen sources which contribute to the nutrient levels of Wisconsin surface waters – many of which are now eutrophic – reported by the Water Subcommittee of the Wisconsin Natural Resources Committee of State Agencies in 1967 [10]. According to this report, 42% of the nitrogen reaching Wisconsin surface waters originates in groundwater; 30% originates in urban and private sewage systems; about 10% is due to runoff from manured crop land; 9% is due to nitrogen in precipitation falling on water areas. This suggests strongly that fertilizer is a major source of the nitrate which contributes to the eutrophication of Wisconsin surface waters. Similar calculations can be derived from the recent Lake Erie Report of the Federal Water Pollution Commission [11]. The report estimates that the lake receives about 75 million pounds of nitrogen per year from the surrounding farmlands. Municipal sewage contributes about 90 million pounds of nitrogen per year. Thus urban and agricultural sources contribute about equally to the nitrogen pollution of Lake Erie. A separate calculation, based on the expected losses of fertilizer nitrogen leaching from the crop land of the Lake Erie drainage basin [12] yields a similar result. These observations suggest that about one-half of the nitrate which pollutes such surface waters (which are in areas that include major urban sources of sewage) is derived from fertilizer nitrogen.

It has been claimed, in the past, by some investigators [13] that high nitrate levels in groundwater wells in rural areas is nearly always due to contamination from sewage systems or livestock feedlots. Recently this problem has been investigated in considerable detail in an intensely farmed irrigated area of the South Platte River valley in Colorado [14]. This study confirmed that groundwater under feedlots was usually contaminated with very high concentrations of nitrate. However, high nitrate levels (about 10–30 ppm of nitrate-N) were also found in groundwater beneath irrigated fields, and it was calculated that 25–30 lb of nitrogen per acre drained from these fields annually. The report concludes:

The ratio of irrigated land to that in feedlots for the study area was approximately 200 to 1. Therefore, although much larger amounts of nitrate per unit area were usually present under feedlots, indications were that irrigated lands contributed more total nitrate to groundwater. Feedlots, however, are usually located near the homestead and may have a pronounced effect on the water quality of domestic wells.

The Colorado study provides an important way to distinguish between pollution due to feedlots and fertilizer: in the former, nitrate is always accompanied by comparable – or greater – concentrations of ammonia and by organic materials. Unfortunately, such data are not available from most of the older studies which claim that ground-water contamination is due to feedlots and not to fertilizer [15]. Data on the nitrate content of groundwater in other parts of the country support the general conclusion that fertilizer is a major source of this contaminant. Thus, McHarg's studies of groundwaters in Minnesota [16] show that nitrate-nitrogen levels of the order of 10 ppm are common, that nitrate originates at the surface of the soil, and is especially concentrated in groundwater under intensely farmed and fertilized areas. In at least one Minnesota town, Elgin, the city's water supply has exceeded the 10 ppm public health limit, and has had to be replaced [17].

The heavily farmed areas of California probably represent the most severe public health hazard from nitrate originating in fertilizer. These agricultural areas are heavily farmed, usually under irrigation, with a very intense and increasing utilization of nitrogen fertilizer. In a typical situation in the San Joaquin Valley it was found that fields irrigated with water containing 1.7 ppm of nitrate-nitrogen yielded, in the sub-surface drainage water, nitrate levels which averaged (over a 4-yr period, 1962–66) 44.5 ppm of nitrate-nitrogen. An average of 167 lbs acre^{-1} yr^{-1} of fertilizer nitrogen was supplied to the soil [18]. Another recent study [19] shows that in 4 different cropping systems an average of 36% of the applied fertilizer nitrogen appears in the drainage water. The nitrate concentrations of drainage waters from different areas generally parallel the amount of fertilizer nitrogen supplied.

The irrigation drainage water in this area is at present collected by the San Joaquin River which discharges into San Francisco Bay. The nitrate content of the river water has now become so high as to induce serious water pollution, due to eutrophica-tion, both in the river itself and in reaches of San Francisco Bay. It is evident from the above that fertilizer nitrogen contributes significantly to this environmental deteriora-tion.

Serious health hazards have arisen in the San Joaquin Valley and in Southern California because of the high nitrate levels of wells derived from groundwater supplies. A study of 800 wells in Southern California in 1960–61 [20] showed that in 172 wells water exceeded 5 ppm of nitrate-nitrogen and in 88 wells nitrate-nitrogen exceeded 10 ppm (the public health limit for potable water). This study provides explicit evidence that the nitrate contamination was not due to organic sources such as sewage or manure. While nitrate-nitrogen levels were in the range 1–10 ppm, the concurrent organic nitrogen levels were of the order of 0.3 ppm and am-monia was in the range of 0.1–0.5 ppm. From the Colorado studies described earlier it is known that groundwater contaminated by organic sources invariably exhibits ammonia and organic nitrogen levels that are in excess of the concurrent nitrate levels. Hence it must be concluded that organic sources do not contribute significantly to the nitrate contamination of wells in Southern California, so that the contributing source must be fertilizer nitrogen. In certain areas of California nitrate levels of well water is so high that public health officials have been required to warn physicians against the use of such water in infant feeding.

It appears evident from these results that in heavily fertilized areas such as California, fertilizer nitrogen is the chief source of two environmental hazards: eutrophication of surface waters and unacceptable levels of nitrate in potable water.

5. Excessive Nitrate in Foods

Under natural circumstances, before the intrusion of modern agriculture, nearly all of the soil nitrate which nourishes the plant is converted to protein and other organic plant constituents; in most plants relatively little of the observed nitrogen accumulates as nitrate. However, at the heavy rates of chemical fertilization now widely used in the U.S., this situation has been changed; plants grown on soil heavily fertilized with nitrate contain much-increased amounts of nitrate. For example, a lettuce crop grown on Missouri soil without added nitrogen contained about 0.1% of nitrate-nitrogen; given 100 lb acre^{-1} of nitrate fertilizer the lettuce nitrate content increased to 0.3% and at about 400 lb of nitrogen fertilizer per acre, the nitrate content of the crop reached a maximum of 0.6% [21]. Other possible causes of nitrate accumulation in crops are weather conditions, light intensity, and time of harvest.

In contrast to the well-reported instances of methemoglobinemia arising from nitrate-polluted drinking water, there are, to my knowledge, in *the United States* no clinical reports of methemoglobinemia relating to nitrate levels of foods. However, in the last five years, European pediatricians have reported a sufficient number of cases of methemoglobinemia associated with ingestion of spinach to warrant a recent general review [22]. In these cases the spinach preparations fed to the affected infants were found to contain sufficient *nitrite* to cause the poisoning. However, *fresh* spinach, even though often high in *nitrate*, contains only insignificant amounts of nitrite, so that the toxic nitrite levels must have developed in the food somewhere between harvesting and use. This process has now been explained by the work of Schuphan [23]. When nitrate-rich spinach preparations are exposed to certain common air-borne bacteria, the latter can quickly convert the nitrate to toxic amounts of nitrite. This may happen when a jar of spinach baby food is opened to the air and kept for a day or two (even in the refrigerator) before using. Bacterial nitrite production may also occur when frozen spinach is kept in the thawed condition for a day or two, either during storage of the original package, or after the package has been opened. Even in the absence of bacterial action, when spinach is transported in tightly packed containers, enzymes in the spinach itself may convert nitrate to nitrite.

On the basis of these generalizations Simon [22] has made the following recommendations regarding the use of spinach for infant feeding:

(1) Prepared spinach must not be kept at room temperature.
(2) Too much fertilizer should not be used for growing spinach. (This is discussed further below.)
(3) Spinach used for infant feeding must not contain more than 300 mg nitrate per kg* because

* mg kg^{-1} is equivalent to ppm.

of the danger of possible mistakes in storage, for nitrite in toxic amounts can only be formed if there is a high nitrate content.

(4) Spinach should not be given to infants in the first 3 months of life, for at this age nitrate can be reduced in the upper parts of the gastrointestinal tract, and because of the lowered diaphorase activity there is then increased susceptibility to methemoglobinemia.

These guidelines provide a useful background for evaluating the significance of observed nitrate levels of foods.

The most detailed study of nitrate in baby foods in North America appears to be that of Kamm *et al.* [24]. They report average values based on the analysis of a total of 37 samples of commercial baby foods in Canada (see Table I).

Baby foods not containing vegetables gave considerably lower nitrate values. A less complete study of baby foods reported earlier by the Missouri Agricultural Experiment Station [21] gave results similar to those cited above.

The most recent baby food analyses known to me are those performed in January, 1968 by W. Brabent, Superintendent and M. Boulerice, Chemist of the Department of Health of the City of Montreal [25]. Analyzing single commercial baby food samples, they found: in spinach, 642 ppm nitrate; in beets, 523 ppm nitrate; in squash, 295 ppm nitrate.

It is clear from these results that infants fed on commercial baby foods of the types analyzed in Canada and the United States in recent years are very likely to exceed, in their diet, the nitrate level recommended by Simon as an acceptable limit.

The European pediatric literature also provides a useful insight into the conditions under which nitrate contained in baby food (specifically, spinach) may be hazardous to infants. It should be recalled that the danger is not from nitrate itself, but from *nitrite* which is readily formed from nitrate under reducing conditions. Nitrate can be converted to nitrite in several different ways:

(1) The infant's intestinal tract normally contains bacteria which can convert nitrate to nitrite. In a normal infant most of the nitrate in food is absorbed from the digestive tract before it reaches the intestine, thus minimizing the hazard. Howevert during digestive disturbances, bacteria from the intestine may reach the upper portion, of the digestive tract where they come in contact with nitrate and produce nitrite. Is

TABLE I

	Average ppm of nitrate	No. of samples analyzed
Mixed vegetables	88	2
Carrots	101	8
Green beans	163	3
Garden vegetables	180	5
Graham crackers	211	1
Squash	282	5
Wax beans	444	2
Beets	977	6
Spinach	1373	5

is also possible that infants may become infected with abnormal strains of intestinal bacteria which are particularly active in converting nitrate to nitrite. (A 1962 epidemic of infant methemoglobinemia in a French hospital was traced to such an effect [26].)

(2) Fresh spinach may, during transport and storage, convert nitrate to nitrite by enzymatic action. Thus Phillips [27] reports that analyses of fresh spinach with an initial nitrate content of 2314 ppm showed a complete conversion of nitrate to nitrite after 21 days of refrigerator storage; in 14 days two-fifths of the nitrate was converted to nitrite. This effect is not uniform and is related to the original nitrate content of the spinach. Similarly, Schuphan [23] reports that toxic levels of nitrite are produced when nitrate-rich spinach is tightly packed in crates and transported and stored over a 4-day period.

(3) When frozen spinach is thawed, bacteria originally present in the spinach may convert nitrate to nitrite. Thus Phillips [27] reports that 3 samples of frozen spinach allowed to thaw over a 39-hr period contained 99, 63, and 244 ppm of *nitrite* respectively. Sinios and Wodsak [28] report an instance in which frozen spinach, thawed and stored at 37 °C for 48 hr contained an average of about 400 ppm of nitrite. It should be noted that frozen spinach might readily be subjected to a thawing period of this duration during storage accidents.

(4) When a sterile jar of baby food spinach puree is opened to the air, bacteria capable of converting nitrate to nitrite may enter the food. Phillips [27] tested several jars of baby food in this way and reported no significant increase in nitrite. However, it is obvious that this effect depends considerably on the degree of local sanitation (i.e., prevalence of dust and contamination from intestinal bacteria) so that one cannot rely on the safety of a jar of nitrate-rich baby food once it has been opened. In a number of the European clinical cases of infant methemoglobinemia, it is evident that nitrite was produced in stored spinach puree by bacterial action (see Simon [22]). This accounts for the warning by European pediatricians against the use of stored containers of baby food spinach.

From the foregoing it is evident that at least several common baby food preparations, spinach, squash, beets, and wax beans, may contain nitrate levels of 300 ppm or more, which, on the basis of European clinical reports, can lead to methemoglobinemia. While such a high level of nitrate in the original food is a *necessary* condition for the occurrence of methemoglobinemia, it is not a *sufficient* condition, for the nitrate must be converted to nitrite, by one of the means described above, before reaction with hemoglobin occurs. Thus, eating nitrate-rich food creates a definite risk of methemoglobinemia. This risk can be reduced by avoiding the feeding of such foods to infants suffering from a gastrointestinal upset, by avoiding the use of frozen foods which have been thawed for a day or two, and by avoiding the use of commercial baby foods (i.e., of the suspect vegetables) which have been stored after being opened. If the original vegetable is low in nitrate, all of these risks can be avoided.

Schuphan [23] has reported observations which tie the production of toxic amounts of nitrite in spinach directly to fertilizer practice. He obtained fresh spinach harvested from land treated with different amounts of inorganic nitrogen fertilizer and studied the conversion of nitrate to nitrite during a 4-day period of transport and storage. In the case of spinach fertilized with 320 kgm of nitrogen per hectare, the leaves contained about 3500 mgm of nitrate per 100 gm dry weight and no nitrite at harvest; after the 4-day period the spinach contained about 360 mgm of nitrite per 100 gm dry weight and the nitrate level had fallen about 30%. Spinach fertilized with 80 kgm of nitrogen per hectare contained at harvest about 500 mgm of nitrate per 100 gm and no nitrite; after transport and storage no changes in either nitrate or nitrite content were observed. On the basis of these data Schuphan concludes that the medical hazard of spinach in infant feeding arises from the excessive use of nitrogen fertilizer. Simon [22] ascribes the recent appearance of infant methemoglobinemia to the shift, especially in Germany, from spinach production by small market gardeners to large-scale 'industrial' farms, which, according to him, are "totally lacking in experience with fertilization".

Unfortunately, there appear to be no detailed studies of the recent trends in the actual levels of nitrogen fertilizer used in the production of specific vegetable crops in the United States, nor comparable year-to-year studies of the resultant nitrate levels in the crops. Jackson *et al.* [29] have made a single comparison of nitrate contents of various vegetable crops grown in 1964 with data for 1907 crops reported by Richardson [30] and find no significant differences. However given the differences in the geographical origins of the 1907 and 1964 crops, lack of information as to the fertilizer practices actually used in growing the crops, and differences in methodology, no firm conclusions can be based on this single comparison.

In contrast, indirect evidence does indicate that the nitrate contents of vegetables grown in the United States have been increasing and that this can be traced to more intensive use of nitrogen fertilizer. Thus Johnson [31] reports that the nitrate content of tomatoes is directly related to the amount of nitrogen fertilizer used. At 328 lb of nitrogen per acre, tomatoes contained 39 ppm of nitrate; with 408 lb of nitrogen per acre, nitrate rose to 57 ppm; at 488 lb of nitrogen per acre, nitrate was 97 ppm; at 568 lb of nitrogen per acre, nitrate was 144 ppm. Johnson found that tomatoes containing 53–78 ppm of nitrate were corrosive to the tinplate of canning containers. Since nitrate-induced corrosion of tinplate by canned vegetables is a phenomenon which has been of concern to canners only in the last few years, we can infer that recent increases in the use of nitrogen fertilizer are indeed causing comparable increases in the nitrate content of commercial vegetable crops. The increasing importance of nitrate poisoning in livestock, which has been widely noted, also suggests that the nitrate levels of crops which are used in livestock feeds have been increasing in recent years.

6. Nitrogen Compounds in the Air

That the atmosphere now contains significant amounts of ammonia and nitrogen oxides is revealed by extensive analyses of the nitrogen content of rain and snow [32]. These compounds are washed out of the atmosphere by precipitation and in the process nitrate (and very little nitrite) is formed both from nitrogen oxides and probably

by oxidation of ammonia. Formation of ammonia from nitrogen gas in the atmosphere appears to be highly unlikely. The observed concentration of nitrogen oxides
in the air cannot be accounted for by lightning discharges [33]; however, it is possible
that some photochemical fixation of nitrogen gas does occur. Generally, natural
atmospheric processes cannot account for the observed concentrations of ammonia
and nitrate in precipitation. Most of these materials clearly enter the atmosphere
from the land. At the coastline and over the ocean their concentrations in precipitation are very low, but over land masses appreciable concentrations of ammonia and
nitrate are found in collected precipitation.

Regional and seasonal variations in the inorganic nitrogen content of precipitation
are quite striking and provide useful clues as to the origins of these materials. Thus,
at Frankfurt-on-Main, Germany, two seasonal maxima in atmospheric nitrogen dioxide are observed. One of these, in April, appears to be due to biological processes
in the soil; the second maximum, in December and January, is associated with fuel
combustion. In contrast, at a rural station in Germany, only a single maximum, in
April, due to soil activity, was observed [34].

In order to evaluate the significance of technological intrusions in this segment of
the nitrogen cycle, we need to consider the possible effects of two activities on the
concentration of nitrogen compounds in precipitation – nitrogen fixation due to
automobiles and certain industrial activities, and fertilizer. As indicated earlier, it is
well-known that automobile exhausts and industrial combustion processes emit
significant amounts of nitrogen oxides and that part of the nitrogen present in the
soil may be released into the air as ammonia and nitrogen oxides. Regional variations
in the nitrate content of precipitation over the U.S. suggested to Junge [35] the influence of both of these technological effects. He observed high concentration of nitrate
over heavily populated and industrialized regions which appear to reflect nitrogen

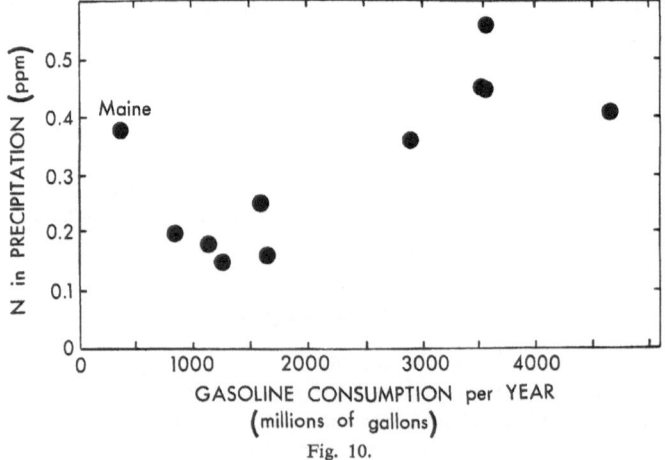

Fig. 10.

fixation in automobile engines and combustion processes. He also notes high seasonal concentration of ammonia and nitrate over agricultural regions (California, Texas, the Corn Belt) which appear to coincide with the time of intensive use of fertilizer.

Recently Lodge *et al.* [36] have reported on the inorganic nitrogen concentration in precipitation from various regions of the U.S. for the period 1960–66. When these results are compared with the regional use of gasoline (as an index of automotive activity) in areas which use relatively little nitrogen fertilizer, the results shown in Figure 10 are obtained. It is evident that the observed inorganic nitrogen content of precipitation correlates well with the regional use of gasoline.

7. The Nitrogen Cycle Outside the U.S.

This discussion has necessarily concentrated on the situation in the U.S., where the needed data are, though very inadequate, relatively available. That the fertilizer problem exists in European countries is evident from the German experience with methemoglobinemia from food-borne nitrate, and from well-known evidence of rapidly increasing eutrophication of surface waters. Japanese reports, where fertilizer use is very intensive, also are indicative of a severe eutrophication problem in recent years. Special problems must be anticipated in tropical and arid regions. That fertilizer practice results in a serious intrusion on the nitrogen cycle in arid regions is evident from Yaalon's recent report [37] on the inorganic nitrogen content of precipitation in Israel. His data for 1962–63 were about one order of magnitude above observations made in 1922–24. This increase in the inorganic nitrogen content of precipitation is attributed to the sharp increase in the use of fertilizer in Israel during the last 40 yr. In fact, Yaalon reports that the total quantity of inorganic nitrogen now deposited annually by precipitation in Israel is about equal to the amount of nitrogen deposited on the land as fertilizer. In effect, the total mass of inorganic nitrogen artificially introduced into the cycle is now carried by the atmosphere.

8. Some Conclusions

This paper has been an inquiry into the state of the nitrogen cycle, chiefly in the U.S., as it has been affected by recent technological intrusions. I believe that the evidence supports the following conclusions:

(1) The natural nitrogen cycle in the U.S. has been stressed by the growth of urban populations in the last generation. However, the stress on the cycle due to the technological introductions of inorganic forms of nitrogen is considerably more severe. This is true because (a) the technological stress has grown much more rapidly than the biological one, and is now about equal in magnitude to the nutritional cycle as a whole; (b) the technological stress intrudes on the natural cycle at its most vulnerable point by introducing into it oxidized forms of nitrogen, which can lead to environmental hazards.

(2) There are two main technological stresses on the nitrogen cycle: the introduction of inorganic nitrogen fertilizers into the soil, and the introduction into the air of nitrogen oxides due to automobiles and other combustion processes.

(3) The massive addition of inorganic nitrogen fertilizer to the soil follows a long period in which the organic nitrogen of the soil has been reduced so that nutrient uptake has become relatively inefficient. With the continued increase of fertilizer application, inorganic nitrogen – chiefly nitrate – has been leached from the soil, entering groundwater, and eventually, surface waters and wells.

(4) In heavily farmed areas, nitrate due to fertilizer makes a major contribution to the inorganic nitrogen content of surface waters, and in many places is responsible for water pollution through eutrophic stimulation of algal growth. This process is now a serious source of environmental pollution.

(5) In heavily farmed areas, the nitrate level of surface waters and wells often exceeds the public health standards for acceptable potable water, resulting in a risk to human health from methemoglobinemia.

(6) Some vegetable products in the United States and Canada, including certain baby foods, with a significant frequency, exceed the nitrate levels recommended for infant feeding by pediatricians. European studies indicate that excess nitrate levels in such products are usually the result of intensive use of nitrogen fertilizer.

(7) The inorganic nitrogen content of precipitation has increased noticeably in the last 25–30 yr due to nitrogen oxides released from the soil and produced by automobiles and other combustion processes.

In my view these conclusions confront us with a number of problems which are large in their magnitude, difficult in their complexity, and grave in their import for the nation. In sum, we have in the United States, thrown the nitrogen cycle seriously out of balance. As a result, the oxidized forms of nitrogen, which are in nature maintained at low, steady-state concentrations in air, water, plants and animals have been elevated to levels which threaten the integrity of major processes in the ecosystem, and vital biological processes in animals and man. The present stress on the nitrogen cycle has already produced important environmental hazards and carries the risk of equally serious medical hazards. Clearly, corrective measures are urgent.

If correction of the present imbalance in the nitrogen cycle required only the development of the necessary technical steps, the problem would be difficult enough. Unfortunately, these difficulties are enormously magnified by the fact that the practices which have produced the imbalance are now deeply embedded in the nation's economy. This is evident from their origin in major economic activities: agriculture, industry and transport, and urbanization. The seriousness of the issues which arise from the impact of these activities on the nitrogen balance is perhaps best illustrated by the problem of water pollution from eutrophication. As I have already indicated, urban wastes and farmland drainage are probably about equally responsible for the eutrophication problem in many areas of the nation. Technological means are available for controlling the input of inorganic salts from municipal and industrial waste; essentially this involves the installation of some type of desalting process as a tertiary stage in treatment plants. However, farmland drainage is not so conveniently localized, but reaches surface waters from every shoreline point. There is no conceivable means of recovering such drainage materials before they reach the water. Similarly, there is little possibility of removing excess nitrate from food. I believe that these difficulties will eventually require the *limitation* of the current high rate of use of inorganic nitrogen fertilizer. I hardly need emphasize the explosive consequences of imposing such a restriction on the current farm economy. Intensive use of fertilizer has become a major source of farm income, for in the last 25 yr the costs of land, labor, and machinery have all risen considerably – but the cost of fertilizer has dropped. The farmer's income has become crucially dependent on the intensive use of fertilizer, particularly nitrogen. If, as I believe, it becomes necessary to limit the use of inorganic nitrogen fertilizer, the present system of farming is faced with a massive dislocation.

Various palliative measures are possible: the use of slow-release fertilizers, more careful seasonal timing of fertilizer applications, precise determination of the amount and rate of fertilizer application. However, such measures do not get at the root of the problem – the continuing depletion of the organic nitrogen content of the soil, and with it, of the soil's efficiency in transferring nutrient to the crop. Attention to this fundamental problem is essential, I believe, if we are to attain a long-term equilibrium in the nitrogen cycle. This will require a serious revision in our approach to agricultural production – for example, the development of new ways to return organic matter to the soil (which would, of course, at the same time alleviate water pollution due to urban wastes), and more intensive use of soil-building pasturage systems. In turn, such an approach will confront us with major economic difficulties for it will require a relatively slow yield from agricultural investment, in contrast with present efforts to find more rapid ways to regain the farmer's capital investment. This approach will also confront the economic inertia due to the present investment of the chemical industry in fertilizer production. At the present time the nitrogen fertilizer industry is in a state of over-capacity and is seeking new ways to sell its product – for example, large scale use of nitrogen to fertilize timber crops. If this is done our present environmental problems will become worse.

What can we do to resolve these issues? Clearly much more needs to be done. Most important is the lack of detailed, continuing surveys of the nitrate and nitrite content of surface waters and their relation to contributions from fertilizer and organic wastes. Although the frequency of such surveys has begun to increase recently, they are usually not sufficiently detailed to yield an assessment of the added nitrogen among the various possible sources. Also important are continuing national surveys of nitrogen in rain and snow – in effect a monitoring system which would parallel the very effective national survey of radioactivity operated by the U.S. Public Health Service, Particularly urgent, it seems to me, is the need for detailed national surveys of nitrate and nitrite in food and drinking water, especially as they affect infants. With such information in hand it should be possible to determine the extent of the hazard from excess nitrate, both to human health, and to the stability of the self-purifying systems in surface waters, and to estimate the degree to which such hazards arise from agriculture, from urban and industrial wastes, and from the emissions of combustion plants and automotive engines.

We know from our experience with the fallout problem that even these actions will be insufficient to resolve the issue. When the fallout controversy had progressed to the point that the harmful consequences of fallout radioactivity was widely acknowledged, the conflict moved into the arena of politics. Given the acknowledged fact that fallout radiation could cause some number of genetic defects and cancers in the populations, it could nevertheless be argued that this risk was justified by the over-riding importance of developing nuclear weapons to secure the nation's military strength. There is, of course, no scientific means for resolving such a conflict. A choice which balances some number of leukemia cases against the development of a new nuclear weapon must reflect the value which we place on human life, and our belief in the wisdom and morality of relying on nuclear weapons for the security of the nation. These decisions are, necessarily, moral and political judgments. They comprised the substance of the great public debate that led to the adoption of the Nuclear Test Ban Treaty.

We can expect the same kind of public debate in connection with the artificial intrusion of nitrogen into the biosphere. Suggestions, such as those which I have made here, that we urgently need more information on the problem, are likely to confront the inertia generated by past acceptance of our present agricultural practices. If, as I expect, they will, new data finally make it clear that we need to restrict the use of inorganic nitrogen fertilizer if hazards to health and to the integrity of our surface waters are to be avoided, other objections will be raised. The world's need for food is acute and will increase. Intensive use of nitrogen fertilizer is clearly responsible for a good part of our present level of agricultural productivity. Shall we worsen world famine in order to protect the health of some portion of our infant population and the integrity of our waste-disposal systems?

There will be those who, perceiving these enormous economic, political and moral issues, will prefer to turn away from scientific discussion and public debate. The scientists who confronted the fallout problem, heard such counsel – and rejected it. They chose to speak out, in the profound conviction that the decisions were not theirs to make, but belonged to the people as a whole.

There is a unique relationship between the scientist's social responsibilities and the general duties of citizenship. If the scientist, directly or by inferences from his actions, lays claim to a special responsibility for the resolution of the policy issues which relate to technology, he may, in effect, prevent others from performing their own political duties. If the scientist fails in his duty to inform citizens, they are precluded from the gravest acts of citizenship and lose their right of conscience.

Every major advance in the technological competence of man has enforced revolutionary changes in the economic and political structure of society. The present age of technology is no exception to this rule of history. We already know the enormous benefits it can bestow, and we have begun to perceive its frightful threats. The political crisis generated by this knowledge is upon us.

Science can reveal the depth of this crisis, but only social action can resolve it.

Science can now serve society by exposing the crisis of modern technology to the judgment of all mankind. Only this judgment can determine whether the knowledge that science has given us shall destroy humanity or advance the welfare of man.

Acknowledgment

This investigation was supported by grant P 10 ES 00139 from the Public Health Service, Department of Health, Education and Welfare through the Center for the Biology of Natural Systems, Washington University, St. Louis.

References

1. Smith, G. E.: Sanborn Field, Bulletin 458, University of Missouri, College of Agriculture, Agricultural Experiment Station, Columbia, Missouri (1942).
2. Plant Food Review, Winter (1965).
3. Allison, F. E.: *Advances in Agron.* **18**, 219 (1966).
4. Larson, T. E. and Larson, B. O.: *Quality of Surface Waters in Illinois*, Illinois State Water Survey, Urbana, Illinois (1957); *Interim Report on the Presence of Nitrate in Illinois Surface Waters*, Illinois State Water Survey, Urbana, Illinois, November (1968).
5. Feth, J. H.: *Water Resources Res.* **2**, 41 (1966).
6. *Eutrophication – A Review*, California State Water Quality Control Board, Publication #34, (1967).
7. Larson, T. E.: personal communication (1968).
8. U.S. Geological Survey, Water Supply Papers Nos. 1198, 1251, 1291, 1351, 1401, 1451, 1521, 1572, 1643, 1743, 1883, 1943, 1949 and USGS Water Quality Records, Nebraska, North Dakota, South Dakota (1964–65).
9. National Water Quality Network, Annual Compilation of Data 1957–63, U.S. Department, H.E.W.
10. Corey, R. B.: *Excessive Water Fertilization*, Report to the Water Subcommittee, Natural Resources Committee of State Agencies, Madison, Wisconsin (1967).
11. Lake Erie Report, Federal Water Pollution Control Administration, August (1968).
12. Commoner, B.: *The Killing of a Great Lake*, World Book Year Book, Field Enterprises Education Corp., Chicago (1968).
13. See for example, Smith, G. E.: *Nitrate Problems in Plants and Water Supplies in Missouri*, Contr. No. 2830, Mo. Agricultural Exp. Station (1964).
14. Stewart, B. A., Viets, Jr., F. G., and Hutchinson, G. L.: *J. Soil Water Conserv.* **23**, 12 (1968).
15. See for example, Engberg, R. A.: Nebraska Water Survey Paper 21, University of Nebraska, Lincoln, Nebr. (1967) and Reference [13].
16. McHarg, I.: Report to Twin-Cities Metropolitan Council (1968).
17. Rochester (Minn.) Post-Bulletin, November 21, p. 17 (1967).
18. Doneen, L. D.: *Effects of Soil Salinity and Nitrates on Tile Drainage in San Joaquin Valley, California*, Water Sci. and Eng. Paper 4002. Sacramento, California (1966); see also, *San Joaquin Master Drain*, Appendix, Part C., Federal Water Pollution Control Administration, Southwest Region (1968).
19. Johnston, W. R., Ittihadieh, F., Daum, R. M., and Pillsbury, A. F.: *Proc. Soil Sci. Soc.* p. 287 (1965).
20. Occurrence of Nitrate in Groundwater Supplies in Southern California, Bureau of Sanitary Engineering, California State Dept. of Health, February (1963).
21. Brown, J. R. and Smith, G. E.: *Nitrate Accumulation in Vegetable Crops as Influenced by Soil Fertility Practices*, Res. Bulletin 920, University of Missouri Agricultural Exp. Station (1967).
22. Simon, C.: 'L'intoxication par les nitrites après ingestion d'épinards', *Arch. Fr. Pediat.* **23**, 231 (1966); 'Nitrate Poisoning from Spinach', *The Lancet* **1**, 872 (1966); also Simon *et al.*: *Z. Kinderheilk.* **91**, 124 (1964).

23. Schuphan, von, W.: 'Der Nitratgehalt von Spinat (Spinacia oleracea L.) in Beziehung zur Methämoglobinämie der Säuglinge', *Z. Ernährungswiss.* **5**, 207 (1965).
24. Kamm, L., McKeown, G. C., and Smith, D. M.: 'New Colorimetric Method for the Determination of the Nitrate and Nitrite Content of Baby Foods', *A.O.A.C.* **48**, 892 (1965).
25. Boulerice, M.: personal communication, Dept of Health, City of Montreal (1968).
26. Fandre, M., Coffin, R., Dropsy, G., and Bergel, J. P.: Epidémie de gastroentérite infantile à Escherichia coli O 127 B8 avec cyanose méthémoglobinémique', *Soc. Péd. Fr.*, Réun. Paris, **19**, 1129 (1962).
27. Phillips, W. E. J.: 'Changes in the Nitrate and Nitrite Contents of Fresh and Processed Spinach During Storage', *Agr. Food Chem.* **10**, 88 (1968).
28. Sinios, A. and Wodsak, W.: 'Die Spinatvergiftung des Säuglings', *Deut. Med. Wochschr.* **90**, 1956 (1965).
29. Jackson, W. A., Steel, J. S., and Boswell, V. R.: 'Nitrates in Edible Vegetables and Vegetable Products', *Am. Soc. Hort. Sci.* **90**, 349 (1967).
30. Richardson, W. D.: *J. Am. Chem. Soc.* **29**, 1747 (1907).
31. Johnson, J. H.: 'Internal Can Corrosion Due to High Nitrate Content of Canned Vegetables', *Proc. Flor. State Hort. Soc.* **79**, 239 (1966).
32. Eriksson, E.: *Tellus* **4**, 215, 271 (1952).
33. Gambell, A. W. and Fisher, D. W.: *J. Geophys. Res.* **69**, 4203 (1964).
34. Georgii, H.: *J. Geophys. Res.* **68**, 3963 (1963).
35. Junge, C. E.: *Trans. Amer. Geophys. Union* **39**, 241 (1958).
36. Lodge, J. P., Jr.: *Chemistry of United States Precipitation*, Nat. Center for Atmospheric Research, Boulder, Colorado (1968).
37. Yaalon, D. H.: *Tellus*, **16**, 200 (1964).

For Further Reading

1. Barry Commoner, *Science and Survival*, Viking Press, New York, 1966.
2. W. V. Bartholomew (ed.), *Soil Nitrogen*, American Society of Agronomy, No. 10, Madison, Wis., 1966.
3. Barry Commoner, *The Closing Circle*, Knopf, New York, 1971.

THE DYNAMICS OF NITROGEN TRANSFORMATIONS
IN THE SOIL*

D. R. KEENEY and W. R. GARDNER

Dept. of Soil Science, University of Wisconsin, Madison, Wisc., U.S.A.

Abstract. As the nation's agriculture becomes more intensive and the need to use the soil as a waste disposal system increases, contaminations of ground water aquifers with nutrients, particularly nitrate, can be expected to become a serious and widespread problem. Already surface water supplies are showing the signs of excessive fertility while reports of high nitrate wells increase in frequency. Nitrogen compounds in surface waters often are associated with enhanced eutrophication rates while high nitrate in waters consumed by humans or animals may give rise to health problems.

1. Introduction

Concern over the pollution of ground water aquifers by nitrate has been expressed in many states. The recent literature contains references to problems in California, Missouri and Colorado while high nitrate wells were reported in Iowa, Wisconsin and Minnesota over two decades ago. The nitrogen cycle, particularly the soil phase, has been the subject of intensive study for many years. Nitrogen undergoes complex transformations between its various forms by means of chemical, physical, and biological processes. Management of this complex system and control of nitrate concentrations in ground and surface waters requires integration of research effort.

2. Environmental Problems Associated with High Nitrates in Ground Water

The most critical problem associated with nitrate levels in ground water aquifers is the possible deleterious effects on human and animal health. With animals, nitrate toxicity can result in abortions, lowered productivity, etc. With humans, only infants under six months of age not on solid foods apparently suffer lethal effects [42], and the problem does not appear to be serious at this time. Nitrate toxity is caused by microbial reduction of NO_3^- to NO_2^- in the gut, and reaction of NO^- with ferrous iron in blood hemoglobin to ferric iron giving methemoglobin, which cannot transport oxygen.

High nitrates in irrigation water can be detrimental to agriculture. Quality and yield of crops such as grapes and sugar beets can be reduced if excess amounts of N are applied at the wrong time [40]. Industrial processes also may be adversely influenced by high N concentrations in the water.

Nitrate in ground waters can contribute to the nutrient load, and hence eutrophi-

* Contribution from the Department of Soil Science, University of Wisconsin, Madison, 53706, and published with the approval of the Director, Research Division of the College of Agricultural and Life Sciences. Supported in part by a grant from the Tennessee Valley Authority.

cation of surface waters, since ground water may become surface water due to 'inter-flow'. Brezonik and Lee [9] have estimated that ground water nitrate contributes 50% of the N in Lake Mendota sediments and waters. Little data are available for streams, but the fact that nitrate often occurs in streams during winter months when runoff is negligible indicates that ground water nitrate is a factor in stream pollution [16].

3. Sources of Nitrogen

The amount of N required to bring a soil percolate to the 10 ppm 'critical' NO_3-N level is 2.27 lb per acre-in. If, e.g., deep percolation out of the root zone is 6 in. yr^{-1}, only 13.6 lb $acre^{-1}$ of N as nitrate would be required to give the critical concentration. Thus we are dealing with a problem which must account for a small fraction of the total available N in an agricultural situation.

Nitrogen entering the soil undergoes various transformations depending upon the form of N added and the physical, chemical and environmental factors operating in the soil.

Inputs of N can be grouped under precipitation, organic wastes and plant debris, fertilizers, and N fixed biologically, with the tacit understanding that N in soil organic matter must also be considered as a potential source of nitrate in percolates.

Corey et al. [13] have estimated the sources of available N in Wisconsin cultivated soils (Table I). These data would differ among states as well as regionally and locally within the state. However, they indicate the relatively small input of fertilizer N in relation to native soil N (O.M. decomposition) and animal waste disposal.

Improved technology and increased competition have dramatically lowered N fertilizer prices recently, and the amount of fertilizer N used has increased propor-tionally. In view of the 'critical' 2.3 lb $acre^{-1}$ in. of percolate value, contributions of fertilizer N to the nitrate levels in certain shallow ground water aquifers may be greater than found in some investigations [36, 40].

Disposal of organic wastes (sewage, manure) has been implicated as a causal factor of high ground water nitrates [32, 36, 38, 40]. Disposal of high concentrations of

TABLE I

Sources and estimated amounts of available
N in cultivated soils in Wisconsin (1965)

Source	Available N	
	lbs A^{-1}	%
Fertilizer	10	9
Legumes	12	10
Precipitation	8	7
O.M. decomposition	45	38
Manure	42	36
Total	117	100

nitrogenous wastes in situations where aerobic conditions occur in at least part of the profile and provisions for adequate removal of nitrate by plant growth are not made would almost certainly lead to zones of high nitrate waters. In some studies [32, 38] ammonium and nitrite have also been found in relatively high levels under feedlots, and Hutchinson and Viets [23] have reported high amounts of ammonia absorption by surface waters near cattle feedlots.

While the absolute amounts of nitrogen in precipitation are not large [16], the seasonal fluctuations and relative yearly constancy of precipitation nitrogen indicates that this source must be evaluated, particularly if undisturbed ecosystems are included in a watershed evaluation.

Biological N fixation occurs through two pathways; fixation of N by free-living soil bacteria and algae and by bacteria living in symbiosis with higher plants [24, 31]. The best estimates indicate that the contribution to the N balance by free-living microorganisms is of little significance agronomically. Relatively large amounts of N (10–200 lb of N per acre per year) may be fixed by the genus Rhizobium living in association with leguminous plants. N fixation by bacteria living with non-legumes (predominately angiosperms) has been found to be an important source of N for these plants. The importance of N fixation in contributing to the N status of undisturbed soils has been pointed out by Stout and Burau [40]. Because biologically fixed N is largely retained in plants, transformations of these forms of N is viewed much the same as organic wastes; only the estimates of the amount of N fixed are difficult.

Many articles stressing N pollution of natural waters have tended to overlook the contribution from soil organic matter. If a soil contains 0.2% organic N and 1–3% yr^{-1} is mineralized [5] from 20 to 60 lb of N $acre^{-1}$ is released. This release is stimulated markedly by cultivation, and cultivation of soils has been implicated as providing the bulk of the nitrate in some soil profiles [38, 40]. Considering the slow movement of percolate waters in many areas, the high nitrate levels in some aquifers may simply be the reflection of the onset of farming a century or more ago.

The comprehensive review by Feth [16] indicates that geologic sources of nitrate have also largely been ignored. Nitrates occur in cave, caliche and playa deposits [36], and most rocks contain N. In fact, nearly 98% of the world's N is contained in fundamental rocks [37]. Carbonate rocks also contain nitrate [16]. Analysis of a large number of limestone samples from Wisconsin [27] has shown that they may contain up to 15 ppm nitrate – N. Feth [16] pointed out the fact that limestone terranes seem to be favored habitats of water high in nitrate, and it would appear that geologic contributions of nitrate should be considered when evaluating sources of nitrate in ground waters.

Recently Kohl et al. [28] attempted to use slight differences in the natural isotopic composition of soil N, atmospheric N and fertilizer N to estimate the contribution of fertilizer N to the nitrate levels of Sangamon River and Lake Decatur, Illinois. A detailed discussion of this approach is outside the scope of this review. It has been criticized by a number of soil scientists with expertise in the use of ^{15}N [22] and recent experimental evidence [8, 14] would indicate that the background variations in the natural ^{15}N content of nitrate-N in soils are sufficient to invalidate the approach of Kohl *et al.* [28]. These variations are due to the isotopic differences in the several sources of nitrifiable nitrogen and to isotopic discriminations that occur during every fractional conversion of one form of nitrogen to another. These effects can be cumulative or can cancel one another out, and will vary widely in time and space. Add to this the difficulty of assessing the N cycle and hydrology on large watersheds and one realizes that research on evaluating the actual contribution of N fertilizers to surface or groundwater requires the use of ^{15}N labeled or depleted fertilizers [14].

4. The Nitrogen Cycle and the Nitrate Budget of the Soil Profile

In soils, the predominant storage reservoir of N is soil organic matter (Figure 1). Breakdown of plant and animal debris during microbiological action, and immobilization of inorganic forms of N in living microbial tissue continually adds to soil organic N, while microbial breakdown of organic N compounds continually depletes soil organic N [3]. Considering the dynamic nature of these processes, the relative stability of soil organic N is quite remarkable. The reasons for organic N stability are numerous, and include formation of resistant heterocyclic N compounds, clay-organic matter complexes, and lack of sufficient carbonaceous energy material for complete breakdown [6].

The amount of N bound in living microbial cells at a given time is usually estimated

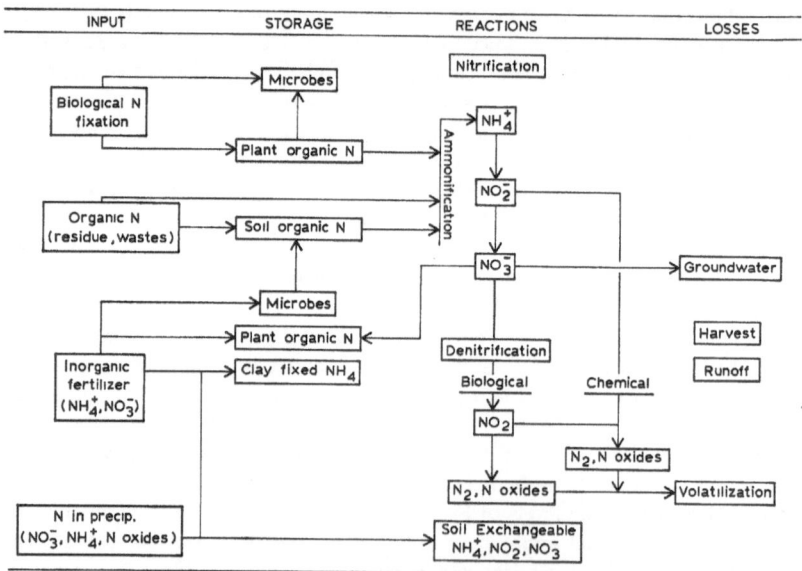

Fig. 1.   Detailed nitrogen cycle.

at less than 1 % of the total N [6]. Clay fixed NH_4, which is relatively unavailable for plant uptake or microbial oxidation, can constitute from 1 to 25 % of the total soil N [6, 30].

Ammonification of organic forms of N by heterotrophs occurs under a wide range of pH, temperature and moisture conditions [3, 11]. In contrast, autotrophic nitrification ($NH_4 \rightarrow NO_2 \rightarrow NO_3$) is easily inhibited by unfavorable conditions [1]. Several heterotrophic nitrifiers have been identified, but their importance in nitrification in soils is not known [1]. However, heterotrophic nitrifiers may be important in nitrate formation in organic-rich environments such as composts and manure piles.

The environmental variables affecting nitrification rates are discussed in several reviews [1, 11, 17, 33]. Briefly, optimal nitrification occurs at 30–35 °C, pH near 7, oxygen concentrations near 20 % and at about two-thirds water saturation. Formation of nitrite by *Nitrosomonas spp.* can be inhibited by a large number of compounds (Alexander, 1965). Oxidation of nitrite to nitrate (*Nitrobacter spp.*) is inhibited by high pH's and free NH_3 giving rise to phytotoxic accumulation of nitrite in some alkaline soils treated with anhydrous NH_3 or urea [1].

Denitrification (reduction of nitrite or nitrate to gaseous forms of N which then can be lost to the atmosphere) can occur by two distinct pathways (Figure 1). Biological reduction of nitrate by soil heterotrophs occurs when insufficient oxygen is available and the organisms thus must use another electron source. In most cases, carbonaceous material for energy must be present, which is the reason significant denitrification does not occur in ground water aquifers. Denitrification by chemical reactions occurs only with nitrite [10]. Nitrite is unstable in acid soils, giving rise to nitrate and N oxides simultaneously. Nitrite also reacts with soil constituents with the fixation of N in organic matter and gaseous N oxide formation [7, 10]. The importance of denitrification reactions in soils is just beginning to be appreciated [43] and this difficult aspect of the N cycle must be more fully understood and evaluated before predictions on N balance sheets, biological nitrate removal from soils, and N fertilizer rates can be predicted accurately [2].

Addition of carbonaceous material to soils stimulates microbial growth. The N requirements of the microbes are met from the inorganic N pool in the soil [3]. These microbes of course decompose, and become part of the soil organic N.

N is the most abundant of the nutrients in plants that must come from the soil or fertilizers [41]. Most crops take up N roughly in proportion to their yield, and as such effectively immobilize N from leaching loss, as evidenced by the fact that fallow soils often have higher profile nitrate contents than cropped soils [32, 38]. Many agronomic crops effectively remove nitrate from several feet of the profile, and use of deep rooted crops such as alfalfa in rotation with shallow rooted crops has been advocated to reduce nitrate levels in soils [38].

5. Nitrate Leaching from the Soil Profile and Transport to Surface or Ground Water

Nitrates are particularly mobile and move with the percolating water to the water table. Both the concentration of the percolate and the rate of leaching depend heavily upon the climate (or irrigation regime) as well as the amount of nitrogen applied as fertilizer. The seasonal nature of evaporation from a land surface, whether vegetated or bare, results in an intermittent leaching of most soil profiles. This leaching often results in annual maxima and minima in the concentration of soluble ions in the soil solution [39]. In general, there is not sufficient mixing due to diffusion and dispersion to obliterate these peaks before they reach the water table. The vertical distance between peaks is given by the net percolation for any one season divided by the soil water content between peaks. The speed and direction which the nitrate moves below the water table is determined by the ground water flow regime [18]. The precision with which the ground water flow can be mapped depends largely upon the amount of effort expended since the mechanism of flow is well known. Random sampling of the ground water without regard to the flow paths can be expected to obscure the picture. If the ground water flow net and the permeability are determined with even a modest degree of precision it should be possible to predict residence times for peak concentrations of nitrates or other solutes and also to predict the actual path from the point at which they enter the ground water to the point where they return in the surface water.

6. Research Needs

Concern has been expressed that the additional burden of nitrogen from man's activities (wastes, auto exhaust emissions, fertilizers) will increase the nitrate load of our waters to the point that surface waters are eutrophied and subsurface waters are toxic [12]. This thesis has resulted in a veritable flurry of panels, hearings, conferences, commissions and reviews (see References section). The general theme of these reports is that a problem may or may not exist, but that with the information at hand, no definite conclusions can be reached. Scientists and the funding agencies have reacted predictably to the situation by paying increased attention to the environmental aspects of the nitrogen cycle. Unfortunately, to date, little imaginative research has been applied to the problem. Most technical reports (of which several dozen have appeared since 1969) have concerned monitoring efforts in field situations where little control of the environment exists. These reports, while useful, can in no way answer the critical question of sources of nitrate contamination of the groundwater. They generally do not take into consideration the nitrogen cycle or the hydrology of the area in question.

Several processes of the nitrogen cycle remain to be quantitated. Of these, nitrogen fixation is one of the inputs of which little information is available. Much more information on the primary output, denitrification, also is needed. For example, recent research has shown that denitrification in lake sediments may remove much of the nitrate entering the lake via seepage [26].

Interdisciplinary effort is required to develop the ability to manage the nitrogen cycle through wise use of nitrogen fertilization and waste disposal practices to minimize nitrate accumulation in the environment. This research must develop quantitative mathematical expressions for the principal nitrogen input, transformation and output rates which determine the concentration of nitrate in the soil solution and combine these models with the equations describing the movement of soil water and solutes through the soil profile to the groundwater aquifer or stream. An example of the approach needed is given in a recent report on systems analysis of the terrestrial nitrogen cycle [15].

References

1. Alexander, M.: 'Nitrification', *Agron.* **10**, 307–343 (1965).
2. Allison, F. E.: 'Evaluation of Incoming and Outgoing Processes that Affect Soil Nitrogen', *Agron.* **10**, 573–606 (1965).
3. Bartholomew, W. V.: 'Mineralization and Immobilization of Nitrogen in the Decomposition of Plant and Animal Residues', *Agron.* **10**, 285–306 (1965).
* Bartholomew, W. V. and Kirkham, D.: 'Mathematical Descriptions and Interpretations of Culture induced Soil Nitrogen Changes', *Trans. 7th Intern. Congr. Soil. Sci.*, Madison, Wisc., pp. 471–477 (1960).
* Biggar, J. W. and Corey, R. B.: 'Agricultural Drainage and Eutrophication', *Agr. Sci. Rev.* **5**, 22–28 (1967).
4. Black, T. A., Gardner, W. R., and Tanner, C. B.: 'Water Storage and Drainage under a Row Crop on a Sandy Soil', *Agron. J.* **62**, 48–51 (1970).
5. Bremner, J. M.: 'Organic Nitrogen in Soils', *Agron.* **10**, 93–149 (1965).
6. Bremner, J. M.: 'Nitrogenous Compounds', in A. D. McLaren and G. N. Peterson (eds.), *Soil Biochemistry*, Marcel Dekker, New York, pp. 19–66 (1967).
7. Bremner, J. M. and Nelson, D. W.: 'Chemical Decomposition of Nitrite in Soils', *Intern. 9th Congr. Soil. Sci.* Vol. II, pp. 495–503 (1968).
8. Bremner, J. M. and Tabatabai, M. A.: 'Nitrogen-15 Enrichment of Soils and Soil-Derived Nitrate', *J. Environ. Qual.* **2**, 363–365 (1973).
9. Brezonik, P. L. and Lee, G. F.: 'Denitrification as a Nitrogen Sink in Lake Mendota, Wis.', *J. Environ. Sci. Technol.* **2**, 120–125 (1968).
10. Broadbent, F. E. and Clark, F.: 'Denitrification', *Agron.* **10**, 344–359 (1965).
* Byerly, T. C.: 'Nitrogen Compounds Used in Crop Production', this volume pp. 377–382.
11. Campbell, N. E. R. and Lees, H.: 'The Nitrogen Cycle', in A. D. McLaren and G. N. Peterson (eds.), *Soil Biochemistry*, Marcel Dekker, New York, pp. 194–215 (1967).
12. Commoner, B.: 'Threats to the Integrity of the Nitrogen Cycle: Nitrogen Compounds in Soil, Water, Atmosphere and Precipitation', this volume pp. 341–366.
13. Corey, R. B., Hasler, A. D., Lee, G. F., Schraufnagel, F. H., and Wirth, T. L.: *Excessive Water Fertilization*, Rept., Water Subcommittee, Natural Resources Committee of State Agencies, Madison, Wis., Jan. 31, 54 pp (1967).
* Delwiche, C. C.: 'The Nitrogen Cycle', *Sci. Amer.* **233**, 136–146 (1970).
14. Edwards, A. P.: 'Isotopic Traces Techniques for Identification of Sources of Nitrate Pollution', *J. Environ. Qual.* **2**, 382–387 (1973).
15. Endelman, F. S., Northup, M. L., Keeney, D. R., Boyle, J. R. and Hughes, R. R.: 'A Systems Approach to an Analysis of the Terrestrial Nitrogen Cycle', *J. Environ. Sys.* **2**, 3–19 (1972).
* Environmental Pollution Panel, President's Science Advisory Committee: *Restoring the Quality of our Environment*, U.S. Govt. Printing Office, Washington, D.C. (1965).
16. Feth, J. S.: 'Nitrogen Compounds in Natural Water – A Review', *J. Water Resources Res.* **2**, 41–58 (1966).
17. Frederick. L. R. and Broadbent, F. E.: 'Biological Interactions', in M. H. McVickar, W. P. Martin, I. E. Miles, and H. H. Tucker (eds.), *Agricultural Anhydrous Ammonia*, Agricultural Ammonia Inst., Memphis, Tenn., pp. 189–212 (1966).
18. Gardner, W. R.: 'Movement of Nitrogen in Soil', *Agron.* **10**, 550–572 (1965).
19. Gardner, W.R.: *Water Uptake and Salt Distribution Patterns in Saline Soils*, International Atomic Energy Agency Symposium on the Use of Radioisotopes and Radiation Techniques in Soil Physics and Irrigation Studies, Istanbul, 335–341 (1967).

20. Goldberg, M. C.: 'Sources of Nitrogen in Water Supplies' in T. L. Willrich and G. E. Smith (eds.), *Agricultural Practices and Water Quality*, Iowa State Univ. Press, Ames., pp. 94–124 (1970).

21. Harmsen, G. W. and Kolenbrander, G. J.: 'Soil Inorganic Nitrogen', *Agron.* **10**, 43–92 (1965).

22. Hauck, R. D., Bartholomew, W. V., Bremner, J. M., Cheng, H. H., Edwards, A. P., Keeney, D. R., Legg, J. O., Olsen, S. R., and Porter, L. K.: 'Use of Variations in Natural Nitrogen Isotope Abundance for Environmental Studies: A Questionable Approach', *Science* **177**, 453–454 (1972)

23. Hutchinson, G. L. and Viets, Jr., F. G.: 'Nitrogen Enrichment of Surface Water by Absorption of Ammonia Voltalized from Cattle Feedlots', *Science* **166**, 514–515 (1969).

24. Jensen, H. H.: 'Nonsymbiotic Nitrogen Fixation', *Agron.* **10**, 436–480 (1965).

25. Keeney, D. R.: 'Nitrates in Plants and Waters', *J. Milk Food Technol.* **33**, 425–432 (1971).

26. Keeney, D. R.: *The Fate of Nitrogen in Aquatic Ecosystems*, Lit. Rev. No. 3, Eutrophication Information Program, Univ. Wis. Resources Center, Madison, 59 p. (NTIS ≠ PB-209 217) (1972).

27. Keeney, D. R. and Chalk, P. M.: 'Nitrate and Ammonium Content of Wisconsin Limestone', *Nature* **217**, 42 (1971).

* Keeney, D. R. and Walsh, L. M.: 'Available Nitrogen in Rural Ecosystems: Sources and Fate', *Hort Science* **7**, 219–223 (1972).

28. Kohl, D. A., Shearer, G. B., and Commoner, B.: 'Fertilizer Nitrogen: Contribution to Nitrate in Surface Water in a Corn Belt Watershed', *Science* **174**, 1331–1334 (1971).

* Martin, W. P., Fenster, W. E., and Hanson, L. D.: 'Fertilizer Management for Pollution Control', in T. L. Wilkich and G. E. Smith (eds.), *Agricultural Practices and Water Quality*, pp. 142–158 (1970).

29. Mortland, M. M. and Wolcott, A. R.: 'Sorption of Inorganic Nitrogen Compounds by Soil Materials', *Agron.* **10**, 150–197 (1965).

* National Academy of Sciences: *Accumulation of Nitrate*, Nat. Acad. Sci., Wash., D.C. (1972).

30. Nommik, H.: 'Ammonium Fixation and Other Reactions Involving a Non-Enzymatic Immobilization of Mineral Nitrogen in Soil', *Agron.* **10**, 198–258 (1965).

31. Nutman, P. S.: 'Symbiotic Nitrogen Fixation', *Agron.* **10**, 360–383 (1965).

32. Olsen, R. J.: *Effect of Various Factors on Movement of Nitrate Nitrogen in Soil Profiles and on Transformations of Soil Nitrogen*, Ph.D. thesis, Department of Soils, Univ. of Wisconsin, Madison, Wis. (1969).

33. Quastel, J. H. and Scholefield, P. G.: 'Biochemistry of Nitrification in Soil', *Bacteriol. Rev.* **15**, 1–53 (1951).

34. Raats, P. A. C. and Scotter, D. R.: 'Dynamically Similar Motion of two Miscible Constituents in Porous Mediums', *J. Water Resources Res.* **4**, 561–568 (1968).

35. Scarsbrook, C. E.: 'Nitrogen Availability', *Agron.* **10**, 481–502 (1965).

36. Smith, G. E.: 'Fertilizer Nutrients as Contaminants in Water Supplies', in N. C. Brady (ed.), *Agriculture and the Quality of our Environment*, Amer. Assn. Adv. Sci., Wash. D.C., pp. 173–186 (1967).

* Stanford, G., Englund, C. B. and Taylor, A. W.: *Fertilizer Use and Water Quality*, ARS 41-168. United States Dept. of Agriculture, Beltsville, Md., (1970).

37. Stevenson, F. J.: 'Origin and Distribution of Nitrogen in Soil', *Agron.* **10**, 1–42 (1965).

* Stevenson, F. J. and Wagner, G. H.: 'Chemistry of Nitrogen in Soils', in *Agricultural Practices and Water Quality*, Iowa State Univ. Press, Ames., pp. 125–141 (1970).

38. Stewart, B. A., Viets, Jr., F. G., and Hutchinson, G. L.: 'Agriculture's Effect on Nitrate Pollution of Ground Water', *J. Soil Water Conserv.* **23**, 13–15 (1968).

39. Stewart, B. A., Viets, F. G., Jr., Hutchinson, G. L., Kemper, W. D., Clark, F. E., Gairbourn, M. L., and Strauch, F.: *Distribution of Nitrates and Other Water Pollutants under Fields and Corrals in the Middle South Platte Valley of Colorado*, USDA-ARS 41-134, (1967).

40. Stout, P. R., and Burau, R. G.: 'The Extent and Significance of Fertilizer Buildup in Soils as Revealed by Vertical Distributions of Nitrogenous Matter between Soils and Underlying Water Reservoirs', in N. C. Brady (ed.), *Agriculture and the Quality of Our Environment*, Amer. Assn. Adv. Sci., Wash., D.C., pp. 283–310 (1967).

* University of Illinois, College of Engineering: *Nitrate and Water Supply: Source and Control*, Proceedings, Twelfth Sanitary Engineering Conference, Engineering Pub. Office, Urbana, (1970).

41. Viets, F. G.: 'The Plant's Need for and Use of Nitrogen', *Agron.* **10**, 503–549 (1965).

* Viets, F. G. and Hageman, R. H.: *Factors Affecting the Accumulation of Nitrate in Soil, Water and Plants*, Agricultural Handbook No. 413, U.S. Govt. Printing Office, Wash., D.C., (1971).
* Vincent, J. M.: 'Environmental Factors in the Fixation of Nitrogen by the Legume', *Agron.* **10**, 384–435 (1965).
* Wadleigh, C. H.: *Wastes in Relation to Agriculture and Forestry*, USDA, Misc. Pub. No. 1065, U.S. Govt. Printing Office, Washington, D.C., (1968).
* Webber, L. R. and Elrick, D. E.: 'The Soil and Lake Eutrophication', in *Proc. 10th Conf. on Great Lakes Resarch*, pp. 404–412 (1967).
* Wheeler, E. W.: 'Fertilizers Are Not Threatening our Environment', *Farm Chemicals* **132**, 45–48 (1969).
* Webber, L. R. (ed.): *Nitrogen in Soil and Water*, Proceedings of a Symposium, Dept. of Soil Science, Ceniv. Guelph, Ontario, Canada, p. 169 (1971).
42. Wright, M. J. and Davison, K. L.: 'Nitrate Accumulation in Crops and Nitrate Poisoning in Animals', *Advance Agron.* **16**, 201–256 (1964).
43. Wullstein, L. H.: 'Soil Nitrogen Volatilization. A Case for Applied Research', *Agron. Sci. Rev.* **5**, 8–13 (1967).

For Further Reading

1. Black, C. A.: *Soil-Plant Relationships*, 2nd edition, John Wiley and Sons, Inc. New York, 1968, pp. 70–152, 405–557.
2. Hine, R. L. (ed.), Water Use: Principles and Guidelines for Planning and Management in Wisconsin, Wisconsin Chapter, Soil Conservation Society of America, Madison, Wis., 1969.
3. McVickar, M. H., W. P. Martin, I. E. Miles and H. H. Tucker (eds.), *Agricultural Anhydrous Ammonia*, American Society of Agronomy, Madison, Wis., 1966.
4. *Yearbook of Agriculture. Soil*, The United States Dept. of Agr., Washington, D.C., 1957.

* Further references, not specifically mentioned in the text.

NITROGEN COMPOUNDS USED IN CROP PRODUCTION

T. C. BYERLY

Cooperative State Research Service, U.S. Department of Agriculture, Washington, D.C., U.S.A.

Abstract. In order to keep pace with the food needs of the world, application of nitrogen in chemical fertilizer must be and is being increased very rapidly, with a doubling time of about 10 yr. The efficiency of use of fertilizer N by crop plants diminishes as rate of application is increased. The N not used by the growing plants may be stored in the rhizosphere, beneath it, percolate to groundwater, or principally, be volatilized to the atmosphere. Efficiency of crop plants in use of N and in its accumulation as nitrate varies widely. Genetic variation is notable; e.g., the chenopods – beets, spinach, and the like – are notable accumulators. Under drought conditions, oats, corn, and other crop plants may also accumulate NO_2. Nitrate in well-water in some areas is presently high; unrelated to fertilizer use.

Research to provide systems for more efficient use of fertilizer N is needed both on grounds of economy and to minimize the accumulation of N from fertilizer in our waters and, as nitrate, in our feed, forage and food plants.

Commoner raises four principal issues. He asserts that:

(1) Nitrate from mineral fertilizer is percolating to groundwater and eventually reaching our lakes and streams and that this nitrate contributes substantially to eutrophication of lakes and streams.

(2) Nitrate content of some vegetables presents a threat to health of young infants.

(3) Nitrate content of drinking water supplies exceeds or approaches tolerance levels in substantial areas of the United States.

(4) Principal dependence for the nitrogen requirements of food crops on nitrogen bound to organic matter would lessen these hazards. With respect to the first issue, surely some nitrate is reaching our waters from mineral fertilizer. More nitrate is likely to reach the environment as nitrogen fertilizer use is increased. At present the principal causes of eutrophication are phosphorus, largely from detergents, and nitrogen from human, plant, and animal wastes.

With respect to nitrate in food plants, nitrate from synthetic fertilizer is only a contributing factor, the principal factors being genetic tendency of chenopods to accumulate nitrate and environmental stresses which accentuate accumulation.

No cases of infant methemoglobinemia attributed to spinach or baby food have been reported in the U.S. It is well known that spinach and beets are nitrate accumulators and that their nitrate content varies widely. While Dr Commoner cites research findings indicating association of high nitrate content in spinach with high nitrate fertilization [1], equally high nitrate levels in spinach were reported in 1907 before synthetic nitrogen compound fertilizers were in use as were found in 1967 [2].

The issue of nitrate in drinking water existed before commercial fertilizer came into general use. Wherever nitrate content of drinking water exceeds or approaches the public health limit tolerance, an improved drinking water supply should be obtained, regardless of origin of the nitrate. Applications of known technology or new technology which may be developed is an obvious alternate to a new supply source for bringing nitrate content of drinking water supplies within acceptable limits.

The fourth issue, that of maximal dependence on nitrogen bound to soil organic matter to supply nutrient requirements of crop plants, is difficult. First, disturbance of soils high in organic matter by tillage releases large amounts of nitrate in the environment. Second, very few soils are likely now or ever to develop needed reservoir capacity of organic-bound nitrogen to support continuous cropping with the high yields now required for food crops.

I do think we have a problem. That problem should be defined. We need to control the entry of nitrates from commercial fertilizers, as well as from sewage, animal wastes, and other sources, into groundwater, lakes, and streams. The degree of control compatible with essential crop production will depend on the improvement of technology through research and the application of that improved technology. This we can and should do. I do not share Dr Commoner's alarm. I see no clear and present danger.

We must increase greatly our use of synthetic nitrogen fertilizers in order to feed the world. The world's population will double by the year 2000 despite pills, intrauterine devices (IUD's), or any other restraint. Most of that increase will take place in the developing countries. And in those countries food production must and can be increased, largely on land now under cultivation [3].

With present technology, world use of commercial nitrogen fertilizers may need to be increased by 5 times over current usage. Nitrogen is not the only factor limiting crop yields; other mineral nutrients, water, pesticides, power machinery, and fuel, are all necessary to obtain the doubled crop yields we must have. The new rice and wheat and corn varieties will give double present yields but only with adequate inputs, and nitrogen is essential.

World use of nitrogen in manufactured fertilizer increased from 4 272 000 metric tons annually in the 1948–52 period to 25 800 000 metric tons in the 1969 period. U.S. use increased from about 1 173 000 metric tons to 6 700 000 – an increase of about 6 times for the world and for the U.S. [4].

A substantial portion of the fertilizer nitrogen was used on cereal crops such as corn, wheat, and rice. Production of all cereals increased about 60 % during the period. Cereal production should be more than double present production in year 2000.

It seems obvious that the large increase in fertilizer nitrogen use which the world must have will lead to some increase in nitrogen in our waters and in our air from this source unless research produces more efficient, economic, technology for fertilizer nitrogen use. Present technology may give us a bushel of soft, white winter wheat with about 10 % protein or 1.0 lb nitrogen per bushel for each 3 lb of fertilizer nitrogen used [5]. Thus, 1 lb of nitrogen applied is in the grain; two go to straw, the rhizosphere, the percolate, the atmosphere. Bread wheat in Decatur County, Kansas, returned about 11 bushels more per acre and about 20 lb more nitrogen in the grain harvested from plots receiving 75 lb. nitrogen per acre than plots receiving no fertilizer nitrogen [6]. And the law of diminishing increments is applicable. In general, the higher the rate of application, the lower the efficiency of nitrogen use [7].

We can double our current corn yields and they should be doubled by year 2000. With our existing technology it may require 300 lb of nitrogen per acre from chemical fertilizer. One Illinois farmer produced 189 bushels per acre in 1965 on a substantial acreage. He used from 265–337 lb of nitrogen [8].

Most plants absorb most of their nitrogen from the soil as nitrate. Soil micro-organisms release nitrogen from organic matter in the soil and convert it to nitrate. Bacteria convert inorganic nitrogen compounds in commercial fertilizer applied to the soil to nitrate. In the presence of sufficient moisture, some nitrate may be leached to groundwater, thence to lakes or streams. This is true both of nitrogen derived from organic matter and for that derived from mineral fertilizer. The amount leached is a complex function of several factors including available nitrate, soil moisture, time of year, temperature, nature of vegetation, and soil composition.

Biological efficiency of use of nitrogen requires that a sufficient supply of available nitrogen be present when needed. Thus, winter wheat yields may be increased by nitrogen applications in fall or early spring. Protein content of the grain will be in-creased only by nitrogen available late in the growing season [6].

Economic efficiency may be highest when nitrogen fertilizer is applied in a single application at the time most convenient to the farmer. This is likely to be fall or spring, outside the growing season. The greater biological efficiency achieved through application as needed by the crop, in most cases, would increase crop production costs and, thus, food costs because of the added costs of fertilizer application.

A great deal is currently being done and more is planned to maintain and enhance the quality of the environment. The role of nitrates in eutrophication is an important, but not the dominant, factor in this problem.

That the people of the world can have an environment of good quality, green with grass and trees, is true because the development and application of improved tech-nology for food production on farms continues to increase output per acre. Increased use of manufactured nitrogen compounds is an essential ingredient of current and prospective farm production technology.

Crops are harvested from less than 300 million acres in the U.S. currently compared to about 360 million acres in 1930. In the world at large and in the United States only modest expansions of harvested acrage will be necessary to feed the world. The FAO Director-General [9] pointed out that, "By and large the developed countries already have more land available for agriculture than they really need". And land not needed for cultivated crops should grow grass and trees. Dr Commoner noted decrease in nitrates in the waters of the upper Missouri and suggested that improved soil conser-vation practices may be a factor. I agree that this may be true. The Great Plains Conservation Program, administered by the Soil Conservation Service, has provided comprehensive conservation treatment for more than 50 million acres in the Great Plains, much of it in the upper Missouri drainage [10].

In December 1968 the United States joined with 49 other nations in a resolution to convoke, not later than 1972, an International Conference on the Problems of the Human Environment. This resolution constituted agenda item 91 of the 23rd Session of the UN General Assembly [11]. It was adopted without objection. The resolution requested the Secretary-General to report to the 24th Session of the General Assembly the main problems which might with particular advantage be considered at such a conference.

Speaking in support of the resolution, the representative of Sweden, Mr Aström, said: "There is no longer any hope of finding new important soils to feed an increasing population. It is rather a question of making already cultivated soils yield richer harvests through higher productivity. Such efforts imply the extensive use of fertilizers, pesticides, the building of huge irrigation systems, and so on." And again, "Substances which in the right place are valuable resources may in another place be harmful. An example of that is the spreading of chemicals as fertilizers on the fields. If these chemicals reach lakes and streams, they become pollutants."

Mr Aström also noted the value of relevant programs for improvement of the environment of the UN Specialized Agencies, the Biosphere Conference held at UNESCO House in September 1968, of the International Council for the Conservation of Nature and Natural Resources, the International Council of Scientific Unions, and the International Biological Program which it sponsors.

U.S. Ambassador Wiggins also spoke in support of the resolution [12].*

The U.S. Department of Agriculture and the cooperating State agricultural experiment stations are currently devoting about 220 scientific man-years (SMY) (full-time equivalent of 220 principal investigators and their assistants and graduate students) to research on plant nutrients. The joint task force on environmental quality [13] estimated need for 55 additional SMY's to do research on plant nutrients by 1977. Ten of these additional SMY's would do research related to control of nitrates reaching our waters from agricultural and other rural sources.

The task force estimated (loc. cit., p. 62) that: "Improved technical information for different soils, under different climates for different crops could avert as much as one-half of the loss from non-beneficial use of this fertilizer."

Even with present technology, ways are known to increase efficiency of nitrogen use. But they are more expensive than current methods and cannot be justified economically at present. People may have to make some choices between food costs and esthetics.

Research is underway in our country and many others on the metabolism of plants. Nitrogen is supplied by biological processes – blue-green algae and free living bacteria as well as those symbiotic with legumes. Legumes are important as food crops. The IBP, in which scientists in many countries, including the U.S., are participating

---

* The UN Conference on the Human Environment met in Stockholm in June 1972. It adopted a Declaration for the Human Environment and recommended to the General Assembly the establishment of a $100 million Environmental Fund to implement recommendations for new international actions to protect and enhance the quality of the Human Environment (cf. Section: 'Further Reading').

includes major programs on the factors affecting the metabolism of nitrogen by plants in production of proteins [14].

In summary, we have reason, in my opinion, to be alert, not to be alarmed, with respect to the role of fertilizer nitrogen in eutrophication and in the accumulation of nitrates in food and feed plants. Increased use of commercial nitrogen fertilizer is necessary. The benefits of such use in feeding the people of the world are unquestionable. More research to produce and adopt more efficient technology for use of commercial nitrogen fertilizer and the application of such technology is needed. More research is needed on the biology and ecology of fixation and metabolism of nitrogen by blue-green algae, free living bacteria, symbiotic bacteria and by other organisms. Much more research is needed on nitrogen in all its roles in all our ecosystems and their effect on human welfare.

Addendum

Kohl et al. [15] reported that about 55% of the nitrate N entering Lake Decatur in Illinois, U.S.A., during the spring of 1970 was derived from fertilizer nitrogen.

Their report was based on measurements of the natural enrichment in ^{15}N of soil nitrogen, fertilizer nitrogen, and the nitrogen of surface water nitrate, as well as on measurements of nitrate concentration.

The validity of these findings was challenged by a group of soil scientists (Hauck et al.) [16]. They argued that the Kohl et al. estimates of organic N being mineralized in the watershed may not have been representative of the entire watershed; that fertilizer nitrogen may lose its identity when it mixes with the pool of organic nitrogen and that the difference in ^{15}N content of fertilizer and soil organic nitrogen is so small as to make discrimination by the procedures used by Kohl et al. improbable. Kohl et al. [17] defended the validity of their results arguing that mixture of fertilizer with the soil organic nitrogen would make the Kohl et al. estimates of fertilizer N contribution to nitrate-N in Lake Decatur conservative.

References

1. Schupan, von, W.: 'Der Nitratgehalt von Spinat (Spinacea oleracea L.) in Beziehung zur Methämoglobinämie der Säuglinge', Z. Ernährungswiss. 5, 207 (1965).
2. Richardson, W. D.: J. Amer. Chem. Soc. 29, 1747 (1907).
   Jackson, W. A., Steel, J. S., and Boswell, V. R.: 'Nitrates in Edible Vegetable Products', Amer. Soc. Hort. Sci. 80, 349 (1967).
3. Boerma, A. H.: 'Food Requirements and Production Possibilities', Sci. Bios. Inf. 18, UNESCO, Paris (1968).
   Byerly, T. C.: 'Benefits and Usefulness of Food Chemicals in Food Production Relative to World Population', in NAS-NRC Publ. 1491, Use of Human Subjects in Safety Evaluation of Food Chemicals', pp. 15–29 (1967).
   Ennis, W. B., Jansen, L. L., Ellis, I. T., and Newson, L. D.: 'Inputs for Pesticides', in The World Food Supply, Vol. III, The White House, Washington, D.C., pp. 130–175 (1967).
4. FAO: Production Yearbook 21, 445. Food and Agriculture Organization, Rome (1967).
5. Leggett, G. E.: 'Relations Between Wheat Yield, Available Moisture, and Available Nitrogen in Eastern Washington Dry Land Areas', Wash. Agr. Expt. St. Bul. 609 (1959).

6. Smith, F. W.: 'Fertilizing Wheat for Profit', *Plant Food Rev.* **10**, 4–6 (1964).
7. Ennis, W. B., Jansen, L. L., Ellis, I. T., and Newson, L. D.: 'Inputs for Pesticides', in *The World Food Supply*, Vol. III, The White House, Washington, D.C., pp. 130–175 (1967).
8. Strohm, J. C. and Ganschon, C.: *The Ford 1968 Almanac*, Golden Press, New York, p. 99 (1968).
9. Boerma, A. H.: 'Food Requirements and Production Possibilities', *Sci. Bios. Inf.* **18**, UNESCO, Paris.
10. Freeman, O. L.: 'Resources in Action Agriculture 2000', U.S. Department of Agriculture, Washington, D.C. (1967).
Freeman, O. L.: 'Agriculture in Transition', Report of the Secretary of Agriculture for 1968, U.S. Department of Agriculture, Washington, D.C. (1969).
11. UN General Assembly, 23rd Session: Resolution adopted by the General Assembly. A/Res/2398 (XXIII), December (1968).
Provisional verbatim record of the seventeen hundred and twenty-second meeting A/PV.1732, December (1968).
12. Wiggins, J. R.: Provisional verbatim record of the seventeen hundred thirty-third meeting. A/PV.1733, December (1968).
13. Byerly, T. C., Acker, D. C., Evans, J. B., Hazen, T. E., Heggestadt, H. E., Schleusener, P. E., Storey, H. C., Hermanson, R., Stubblefield, T. M., Treadway, R. H., Wadleigh, C. H., Yeck, R.G., Brinkley, P. C., Cellman, I., MacKenzie, D., Buckley, J. L., Deevey, Jr., E. S., King, D. R., Porter, R., Bullard, W. E., Geyer, H. G., and Ward, D. J.: 'Environmental Quality: Pollution in Relation to Agriculture', Report of a Joint Task Force of the U.S. Department of Agriculture and the State Universities and Land-Grant Colleges (1968).
14. NAS-NRC: 'Man's Survival in a Changing World', U.S. Participation in the International Biological Program. National Academy of Sciences-National Research Council, Washington, D.C. (1968).
15. Kohl, D. H., Shearer, G. B., and Commoner, B.: 'Fertilizer Nitrogen Contribution to Nitrate in Surface Water in a Cornbelt Watershed', *Science* **174**, 1331–1334 (1971).
16. Hauck, R. D., Bartholomew, W. V., Bremmer, J. M., Broadbent, F. E., Cheng, H. H., Edwards, A. P., Keeney, D. R., Legg, J. O., Olsen, S. R., and Porter, L. K.: 'Use of Variation in Natural Nitrogen Isotope Abundance for Environmental Studies: A Questionable Approach', *Science* **177**, 453–454 (1972).
17. Kohl, D. H., Shearer, G. B., and Commoner, B.: Ibid, *Science* **177**, 454–456 (1972).

For Further Reading

1. Nyle C. Brady (ed.), *Agriculture and the Quality of Our Environment*. A Symposium Presented at the 133rd Meeting of the American Association for the Advancement of Science, December 1966.
2. Wastes in Relation to Agriculture and Forestry. Cecil H. Wadleigh, Director, Soil & Water Conservation Research Division, Agricultural Research Service, U.S. Department of Agriculture. Miscellaneous Publication No. 1065, USDA, March 1968.
3. Clifford M. Hardin (ed.), *Overcoming World Hunger*. The American Assembly, Columbia University. Published by Prentice-Hall International, Inc., Englewood Cliffs, N.J., 1969.
4. Cleaning Our Environment: The Chemical Basis for Action. A Report by the Subcommittee on Environmental Improvement, Committee on Chemistry and Public Affairs, American Chemical Society, 1969.
5. Viets, Jr., F. G. and Hageman, R. H.: 1971, *Factors Affecting the Accumulation of Nitrate in Soil, Water, and Plants*, Ag. Handbook 413, U.S. Department of Agriculture, Washington, D.C. 20250.
6. Committee on Nitrate Accumulation: 1972, *Accumulations of Nitrate*, National Academy of Sciences, Washington, D.C. 20418.
7. U.N. General Assembly Report of the United Nations Conference on the Human Environment held at Stockholm, 5–16 June, 1972. A/CONF.48/14, July 3, 1972.

MAN-INDUCED EUTROPHICATION OF LAKES

ARTHUR D. HASLER

University of Wisconsin, Laboratory of Limnology, Madison, Wis., U.S.A.

1. Introduction

Many lakes the world over are becoming less desirable places on which to live because of nutrient wastes pouring into them from a man-changed environment. Man's
activities, which introduce excess nutrients to lakes, streams, and estuaries are rapidly
accelerating the process of cultural eutrophication. [10] Excessive enrichment, brought
about by population and industrial growth, intensified agriculture, river-basin development, recreational use of public waters, and domestic and industrial exploitation
of shore properties, accelerates the deterioration of waters. The process causes changes
in plant and animal life which usually interfere with multiple uses of waters, reduce
their aesthetic qualities and economic value, and threaten the destruction of precious
water resources. Overwhelming excessive scums of blue-green algae and aquatic
plants choke the open water, makes the water turbid and nonpotable. They die, rot
and repel human residents with repugnant odors. Organic matter from this crop sinks
and consumes the deep-water oxygen vital for fish and other animal life.

Under natural conditions lakes proceed toward geological extinction at varying
rates through eutrophication or bog formation. Many lakes, in unpopulated temperate zones, and lying in sandy granite drainage basins are still pristine and clear
(oligotrophic) even though 10000 yr have elapsed since the glacier formed them. Other
lakes in the same area, such as shallow bog lakes which were likewise shaped by
grinding ice during the same glacial epoch, are already extinct. They are grown over
with mats of sphagnum moss interspersed with orchids and pitcher plants. Brown
colored water lies below the mat which deteriorates and slowly fills in the basin. In
some, the terminal stages of bog formation are evident because these former lakes are
now covered with shrubs, tamarack and black spruce forests. This type of extinction
is not eutrophication. How this succession or continuum proceeds from open lake to
forest is too complex to be developed in this brief essay.

Archeological studies by G. E. Hutchinson and R. Patrick of cores of lake sediments of the Italian lake, *Lago di Monterosi*, reveal that the Romans by constructing
roads inadvertently increased the nutrient drainage of a landscape by cutting the
trees and exposing limestone strata. The erosion from these nutrient richer strata was
followed by a eutrophic period in the lake's history as recognized by the kinds of
diatoms found in the cores. E. S. Deevey, Yale Univ., also recognized prehistoric
changes of climate and rate of eutrophication in Linsley Pond, Conn., which are
correlated with the fossils in the cores. He determined the abundance and variety of
microscopic organisms, plankton crustacea and insect larvae in different strata.

S. Fred Singer (ed.), The Changing Global Environment, 383–399. All rights reserved.
Copyright © 1975 by D. Reidel Publishing Company, Dordrecht–Holland.

The rate of eutrophication of lakes in geological time can often be predicted by examining the soil and vegetation of its drainage basin. If the drainage area is large, the vegetation pristine and the soil rich and erodable, the lake water will be rich in algae and fish; if poor, the water will produce little and will retain its clarity because of low algae count and high aesthetic quality.

2. The Algal Community (Phytoplankton)

Algae are microscopic one-celled plants which require, for their growth, the same nutrients as do garden flowers and lawns. If fertilized richly in spring and summer they flourish; if impoverished they grow sparsely. A community of free floating algae is much richer and more diverse in kinds of species than is any garden.

Pure cultures of algae grown in flasks and chemostats in the laboratory, multiply and grow rapidly if nitrogen-and phosphorus-bearing chemicals are added; the algae in similar cultures grow even more luxuriantly if only small amounts of sewage effluent are added containing the same amounts of nitrogen and phosphorus as in the original above, hence demonstrating that in addition to P and N there are ingredients (probably vitamins and growth hormones) in sewage which promote growth.

A nutrient-poor temperate zone lake will be clear; hence one might collect, with a cone shaped, fine meshed, silk net, thousands of minuscule algae cells at depths of 150 ft and more. In a nutrient-rich lake, on the other hand, the high numbers of algae, the food of protozoa, rotifers and waterfleas will lend a greenish cast to the surface water, restrict the penetration of sunlight and therefore limit photosynthesizing algae as well as rooted aquatic plants to the shallower depths.

The upshot of eutrophicated (nutrient-enriched) conditions has adverse ramifications. When the enriched conditions are owing to man-made effluents, the algae grow so profusely that the waterfleas (the basic food of all larval fishes) cannot consume the algae fast enough to reduce their numbers significantly; hence abnormal amounts die uneaten.

The biological communities of a lake become upset when bacteria are unable to convert dead organic matter into plant and animal food.

Not only is oxygen, in the deep cool water, exhausted by organic products but hydrogen sulfide (rotten egg gas) accumulates to poisonous levels.

The finale in these despoiled depths is the demise of all noble fishes, e.g. whitefish, trout and cisco which demand oxygen rich water depths for life. Moreover, some noble fishes such as cisco spawn in the fall – their eggs must incubate throughout winter, but an enriched lake having lost its oxygen in the deep layers (under the ice) cannot nourish the eggs for hatching in the spring.

3. Sources of Nutrients

Phosphate additions appear to be one of the major factors in pollution of European and North American lakes, although the rate at which nutrients pass through chemical and biological cycles is also important. Sources of plant nutrients are principally from human sewage and industrial wastes, including the phosphate-rich detergents as can be seen from Table I. Drainage from farmland is second in importance as a nutrient source in temperate zones, where farm manure spread on frozen ground in winter is flushed into streams during spring thaws and rains. A shocking statistic which points up the gravity of our contemporary situation is that farm animals in the Midwest alone provide unsewered and untreated excrement which is equivalent to that from a population of 350 million people. Also, it is surprising that substantial quantities of nitrates of combustion engine and smokestack origin augment these sources. City streets also provide a source of phosphates and nitrates that has to be dealt with.

TABLE I

Summary of estimated nitrogen and phosphorus reaching Wisconsin surface waters

Source	N Lb yr^{-1}	P	N (% of total)	P
Municipal treatment facilities	20000000	7000000	24.5	55.7
Private sewage systems	4800000	280000	5.9	2.2
Industrial wastes[a]	1500000	100000	1.8	0.8
Rural sources				
Manured lands	8110000	2700000	9.9	21.5
Other cropland	576000	384000	0.7	3.1
Forest land	435000	43500	0.5	0.3
Pasture, woodlot and other lands	540000	360000	0.7	2.9
Ground water	34300000	285000	42.0	2.3
Urban runoff	4450000	1250000	5.5	10.0
Precipitation on water areas	6950000	155000	8.5	1.2
Total	81661000	12557500	100.0	100.0

[a] Excludes industrial wastes that discharge to municipal systems. Table does not include contributions from aquatic nitrogen fixation, waterfowl, chemical deicers and wetland drainage.

In toto the results of man-induced eutrophication are catastrophic as noted in the case history of Lake Zürich, Switzerland, where all noble, deep-water fishes, which had provided gourmet specimens for generations, disappeared within 20 yr after sewage from the surrounding villages was changed from the outdoor privy type to flush toilets.

The Zürichsee, a lake in the foothills of the Alps, offers a sad example of the effects of sewage effluent. It is composed of two distinct basins, the Obersee (50 m) and the Untersee (141 m), separated only by a narrow passage. In the past five decades the deeper of the two, at one time a decidedly clear and oligotrophic lake, has become strongly eutrophic, owing to urban effluents originating from a group of small communities totaling about 110000 people. The shallower of the two received no major urban drainage and retained its oligotrophic characteristics for a longer period. Thus we have an experimental and reference lake side by side.

4. History of Fishing

Minder [18] observed that hand in hand with domestic fertilization the Zürichsee changed from a whitefish (coregonid) lake to a coarse fish lake. In fact, the trout, *Salmo salvelinus*, and a whitefish, *Coregonus exignus*, disappeared from the Untersee and are no longer common in the Obersee; restocking has not been successful. An upsurge of cyprinid fishes (minnow-carp family), chiefly *Alburnus lucidus*, has been striking; mass harvesting of this species has been practiced recently, while *Albramis brama* and *Leuciscus rutilus* have also become abundant with the progressive eutrophication.

5. History of Plankton Succession

Minder is convinced that the decided increase of plankton is not an expression of a natural ripening process, but is owing to plant nutrients, principally P and N, from domestic sources. The diatom *Tabellaria fenestra* appeared, explosively in 1896. Two years later occurred an eruption of the blue-green alga, *Oscillatoria rubescens*. The latter had been known from the eutrophic Murtenersee for 70 yr, but otherwise had not been recorded elsewhere in Switzerland, except that it was reported from the Baldeggersee in 1894 where it has appeared in spring and winter. It had not been seen in the Zürichsee plankton until the 1896 eruption when it replaced the usually dominant *Fragilaria capucina*. When Minder [17] studied the lake during 1920–24, *Oscillatoria rubescens* appeared in quantities in the surface plankton, with a maximum in fall and winter. Oscillatoria generally flourished in the deeper water of the lake in summer. In 1936 Minder [18] observed a red scum of it over most of the lake. An odor of fish oil is frequently noticeable in summer. There were 1.75 gm wet weight of algae per liter, chiefly Oscillatoria, on 5 May 1899.

Further evidence for recent sudden increase in biological productivity can be found in bottom sediment studies. These demonstrate that Tabellaria occurs in only the most recent layers, its appearance coinciding with the period of Schröter's observations. Moreover, the modern layers are laminated, at least in the deeper parts of the lake, and everywhere are darker than the underlying sediment. The darkening and the laminated character of the sediments are especially pronounced from 1896 onward, the date being determined by counting the seasonal laminae.

Minder [18] cites some comparative plankton analyses on the Untersee and Obersee: first, in no series were the biocoenoses of the two sections identical; second, *Oscillatoria rubescens* was never found in the Obersee and Tabellaria, very infrequently; third, quantitatively the entire plankton of the Untersee was vastly richer; and fourth, plankton quantities are greater in the Untersee downstream from the town of Rapperswil where most of the sewage enters. Minder also observed the rotifer *Keratella quadrata* as appearing first in 1900. *Bosmina longirostris* largely replaced *B. coregoni* after 1911. Since 1920 one of the Ulothricales has become common in summer.

6. History of Other Limnological Factors

Chemical analyses which prove that certain elements of domestic sewage origin have accumulated markedly are cited by Minder [16], who shows there has been a gradual increase of chloride ion over a relatively few decades. Analyses in 1888 showed the water contained 1.3 mg Cl l^{-1}, by 1916 it had risen to 4.9 mg l^{-1}. The organic matter as measured by loss on ignition also rose from 9 mg l^{-1} in 1888 to 20 in 1914. Vollenweider [23] gives as a rule of thumb 0.2–0.5 gm m^{-2} yr^{-1} of P and 5–10 gm m^{-2} yr^{-1} of N as the levels of these nutrients which are associated with nuisance blooms of algae in European lakes.

Minder [19] gives comparisons of the changes in transparency (average of 100 readings, see Table II).

TABLE II

	Maximum disc reading	Minimum disc reading
Before 1910	16.8	3.1
1905–1910	10.0	2.1
1914–1928	10.0	1.4

It is significant also that O_2 values in the deep water have decreased in the last four decades [20]. Midsummer values at 100 m were nearly 100% saturation from 1910–1930; from 1930–42, however, they were as low as 9% saturation but averaged about 50%.

7. Man-Made Lakes

Lakes are more adversely affected by sewage effluent than are flowing streams chiefly because flowing water is not conducive to algae growth although diatoms do grow on bottom stones and large aquatic plants grow profusely if the water is not too swift or turbid. Hence, the diversion of sewage around lakes and into streams is the lesser evil. Nevertheless, while alleviating the lake problem it places an increasing burden upon the stream's biological system and upon the communities downstream which must purify it. In modern times most large streams have man-made dams for impounding the water and whose outflow provides energy for hydro-electric power. If such a reservoir receives sewage, the quality of the water deteriorates as does a lake, and the cost of water purification and odor control rises for downstream communities. There is therefore a finite limit. Permissible levels of impurities will have to decrease, hence the technological improvement to obtain more complete nutrient removal must become more efficient.

8. Limnological Features of Eutrophicated Lakes

Lakes are wasteful of phosphorus when phosphorus is added [7] and Einsele and Hasler [8] and Edmondson [5]. It is soon incorporated into the phytoplankton and macrophytes and carried to the bottom where very little of it ever returns to solution. Hutchinson and Bowen [11] added the radionuclide 32 P to a lake in Connecticut (Linsley Pond). Within 23 hr 57% was absorbed by algae or chemically removed and lost to deeper water (below 6 m), while 28% was taken up by soils and aquatic plants in the shallow zones (0–3 m).

That phosphorus is lost from the system is proved by Edmondson's [5, 6] beautiful data on Lake Washington (Seattle) Figure 1a, where phosphorus continued to build up as the sewage volume increased but returned almost to natural or pre-sewage levels after diversion of domestic sewage from Lake Washington.

Figure 1b shows the data obtained by Edmondson and his associates during their continuing long-term investigation of Lake Washington at Seattle. Changes in chlorophyll concentration in the epilimnion correspond with the trend of phosphate ac-

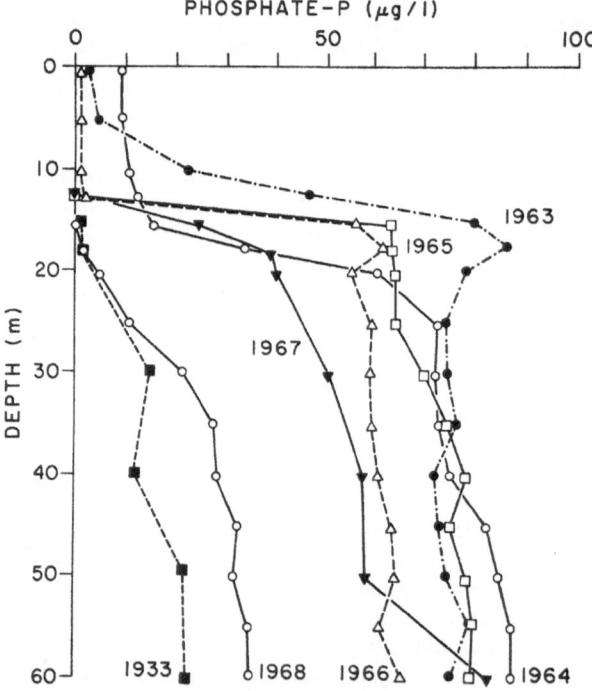

Fig. 1a. Depth distribution of phosphate P in Lake Washington before sewage inflow (1933), during (to 1964) and after diversion (1965–68).

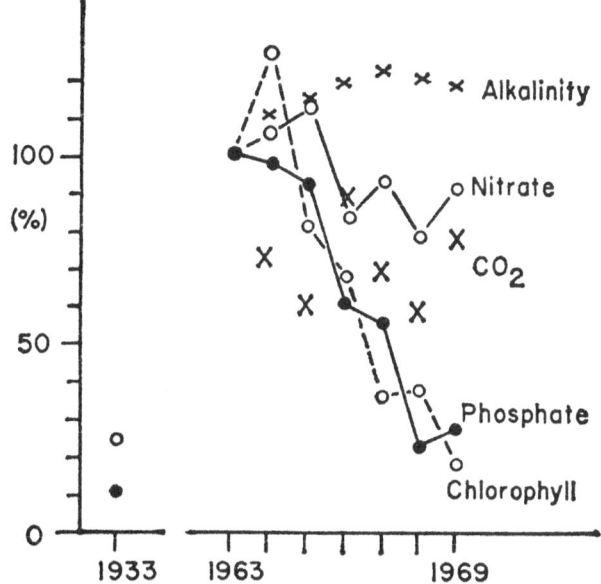

Fig. 1b. Mean winter (January to April) values in surface water of phosphate-phosphorus and nitrate-nitrogen, and mean summer (July and August) values of chlorophyll in surface phytoplankton. The 1963 values, plotted as 100%, were (in micrograms per liter): P, 57; N, 428; and chlorophyll, 38. Unconnected points show winter means (January and February) of bicarbonate alkalinity and free carbon dioxide in surface water (25.3 and 3.2 mg l^{-1} in 1963).

cumulation, reduced transparency, and increase in the O_2 deficit, and are all therefore expressions of galloping eutrophication.

Findenegg [9] (Figure 2) has used the ^{14}C method (radioactive carbon) to evaluate degree of eutrophication. It has the advantage that it measures the rate at which energy is converted into carbohydrate in the photosynthetic process. A eutrophic lake fixes carbon principally in the near surface water because its turbidity prevents light from supplying essential energy to algae in the deeper layers. Higher nutrients in a eutrophic lake also serve to stimulate higher rates of ^{14}C fixation in the surface waters.

Dr Findenegg's example demonstrates a general principle in biology that in senility the older animal often consumes as much food as when he was young, but utilizes it less efficiently. In jargon we say, "He spins his wheels!". The four examples show how with increasing levels of eutrophication there is a steady increase in carbon fixed, but as turbidity rises (4th example), photosynthesis is restricted to the surface waters, hence less depth can be used for production and the utilization drops off – frankly the lake is overfed and obese, perhaps also physiologically senile and – it is spinning its wheels.

Reduction in transparency (by Secchi disc measurement) of Lakes Washington and

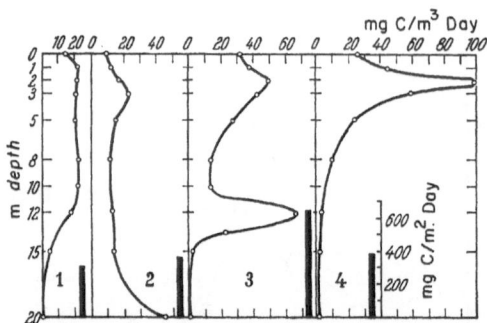

Fig. 2.   The production of carbon by photosynthesizing algae in four alpine lakes (1. Millstätter, 2. Klopeiner, 3, Wörther, and 4. lower part of the Lake of Constance) expressed as mg C m^{-3} day^{-1}. The columns give the total production below 1 m^2 in mg C m^{-2} day^{-1}. Note the restriction of production to the surface waters in the most eutrophic lake, No. 4.

Zürich appear to be characteristic for rapidly eutrophicated oligotrophic lakes. Hyper-eutrophication of a natural eutrophic lake, Lake Mendota, Wisconsin, has not changed the average transparency nor the average hypolimnetic deficit. However, other conspicuous characteristics are observed such as loss of cisco, decrease in diversity of macrophytes, but an increase in quantity of one species. *Myriophyllum spicatum*, which comprises over 90% of the macrophyte population [1969].

9. Phytoplankton

Comparisons of algae diversity of oligotrophic Trout Lake and eutrophic Lake Mendota show the latter to have fewer species but the size of the organisms is considerably larger indicating higher levels of production than in the oligotrophic lake [21].

Often the low species diversity of the phytoplankton in eutrophic lakes is a result of high populations of blue-green algae such as *Aphanizomenon flos aqua* and *Anabaena spiroides* in Lake Mendota. In some seasons *Fragillaria crotonensis* and *Stephanodiscus astrae* become dominant.

In many eutrophicated northeastern U.S. lakes rooted aquatic plants Myriophyllum and Ceratophyllum become festooned with the filamentous alga *Cladophora* and form dense mats in shallow areas.

10. Great Lakes

Until recently it was thought that eutrophication would not be a major problem in large lakes because of the vast diluting effect of their size. However, evidence is accumulating that indicates eutrophication is occurring in the lower Great Lakes. Furthermore, the undesirable changes in the biota appear to have been initiated in relatively recent years. Charles C. Davis [3], utilizing long-term records from Lake

Erie, has observed both qualitative and quantitative changes in the phytoplankton of that large body of water owing to cultural eutrophication. Total numbers of phytoplankton have increased more than threefold since 1920, while the dominant genera have changed from *Asterionella* and *Synedra* to *Melosira*, *Fragillaria*, and *Stephanodiscus*.

Other biological changes usually associated with the eutrophication process in small lakes have also been observed in the Great Lakes. Alfred M. Beeton [1] recently summarized the literature pertaining to the trophic status of the Great Lakes in terms of their biological and physiochemical characteristics and indicated that, of the five lakes, Lake Erie has undergone the most noticeable changes due to eutrophication. In terms of annual harvests, commercially valuable species of fish, such as the lake herring or cisco, sauger, walleye, and blue pike, have been replaced by less desirable species, such as the freshwater drum or sheepshead, carp, and smelt. Similarly, in the organisms living in the bottom sediments of Lake Erie drastic changes in species composition have been observed. Where formerly the mayfly nymph *Hexagenia* was abundant to the extent of 500 organisms per square meter, it presently occurs at levels of 5 and less per square meter. Chironomid midges and tubificid worms now are dominant members in this community.

11. What Can Be Done to Reduce the Galloping Rate of Eutrophication?

The deterioration of our lakes proceeds at such a galloping pace that there is insufficient time to raise an enlightened younger generation which could cope with the causes of eutrophication; hence every effort must be undertaken to convince government officials and voters that action, even though expensive, must be taken immediately to avoid catastrophe. In order to obtain positive returns, time operates negatively against delay. To insure a brighter future, universities, colleges, churches, service clubs, the press, radio and television must acquire a knowledge of the causes, prevention, and cure and begin without delay and help to disseminate factual information.

Provided they are given the facts, preachers and rabbis could preach sin against the environment as convincingly from the pulpit as they preach sin against the soul.

Decision makers, such as legislators, state, county, and village officials, decision planners such as architects, decision formulators such as lawyers and judges, decision executors such as realtors, engineers and contractors must all receive enlightenment about the implication and possible perturbations of the landscape whose environmental health influences the well being of the lake into which the land's effluents flow.

I urge that every educational body, in every community, organize week-long intensive clinics, seminars or working groups to which experienced limnologists and ecologists are invited as teachers, lecturers, demonstrators and guides. I urge journalists and editors, television and radio directors to send their personnel to clinics, or to meet with experienced ecologists in order to obtain facts and illustrations, the basis for which readable and effective articles can be written and programs prepared. They should be taken on field trips to areas where the problem can be demonstrated at first hand in order to demonstrate to the writer or programmer the reality of a eutrophied lake, capture their enthusiasm and stimulate their originality toward the preparation of dynamic and imaginative programs.

The processes of eutrophication are too rapid to risk delay in taking legal action. In applying new concepts of water law to the alleviation of eutrophication there is a need for proper zoning ordinances and forthright public initiative in modernizing the law when the scientific data, even if not complete, suggest action. A new law in Wisconsin requires a 1000-ft setback for all cottages and buildings on lakes and a 300-ft setback on streams, together with stricter specifications for septic tank construction depending upon soil permeability.

I would urge legislative lawyers to draft critical legislation for water usage, in regions where the legal procedures are inadequate and encourage them to draft laws which will provide adequate protection of a lake or reservoir from perturbations.

In Wisconsin the late Prof. Jacob Beuscher, through his association with ecologists and landscape architects, drafted unique aforementioned zoning legislation for Wisconsin which now has been passed (Wisconsin Water Resources Law, 1965). Hence, if a dwelling or resort is planned it would have to meet exact specifications for sewage disposal so that no effluent could seep into the water. Some soils are less able to absorb effluents than others, hence the setback of a planned hotel or dwelling must be at the extreme end of 1000 ft if the soil has poor drainage qualities. His legislative bill also specified beauty for this zoned corridor as a quality to be preserved. Vilas County, rich in lakes, prohibits cutting the natural vegetation from more than 10% of the shoreline fronting a property.

12. Social, Legal, and Economic Aspects

In the preceding discussion emphasis was placed upon the effectiveness of various management procedures. Of equal importance are comprehensive economic analyses of new approaches to management. Included should be studies to develop methods to quantify costs and benefits and to analyze public opinion so that the management programs developed are acceptable to society.

Beuscher [2] writes:

Since resource management requires not only scientific knowledge and techniques but also governmental and legal structures by which desired management can be achieved, the entire field of legal and governmental structure is a necessary research area. Wisconsin's assertion that it is trustee of all navigable waters of the state is one of the strongest examples of a state's assertion of its right and duty to protect public interests in natural resources, and this could form a basis for a variety of strong regulatory policies. Potential conflicts between the asserted trusteeship and the rights of private littoral and riparian owners exist, however, and should be investigated as a guide both to the potentials for regulation and to the limitations on regulation without compensation.

Zoning is one type of regulatory action which is likely to be of significant value in attacking the problems associated with inland lakes. As for other water resources, the Wisconsin Legislature has acted and the counties are presently required by law to enact river and lakeshore zoning ordinances. Research is needed to review the powers of the counties under present statutes, especially noting where powers which seem necessary to accomplish desired regulation are lacking or unclear. Creative proposals and careful analyses are needed concerning the present procedures for administration of zoning by the counties. Projects in various areas of scientific research could be undertaken with a view toward producing facts and testing procedures which the counties might employ to guide and defend their regulation of stream and lakeshore lands.

Owing to the traditional lack of compensation for regulations imposed, zoning is a limited device for the control of lakeshore lands. Imaginative legal research is needed on a broad range of new control devices, such as compensable regulation and partial condemnation, some of which are being tried in some parts of the country. Finally, the powers of all levels of government and the potential powers of private groups and resource control and management corporations should be analyzed as means toward proposing more systematic and creative methods of management than exist presently.

13. Examples of Success

Our knowledge of what causes eutrophication is already sufficiently good that firm and effective precautions can be recommended. They may be expensive to achieve, but the predictive facts are at hand. While improvement in methods can be made more economical, it is not a lack of knowledge which prevents us from action. Three case histories are at hand:

13.1. LAKE MONONA

Complaints about the unpleasant odors arising from Madison's Lake Monona were published in the newspapers as early as 1850. Sewage effluents were impugned as the villain in 1885 when a consultant J. Nader advised "...that the lakes were not properly used as receptacles for sewage in the crude state". In 1895 a $1\frac{1}{2}$ mill sewer tax was imposed on assessed valuation, but the sewage treatment plant built from these funds failed in 1898. Septic tanks and cinder filter beds were then constructed but reached capacity in 1906, and it was not until 1914 that a modern sewage treatment plant was constructed. Its effluent entered Lake Monona and its fertility continued to feed the algae and weeds. The process of eutrophication accelerated on Lake Monona and in 1920 its city council minutes read:

Winds ... drive detached masses of putrefying algae onto shores ... if stirred with a stick, look like human excrement and smell exactly like odors from a foul and neglected pig sty.

In 1921 consideration was given to piping the effluent to the Wisconsin River but another plant was built below Lake Monona in 1928. One half of Madison's effluent then first passed into Lake Waubesa; later (1936) all of it. In spite of heavy applications of the algal poison $CuSO_4$ to the lake to halt the growth of well fed algae, the build-up of offensive and obnoxious odors in Waubesa and Kegonsa worsened. In desperation the Lewis Anti-Pollution Law was introduced in 1941 but was vetoed by Governor Julius P. Heil because of conflicting issues on whether sewage or rural run-off was the culprit.

In 1942 the Burke Plant which had been discontinued in the '30's was reopened to accommodate the military needs of Truax Field. The need for more copper sulfate during this period is obvious from the graph. In 1943 the Lewis bill was passed to take effect one year after the war.

In actuality the effluent did not by-pass the lakes until 1958. During this century of time, buck passing, economy measures, false information from communications, unconclusive action, and lack of cooperation between government and citizens hampered progress.

The fact that copper sulfate treatment of the lake to curb the burgeoning algae growths dropped from freight car load quantities to minimal local treatments is proof that even though agricultural drainage unfortunately continues, the diversion of city sewage produced a change for the better.

13.2. Lake washington

Lake Washington was used for the disposal of first raw sewage and later treated sewage from the City of Seattle. About 1930, the last major source of raw sewage was removed from the lake, but up until 1959 untreated sewage still entered the lake in relatively small quantities through storm sewer overflows at times of heavy rainfall as well as from seepage from septic tanks whose effluent trickled through or across the ground into small streams entering the lake. In 1959 there were 10 sewage treatment plants, serving 64000 people, putting treated effluent into Lake Washington.

While biologists and engineers warned the community of the impending doom of the lake, it took the dense blooms of *Oscillatoria rubescens*, the same lavender colored alga that produced nuisances in Lake Zürich, to awaken the citizens to the reality of these warnings. While some argued as they did in Madison that the run-off from fertile land was causing the nuisances, others contended that the major sources (city sewage) could be diverted. Radio and television debates were heard and viewed. 5000 women of the League of Women Voters knocked on doors in the campaign. In addition citizen's groups held meetings culminating in sufficient public opinion to support the city officials in a bonding campaign amounting to an anticipated expenditure of $121000000 for the diversion of 50000000 gall of sewage around Lake Washington and to empty the sewers into Puget Sound. While the sewers are not yet entirely complete a major part of the sewage has now been diverted. Already the quality of the water has improved noticeably, and the measurements of clarity are improving.

13.3. Lake tahoe

In order to protect the pristine beauty and crystal clarity of Lake Tahoe several sanitary engineers [15] made a study of the sewage disposal problems of Lake Tahoe and published a comprehensive study in 1963. They described the problem and projected the rate of growth of the communities and tourist facilities whose sewage from treatment plants and septic tanks is entering this beautiful mountain lake. The lake's great size and immense depth meant that it could absorb some sewage without showing general signs of deterioration yet at the sewage outlets objectionable growths of green algae accumulated on the stones. Hence it was only a matter of time before the increased sewage load from a skyrocketing population would change this clearest of all North American lakes to a lake of lesser esthetic value.

In spite of the 70 odd governmental units in Nevada and California which surround the lake, this initial limnological and engineering evaluation inspired the creation of citizen-governmental action committees and associations which went into action. Already South Tahoe (1968) has built a $19 000 000 sewage treatment plant in which the treated effluent is pumped over a 7500 ft pass to a reservoir, over the mountain, where the water will be used for irrigation, and a fertile water at that! Some $10 000 000 of the total budget was in federal grants acquired by the South Tahoe Public Utility District to help offset this cost – a demonstration of cooperative action of federal and local government in solving a local and national problem. Acknowledging the imminent danger of despoiling this esthetic and financial resource, other communities are facing reality in an action program. A small community of Round Hill in Nevada with only 42 voting citizens, albeit many are owners of gaming houses, has bonded itself for 5.8 mill to treat its sewage and pump it out of the basin. We can only hope that the hotel and residential sewage from other parts of this once gin-clear lake can be similarly diverted in time to avoid the certain despoilation of Sierra Nevada's most magnificent landscape gem.

European lakes, Schliersee and Tegernsee in Germany's Bavarian Alps were eutrophicated by hotel sewage, but are now slowly reverting to more tolerable conditions following a diversion of sewage to the outlets [13]. Lake d'Annecy, France (near Lake Geneva) is following suit [12]. All lakes, from which sewage has been diverted, have shown improvement (see case for Lakes Washington and Monona, Figures 1a and 3) therefore demonstrating that ingredients in sewage contribute greatly to eutrophication and if withdrawn, improvement sets in. These facts negate the argument "Why divert sewage at great cost if rainwater and rural drainage is so rich in nitrogen?"

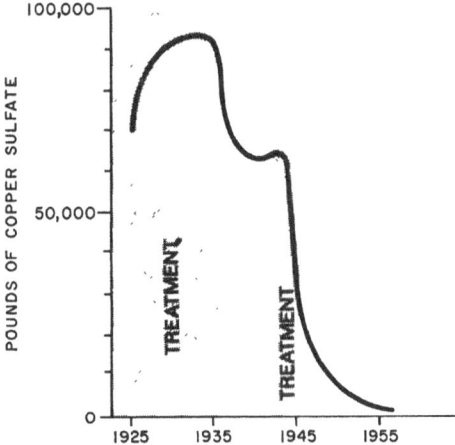

Fig. 3. The amount of copper sulfate required to control alga nuisances in Lake Monona near Madison, Wisconsin, in the period 1925–56. The amount used is a function both of the abundance of algae and the degree of public disturbance; thus small differences are probably unimportant, but large changes are significant. The main trend is shown by the solid line.

The 'healing' is, of course, more rapid in lakes in oligotrophic and high rainfall land-scapes. The Madison, Wis., lakes were naturally eutrophic, nevertheless the hyper-eutrophic and repulsive conditions do become milder after diversion. In lakes Wau-besa and Kegonsa after diversion several species of algae replaced the highly eutrophic single species blooms.

14. Harvesting and Utilization of Excess Crops of Plants and Fish

Machinery, some still in the idea stage, is needed for harvesting large aquatic plants. Removal of this crop from the lake along with its phosphorus and nitrogen containing organic matter impoverishes the water of nutrients. It also improves the esthetics, opens the water area to boating and swimming and creates better shoreline sanitation. More research needs to be done to find a commercial product for aquatic plants and to utilize them for a useful purpose. Eutrophicated lakes produce large crops of fish which should also be harvested more intensively, by commercial fishermen if neces-sary, because the unharvested fish dying of old age decompose, adding nutrients to the already over-rich environment.

In Lake Mendota in late 1966, 40 lb acre^{-1} of carp alone were harvested in a single seine haul. Yields of 250 lb acre^{-1} yr^{-1} of fish of all species could be harvested easily from this lake without damaging the fishery. In terms of nitrogen and phosphorus 1000 lb of rotting fish would yield to the lake 25 lb of N and 2 lb of P.

15. Chemical Control of Nuisance Growths

Chemicals which poison unwanted aquatic plants and algae have deleterious effects in and around treated areas. Moreover, the killed weeds rot and add to the nutrient supply. This is a bad conservation practice because no good is accomplished. Chemi-cals distort the structure of multi-species aquatic communities and hence are less useful in lakes than they are in agriculture, where weeds are to be eradicated from a crop of a single species such as wheat. Herbicide use cannot be justified in a lake ecosystem. In addition, chemicals are more difficult to manage than on land for they are soon drifted to other areas. The toxic actions of these chemicals on other species in the lake have not been tested, nor have the possible insidious side effects of sublethal actions over longer periods. At present, the use of chemicals to combat algae blooms or rooted aquatic vegetation can be no more than a palliative. They should be used in aquatic ecosystems only as a last resort. What is eradicated is sure to be replaced by something else that may be more difficult to poison.

16. Utilization of Sewage and Farm Manure

Highly valuable fertilizers are found in sewage (Chicago had 30 tons P per day effluent in 1960). In fact each human produces 1.5 to 4 lb yr^{-1}.

After sewage has undergone secondary treatment it still contains phosphorus. The average P content of sewage (secondary) is 8 mg^{-1}. Hormones, vitamins and growth substances are also fertilizing ingredients. Moreover phosphate rich detergents now added are not entirely removed. In fact, secondary treatment in most treatment plants removes only about 80 % of the P and high costs deter removing more. An increasing human population adds to the total residual left in sewage effluent after treatment. "We seem to be on a treadmill," comments G. A. Rohlich, an eminent engineer, who states further that in spite of new advances we are not much further ahead of the problem than at the turn of the century.

The price of clean water may rise to a point where we may have to insist upon and want to afford evaporation of the effluents in order to obtain a dry solid and distilled water, as is done in some types of desalting techniques. Secondary treatment of sewage does not remove organic growth factors but probably produces them, hence the evaporation looms as an essential though expensive treatment.

In temperate climates of North America it is customary to scatter farm manure on the frozen land. Large amounts of valuable fertilizer are flushed into streams and lakes during early spring thaws and spring rains. Because a cow produces 6 lb nitrogen and 1.5 lb phosphorus yr^{-1} it is clear that this source is important. Modernization of the European method of fluidizing dairy cow manure, storing it in huge tanks and distributing it with a 'honey wagon' as soon as the soil can absorb it, is now being recommended. However, economic limitations inhibit progress in converting to a more efficient method of manuring. Fortunately, forest and agricultural soils have a remarkable tenacity for phosphorus. Agricultural and forest crops could profit from the fertilizers from our domestic wastes, but the technology for processing and distributing it are still very expensive when compared with the cheapness of sacked artificial fertilizer.

The volume and weight of dried sewage to be disposed of will have staggering proportions. Settlings from primary treatment of sewage abound near every city, but a farmer can buy and distribute sacked fertilizer cheaper than he can haul dried sewage sludge which is available free of charge. We are too affluent to be able to afford the use of our 'night soil'.

The City of Milwaukee markets in paper bags a dried sludge called Milorganite from its primary settlings which is rich in organic matter. Every city could do this. It piles up and presents a disposal problem because Milwaukee's product satisfies the available market for this product.

17. Benefits of Guided Eutrophication

All eutrophication is not necessarily bad. Well planned enrichment could increase the production of food organisms for fish and hence raise the protein productivity of a natural or man-made lake. Because of the complexity of interactions at various depths and seasons, more knowledge than we now have is needed before we can guide eutrophication and harvest the fish without exceeding the fertility levels that destroy the esthetics of a lake.

18. Predictions

Predicting the consequences of eutrophication would be highly desirable for decision-makers. Systems analysis employs new techniques for constructing mathematical models of a drainage basin to make it possible to evaluate changes which might take place as various eutrophicating factors occur and hence is a powerful tool in dealing with these complex problems in which multifactor cause and effect are involved.

19. In Summary

It is now of greatest urgency to prevent further damage to water resources and to take corrective steps to reverse present damages. Suggested preventive and corrective measures include removing nutrients from municipal, industrial, and agricultural wastes, diversion of treated effluents from lakes, harvesting algae, aquatic plants and fish from lakes in order to help impoverish the water and to improve esthetic qualities; establish regulations for shoreland corridors in order to protect lakes from further damage.

References

1. Beeton, A. M.: 'Eutrophication of the St. Lawrence Great Lakes', *Limnology Oceanography* **10**, 240–254 (1965).
2. Beuscher, J.: in *International Symposium on Eutrophication – Eutrophication: causes, consequences, correctives* (University of Wisconsin, Madison, June, 1967). National Academy of Sciences, National Research Council, Washington, D.C. (1969).
3. Davis, C. C.: 'Evidence for Eutrophication of Lake Erie from Phytoplankton Records', *Limnology Oceanography* **9**, 275–283 (1964).
4. Edmondson, W. T.: 'Water-Quality Management and Lake Eutrophication: The Lake Washington Case', reprinted from *Water Resources Management and Public Policy*, (ed. by Thomas H. Campbell and Robert O. Sylvester), University of Washington Press, Seattle, pp. 139–178 (1968).
5. Edmondson, W. T.: Eutrophication in North America, in *Eutrophication NAS Symposium*, pp. 124–149 (1969).
6. Edmondson, W. T.: 'Phosphorus, Nitrogen, and Algae in Lake Washington after Diversion of Sewage', *Science* **169**, 690–691 (1970).
7. Einsele, W.: 'Über die Beziehungen des Eisenkreislaufs zum Phosphatkreislauf in eutrophen seen', *Arch. Hydrobiol.* **29**, 664–686 (1936).
8. Einsele, W. and Hasler, A. D.: 'Fertilization for Increasing Productivity of Natural Inland Waters', *Trans. 13th N. Amer. Wildlife Conf.*, pp. 527–554 (1948).
9. Findenegg, I.: 'Bestimmung des Trophiegrades von Seen nach der Radiocarbonmethode', *Naturwissenschaften* **51**, 368–369 (1964).
10. Hasler, Arthur D.: 'Eutrophication of Lakes by Domestic Drainage', *Ecology* **28**, 383–395 (1947).
11. Hutchinson, G. E. and Bowen, V. T.: 'A Direct Demonstration of the Phosphorus Cycle in a Small Lake', *Proc. Nat. Acad. Sci. U.S.* **33**, 148–153 (1947).
12. Laurent, P. J.: *Station Centrale d'Hydrobiologie*, personal communication (1970).
13. Liebmann, H.: personal communication (1970).
14. Lind, C. T. and Cottam, G.: 'The Submerged Aquatics of University Bay: A Study in Eutrophication', *Amer. Midl. Nat.* **81**, 353–369 (1969).
15. McGauhey, P. H., Eliassen, R., Rohlich, G. A., Ludwig, H. F., and Pearson, E. A.: 'Comprehensive Study of Protection of Water Resources of Lake Tahoe', to Lake Tahoe Area Council Engineering-Sciences, Inc., Arcadia, Calif. (1963).

16. Minder, Leo: 'Zur Hydrophysik des Zürich und Walensees, nebst Beitrag zur Hydrochemie und Hydrobakteriologie des Zürichsees', *Archw. Hydrobiol.* **12**, 122–194 (1918).
17. Minder, Leo: 'Biologisch-chemische Untersuchungen im Zürichsee', *Rev. Hydrologie* **3**, 1–69 (1926).
18. Minder, Leo: 'Der Zürichsee als Eutrophierungsphänomen. Summarische Ergebnisse aus fünfzig Jahren Zürichseeforschung', *Geologie Meere Binnengewasser* **2**, 284–299 (1938).
19. Minder, Leo: *Der Zurichsee im Lichte der Seetypenlehre*, Naturforschenden Gesellschaft in Zürich (1943).
20. Minder, Leo: 'Neuere Untersuchungen über den Sauerstoffgehalt und die Eutrophie des Zürichsees', *Archw. Hydrobiol.* **40**, 279–301 (1943).
21. Sager, P: 'Species Diversity in Lacustrine Phytoplankton. I. The Components of the Index of Diversity from Shannon's Formula', *Amer. Natur.* **103**, **929**, 51–59 (1969).
22. Stewart, K. M.: *Physical Limnology of Some Madison Lakes*, Ph.D. Thesis, University of Wisconsin (1965).
23. Vollenweider, R. A.: *Scientific Fundamentals of the Eutrophication of Lakes and Flowing Waters, With Particular Reference to Nitrogen and Phosphorus as Factors in Eutrophication*, Technical Rept. DAS/CSI/6827, Organization for Economic Co-operation and Development (OECD), Paris (1968).

For Further Reading

On the causes, consequences, and corrective measures of worldwide eutrophication:

Proceedings of International Symposium on Eutrophication – *Eutrophication: causes, consequences, correctives* (Univ. of Wisconsin, Madison, June, 1967), National Academy of Sciences, National Research Council, Washington, D.C., 1969.
Likens, G. E. (ed.): 1972. *Nutrients and Eutrophication, The Limiting-Nutrient Controversy*, Special Symposium, Vol. 1, Amer. Soc. Limnology+Oceanogr., Inc., Allen Press.

ABOUT THE AUTHORS

S. FRED SINGER received his Ph.D. in physics from Princeton University in 1948. He has alternated between universities and government service and now is Professor of Environmental Sciences at the University of Virginia in Charlottesville, Va. Until 1970 he held the position of Deputy Assistant Secretary, U.S. Department of the Interior. Following his earlier association with the U.S. Weather Bureau as Director of the National Weather Satellite Center, he received the Gold Medal Award for Distinguished Federal Service. He has served as Dean of Environmental Sciences at the University of Miami. His scientific interests have been in atmospheric physics, space physics, oceanography, and the early history and development of the earth-moon system. His current professional concerns include a wide range of problems, including energy, ecological and environmental concerns. He is Chairman of the Committee on Environmental Quality of the American Geophysical Union.

WILLIAM W. KELLOGG is a Senior Scientist at the National Center for Atmospheric Research (NCAR) in Boulder, Colorado. He received his Ph.D. from UCLA in meteorology. Following military service as a pilot-weather officer, he served as assistant professor at UCLA. He was head of the Planetary Sciences Department of the Rand Corporation and became Director of the Laboratory of Atmospheric Science of NCAR in 1964. He has served on committees of the International Association of Meteorology and Atmospheric Physics (IAMAP), and other international scientific groups. He has been a member of several National Academy of Sciences panels and boards, and also served on advisory committees of Air Force, NASA and EPA. He is currently President of the American Meteorological Society and of the Meteorology Section of the American Geophysical Union.

FRANCIS S. JOHNSON is Director, Center for Advanced Studies of the University of Texas at Dallas, Texas. He received his Ph.D. in meteorology from the University of California, Los Angeles, and has done extensive research in meteorology and space physics, with particular emphasis on the upper atmosphere. He is a member of numerous committees of the National Academy of Sciences, including the Space Science Board and the Advisory Committee to the Environmental Science Services Administration. He is a member of the Executive Committee of the International Association of Geomagnetism and Aeronomy, and the English Secretary of Commission IV of the International Union of Radio Science. He has served as Associate Editor of the *Journal of Geophysical Research*. His most recent research interests include the development of planetary atmospheres.

FREDERICK D. SISLER is a marine microbiologist who received his Ph.D. from the Scripps Institution of Oceanography in La Jolla, Calif., in 1949. He has held positions in private industry and government laboratories including the Federal Water Pollution Control Administration. He is presently with the U.S. Environmental Protection Agency. His original research has included such topics as biochemical fuel cells, organic matter in meteorites, and oceanographic instrumentation.

SYUKURO MANABE received his Ph.D. in meteorology from Tokyo University in 1958, and since then has served as Research Meteorologist with the U.S. Weather Bureau. Since 1962 he has been a member of the Geophysical Fluid Dynamics Laboratory, now located in Princeton, N.J., and concerned with the simulation of atmospheric circulation problems by means of high speed electronic computers.

WILLARD F. LIBBY received degrees in chemistry from the University of California, Berkeley, and in addition holds a number of honorary doctorates. He teaches at the University of California, Los Angeles, where he is Director of the Institute of Geophysics, and at the University of Colorado. He has served a five year-term as Commissioner of the U.S. Atomic Energy Commission. He is a Nobel

Laureate in Chemistry (1960) in recognition of his work on natural radiocarbon and its application to the dating of ancient artifacts. He has received many honors and awards, including the Willard Gibbs Medal of the American Chemical Society, the Day Medal of the Geological Society of America, and the Albert Einstein Medal.

RAINER BERGER received his Ph.D. in organic isotope chemistry at the University of Illinois in 1960. He is Professor of Geography, Geophysics and Anthropology at the University of California, Los Angeles, having been affiliated with aerospace firms in California. His interests now include the application of radiocarbon dating to archeology and anthropology.

LOUIS S. JAFFE is a chemist and physiologist with degrees from Brooklyn College and Columbia University. He is Associate Clinical Professor of Epidemiology and Environmental Health, The George Washington University School of Medicine, where he teaches courses on air pollution. He also serves as a consultant in the areas of environmental health, pollution, methods of measurement, and related fields. For the past 30 yr, he held various responsible government positions in these areas. Most recently, he was concerned with developing air quality criteria for the Department of Health, Education and Welfare. He is the author of many publications concerning atmospheric pollution and environmental health and has been particularly concerned with the problems of ozone and carbon monoxide.

ELMER ROBINSON is Professor (air pollution) and research meteorologist at Washington State University, Pullman, Wash. Mr Robinson graduated from the University of California at Los Angeles in 1948 with an M.A. degree in Meteorology. In 1948 he joined Stanford Research Institute and became active in research programs dealing with air pollution, cloud and fog physics, and atmospheric chemistry. He served as Senior Meteorologist and Chairman Environmental Research Department. During the period 1957–1960 he served as Chief, Air Analysis Section, Bay Area Air Pollution Control District, San Francisco, California.

ROBERT C. ROBBINS, Senior Physical Chemist, Environmental Research Department, Stanford Research Institute. He received his Ph.D. from the University of Delaware in Physical Chemistry in 1953. Since 1954 he has been active in various fields of atmospheric research at Stanford Research Institute. His special interests have included air pollution, photochemistry, atmospheric chemistry, aerosol chemistry and physics, and atmospheric carbon monoxide research. He served as a meteorologist with the U. S. Navy in WW II and has also been with the Du Pont Co.

REID A. BRYSON received his Ph.D. from the University of Chicago in meteorology and has taught mainly at the University of Wisconsin-Madison where he established the Department of Meteorology and now serves as Professor of Meteorology and Geography, and the Director of the Institute for Environmental Studies. He is a Fellow of the American Meteorology Society. His special interest is world climatology with an emphasis on paleoclimatology. Field experience in many parts of the world account for a number of years of his career.

WAYNE M. WENDLAND received his Ph.D. from the University of Wisconsin-Madison in meteorology and is presently Assistant Professor of Geography and Meteorology. He served eight years as a forecaster in the U.S. Air Force. His areas of interest are historical climatology and the relation between climate and man.

J. MURRAY MITCHELL received his Ph.D. in Meteorology at the University of Pennsylvania in 1960. He has been Research Meteorologist with U.S. Weather Bureau since 1955. He has been appointed as a Visiting Lecturer at a number of institutions, is a member of the U.S. National Committee, International Association for Quaternary Research, and a member of committees of the American Meteorological Society and American Geophysical Union concerned with paleoclimatology and climate change. He has served as Associate Editor of the *Journal of Applied Meteorology* and is presently Editor of the Meteorological Monograph Series.

VINCENT J. SCHAEFER is Director of the Atmospheric Science Center of the State University of New York at Albany. For most of his career he has been associated with the General Electric Research

Laboratory and principally with Dr Irving Langmuir, with whom he is co-inventor of the artificial-fog smoke-screen generator. He is also the discoverer of the dry ice seeding technique for cloud modification. He has received an honorary doctorate from the University of Notre Dame and the Robert M. Losey Award of the Institute of Aeronautical Sciences.

HELMUT E. LANDSBERG received his Ph.D. in geophysics and meteorology at the University of Frank-furt. He then joined the faculty of Pennsylvania State University and became Associate Professor at the University of Chicago in 1942. In 1954, Dr Landsberg became Director of Climatology of the U.S. Weather Bureau, and when the Environmental Science Services Administration (now NOAA) was formed, he became the first Director of the Environmental Data Service. He is Research Professor and Chairman of Meteorology at the University of Maryland, a member of the National Academy of Engineering, and a fellow of the American Academy of Arts and Sciences, the American Geophysical Union (of which he was President in 1968–1970), and other scientific societies.

HUGH W. ELLSAESSER is a senior scientist at the University of California's Lawrence Livermore Laboratory where he has performed meteorological research and consultation since 1963. He earned his meteorological degrees at UCLA (M.A. 1947) and the University of Chicago (ph.D. 1964). He served over twenty years as a weather officer with the USAF Air Weather Service, including assignments as Air Force Hurricane Officer in Miami, Florida. Since 1969 he has been deeply concerned by dramatic and unsubstantiated claims relating to man's 'fouling of his nest'.

EDWARD D. GOLDBERG received his Ph.D. in chemistry from the University of Chicago in 1949. He then joined the Scripps Institution of Oceanography where he now serves as full Professor. His scientific interests include radio chemistry of marine waters and sediments but extend into problems of planetary atmospheres. He is active in professional societies and Editor of three journals. He was elected Vice President of the Section of Volcanology in the American Geophysical Union.

GEORGE M. WOODWELL received his Ph.D. from Duke University in 1958. He now serves as Senior Ecologist in the Biology Department of the Brookhaven National Laboratory on Long Island, New York. In addition, he holds an adjunct appointment as Lecturer in Ecology at Yale University. His research interests have focused on the ecological effects of ionizing radiation and on the movement of nutrient elements and toxic materials through the various biological, geological, and chemical cycles. His more recent interests include the cycling of DDT over the surface of the earth. His society member-ships include not only scientific societies but also the Sierra Club and Nature Conservancy. He is a member of the Board of Trustees of the Environmental Defense Fund, designed to use the courts to restrict the use of persistent pesticides.

PAUL P. CRAIG is Deputy Director of the Office of Energy R & D Policy of the National Science Foundation. He received a Ph.D. in physics from the California Institute of Technology. Before joining NSF in 1971, he worked at the Brookhaven National Laboratory and at the Los Alamos Scientific Laboratory. His fields of expertise include energy research and technology, environmental science, material sciences, and low temperature physics.

HORTON A. JOHNSON is Professor of Pathology at the Indiana University School of Medicine. He received his medical degree from Columbia University in 1953 and interned at the University of Michigan. Most of his professional career was spent at the Brookhaven National Laboratory, Long Island, N.Y., investigating radiation injury and cell proliferation kinetics.

BOSTWICK H. KETCHUM received a Ph.D. in biology from Harvard University in 1938. Most of his professional career has been spent at the Woods Hole Oceanographic Institution. He is presently on leave at the National Science Foundation, serving as Section Head, Ecology and Systematic Biology. He is involved in many professional societies and in editorial activities. He serves as a Consultant to the U.S. Public Health Service and is a member of advisory committees of the National Academy of Sciences and other scientific organizations. His range of interest includes marine ecology, physiolo-gy of algae, nutrient cycles in the ocean, and the special problems of estuaries.

BENGT LUNDHOLM is Executive Member of the Ecological Research Committee at the Swedish Natural

Science Research Council in Stockholm. He is also Chairman of the ad hoc Committee on the Human Environment which was set up by the International Council of Scientific Unions in 1968. When the International Biological Program created a special working group on global baseline stations, he was chosen as Chairman. He graduated from the University of Upsala and spent 3 yr in east and south Africa studying mammals. During 1964–68 he was Secretary-General in the Swedish Royal Commission on Natural Resources which laid the basis for the present Swedish policy concerning the environment.

BARRY COMMONER received his Ph.D. from Harvard University in 1941. Following military service, he joined the faculty of Washington University in St. Louis, becoming Chairman of the Department of Botany in 1965 and Director of the Center for the Biology of Natural Systems. He has received many honors including honorary degrees and the Newcomb Cleveland Prize of the American Association for the Advancement of Science. He has served as Chairman of the AAAS Committee on Science in the Promotion of Human Welfare and, in 1967, was elected to the AAAS Board of Directors. He has been actively investigating fundamental problems relating to the physio-chemical basis of biological processes, including pioneer studies on free radicals and on virus replication. He has also developed a deep interest in the interaction between science and social problems; his book *Science and Survival* deals with the threats to human survival from technological changes.

D. R. KEENEY is Associate Professor, Department of Soil Science, University of Wisconsin, Madison. He received his B.S. (1959) from Iowa State University, his M.S. (1961) from the University of Wisconsin, and his Ph.D. (1965) from Iowa State University. He joined the University of Wisconsin staff in 1966 after a year's postdoctorate study at Iowa State. Dr Keeney's research deals primarily with methods for determining various forms of nitrogen and with nitrogen transformations in soils, lake sediments and waters.

W. R. GARDNER is Professor, Department of Soil Science, University of Wisconsin, Madison. He received his B.S. (1949) from Utah State University and his M.S. (1951) and Ph.D. (1953) from Iowa State University. Before joining the University of Wisconsin staff in 1966, he served for 15 yr as physicist, Agricultural Research Service, U.S. Department of Agriculture, Salinity Laboratory, Riverside, Calif. He has pioneered investigations on water and solute movement through soils and the ion transport phenomena in plants. He also is actively conducting research on micrometeorology, hydrology, and the movement of pollutants through soil.

THEODORE D. BYERLY is Assistant Director of Science and Education of the U.S. Department of Agriculture. He received his Ph.D. from the University of Iowa in 1926 and has been associated with universities and agricultural research throughout his career. He has received many honors and awards including the Department of Agriculture Distinguished Service Award. He served as Chairman of the Division of Biology and Agriculture of the National Academy of Sciences/National Research Council.

ARTHUR C. HASLER received his Ph.D. from the University of Wisconsin in 1937 and is associated with that institution as Professor of Zoology and Director of the Laboratory of Limnology. He has been the recipient of many honors including election to the National Academy of Sciences, honorary doctorates, President of the American Society of Limnology and Oceanography, President of the International Association for Ecology and Director of The Institute of Ecology. His successful research career involves experimental limnology and deciphering the environmental cues which guide migrating salmon.

INDEX OF NAMES

INDEX OF SUBJECTS